Lecture Notes in Artificial Intelligence   5208

Edited by R. Goebel, J. Siekmann, and W. Wahlster

Subseries of Lecture Notes in Computer Science

Helmut Prendinger   James Lester
Mitsuru Ishizuka (Eds.)

Intelligent Virtual Agents

8th International Conference, IVA 2008
Tokyo, Japan, September 1-3, 2008
Proceedings

 Springer

Series Editors

Randy Goebel, University of Alberta, Edmonton, Canada
Jörg Siekmann, University of Saarland, Saarbrücken, Germany
Wolfgang Wahlster, DFKI and University of Saarland, Saarbrücken, Germany

Volume Editors

Helmut Prendinger
National Institute of Informatics
Digital Contents and Media Sciences Research Division
2-1-2 Hitotsubashi, Chiyoda-ku, 101-8430 Tokyo, Japan
E-mail: helmut@nii.ac.jp

James Lester
North Carolina State University, Department of Computer Science
Raleigh, NC 27695, USA
E-mail: lester@ncsu.edu

Mitsuru Ishizuka
University of Tokyo, Department of Information and Communication Engineering
7-3-1, Hongo, Bunkyo-ku, Tokyo 113-8656, Japan
E-mail: ishizuka@i.u-tokyo.ac.jp

Library of Congress Control Number: 2008933373

CR Subject Classification (1998): I.2.11, H.5.1, H.4, K.3

LNCS Sublibrary: SL 7 – Artificial Intelligence

ISSN   0302-9743
ISBN-10   3-540-85482-7 Springer Berlin Heidelberg New York
ISBN-13   978-3-540-85482-1 Springer Berlin Heidelberg New York

This work is subject to copyright. All rights are reserved, whether the whole or part of the material is concerned, specifically the rights of translation, reprinting, re-use of illustrations, recitation, broadcasting, reproduction on microfilms or in any other way, and storage in data banks. Duplication of this publication or parts thereof is permitted only under the provisions of the German Copyright Law of September 9, 1965, in its current version, and permission for use must always be obtained from Springer. Violations are liable to prosecution under the German Copyright Law.

Springer is a part of Springer Science+Business Media

springer.com

© Springer-Verlag Berlin Heidelberg 2008
Printed in Germany

Typesetting: Camera-ready by author, data conversion by Scientific Publishing Services, Chennai, India
Printed on acid-free paper   SPIN: 12464449   06/3180   5 4 3 2 1 0

Preface

Welcome to the Proceedings of the 8th International Conference on Intelligent Virtual Agents, which was held on September 1–3, 2008 in Tokyo, Japan. Intelligent virtual agents (IVAs) are autonomous, graphically embodied agents in a virtual environment that are able to interact intelligently with human users, other IVAs, and their environment. The IVA conference series is the major annual meeting of the intelligent virtual agents community, attracting interdisciplinary minded researchers and practitioners from embodied cognitive modeling, artificial intelligence, computer graphics, animation, virtual worlds, games, natural language processing, and human–computer interaction.

The origin of the IVA conferences dates from a successful workshop on Intelligent Virtual Environments held in Brighton, UK, at the 13th European Conference on Artificial Intelligence (ECAI 2008). This workshop was followed by a second one held in Salford in Manchester, UK in 1999. Subsequent events took place in Madrid, Spain in 2001, Irsee, Germany 2003 and Kos, Greece in 2005. Starting in 2006, IVA moved from being a biennial to an annual event and became a full-fledged international conference, held in Marina del Rey, California, USA in 2006, and Paris, France in 2007. From 2005, IVA also hosted the Gathering of Animated Lifelike Agents (GALA), an annual festival to showcase the latest animated lifelike agents created by university students and academic or industrial research groups.

IVA 2008 was the first time that IVA was organized in Asia and we are happy to report that a large number of papers were submitted. IVA 2008 received 99 submissions from Europe, the Americas, and Asia. Out of 71 long paper submissions, only 18 were accepted, and an additional 22 were accepted as short papers, or as posters (30). Out of 23 short paper submissions, we could only accept 6. A further 15 were accepted as posters. In total, there were 28 short papers and 42 poster papers accepted.

IVA 2008 was hosted by the National Institute of Informatics, Tokyo. We would like to thank the members of the Program Committee for providing insightful reviews, and the members of the Senior Program Committee for selecting the best papers. We thank Hugo Hernault for preparing the IVA 2008 website and Ulrich Apel for providing the content on Tokyo and the institute. A special thanks goes to two people: Shigeko Tokuda for her excellent work in preparing and running the conference, and Boris Brandherm for his superb and efficient management of the submitted papers and preparation of the proceedings. We would also like to mention that the continued and active support of the IVA Steering Committee at each stage of the conference preparation process is greatly appreciated.

Last but not least, we would like to thank the authors for submitting their high-quality work to IVA 2008. We hope the readers will enjoy the papers in this volume, and look forward to future IVA conferences.

June 2008

Helmut Prendinger
James Lester
Mitsuru Ishizuka

Organization

Conference Chairs

Helmut Prendinger   National Institute of Informatics, Japan
James Lester   North Carolina State University, USA
Mitsuru Ishizuka   The University of Tokyo, Japan

Senior Program Committee

Elisabeth André   University of Augsburg, Germany
Ruth Aylett   Heriot-Watt University, UK
Marc Cavazza   University of Teesside, UK
Jonathan Gratch   University of Southern California, USA
Stefan Kopp   Bielefeld University, Germany
Jean-Claude Martin   LIMSI-CNRS, France
Patrick Olivier   University of Newcastle upon Tyne, UK
Catherine Pelachaud   University of Paris 8, INRIA, France
Seiji Yamada   National Institute of Informatics, Japan

Best Paper Chair

W. Lewis Johnson   Alelo, Inc., USA

Publicity Chair

Anton Nijholt   University of Twente, The Netherlands

Submissions Chair

Boris Brandherm   National Institute of Informatics, Japan

Local Organizing Committee

Helmut Prendinger   National Institute of Informatics (Local Organization Chair)
Ulrich Apel   National Institute of Informatics
Boris Brandherm   National Institute of Informatics
Hugo Hernault   The University of Tokyo
Shigeko Tokuda   National Institute of Informatics
Sebastian Ullrich   RWTH Aachen University

Invited Speakers

Hiroshi Ishiguro   Osaka University, Japan
Ted Tagami   Millions of Us, USA

Program Committee

Jan Allbeck
Jens Allwood
Norman Badler
Jeremy Bailenson
Amy Baylor
Christian Becker-Asano
Timothy Bickmore
Gaspard Breton
Stéphanie Buisine
Felix Burkhardt
Justine Cassell
Jeffrey Cohn
Angélica de Antonio
Fiorella de Rosis
Zhigang Deng
Doron Friedman
Patrick Gebhard
Marco Gillies
Dirk Heylen
Tetsunari Inamura
Katherine Isbister
Ido Iurgel
Kostas Karpouzis
Michael Kipp
Yasuhiko Kitamura
Takanori Komatsu
Nicole Krämer
Brigitte Krenn

Michael Kruppa
Steve Maddock
Wenji Mao
Andrew Marriott
Stacy Marsella
Hideyuki Nakanishi
Yukiko Nakano
Michael Neff
Toyoaki Nishida
Pierre-Yves Oudeyer
Ana Paiva
Igor Pandzic
Maja Pantic
Sylvie Pesty
Christopher Peters
Paolo Petta
Paul Piwek
Rui Prada
Matthias Rehm
Mark Riedl
Brent Rossen
Martin Rumpler
Zsófia Ruttkay
Marc Schröder
Candy Sidner
Daniel Thalmann
Kristinn Thórisson

IVA Steering Committee

Ruth Aylett   Heriot-Watt University, UK
Jonathan Gratch   University of Southern California, USA
Stefan Kopp   Bielefeld University, Germany
Patrick Olivier   University of Newcastle upon Tyne, UK
Catherine Pelachaud   University of Paris 8, INRIA, France

Held in Cooperation with

The American Association of Artificial Intelligence (AAAI)
The European Association for Computer Graphics (EG)
The Association for Computing Machinery (ACM)
Special Interest Group on Artificial Intelligence (SIGART)
Special Interest Group on Computer-Human Interaction (SIGCHI)

Table of Contents

Emotion and Empathy

The Relation between Gaze Behavior and the Attribution of Emotion:
An Empirical Study 1
    Brent Lance and Stacy C. Marsella

Affect Simulation with Primary and Secondary Emotions 15
    Christian Becker-Asano and Ipke Wachsmuth

User Study of AffectIM, an Emotionally Intelligent Instant Messaging
System 29
    Alena Neviarouskaya, Helmut Prendinger, and Mitsuru Ishizuka

Expressions of Empathy in ECAs 37
    Radoslaw Niewiadomski, Magalie Ochs, and Catherine Pelachaud

Narrative and Augmented Reality

Archetype-Driven Character Dialogue Generation for Interactive
Narrative 45
    Jonathan P. Rowe, Eun Young Ha, and James C. Lester

Towards a Narrative Mind: The Creation of Coherent Life Stories for
Believable Virtual Agents 59
    Wan Ching Ho and Kerstin Dautenhahn

Emergent Narrative as a Novel Framework for Massively Collaborative
Authoring 73
    Michael Kriegel and Ruth Aylett

Towards Real-Time Authoring of Believable Agents in Interactive
Narrative 81
    Martin van Velsen

Male Bodily Responses during an Interaction with a Virtual Woman ... 89
    Xueni Pan, Marco Gillies, and Mel Slater

A Virtual Agent for a Cooking Navigation System Using Augmented
Reality 97
    Kenzaburo Miyawaki and Mutsuo Sano

Conversation and Negotiation

Social Perception and Steering for Online Avatars 104
    Claudio Pedica and Hannes Vilhjálmsson

Multi-party, Multi-issue, Multi-strategy Negotiation for Multi-modal
Virtual Agents.. 117
  *David Traum, Stacy C. Marsella, Jonathan Gratch, Jina Lee, and
  Arno Hartholt*

A Granular Architecture for Dynamic Realtime Dialogue.............. 131
  Kristinn R. Thórisson and Gudny Ragna Jonsdottir

Generating Dialogues for Virtual Agents Using Nested Textual
Coherence Relations... 139
  *Hugo Hernault, Paul Piwek, Helmut Prendinger, and
  Mitsuru Ishizuka*

Integrating Planning and Dialogue in a Lifestyle Agent 146
  *Cameron Smith, Marc Cavazza, Daniel Charlton, Li Zhang,
  Markku Turunen, and Jaakko Hakulinen*

Audio Analysis of Human/Virtual-Human Interaction 154
  Harold Rodriguez, Diane Beck, David Lind, and Benjamin Lok

Nonverbal Behavior

Learning Smooth, Human-Like Turntaking in Realtime Dialogue 162
  Gudny Ragna Jonsdottir, Kristinn R. Thórisson, and Eric Nivel

Predicting Listener Backchannels: A Probabilistic Multimodal
Approach 176
  Louis-Philippe Morency, Iwan de Kok, and Jonathan Gratch

IGaze: Studying Reactive Gaze Behavior in Semi-immersive
Human-Avatar Interactions 191
  Michael Kipp and Patrick Gebhard

Estimating User's Conversational Engagement Based on Gaze
Behaviors 200
  Ryo Ishii and Yukiko I. Nakano

The Effects of Agent Nonverbal Communication on Procedural and
Attitudinal Learning Outcomes 208
  Amy L. Baylor and Soyoung Kim

Evaluating Data-Driven Style Transformation for Gesturing Embodied
Agents 215
  Alexis Heloir, Michael Kipp, Sylvie Gibet, and Nicolas Courty

Models of Culture and Personality

Culture-Specific First Meeting Encounters between Virtual Agents .. 223
  *Matthias Rehm, Yukiko Nakano, Elisabeth André, and
  Toyoaki Nishida*

Virtual Humans Elicit Skin-Tone Bias Consistent with Real-World
Skin-Tone Biases... 237
  *Brent Rossen, Kyle Johnsen, Adeline Deladisma, Scott Lind, and
  Benjamin Lok*

Cross-Cultural Evaluations of Avatar Facial Expressions Designed by
Western Designers 245
  Tomoko Koda, Matthias Rehm, and Elisabeth André

Agreeable People Like Agreeable Virtual Humans 253
  Sin-Hwa Kang, Jonathan Gratch, Ning Wang, and James H. Watt

A Listening Agent Exhibiting Variable Behaviour 262
  Elisabetta Bevacqua, Maurizio Mancini, and Catherine Pelachaud

Markup and Representation Languages

The Next Step towards a Function Markup Language................... 270
  *Dirk Heylen, Stefan Kopp, Stacy C. Marsella,
  Catherine Pelachaud, and Hannes Vilhjálmsson*

Extending MPML3D to Second Life 281
  *Sebastian Ullrich, Klaus Bruegmann, Helmut Prendinger, and
  Mitsuru Ishizuka*

An Extension of MPML with Emotion Recognition Functions
Attached.. 289
  Xia Mao, Zheng Li, and Haiyan Bao

Architectures for Robotic Agents

ITACO: Effects to Interactions by Relationships between Humans and
Artifacts 296
  Kohei Ogawa and Tetsuo Ono

Teaching a Pet Robot through Virtual Games 308
  Anja Austermann and Seiji Yamada

Cognitive Architectures

Modeling Self-deception within a Decision-Theoretic Framework 322
  Jonathan Y. Ito, David V. Pynadath, and Stacy C. Marsella

Modeling Appraisal in Theory of Mind Reasoning 334
  Mei Si, Stacy C. Marsella, and David V. Pynadath

Improving Adaptiveness in Autonomous Characters 348
  Mei Yii Lim, João Dias, Ruth Aylett, and Ana Paiva

The Embodiment of a DUAL/AMBR Based Cognitive Model in the
RASCALLI Multi-agent Platform 356
  Stefan Kostadinov and Maurice Grinberg

BDI Model-Based Crowd Simulation 364
  *Kenta Cho, Naoki Iketani, Masaaki Kikuchi, Keisuke Nishimura,
  Hisashi Hayashi, and Masanori Hattori*

The Mood and Memory of Believable Adaptable Socially Intelligent
Characters 372
  Mark Burkitt and Daniela M. Romano

Agents for Healthcare and Training

Visualizing the Importance of Medical Recommendations with
Conversational Agents 380
  Gersende Georg, Marc Cavazza, and Catherine Pelachaud

Evaluation of Justina: A Virtual Patient with PTSD 394
  *Patrick Kenny, Thomas D. Parsons, Jonathan Gratch, and
  Albert A. Rizzo*

Elbows Higher! Performing, Observing and Correcting Exercises by a
Virtual Trainer 409
  Zsófia Ruttkay and Herwin van Welbergen

A Virtual Therapist That Responds Empathically to Your Answers 417
  Matthijs Pontier and Ghazanfar F. Siddiqui

Agents in Games, Museums and Virtual Worlds

IDEAS4Games: Building Expressive Virtual Characters for Computer
Games 426
  *Patrick Gebhard, Marc Schröder, Marcela Charfuelan,
  Christoph Endres, Michael Kipp, Sathish Pammi,
  Martin Rumpler, and Oytun Türk*

Context-Aware Agents to Guide Visitors in Museums 441
  Ichiro Satoh

Virtual Institutions: Normative Environments Facilitating Imitation
Learning in Virtual Agents 456
  Anton Bogdanovych, Simeon Simoff, and Marc Esteva

Posters

Enculturating Conversational Agents Based on a Comparative Corpus
Study 465
  *Afia Akhter Lipi, Yuji Yamaoka, Matthias Rehm, and
  Yukiko I. Nakano*

The Reactive-Causal Architecture: Towards Development of Believable
Agents 468
    Ali Orhan Aydın, Mehmet Ali Orgun, and Abhaya Nayak

Gesture Recognition in Flow in the Context of Virtual Theater 470
    Ronan Billon, Alexis Nédélec, and Jacques Tisseau

Automatic Generation of Conversational Behavior for Multiple
Embodied Virtual Characters: The Rules and Models behind Our
System 472
    Werner Breitfuss, Helmut Prendinger, and Mitsuru Ishizuka

Implementing Social Filter Rules in a Dialogue Manager Using
Statecharts.. 474
    Jenny Brusk

Towards Realistic Real Time Speech-Driven Facial Animation 476
    Aleksandra Cerekovic, Goranka Zoric, Karlo Smid, and
    Igor S. Pandzic

Avatar Customization and Emotions in MMORPGs 479
    Shun-an Chung and Jim Jiunde Lee

Impact of the Agent's Localization in Human-Computer Conversational
Interaction 481
    Aurélie Cousseau and François Le Pichon

Evolving Expression of Emotions in Virtual Humans Using Lights and
Pixels 484
    Celso de Melo and Jonathan Gratch

Motivations and Personality Traits in Decision-Making 486
    Etienne de Sevin

A Flexible Behavioral Planner in Real-Time 488
    Etienne de Sevin

Face to Face Interaction with an Intelligent Virtual Agent: The Effect
on Learning Tactical Picture Compilation 490
    Willem A. van Doesburg, Rosemarijn Looije,
    Willem A. Melder, and Mark A. Neerincx

Creating and Scripting Second Life Bots Using MPML3D 492
    Birgit Endrass, Helmut Prendinger, Elisabeth André, and
    Mitsuru Ishizuka

Piavca: A Framework for Heterogeneous Interactions with Virtual
Characters 494
    Marco Gillies, Xueni Pan, and Mel Slater

Interpersonal Impressions of Agents for Developing Intelligent
Systems 496
   Kaoru Sumi and Mizue Nagata

Comparing an On-Screen Agent with a Robotic Agent in
Non-Face-to-Face Interactions 498
   Takanori Komatsu and Yukari Abe

Sustainability and Predictability in a Lasting Human–Agent
Interaction 505
   Toshiyuki Kondo, Daisuke Hirakawa, and Takayuki Nozawa

Social Effects of Virtual Assistants. A Review of Empirical Results
with Regard to Communication 507
   Nicole C. Krämer

SoNa: A Multi-agent System to Support Human Navigation in a
Community, Based on Social Network Analysis 509
   Shizuka Kumokawa, Victor V. Kryssanov, and Hitoshi Ogawa

Animating Unstructured 3D Hand Models 511
   Jituo Li, Li Bai, and Yangsheng Wang

Verification of Expressiveness of Procedural Parameters for Generating
Emotional Motions 514
   Yueh-Hung Lin, Chia-Yang Liu, Hung-Wei Lee,
   Shwu-Lih Huang, and Tsai-Yen Li

Individualised Product Portrayals in the Usability of a 3D Embodied
Conversational Agent in an eBanking Scenario 516
   Alexandra Matthews, Nicholas Anderson, James Anderson, and
   Mervyn Jack

Multi-agent Negotiation System in Electronic Environments 518
   Dorin Militaru

A Study of the Use of a Virtual Agent in an Ambient Intelligence
Environment 520
   Germán Montoro, Pablo A. Haya, Sandra Baldassarri,
   Eva Cerezo, and Francisco José Serón

Automatic Torso Engagement for Gesturing Characters 522
   Michael Neff

Modeling the Dynamics of Virtual Agent's Social Relations 524
   Magalie Ochs, Nicolas Sabouret, and Vincent Corruble

Proposal of an Artificial Emotion Expression System Based on External
Stimulus and Emotional Elements 526
   Seungwon Oh and Minsoo Hahn

A Reactive Architecture Integrating an Associative Memory for
Sensory-Driven Intelligent Behavior 528
   David Panzoli, Hervé Luga, and Yves Duthen

Social Responses to Virtual Humans: Automatic Over-Reliance on the
"Human" Category 530
   Sung Park and Richard Catrambone

Adaptive Self-feeding Natural Language Generator Engine 533
   Jovan David Rebolledo Méndez and Kenji Nakayama

A Model of Motivation for Virtual-Worlds Avatars 535
   Genaro Rebolledo-Mendez, David Burden, and Sara de Freitas

Towards Virtual Emotions and Emergence of Social Behaviour 537
   Dirk M. Reichardt

Using Virtual Agents for the Teaching of Requirements Elicitation in
GSD 539
   Miguel Romero, Aurora Vizcaíno, and Mario Piattini

A Virtual Agent's Behavior Selection by Using Actions for Focusing of
Attention 541
   Haris Supic

Emergent Narrative and Late Commitment 543
   Ivo Swartjes, Edze Kruizinga, Mariët Theune, and Dirk Heylen

"I Would Like to Trust It but" Perceived Credibility of Embodied
Social Agents: A Proposal for a Research Framework 545
   Federico Tajariol, Valérie Maffiolo, and Gaspard Breton

Acceptable Dialogue Start Supporting Agent for Avatar-Mediated
Multi-tasking Online Communication 547
   Takahiro Tanaka, Kyouhei Matsumura, and Kinya Fujita

Do You Know How I Feel? Evaluating Emotional Display of Primary
and Secondary Emotions 548
   Julia Tolksdorf, Christian Becker-Asano, and Stefan Kopp

Comparing Emotional vs. Envelope Feedback for ECAs 550
   *Astrid von der Pütten, Christian Reipen, Antje Wiedmann,
   Stefan Kopp, and Nicole C. Krämer*

Intelligent Agents Living in Social Virtual Environments – Bringing
Max into Second Life 552
   Erik Weitnauer, Nick M. Thomas, Felix Rabe, and Stefan Kopp

Author Index 555

The Relation between Gaze Behavior and the Attribution of Emotion: An Empirical Study

Brent Lance and Stacy C. Marsella

University of Southern California
Information Sciences Institute
4676 Admiralty Way Suite 1001
Marina Del Rey, CA 90034
{lance,marsella}@isi.edu

Abstract. Real-time virtual humans are less believable than hand-animated characters, particularly in the way they perform gaze. In this paper, we provide the results of an empirical study that explores an observer's attribution of emotional state to gaze. We have taken a set of low-level gaze behaviors culled from the nonverbal behavior literature; combined these behaviors based on a dimensional model of emotion; and then generated animations of these behaviors using our gaze model based on the Gaze Warping Transformation (GWT) [9], [10]. Then, subjects judged the animations displaying these behaviors. The results, while preliminary, demonstrate that the emotional state attributed to gaze behaviors can be predicted using a dimensional model of emotion; and show the utility of the GWT gaze model in performing bottom-up behavior studies.

Keywords: Gaze, Nonverbal Behavior, Emotional Expression, Character Animation, Procedural Animation, Motion Capture, Posture, Virtual Agent.

1 Introduction

Animated characters in feature films function at a high level of believability, appear to come alive, and successfully engage the film's audience; as do characters in many video games, although arguably to a lesser extent. Unfortunately, virtual embodied agents struggle to achieve this goal. However, the animation methods, used to create the film and video game characters are expensive and time consuming, and only allow for limited interaction in dynamic environments, making them unsuitable for the development of virtual embodied agents. Instead, real-time animation systems that express believable behavior are necessary. Our specific interest is in gaze behavior, which is expressive not only in terms of where the gaze is directed, but also in how the gaze is performed, its physical manner. As such, the goal of this paper is to find a model that maps between emotion and the physical manner of gaze. The purpose of this mapping is to allow for the generation of believable, emotionally expressive gaze shifts in an interactive virtual human, while using the minimum amount of motion capture data necessary to maintain realistic physical gaze manner.

We present an approach to this problem that consists of an exploratory empirical study of the mapping between a set of gaze behaviors and the emotional content attributed to gaze shifts performing those behaviors by observers. This study is similar to the "reverse engineering" approach used by Grammer et al. [7], to study emotional state and facial expression. In this context, "reverse engineering" is used to mean a bottom-up approach where nonverbal behavior expressions are generated through the combination of low-level physical behaviors, and then displayed to subjects who rate the expression on its emotional content. Specifically, Grammer et al. [7], use Poser to generate random facial expressions from the space of all possible combinations of FACS Action Units. Users then evaluated the resulting expressions using a circumplex model of emotion.

Similarly, we found a model that describes the mapping between gaze behaviors and the attribution of emotion to gaze shifts displaying those behaviors by first determining how the model describes emotion. We used two different representations of emotion, a set of emotional categories, such as anger or fear, and the PAD dimensional model of emotion [13]. Then we determined the space of possible gazes and the physical manners which they perform. To do this, we have culled a set of low-level, composable gaze behaviors from the nonverbal behavior literature, such as a bowed head during gaze. We then generated all possible gazes allowed by our space of low-level behaviors using our model of expressive gaze manner based on the Gaze Warping Transformation (GWT) [9], [10].

We use this model because it is capable of displaying an arbitrary selection of gaze behaviors while directed towards an arbitrary target with a minimal library of motion capture data. In [9], we first described and evaluated the Gaze Warping Transformation (GWT) a method for producing emotionally expressive head and torso movement during gaze shifts. We then provided a neuroscience-based eye model, and integrated it with GWTs [10].

Finally, we collected data to determine what emotional states subjects attributed to animated characters displaying these behaviors during gaze shifts. As a result of this reverse engineering study, we were able to demonstrate that composition of these low-level gaze behaviors preserved the PAD dimensional ratings. These results, while promising, are still preliminary. However, the study clearly demonstrates the utility of the GWT as a research tool beyond generating animations, and points out several areas for future research.

While these results have the most application to our GWT-based gaze model, any procedural gaze model with sufficient control over the animation curves used to generate gaze shifts should be able to take advantage of this mapping.

2 Related Work

There have been many implementations of gazing behaviors in real-time applications such as embodied virtual agents. Several of these gaze implementations in virtual characters are based on communicative signals (e.g. [2], [16]). Other gaze models have been developed for agents that perform tasks in addition to dialog, such as [6], [17]. There are also models of resting gaze, which simulate eye behavior when the eye is not performing any tasks [4] [11]. Additionally, there are attention-based models of gaze that perform eye movements based on models of attention and saliency [18], [19].

There are several trends which can be seen in these implementations of gaze. First, the models focus on when and where the character looks, not on how the gaze shift occurs. Second, these models, with few exceptions, focus on communicative or task-related gaze behaviors, not on how gaze reveals emotional state.

In addition to the previous research on implementing models of nonverbal gazing behavior, there has been recent work focused on the manipulation of parameters describing the way in which general movement is performed. This concept is referred to as manner or style. This research can provide methods for manipulating the way in which movements are performed, or to obtain the style from one movement and transfer it to another [1], [3], [22]. This research was inspirational to the development of the Gaze Warping Transformation, but does not deal with the constraints specific to gaze movement, nor does it identify specific styles and their expressive meaning, which is the purpose of this study.

3 Expressive Gaze Model

We used our previous work on gaze to generate the gaze shifts for this study. Our gaze model combines two parts: first, a parameterization called the Gaze Warping Transformation (GWT), that generates emotionally expressive head and torso movement during gaze shifts [9]. The GWT is a set of parameters that transforms an emotionally neutral gaze shift towards a target into an emotionally expressive gaze shift directed at the same target. A small number of GWTs can then produce gazes displaying varying emotional content directed towards arbitrary targets.

The second part is a procedural model of eye movement based on stereotypical eye movements described in the visual neuroscience literature [10]. The procedural eye movement is layered framewise onto the GWT-generated body movement. Emotion is expressed using the GWT, while the procedural eye model ensures realistic motion.

3.1 Gaze Warping Transformation

A Gaze Warping Transformation, or GWT, is found by obtaining two motion captures of gaze shifts directed from the same start point to the same target, one emotionally expressive, the other emotionally neutral, and finding a set of warping parameters that would convert the animation curve representing each degree of freedom in the emotionally neutral animation into the animation curve for the corresponding degree of freedom in the emotionally expressive movement [9].

This works by transforming the keyframes of an animation curve. The keyframes of an animation are a subset of that animation's frames, such that the values of the motion curves for intermediate frames are found by interpolating between the keyframes. We select the keyframes for each gaze by aligning it to a "stereotypical" gaze shift with known keyframe locations [10]. The gazes are aligned using the ratio of movement that occurred by each frame to that throughout the entire curve [1].

The result of this is a set of keyframes $x(t)$, defined as a set of value, frame pairs, (x_i, t_i). These keyframes are transformed to those of a new motion $x'(t')$, defined as the set of pairs (x_i', t_i') through the use of two functions [21]. The first function, given a frame in the emotional curve t_i', calculates the frame t_i in the neutral motion curve to obtain the corresponding amplitude x_i. For the GWT, we use the function

$$t_i = g(t_i'), \qquad (1)$$

$$g(t_i') = c(t_i') * (t_i' - t_{i-1}'), \qquad (2)$$

where given a frame time in the emotional movement t_i', $g(t)$ determines the corresponding frame t_i in the neutral movement through a scaling parameter $c(t_i')$, which scales the time span between two adjacent keyframes. The second function is

$$x'(t_i') = x(t_i) + b(t_i), \qquad (3)$$

where $b(t_i)$ is a spatial offset parameter that transforms the neutral curve amplitude $x(t_i)$ into the corresponding emotional amplitude $x'(t_i')$. The final GWT is an $m * n$ set of (c, b) pairs, where m is the number of degrees of freedom in the animated body, and n is the number of keyframes in the animation.

As the GWT is based on a technique of simple geometric transformations [21], the generated animations can move outside the physical limits of a human body. To solve this, we use an inverse kinematics system implemented using nonlinear optimization. This system simulates a rigid skeleton, keeping our animated movement within the limits of the human body [10].

Table 1. List of Gaze Types

Gaze Type
Eye-Only Gaze Shift
Eye-Head Gaze Shift
Eye-Head-Body Gaze Shift
Head-Only Movement
Head-Body Movement

3.2 Procedural Model of Eye Movement

In addition to the GWT, which describes head and torso movement during gaze shifts, we developed an integrated procedural model of eye movement [10]. This model of eye movement is based on the visual neuroscience literature, specifically on research describing the different movements eyes perform during gaze, and the way in which eye movement and head movement are integrated during gaze shifts [12]. It generates several classes of gaze shifts (Table 1) using the following building blocks:

- Saccades. The saccade is a very rapid, highly-stereotyped eye movement which rotates the eye from its initial position directly to the target;
- Vestibulo-Ocular Reflex (VOR). Through the VOR, the eyes rotate within their orbit so that the gaze maintains the same target while the head moves. It produces the Head-Only and Head-Body movements; and
- Combined Eye-Head Movement. This is used to integrate eye movement and head-torso movement, and generates the Eye-Head and Eye-Head-Body gaze shifts;

4 Approach

We performed an empirical study to determine a preliminary mapping between a space of gaze behaviors and emotion attributed to the gaze behaviors by subjects. To obtain this mapping, we first selected appropriate emotional models and the space of gaze behaviors to map between. To determine the mapping between a particular gaze and the attribution of emotion to that gaze, we use a "reverse engineering" approach [7]. Specifically, we generate all the gazes allowed by our space of gaze behaviors, and collect data of subjects attributing emotion to these gaze shifts.

4.1 Structure of Model

Selected Emotion Model. There are many potential models of emotion we could have mapped to the gaze behaviors. We selected two: the first is the PAD model [13]; a model of emotion that views emotion as a space described with a three dimensions: pleasure / displeasure, arousal / non-arousal, and dominance / submissiveness.

The categories of emotion, such as anger or happiness, are represented in this model by subregions in the space defined by the emotional dimensions. For example, anger is defined as negative valence, high arousal, and high dominance, while fear is defined as negative valence, high arousal, and low dominance.

We are also using a categorization of emotion to map gaze behaviors to a set of intuitive emotional descriptors. Rather than using an existing categorical model, this categorization is derived from observer responses to the animations.

Table 2. Gaze Behaviors

Hypothesized Behaviors
Head Raised
Head Bowed
Faster Velocity
Slower Velocity
Torso Raised
Torso Bowed

Selected Gaze Behavior. In addition to the emotional model, we had to determine a space of gaze behaviors, due to the lack of a descriptive set of known gaze behaviors analogous to the FACS system. We identified a set of "emotional behaviors" from the psychology and arts literature that are likely to be used to reveal emotional state. This set of behavior guidelines can be seen in Table 2.

These guidelines are simplifications of the actual literature [5, 8]. Our guidelines are that users will view the character as more dominant when its head is turned upwards than when its head is turned downwards [15], that the perception of arousal is strongly related to velocity [9], and that vertical posture of the body will display emotional pleasure [20]. While there are many alternative gaze behaviors that could also be modeled using the GWT, such as subtle variations in dynamics, or wider variations on posture, this limited set provides a starting point for this research.

4.2 Motion Capture Collection

For the head and torso behaviors, we asked the actor to perform "raised," "neutral," and "bowed" versions of the behavior, and collected data from the resulting movement. We also collected "fast," "neutral," and "slow" velocity movements. However, the "raised" torso posture was indistinguishable from the neutral torso posture, due to the limitations of the motion tracking system we used, resulting in the set of physical behaviors shown in Table 3. All captured gaze shifts consisted of the desired behavior being displayed in a gaze aversion that started gazing straight ahead in a neutral position and posture, and ended gazing 30 degrees to the right displaying the intended gaze behavior. From this motion data, we produced eight behavior GWTs, one for each behavior listed in Table 3.

We also collected motion capture of the different gaze types (Table 1), and produced GWTs for each gaze type as well as the gaze behaviors. The gaze types were captured as gaze aversions that began gazing straight ahead and ended gazing 30° to the right, and gaze attractions that began 30 degrees to the right and ended gazing straight ahead. This resulted in 10 GWTs – one aversive and one attractive gaze shift for each of the different types of gaze in Table 1.

Table 3. Discretization of Gaze Behaviors

Behavior Dimension	Possible Values
Head Posture	Raised, Neutral, Bowed
Torso Posture	Neutral, Bowed
Movement Velocity	Fast, Neutral, Slow

4.3 Animation Generation

From these 8 GWTs representing the discretized physical behaviors (Table 3) and 10 GWTs representing the various gaze types (Table 1), we generated 150 animations for use in our empirical bottom-up study. We combined the gaze behaviors in Table 3 in all possible ways, leaving out combinations of a raised head with bowed torso due to the physical implausibility of the behavior, resulting in 15 total behavior combinations. Then, these combined gaze behaviors were applied to the 10 gaze type GWTs, resulting in 150 GWTs. Finally, to generate the animations, we applied these 150 GWTs to neutral gaze shifts, with the resulting output rendered using Maya. These animations can be seen at:
http://www.isi.edu/~marsella/students/lance/iva08.html

4.4 Category Formation

In order to determine the categories for our primary experiment, and obtain a picture of how well the animated gaze behaviors covered the emotional space defined by the emotion models, we performed a preliminary category formation study.

Approach. 31 informally selected people each viewed 20 animations randomly selected from the set of 150 animations with no duplicates, giving us 620 views, or approximately 4 per animation, and provided an open-ended written response to the question "What emotional state is the character displaying?" We then categorized the affective responses based on the hierarchical model of emotion described in [14].

Results. We used the hierarchical model as a sorting guideline, to divide the individual responses into ten categories (Table 4); for example categorizing "expression of contempt" as Contempt, or "terrified" as Fear. However, we utilized additional categories not described by the hierarchical model. After categorizing the responses, we then selected categories where at least one video had 50% of the subjects rate it with that category. We then discarded those categories that were related to attention, discarding responses such as "change in attention," "displaying strong interest," and "distracted." Finally, we discarded the responses indicating "uncertainty," as we were concerned that it would be applied when the subject was uncertain of the character's state, not when the character was displaying uncertainty.

Table 4. Emotional Categories

Emotional Categories
Anger
Contempt
Disbelief
Excitement
Fear
Flirtatious
Guilt
Sadness
Secretive
Surprise

4.5 Emotional Attribution Experiment

After selecting the low-level behaviors, generating the animations, and setting the emotional categories, we performed the empirical study. The animations were placed online, and subjects rated the animation in two ways: first by selecting the emotional category (Table 4) that most closely approximated the emotion that they perceived in the animation, and second by locating the animation's perceived emotion along the emotional dimensions of the PAD model. One hundred subjects selected through social networking rated fifteen unique, randomly selected animations each, resulting in ten ratings for each of the 150 animations. Subjects rated the animation's location within the PAD model by using five-point Likert scales to indicate their agreement with two statements representing each dimension, seen in Table 5. The Likert scales were 1 = Strongly Disagree, 2 = Disagree, 3 = N/A, 4 = Agree, 5 = Strongly Agree. Emotional categories and rating statements were displayed in random order.

Table 5. Emotional Dimension Rating Scales

Emotional Dimension	Rating Statement
High Dominance	The character is dominant.
Low Dominance	The character is submissive.
High Arousal	The character is agitated.
Low Arousal	The character is relaxed.
High Valence	The character is pleased.
Low Valence	The character is displeased.

5 Results

We uncovered the mapping between emotion models and physical behaviors, in order to answer the following questions:

1. How did the subjects rate gaze shifts containing the low-level gaze behaviors in Table 3 along the PAD dimensions?
2. Does composition of low-level gaze behaviors in Table 3 preserve the PAD dimensions? For example, if a gaze shift displays low Dominance and low Pleasure behaviors, are low Dominance and low Pleasure attributed to it?
3. How did the subjects rate gaze shifts containing the low-level gaze behaviors in Table 3 using the emotional categories in Table 4?

While we had originally intended to find which of the 150 individual animations varied across emotional state, ten ratings per animation was too few to perform a reliable statistical analysis. Instead, we combined the gazes across gaze type (Table 1), giving us 50 ratings for each of the 15 combinations of gaze behaviors.

5.1 Dimensional Results

How reliable were the dimensional ratings scales?
Before exploring the dimensional results, we tested how well our dimensional rating scales measured the emotional dimensions they were intended to by calculating the correlation and Cronbach's Alpha between each pair of rating scales from Table 5.

The Pleased and inverted Displeased scales performed well. The correlation between the two was 0.615, and the standardized Alpha score indicating scale reliability was high, with $\alpha = 0.7610$, ($\alpha > 0.7$ is considered a reliable scale). Dominant and inverted Submission also did well, with a correlation of 0.6649, and a high Alpha ($\alpha = 0.7987$). Therefore, we averaged Pleased and inverted Displeased into one Pleasure scale, and combined Dominant and inverted Submission into one Dominance scale.] Correlations between the Dominance and Pleasure scales were low, (0.1569), indicating no overlap.

However using the ratings of Relaxed and Agitated as a scale for Arousal was less reliable, as both correlation (0.3745) and Alpha ($\alpha = 0.5449$) were low. In addition, correlations between Relaxed and Pleased (0.5353) and between Agitated and Displeased (0.4889) were higher than between Relaxed and Agitated. There are several possible explanations for this, and further research will be necessary to determine the actual reason, but for the remainder of this paper, we will be using the two scales separately as Relaxed and Agitated.] As we used 5-point Likert scales, but only animated 3-point scales of physical behavior, we condensed the collected data into 3-point scales by combining "Strongly Disagree" and "Disagree", as well as "Strongly Agree" and "Agree", leaving Neutral ratings unchanged.

How did the subjects rate gaze shifts containing the low-level gaze behaviors in Table 3 along the PAD dimensions?
To answer this question, we performed a series of MANOVAs (multivariate analysis of variance) and t-tests to determine whether or not the mean emotion dimensions ratings differed across to the low level behaviors found in Table 3. Results of this analysis can be seen in Table 6.

Table 6. Significant Relationships between PAD Dimension and Gaze Behaviors

Emotional Dimension	Head	Body	Velocity
High Dominance	Raised	Bowed	Fast
Low Dominance	Bowed	Neutral	Non-Fast
Relaxed		Bowed	
Agitated	Non-Bowed		Fast
High Pleasure	Neutral	Bowed	
Low Pleasure	Non-Neutral	Neutral	

Four MANOVAs were performed, each with one dimension (Pleasure, Agitation, Relaxation, or Dominance) as the dependent variable, and Head Orientation, Torso Orientation, Velocity, and Subject as the independent variables, while testing for second degree factorial interaction between the independent variables.

The MANOVA results for Dominance showed significant effects ($N = 1500$, $DF = 18$, $F = 14.51$, $p < .001$) for head orientation ($F = 24.08$, $p < .001$), torso orientation ($F = 82.55$, $p < .001$), and velocity ($F = 7.38$, $p < .001$), with a significant interaction between head and torso orientation ($F = 6.47$, $p < .05$). The t-tests showed clear differences between group means, with raised head corresponding to higher Dominance, and bowed head with lower. In addition the t-tests revealed that a bowed posture was rated higher than a neutral posture, and that the Dominance rating for fast was higher than for slow or for neutral (all significant to $p < .01$).

The Relaxed results showed significant differences ($N = 1500$, $DF = 18$, $F = 1.89$, $p < .05$) across the torso orientation ($F = 11.41$, $p < .001$) and the velocity ($F = 3.78$, $p < .05$), with a significant interaction effect between torso and velocity ($F = 3.68$, $p < .05$); and the t-tests revealed that a bowed body was rated more Relaxed than a neutral body ($p < .01$). However, the t-tests did not reveal useful information about the velocity, indicating that the significant difference found by the MANOVA was likely related to the interaction between torso and velocity.

The MANOVA for Agitation found significant differences (N = 1500, DF = 18, $F = 4.60$, $p < .001$) across the head orientation ($F = 19.61$, $p < .001$), the velocity ($F = 6.04$, $p < .01$), and the subject ($F = 17.12$, $p < .001$), and a significant interaction effect between the head and the velocity ($F = 7.17$, $p < .05$). The t-tests showed that raised and neutral head were rated as significantly more Agitated than bowed head, and the rating for high velocity was higher than for slow or neutral ($p < .05$).

Finally, the analysis revealed that Pleasure significantly differed ($N = 1500$, $DF = 18$, $F = 5.93$, $p < .001$) across both the vertical orientation of the head ($F = 6.58$, $p < .05$) and the torso ($F = 77.57$, $p < .001$), with no significant interaction effects. The ratings for Pleasure also differed significantly ($F = 4.06$, $p < .05$) across subject. T-tests ($p < .01$) showed that the Pleasure rating for a neutral head orientation was significantly higher than those for bowed and raised head orientations, and that a bowed posture was rated higher than a neutral posture.

Does composition of low-level gaze behaviors preserve the PAD dimensions?

In order to determine whether the low-level behaviors can be combined according to the PAD model of emotion, we performed a second analysis. We performed six MANOVAs, each using an emotional dimension (High Dominance, Low Dominance, Relaxed, Agitated, High Pleasure, and Low Pleasure) as the dependent variable. We then used the number of behaviors associated with that emotional dimension, and the subject as the two independent variables. This tested whether or not gaze shifts displaying different numbers of behaviors attributed to a specific emotional dimension would have different values attributed to them. The results of this analysis showed that mean attributed ratings for an emotional dimension increased as the number of gaze behaviors associated with that emotional dimension increased, as seen in Figure 1. This indicates that physical gaze behaviors, when combined according to PAD dimensions will be rated as predicted by the combined behaviors.

The specific results for dominance show significant differences ($N = 1500$, $DF = 6$, $F = 32.24$, $p < .01$) across the number of both low ($F = 14.17$, $p < .001$) and high ($F = 26.39$, $p < .001$) dominance behaviors displayed in a gaze shift, and a significant interaction effect ($F = 6.93$, $p < .01$) between low and high dominance. T-tests showed that as the number of dominance gaze behaviors increased, the rating of dominance significantly increased ($p < .01$) for high dominance behaviors, and significantly decreased ($p < .05$) for low dominance behaviors.

The MANOVA for Agitated revealed significant differences ($N = 1500$, $DF = 3$, $F = 18.31$, $p < .001$) across the number of behaviors displayed in a shift, and showed significant differences across subjects ($F = 20.50$, $p < .001$), with no interaction effects. T-tests demonstrated that gaze shifts with no Agitated behaviors were rated significantly less agitated than those with 1 or 2 Agitated behaviors ($p < .01$).

Both low and high pleasure showed significant differences across the number of behaviors ($N = 1500$, $DF = 3$, $F = 22.96$, $p < .001$), although there were also significant differences across subjects ($F = 4.87$, $p < .05$), and no interaction effects. Subsequent t-tests showed that mean ratings of pleasure significantly differed ($p < .01$) across all numbers of pleasure-associated behaviors, and that as low pleasure behaviors increased, pleasure decreased and vice versa for high pleasure behaviors.

As the relaxed dimension only had one behavior associated with it, no further testing was performed.

5.2 Categorical Results

How did the subjects rate gaze shifts containing the low-level gaze behaviors in Table 3 using the emotional categories in Table 4?

To answer this question, we generated a cross tabulation of the 15 combinations of gaze behaviors against the emotional categories (Table 4), and used Pearson's chi squared (X^2) test to examine relationships in the data. We then performed further tests on the residuals to determine which had significant differences.

Table 7. Emotional Categories and Significantly Related Behavior Combinations

Emotional Categories	Significantly Related Behavior Combinations
Contempt	Head Raised, Body Neutral, Velocity Neutral
Excitement	Head Neutral, Body Bowed, Velocity Fast
Fear	Head Neutral, Body Neutral, Velocity Neutral
	Head Neutral, Body Neutral, Velocity Slow
Guilt	Head Bowed, Body Neutral, Velocity Neutral
	Head Bowed, Body Neutral, Velocity Slow
Sadness	Head Bowed, Body Neutral, Velocity Fast
	Head Bowed, Body Neutral, Velocity Neutral

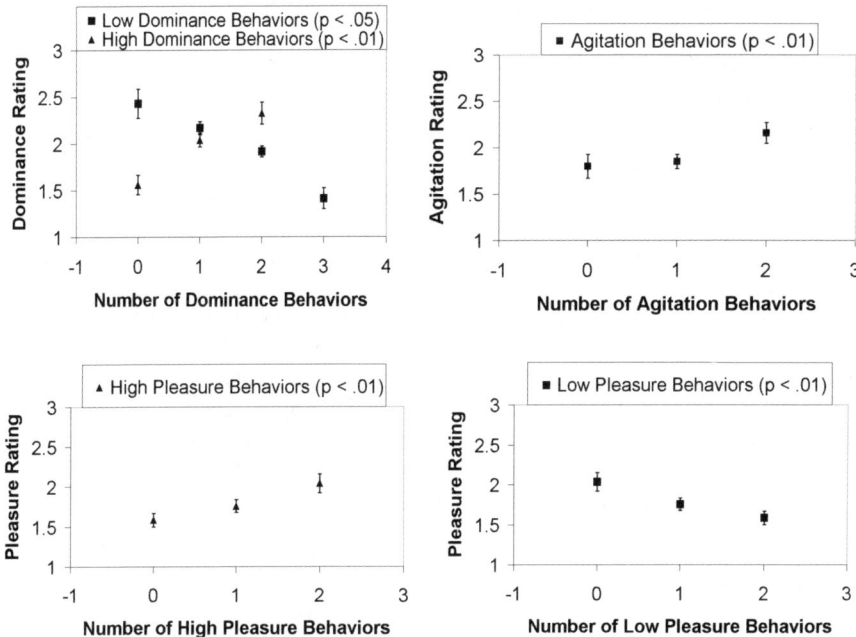

Fig. 1. Plots of Mean Ratings vs. Number of Behaviors for Dominance, Pleasure and Agitation

Results of this analysis can be seen in Table 7. The X^2 test showed that gaze combinations and emotional categories were not randomly related ($N = 1500$, $DF = 126$, $X^2 = 775.817$, $p < .01$). The table rows show behavior combinations with a significant number ($p < .05$) of ratings for that emotional category.

While only 5 of the 15 gaze behavior combinations had significant associations to emotional categories, it was clear through examination of the residuals that further analysis of the relationship between the emotional categories and the low-level behaviors from Table 3 could be useful. For example, while no individual gaze behavior combination was rated significantly high for Flirtatious, all gaze shifts with the bowed head behavior had more Flirtatious ratings than did the gaze shifts without bowed head. To examine this, we generated crosstabs of individual gaze behaviors against emotional categories, and performed additional X^2 tests (Table 8).

Table 8. Significant Relationships between Emotional Categories and Gaze Behaviors

Emotional Category	Head	Torso
Contempt	Raised	Neutral
Excitement		Bowed
Fear		Neutral
Flirtatious	Bowed	
Guilt	Bowed	Neutral
Sadness	Bowed	
Surprise	Neutral	

We found significant interactions between head vertical orientation and emotional categories, ($N = 1500$, $DF = 18$, $X^2 = 329.47$, $p < .001$). Testing the residuals showed that the Contempt category was more likely ($X^2 = 70.35$, $p < .05$) to be attributed to a gaze shift with the head raised behavior, while Flirtatious ($X^2 = 73.41$, $p < .01$), Guilt ($X^2 = 81.33$, $p < .01$), and Sadness ($X^2 = 42.51$, $p < .01$) were all more likely to be attributed to bowed head gaze shifts. Finally, Surprise was significantly less likely ($X^2 = 55.30$, $p < .01$) to be attributed to bowed head gazes. Anger, Disbelief, Excitement, Fear, and Secretive do not relate to head vertical orientation significantly.

Torso posture was not randomly related to emotional category ($N = 1500$, $DF = 9$, $X^2 = 187.49$, $p < .001$). Excitement was more likely to have a bowed torso ($X^2 = 62.94$, $p < .01$), while Contempt ($X^2 = 24.24$, $p < .05$), Fear ($X^2 = 29.19$, $p < .01$), and Guilt ($X^2 = 19.88$, $p < .01$) were attributed more often to neutral torso animations.

We also found, despite our expectations, no strong relationships between the emotional categories and the velocity of the gaze using a crosstab of emotions by velocity. While, the chi squared test showed that emotional category and velocity are not randomly related ($N = 1500$, $DF = 18$, $X^2 = 42.36$, $p < .001$), upon examination of the residuals, no emotional categories significantly differed across velocities.

6 Discussion

As a result of this reverse engineering study, we were able to demonstrate that composition of low-level gaze behaviors in Table 3 preserved the PAD dimensions (Figure 1). In addition, this preservation of PAD dimensions through composition can even be extended to some emotional categories. For example, Guilt can be mapped into the PAD space as low Pleasure, low Arousal, and low Dominance. By combining the behaviors associated with low Dominance, low Pleasure, and Relaxation (Table 6) we generate a movement with a bowed head, neutral torso, and slow velocity. Table 8 reveals that Guilt is attributed to gaze shifts displaying a bowed head, neutral torso, and slow velocity, just as predicted by the PAD model.

We also have a partial mapping between emotional categories and gaze behaviors. For example, in Table 7, we can see that the gaze categories Contempt, Excitement, Guilt, and Sadness are all clearly associated with specific combinations of gaze behaviors, although there is some overlap between Guilt and Sadness. In addition, through examining the relationships between individual low-level gaze behaviors and emotional categories (Table 8), we obtain reinforcement of this mapping, and additional information about how Flirtatious behavior is displayed as well.

This indicates that virtual embodied agents with disparate models of emotion should be able to make use of this mapping between gaze behaviors and attributed emotional state. If the agent uses a categorical model of emotion with emotional categories beyond those used in this study, then by mapping those categories into the PAD space, appropriate gaze behavior may still be generated.

A more thorough description of the low-level behavior components for gaze, similar to FACS for facial expression, would be very valuable to this type of research. While we determined our own space of gaze behaviors for this study, there are other possible ways to structure the gaze behavior space that may provide better results.

7 Conclusion

In this paper, we have provided the results of a reverse engineering study resulting in a preliminary mapping between gaze behaviors and emotional states that could be used with a variety of gaze or emotion models. In addition, we have shown that combining low-level behaviors associated with emotional dimensions in accordance with those dimensions generates a gaze shift that subjects attribute the combined emotional state to. These results, while promising, are still preliminary. However, this study demonstrates the utility of the GWT as a nonverbal behavior research tool, and points towards several directions for future research.

Many of the results of the mapping were not surprising, such as the link between increased dominance and increased vertical orientation of the head. However, there were unexpected results; for example, the link between high Pleasure and a bowed forward body. This indicates the need for a broader selection of gaze behaviors to determine why these unexpected results occurred. This work would also benefit from a more complete exploration of the way in which emotional state is attributed to different combinations of head and eye behavior, as well as a real-time implementation of the gaze mapping in a virtual embodied agent for evaluation.

Acknowledgements. Dr. Bosco Tjan, Dr. Skip Rizzo, and Tiffany Cole provided invaluable assistance, without which this paper would not have been possible. This work was sponsored by the U.S. Army Research, Development, and Engineering Command (RDECOM), and the content does not necessarily reflect the position or the policy of the Government, and no official endorsement should be inferred.

References

1. Amaya, K., Bruderlin, A., Calvert, T.: Emotion From Motion. In: Proceedings of the Conference on Graphical Interface, pp. 222–229 (1996)
2. Bickmore, T., Cassell, J.: Social Dialogue with Embodied Conversational Agents. In: Bernsen, N. (ed.) Natural, Intelligent and Effective Interaction with Multimodal Dialogue Systems. Kluwer Academic Publishers, Dordrecht (2004)
3. Brand, M., Hertzmann, A.: Style Machines. In: Proceedings of SIGGRAPH 2000 (2000)
4. Deng, Z., Lewis, J.P., Neumann, U.: Automated Eye Motion Using Texture Synthesis. IEEE Computer Graphics and Applications 25(2) (2005)
5. Exline, R.: Visual Interaction: The Glances of Power and Preference. In: Weitz, S. (ed.) Nonverbal Communication: Readings with Commentary, Oxford University Press, Oxford (1974)
6. Gillies, M.F.P., Dodgson, N.A.: Eye Movements and Attention for Behavioral Animation. The Journal of Visualization and Computer Animation 13, 287–300 (2002)
7. Grammer, K., Oberzaucher, E.: Reconstruction of Facial Expressions in Embodied Systems. Mitteilungen, ZiF (2006)
8. Kleinke, C.: Gaze and Eye Contact: A Research Review. Psychological Bulletin 100(1) (1986)
9. Lance, B., Marsella, S.: Emotionally Expressive Head and Body Movement During Gaze Shifts. In: Pelachaud, C., et al. (eds.) IVA 2007. LNCS (LNAI), vol. 4722. Springer, Heidelberg (2007)

10. Lance, B., Marsella, S.: A Model of Gaze for the Purpose of Emotional Expression in Virtual Embodied Agents. In: Padgham, Parkes, Müller, Parsons (eds.) Proc. of 7th Int. Conf. on Autonomous Agents and Multiagent Systems (AAMAS 2008) (2008)
11. Lee, S., Badler, J., Badler, N.: Eyes Alive. ACM Transactions on Graphics 21(3) (2002)
12. Leigh, R.J., Zee, D.: The Neurology of Eye Movements, 4th edn. Oxford Press (2006)
13. Mehrabian, A.: Silent Messages: Implicit Communication of Emotions and Attitudes, 2nd edn. Wadsworth Publishing Company (1981)
14. Metts, S., Bowers, J.: Emotion in Interpersonal Communication. In: Knapp, M., Miller, G. (eds.) Handbook of Interpersonal Communication, 2nd edn. Sage, Thousand Oaks (1994)
15. Mignault, A., Chaudhuri, A.: The Many Faces of a Neutral Face: Head Tilt and Perception of Dominance and Emotion. Journal of Nonverbal Behavior 27(2) (2003)
16. Pelachaud, C., Bilvi, M.: Modeling Gaze Behavior for Conversational Agents. In: Rist, T., Aylett, R., Ballin, D., Rickel, J. (eds.) IVA 2003. LNCS (LNAI), vol. 2792. Springer, Heidelberg (2003)
17. Rickel, J., Johnson, W.L.: Animated Agents for Procedural Training in Virtual Reality: Perception, Cognition, and Motor Control. Applied Art. Int. 13(4-5) (1999)
18. Peters, C., Pelachaud, C., Bevacqua, E., Mancini, M.: A Model of Attention and Interest Using Gaze Behavior. In: Panayiotopoulos, T., Gratch, J., Aylett, R.S., Ballin, D., Olivier, P., Rist, T. (eds.) IVA 2005. LNCS (LNAI), vol. 3661. Springer, Heidelberg (2005)
19. Picot, A., Bailly, G., Elisei, F., Raidt, S.: Scrutinizing Natural Scenes: Controlling the Gaze of an Embodied Conversational Agent. In: Pelachaud, C., et al. (eds.) IVA 2007. LNCS (LNAI), vol. 4722. Springer, Heidelberg (2007)
20. Schouwstra, S., Hoogstraten, H.: Head Position and Spinal Position as Determinants of Perceived Emotional State Perceptual and motor skills 81(22), 673–674 (1995)
21. Witkin, A., Popovic, Z.: Motion Warping. In: Proceedings of SIGGRAPH 1995 (1995)
22. Zhao, L., Badler, N.: Acquiring and Validating Motion Qualities from Live Limb Gestures. Graphical Models 67(1) (2005)

Affect Simulation with Primary and Secondary Emotions

Christian Becker-Asano and Ipke Wachsmuth

Faculty of Technology, University of Bielefeld, 33594 Bielefeld, Germany
{cbecker,ipke}@techfak.uni-bielefeld.de

Abstract. In this paper the WASABI[1] Affect Simulation Architecture is introduced, in which a virtual human's cognitive reasoning capabilities are combined with simulated embodiment to achieve the simulation of primary and secondary emotions. In modeling primary emotions we follow the idea of "Core Affect" in combination with a continuous progression of bodily feeling in three-dimensional emotion space (PAD space), that is only subsequently categorized into discrete emotions. In humans, primary emotions are understood as onto-genetically earlier emotions, which directly influence facial expressions. Secondary emotions, in contrast, afford the ability to reason about current events in the light of experiences and expectations. By technically representing aspects of their connotative meaning in PAD space, we not only assure their mood-congruent elicitation, but also combine them with facial expressions, that are concurrently driven by the primary emotions. An empirical study showed that human players in the Skip-Bo scenario judge our virtual human MAX significantly older when secondary emotions are simulated in addition to primary ones.

1 Introduction and Motivation

Researchers in the field of Embodied Conversational Agents (ECAs) [8,29] build anthropomorphic systems, which are employed in different interaction scenarios, that afford communicative abilities of different style and complexity. As these agents comprise an increasing number of sensors as well as actuators together with an increase in expressive capabilities, they need to be able to recognize and produce social cues in face-to-face communication.

One factor in social interaction is the ability to deal with the affective dimension appropriately. Therefore researchers in the growing field of "Affective Computing" [27] discuss ways to derive human affective states from all kinds of intrusive and non-intrusive sensors. With regard to the expressive capabilities of embodied agents the integration of emotional factors influencing bodily expressions is argued for. These bodily expressions include, e.g., facial expression, body posture, and voice inflection and all of them must be modulated in concert to synthesize coherent emotional behavior.

---
[1] [W]ASABI [A]ffect [S]imulation for [A]gents with [B]elievable [I]nteractivity.

With the WASABI architecture we present our attempt to exploit different findings and conceptions of emotion psychology, neurobiology, and developmental psychology in a fully-implemented computational architecture. It is based on the simulation of an emotion dynamics in three-dimensional emotion space, which has proven beneficial to increase the lifelikeness and believability of our virtual human MAX in two different interaction scenarios [3]. With this emotion dynamics simulation, however, we are limited to a class of rather simple emotions, that are similar to Damasio's conception of primary emotions. In the WASABI architecture we use our agent's cognitive reasoning abilities to model the mood-congruent elicitation of secondary emotions as well.

In the following section the psychological as well as neurobiological background is described with respect to the distinction of primary as well as secondary emotions. Subsequently we give an overview of related work in the field of Affective Computing. The WASABI architecture is presented in Section 3. We explain how nine primary emotions together with three secondary emotions namely the prospect-based emotions *hope*, *fears-confirmed*, and *relief* were integrated in such a way that their mood-congruent elicitation can be guaranteed. We conclude by discussing results of a first empirical study on the effect of secondary emotion simulation in the Skip-Bo card game scenario.

2 Background and Related Work

We will first introduce the psychological and neurobiological background before an overview of related work in affective human-computer interaction is given.

2.1 Psychological and Neurobiological Background

According to [33] the "psychological construct" labeled emotion can be broken up into the component of cognitive appraisal, the physiological component of activation and arousal, the component of motor expression, the motivational component, and the component of subjective feeling state. We follow the distinction of a cognitive appraisal component and a physiological component in the computational simulation of affect.[2] We also account for the motivational component, because our agent's reasoning capabilities are modeled according to the belief-desire-intention approach to modeling rational behavior [30]. We further believe, that we first need to realize the dynamic interaction between a cognitive and a physiological component before we can tackle the question of how to computationally realize a subjective feeling state.

Recently, psychologists started to investigate "unconscious processes in emotions" [35] and Ortony at al. discuss levels of processing in "effective functioning" by introducing a distinction between "emotions" and "feelings" [26]. They understand feelings as "readouts of the brain's registration of bodily conditions and changes" whereas "emotions are interpreted feelings" [26, p. 174] and propose

---

[2] The motor expression component has been realized as well, but is not in the scope of this paper (see [3] for details).

three different levels of information processing [26], which are compatible with Scherer's three modes of representation [35].

According to Ortony and colleagues, lower-levels have to contribute in order to experience "hot" emotions such that "cold, rational anger" could be solely the product of the cognitive component "without the concomitant feeling components from lower levels." [26, p. 197] A purely primitive feeling of fear, on the contrary, also lacks the necessary cognitive elaboration to become a full-blown emotion. These psychological considerations are compatible with LeDoux's distinction of a low and a high road of fear elicitation in the brain [20] and Damasio's assumption of bodily responses causing an "emotional body state" [9, p. 138] that is subsequently analyzed in the thought process.

Two classes of emotions. Damasio's neurobiological research on emotions suggests the distinction of at least two classes of emotions, namely, primary and secondary emotions [9].

The term "primary emotions" [9] refers to emotions which are supposed to be inate. They developed during phylogeny to support fast and reactive response behavior in case of immediate danger, i.e. basic behavioral response tendencies like "flight-or-fight" behaviors. In humans, however, the perception of the changed bodily state is combined with the object that initiated it resulting in a "feeling of the emotion" with respect to that particular object [9, p. 132]. Primary emotions are also understood as a prototypical emotion types which can already be ascribed to one year old children [10].

They are comparable to the concept of "Core Affect" [32,14], which is based on the assumption that emotions cannot be identified by distinct categories from the start. "Core Affect" is represented in two-dimensional emotion space of Pleasure/Valence and Arousal, which is sometimes extended by a third dimension [37,31,17] labeled "Control/Dominance/Power" of connotative meaning. Furthermore, Wundt claims that any emotion can be characterized as a continuous progression in such three-dimensional emotion space [37]. The degree of Dominance not only describes the experienced "control" over the emotion or the situational context, but can also belong to an agent's personality traits. The three-dimensional abstract emotion space is often referred to as PAD space.

Secondary emotions like "relief" or "hope" are assumed to arise from higher cognitive processes, based on an ability to evaluate preferences over outcomes and expectations. Accordingly, secondary emotions are acquired during ontogenesis through learning processes in the social context.

Damasio uses the adjective "secondary" to refer to "adult" emotions, which utilize the machinery of primary emotions by influencing the acquisition of "dispositional representations", which are necessary for the elicitation of secondary emotions. These "acquired dispositional representations", however, are believed to be different from the "innate dispositional representations" underlying primary emotions. Furthermore, secondary emotions influence bodily expressions through the same mechanisms as primary emotions.

2.2 Related Work

El-Nasr et al. [13] present FLAME as a formalization of the dynamics of 14 emotions based on fuzzy logic rules, that includes a mood value, which is continuously calculated as the average of all emotion intensities to provide a solution to the problem of conflicting emotions being activated at the same time. The idea of expectations is realized in FLAME by means of learning algorithms based on rewards and punishments. Although it was not integrated into the simulation of a virtual human, the mutual influence of emotion and mood is quite similar to the conception of emotion dynamics in the WASABI architecture.

Marsella and Gratch [23] focus with their "EMA" model of emotions on the dynamics of emotional appraisal. They also argue for a mood value as an addend in the calculation of otherwise equally activated emotional states following the idea of mood-congruent emotions. Their framework for modeling emotions is the first fully implemented, domain-independent architecture for emotional conversational agents.

Marinier and Laird [21] aim to combine the work of Gratch and Marsella [23] with the findings of Damasio [9]. In later publications [22], however, Damasio's work is less central and they follow the ideas of Scherer [34]. Their central idea of "appraisal frames" is based on the EMA model [23] and eleven of Scherer's sixteen appraisal dimensions are modeled for integration in the Soar cognitive architecture. They distinguish an "Active Appraisal Frame", which is the result of a momentary appraisal of a given event, from a "Perceived Appraisal Frame", which results from the combination of the actual mood and emotion frames. Thereby, they claim to account for Damasio's distinction between emotion and feeling—similarly to the conception underlying the WASABI architecture.

Although André et al. [1] start with the distinction of primary and secondary emotions, it is not taken up in their later publications, e.g. [16]. Gebhard [15], recently, uses the PAD space to derive a mood value from emotions resulting from OCC-based appraisal. Three-dimensional emotion spaces similar to PAD space are also used to drive the sociable robot "Kismet" [7] or the humanoid robot WE-4RII [19].

3 The WASABI Architecture

The WASABI architecture conceptualized here builds upon previous work on the simulation of emotion dynamics for the virtual human MAX [2] that has proven to support the agent's believability in two different interaction scenarios [3,6]. It was, however, limited to the simulation of primary emotions.

Accordingly, the WASABI architecture [5] combines bodily emotion dynamics with cognitive appraisal in order to simulate infant-like primary emotions as well as cognitively elaborated, more adult secondary emotions. In the following a suitable specification of the different concepts *emotion* and *mood* is derived from the theoretical background presented above.

Emotions are understood as current states with a specific quality and intensity, which are the outcome of complex neurophysiological processes for communication. The processes include neural activity of the brain as well as physiological responses of the body. One gets aware of one's emotions in two cases: (1) if their awareness likelihood w exceeds a certain threshold (cf. Section 3.3) or (2) if one concentrates on the underlying processes by means of introspection.

Emotions can be classified into primary and secondary ones, but every emotion has either positive or negative valence of a certain value and compared to mood an emotion lasts significantly less long. The differences between primary and secondary emotions are conceptualized as follows:

- Secondary emotions are based on more complex data structures than primary ones. Accordingly, only some general aspects of secondary emotions (such as their valence components) are represented in PAD space.
- The appraisal of secondary emotions depends much more on the situational and social context than that of primary emotions. Thus, secondary emotions are more dependent on the agent's cognitive reasoning abilities.
- The releasers of secondary emotions might be learned based on the history of primary emotions in connection with memories of events, agents and objects.
- The agent's facial expressions of primary emotions (cf. Figure 1) may accompany secondary emotions.
- Secondary emotions also modulate the agent's simulated embodiment.

Mood is understood as a background state with a much simpler affective quality than emotions. It is assumed that bodily responses influence the development of mood over time and that a mood is a diffuse valenced state, i.e. the experiencing individual is likely to be unable to give a clear reason for his or her current mood. Emotions have a fortifying or alleviating effect on an individual's mood, which, in turn, influences the elicitation of emotions [24]. A mood's duration is in general longer than that of emotions.

3.1 Nine Primary Emotions

Primary emotions (PE) are inborn affective states, which are triggered by reflexes in case of potentially harmful stimuli. They result in fast, reactive behavioral responses and, thus, are quite similar to the concept of proto-affect proposed by [26]. According to developmental psychology, young children express their (primary) emotions directly, because they did not yet internalize this process as in the case of adults [18].

In our previous realization of emotion dynamics [2] this direct expression of primary emotions is achieved by implementing five of Ekman's six "basic emotions". In addition, the emotions "bored", "annoyed", and "depressed" as well as the non-emotional state "concentrated" are simulated. Each of the primary emotions is located in PAD space according to Table 1, for which the coordinates are derived from some of the values given in [31, p. 286ff]. The seven facial expressions of MAX corresponding to the eight primary emotions and the neutral state "concentrated" (cf. Table 1) are shown in Figure 1. In case of high

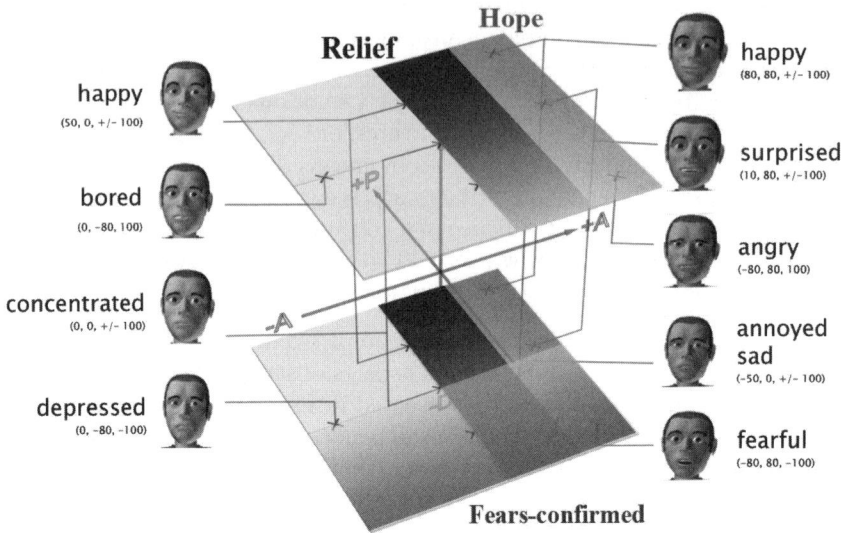

Fig. 1. The nine primary emotions of Table 1 as points together with the three secondary emotions of Table 2 as weighted areas in PAD space

pleasure Ekman's set of "basic emotions" [12] only contains one obviously positive emotion, namely happiness. Thus, in the presented implementation [5] this primary emotion covers the whole area of positive pleasure regardless of arousal or dominance as it is located in PAD space four times altogether.

3.2 Three Secondary Emotions

According to Damasio, the elicitation of secondary emotions involves a "thought process", in which the actual stimulus is evaluated against previously acquired experiences and online generated expectations.

The "prospect-based emotions" cluster of the OCC-model of emotions [25] is considered here to belong to the class of secondary emotions, because their appraisal process includes the evaluation of events against experiences and expectations. This OCC-cluster consists of six emotions, of which *hope*, *fears-confirmed*, and *relief* are simulated in the WASABI architecture.

Hope. Ortony et al. describe *hope* as resulting from the appraisal of a prospective event [25]. If the potential event is considered desirable for oneself, one is likely to be "pleased about the prospect of a desirable event" [25, p. 110]. The calculation of this emotion's awareness likelihood, however, is rather independent from these cognitive processes.

The previous analysis provides the rationale for modeling *hope* in the following way:

– Pleasure: The awareness likelihood of *hope* increases the more pleasurable the agent feels.

Table 1. Primary emotions in PAD space: The five "basic emotions" of [11] are assigned to the corresponding facial expressions modeled in [2] whenever such a mapping is possible (cp. Figure 1) and additionally an individual base intensity i_{pe} is set for each primary emotion (see also Section 3.3)

PE final	Facial expr. (Ekman)	PAD values	base intensity i_{pe}
1. angry	anger (*anger*)	(80, 80, 100)	0.75
2. annoyed	sad (*sadness*)	(-50, 0, 100)	0.75
3. bored	bored (*none*)	(0, -80, 100)	0.75
4. concentrated	neutral (*none*)	(0, 0, ±100)	0.75
5. depressed	sad (*sadness*)	(0, -80, -100)	0.75
6. fearful	fear (*fear*)	(-80, 80, 100)	0.25
7. happy	happy (*happiness*)	(80, 80, ±100) (50, 0, ±100)	0.75
8. sad	sad (*sadness*)	(-50, 0, -100)	0.75
9. surprised	surprised (*surprise*)	(10, 80, ±100)	0.0

- Arousal: With respect to an agent's arousal, *hope* is more likely elicited the higher the agent's arousal value.
- Dominance: The awareness likelihood of *hope* is modeled to be independent of the agent's general level of dominance.

To realize this distribution of awareness likelihood in case of hope, two areas are introduced in Figure 1, one in the high dominance plane and the other in the low dominance plane. In Table 2 the exact values of the four corners of each of the two areas together with the respective base intensity in each corner is given for *hope*.

Fears-confirmed. According to Ortony et al., *fears-confirmed* is elicited when being "displeased about the confirmation of the prospect of an undesirable event." [25, p. 110] With respect to its representation in PAD space the similarity to the primary emotion *fearful* is taken into account and the following decisions are taken:

- Pleasure: The awareness likelihood of *fears-confirmed* increases the less pleasurable the agent feels.
- Arousal: *fears-confirmed* is considered to be independent from the agent's arousal value.
- Dominance: *fears-confirmed* can only be perceived by the agent, when he feels submissive as in the case of *fearful*.

This distribution of awareness likelihood is realized in PAD space (cf. Figure 1) by introducing an area in the low dominance plane (cf. Table 2 for the exact coordinates and intensities).

Relief. The secondary emotion *relief* is described as being experienced whenever one is "pleased about the disconfirmation of the prospect of an undesirable event." [25, p. 110] Taking the similarity with Gehm and Scherer's "content" cluster into account [17], *relief* is represented in PAD space according to the following considerations:

Table 2. The parameters of the secondary emotions *hope*, *fears-confirmed*, and *relief* for representation in PAD space

Area	(PAD values), intensity
HOPE	
high dominance	(100, 0, 100), 0.6; (100, 100, 100), 1.0; (-100, 100, 100), 0.5; (-100, 0, 100), 0.1
low dominance	(100, 0, -100), 0.6; (100, 100, -100), 1.0; (-100, 100, -100), 0.5; (-100, 0, -100), 0.1
FEARS-CONFIRMED	
low dominance	(-100, 100, -100), 1.0; (0, 100, -100), 0.0; (0, -100, -100), 0.0; (-100, -100, -100), 1.0
RELIEF	
high dominance	(100, 0, 100), 1.0; (100, 50, 100), 1.0; (-100, 50, 100), 0.2; (-100, 0, 100), 0.2
low dominance	(100, 0, -100), 1.0; (100, 50, -100), 1.0; (-100, 50, -100), 0.2; (-100, 0, -100), 0.2

- Pleasure: *relief* is more likely to become aware the more pleasurable the agent feels.
- Arousal: Only in case of relatively low arousal levels the agent is assumed to get aware of the emotion *relief*.
- Dominance: The awareness likelihood of *relief* is considered to be independent from the agent's state of dominance.

The awareness likelihood is represented in Figure 1 by two areas, one located in the high dominance plane and the other in the low dominance plane (cf. Table 2).

3.3 Emotion Dynamics and Awareness Likelihood

The implementation of emotion dynamics is based on the assumption that an organisms natural, homeostatic state is characterized by emotional balance, which accompanies an agent's normal level of cognitive processing [36]. Whenever an emotionally relevant internal or external stimulus is detected, however, its valence component serves as an emotional impulse, which disturbs the homeostasis causing certain levels of pleasure and arousal in the emotion module. Furthermore, a dynamic process is started by which these values are continuously driven back to the state of balance (see [3] for details).

The two valences are mathematically mapped into PAD space (cf. Figure 1) and combined with the actual level of Dominance, which is derived from the situational context in the cognition of the architecture. This process results in a course of a reference point in PAD space representing the continuously changing, bodily feeling state from which the awareness likelihood of primary and secondary emotions is derived with an update rate of 25Hz (see also [4,6]).

Awareness likelihood of primary emotions. The awareness likelihood of any of the nine primary emotions pe (cf. Table 1) increases the smaller the distance between the actual PAD values and the primary emotion's PAD values (i.e. d_{pe} in Equation 1). When d_{pe} falls below Φ_{pe} units for a particular primary emotion pe, the calculation of its awareness likelihood w_{pe} is started according to Equation 1 until d_{pe} falls below Δ_{pe} units in which case the likelihood is w_{pe} equals the primary emotion's base intensity i_{pe}.

$$w_{pe} = (1 - \frac{d_{pe} - \Delta_{pe}}{\Phi_{pe} - \Delta_{pe}}) \cdot i_{pe}, \quad with \quad \Phi_{pe} > \Delta_{pe} \ \forall pe \in \{pe_1, \ldots, pe_9\} \quad (1)$$

In Equation 1, Φ_{pe} can be interpreted as the activation threshold and Δ_{pe} as the saturation threshold, which can be adjusted for every primary emotion $pe_n \in \{pe_1, \ldots, pe_9\}$ independently[3]. By setting a primary emotion's base intensity i_{pe} to 0.0 (as in the case of *surprised*, cf. Table 1) it needs to be *triggered* by the cognition before it might gain a non-zero awareness likelihood w_{pe}.

In case of primary emotions that are represented in PAD space more than once (i.e. concentrated, happy, and surprised; cf. Table 1) the representation with the minimum distance to the reference point is considered in Equation 1 for calculation of its awareness likelihood.

Awareness likelihood of secondary emotions. With representing the three secondary emotions in PAD space their mood-congruent elicitation can be assured, because the actual PAD values are also relevant for calculating every secondary emotion's awareness likelihood. In contrast to most primary emotions, all secondary emotion's *base intensities* are set to zero by default (cp. the case of *surprised* above).

Accordingly, every secondary emotion has first to be triggered by a cognitive process, before it gains the potential to get aware to the agent. Furthermore, a secondary emotion's *lifetime* parameter (set to 10.0 by default) together with its *decay function* (set to linear by default) are used to decrease its intensity over time until its *base intensity* of zero is reached again.

In the Skip-Bo card game scenario [6], for example, Max might believe that the opponent may play a card hindering him to fulfill one of his goals and the expectation of an undesirable event is generated. Later, upon perceiving and interpreting the opponent's actions, Max might realize that the opponent fulfilled the expectation by playing that undesired card. In result the secondary emotion *fears-confirmed* is triggered and, at the same time, a negative emotional impulse is sent to the emotion dynamics. If the corresponding PAD values fall into the *fears-confirmed* region (cf. Figure 1), Max will get aware that his *fears are confirmed* with a likelihood that results from linear interpolation between the current intensities in the four corners of the *fears-confirmed* area at the location given by the actual PAD values. Currently, each non-zero likelihood of a secondary emotion lets MAX produce an appropriate verbal expression (cf. Figure 2(a) and Figure 2(b)).

---
[3] The nine primary emotions are indexed according to Table 1.

(a) MAX expresses his *hope* that the human player will play the card with the number seven next by saying "Kannst Du nicht die 7 spielen?" (Can't you play the seven?)

(b) MAX realizes that his *fears* just got *confirmed* and utters "Das hatte ich schon befrchtet!" (I was already afraid of that!)

Fig. 2. MAX expressing his *hope* and realizing that his *fears* got *confirmed*

3.4 Connecting Cognition and Embodiment

In Figure 3 the conceptual distinction of an agent's simulated embodiment and its cognition is presented and the different modules and components of the WASABI architecture are assigned to the corresponding layers.

To the left of Figure 3 the virtual human MAX perceives some (internal or external) stimulus. *Non-conscious appraisal* is realized by directly sending a small positive *emotional impulse* to the *Emotion dynamics* component of the WASABI architecture. This establishes the "low road" [20] of primary emotion elicitation. For example, the presence of visitors in the museum [2] is interpreted as *intrinsically pleasant* following the ideas of Scherer [34].

Another path resulting in *emotional impulses* begins with *conscious appraisal* of the perceived stimulus (cf. Figure 3, top left). This process resides in the *Cognition*, because it is based on the evaluation of goal-conduciveness of an event [34] and can be considered the "high road" of emotion elicitation [20]. Therefore, MAX exploits his BDI-based cognitive reasoning abilities to update his *memory* and generate *expectations*. These deliberative processes not only enable MAX to derive his subjective level of *Dominance* from the situational and social context, but also propose cognitively plausible *secondary emotions*.

These *secondary emotions* are, however, first *filtered* in *PAD space*, before MAX might get *aware* of them (cf. Figure 3, middle). Independent of this "awareness filter", every cognitively plausible *secondary emotion* influences the *Emotion dynamics* component of the WASABI architecture, thereby modulating MAX's *Pleasure* and *Arousal* values, i.e. his simulated *Embodiment*. This influence is achieved by interpreting the valence component of any *secondary emotion* as an *emotional impulse* (cf. Figure 3, left). This way, *secondary* emotions "utilize the

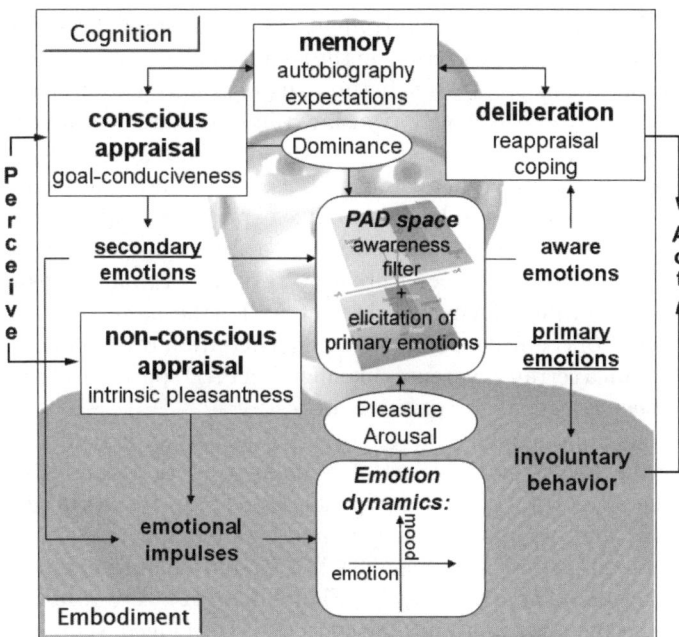

Fig. 3. The conceptual distinction of cognition and embodiment in the WASABI architecture

machinery of primary emotions" [9], because they might result in the elicitation of mood-congruent *primary emotions*, which in the WASABI architecture drive MAX's facial expressions *involuntarily*. Furthermore, as the *Pleasure* and *Arousal* values are incessantly modulating MAX's *involuntary behaviors* (i.e. breathing and eye blinking) as well, even "unaware" *secondary emotions* have an effect on MAX's bodily state and involuntary behavior.

In combination with the actual level of *Dominance*, primary emotions are elicited by means of a distance metric in *PAD space*. As mentioned before, these primary emotions are directly driving MAX's facial expressions. Although this automatism might be considered unnatural for an adult, it has proven applicable and believable in the situational contexts in which MAX was integrated so far.

After the awareness filter has been applied, the resulting set of *aware emotions* consists of primary and secondary emotions together with their respective awareness likelihoods. They are finally subject to further deliberation and reappraisal resulting in different coping behaviors. In the card game scenario the direct vocal and facial expression of negative emotions is sufficient to let the human players play in accordance with the rules.

4 Discussion and Conclusion

We presented the WASABI architecture for mood-congruent simulation of primary and secondary emotions as it is integrated in, and makes use of, the overall cognitive architecture of our Virtual Human MAX. The simulation and direct expression of primary emotions is based on the idea to capture an agent's bodily feeling state as a continuous progression in three-dimensional emotion space (i.e. PAD space), which is only subsequently translated into weighted, primary emotions. Secondary emotions, in contrast, are understood as onto-genetically later types of emotions, which require higher cognitive reasoning abilities and a certain sense of time, in that an agent has to be able to take experiences and expectations into account to generate prospect-based emotions. To also assure mood-congruency of secondary emotions, we roughly capture aspects of their connotative meanings in PAD space as well by introducing weighted areas. Furthermore, to account for the decisive influence of cognitive processes in the elicitation of secondary emotions, they can only gain a certain awareness likelihood in PAD space of the embodiment, after having been triggered by cognitive processes.

Although our approach has not yet been subject to an extended evaluation, some empirical evidence on the effect of secondary emotion simulation could already be gathered. A total of 23 participants played Skip-Bo either against MAX only simulating and expressing primary emotions (similar to the negative empathic condition in [28], n=11) or against MAX additionally simulating and verbally expressing secondary emotions (cf. Figure 2, n=12). As a result, MAX with primary and secondary emotions "in concert" was judged significantly older (mean value 27,5 years, standard deviation 7.5) than MAX with simulated primary emotions alone (mean value 19,8 years, standard deviation 7.7)[4].

In summary, we believe that the WASABI architecture is a helpful model to understand how the dynamic interplay of a human's body and mind together with his past experiences and future expectations sometimes turns "cold" cognitions into "hot" affective states.

References

1. André, E., Klesen, M., Gebhard, P., Allen, S., Rist, T.: Integrating models of personality and emotions into lifelike characters. In: Proceedings International Workshop on Affect in Interactions - Towards a New Generation of Interfaces, pp. 136–149 (1999)
2. Becker, C., Kopp, S., Wachsmuth, I.: Simulating the emotion dynamics of a multimodal conversational agent. In: André, E., Dybkjær, L., Minker, W., Heisterkamp, P. (eds.) ADS 2004. LNCS (LNAI), vol. 3068, pp. 154–165. Springer, Heidelberg (2004)
3. Becker, C., Kopp, S., Wachsmuth, I.: Why emotions should be integrated into conversational agents. In: Nishida, T. (ed.) Conversational Informatics: An Engineering Approach, November 2007, ch. 3, pp. 49–68. Wiley, Chichester (2007)
4. Becker, C., Wachsmuth, I.: Modeling primary and secondary emotions for a believable communication agent. In: Reichardt, D., Levi, P., Meyer, J.-J.C. (eds.) Proceedings of the 1st Workshop on Emotion and Computing, Bremen, pp. 31–34 (2006)

---

[4] A two-tailed t-test assuming unequal variances results in $p = 0.025$. No significant effect of the participants' gender could be found.

5. Becker-Asano, C.: WASABI: Affect Simulation for Agents with Believable Interactivity. PhD thesis, AI Group, University of Bielefeld (to appear, 2008)
6. Becker-Asano, C., Kopp, S., Pfeiffer-Lemann, N., Wachsmuth, I.: Virtual Humans Growing up: From Primary Toward Secondary Emotions. KI Zeitschrift (German Journal of Artificial Intelligence) 1, 23–27 (2008)
7. Breazeal, C.: Emotion and sociable humanoid robots. International Journal of Human-Computer Studies 59, 119–155 (2003)
8. Cassell, J., Sullivan, J., Prevost, S., Churchill, E.: Embodied Conversational Agents. The MIT Press, Cambridge (2000)
9. Damasio, A.: Descartes' Error, Emotion Reason and the Human Brain. Grosset/Putnam (1994)
10. Damasio, A.: Looking for Spinoza: Joy, Sorrow, and the Feeling Brain. Harcourt (2003)
11. Ekman, P.: Facial expressions. In: Handbook of Cognition and Emotion, ch. 16, pp. 301–320. John Wiley, Chichester (1999)
12. Ekman, P., Friesen, W., Ancoli, S.: Facial sings of emotional experience. Journal of Personality and Social Psychology 29, 1125–1134 (1980)
13. El-Nasr, M.S., Yen, J., Ioerger, T.R.: FLAME - Fuzzy Logic Adaptive Model of Emotions. Autonomous Agents and Multi-Agent Systems 3(3), 219–257 (2000)
14. Feldman Barrett, L.: Feeling is perceiving: Core affect and conceptualization in the experience of emotion. In: The unconscious in emotion, ch. 11, pp. 255–284. Guilford Press, New York (2005)
15. Gebhard, P.: ALMA - A Layered Model of Affect. In: Autonomous Agents & Multi Agent Systems, pp. 29–36 (2005)
16. Gebhard, P., Klesen, M., Rist, T.: Coloring multi-character conversations through the expression of emotions. In: André, E., Dybkjær, L., Minker, W., Heisterkamp, P. (eds.) ADS 2004. LNCS (LNAI), vol. 3068, pp. 128–141. Springer, Heidelberg (2004)
17. Gehm, T.L., Scherer, K.R.: Factors determining the dimensions of subjective emotional space. In: Scherer, K.R. (ed.) Facets of Emotion, ch. 5. Lawrence Erlbaum Associates, Mahwah (1988)
18. Holodynski, M., Friedlmeier, W.: Development of Emotions and Emotion Regulation. Springer, Heidelberg (2005)
19. Itoh, K., Miwa, H., Nukariya, Y., Zecca, M., Takanobu, H., Roccella, S., Carrozza, M., Dario, P., Takanishi, A.: Behavior generation of humanoid robots depending on mood. In: 9th International Conference on Intelligent Autonomous Systems (IAS-9), pp. 965–972 (2006)
20. LeDoux, J.: The Emotional Brain. Touchstone. Simon & Schuster (1996)
21. Marinier, R., Laird, J.: Toward a comprehensive computational model of emotions and feelings. In: International Conference on Cognitive Modeling (2004)
22. Marinier, R.P., Laird, J.E.: Computational modeling of mood and feeling from emotion. In: CogSci, pp. 461–466 (2007)
23. Marsella, S., Gratch, J.: EMA: A computational model of appraisal dynamics. In: European Meeting on Cybernetics and Systems Research (2006)
24. Neumann, R., Seibt, B., Strack, F.: The influence of mood on the intensity of emotional responses: Disentangling feeling and knowing. Cognition & Emotion 15, 725–747 (2001)
25. Ortony, A., Clore, G.L., Collins, A.: The Cognitive Structure of Emotions. Cambridge University Press, Cambridge (1988)

26. Ortony, A., Norman, D., Revelle, W.: Affect and proto-affect in effective functioning. In: Fellous, J., Arbib, M. (eds.) Who needs emotions: The brain meets the machine, pp. 173–202. Oxford University Press, Oxford (2005)
27. Picard, R.W.: Affective Computing. The MIT Press, Cambridge (1997)
28. Prendinger, H., Becker, C., Ishizuka, M.: A study in users physiological response to an empathic interface agent. International Journal of Humanoid Robotics 3(3), 371–391 (2006)
29. Prendinger, H., Ishizuka, M. (eds.): Life-Like Characters. Tools, Affective Functions, and Applications. Cognitive Technologies. Springer, Heidelberg (2004)
30. Rao, A., Georgeff, M.: Modeling Rational Agents within a BDI-architecture. In: Allen, J., Fikes, R., Sandewall, E. (eds.) Proc. of the Intl. Conference on Principles of Knowledge Representation and Planning, pp. 473–484. Morgan Kaufmann publishers Inc, San Mateo (1991)
31. Russell, J., Mehrabian, A.: Evidence for a three-factor theory of emotions. Journal of Research in Personality 11(11), 273–294 (1977)
32. Russell, J.A., Feldmann Barrett, L.: Core affect, prototypical emotional episodes, and other things called emotion: Dissecting the elephant. Journal of Personality and Social Psychology 76(5), 805–819 (1999)
33. Scherer, K.R.: On the nature and function of emotion: A component process approach. In: Scherer, K., Ekman, P. (eds.) Approaches to Emotion, pp. 293–317. Lawrence Erlbaum, Mahwah (1984)
34. Scherer, K.R.: Appraisal considered as a process of multilevel sequential checking. In: Scherer, K.R., Schorr, A., Johnstone, T. (eds.) Appraisal Processes in Emotion, ch. 5. Oxford University Press, Oxford (2001)
35. Scherer, K.R.: Unconscious processes in emotion: The bulk of the iceberg. In: Niedenthal, P., Feldman Barrett, L., Winkielman, P. (eds.) The unconscious in emotion. Guilford Press, New York (2005)
36. Sloman, A., Chrisley, R., Scheutz, M.: The architectural basis of affective states and processes. In: Who needs emotions? Oxford University Press, Oxford (2005)
37. Wundt, W.: Vorlesung über die Menschen- und Tierseele. Voss Verlag, Leipzig 1922/1863

User Study of AffectIM, an Emotionally Intelligent Instant Messaging System

Alena Neviarouskaya[1], Helmut Prendinger[2], and Mitsuru Ishizuka[1]

[1] University of Tokyo, Department of Information and Communication Engineering, Japan
`lena@mi.ci.i.u-tokyo.ac.jp,ishizuka@i.u-tokyo.ac.jp`
[2] National Institute of Informatics, Japan
`helmut@nii.ac.jp`

Abstract. Our research addresses the tasks of recognition, interpretation and visualization of affect communicated through text messaging. In order to facilitate sensitive and expressive interaction in computer-mediated communication, we previously introduced a novel syntactical rule-based approach to affect recognition from text. The evaluation of the developed Affect Analysis Model showed promising results regarding its capability to accurately recognize affective information in text from an existing corpus of informal online conversations. To enrich the user's experience in online communication, make it enjoyable, exciting and fun, we implemented a web-based IM application, AffectIM, and endowed it with emotional intelligence by integrating the developed Affect Analysis Model. This paper describes the findings of a twenty-person study conducted with our AffectIM system. The results of the study indicated that automatic emotion recognition function can bring a high level of affective intelligence to the IM application.

Keywords: Affective sensing from text, affective user interface, avatar, emotions, online communication, user study.

1 Introduction and Motivation

The essentialness of emotions to social life is manifested by the rich history of theories and debates about emotions and their nature. Recently, the task of recognition of affective content conveyed through written language is gaining increased attention by researchers interested in studying different kinds of affective phenomena, including sentiment analysis, subjectivity and emotions. In order to analyse affect communicated through written language, researchers in the area of natural language processing proposed a variety of approaches, methodologies and techniques [2,8-11].

Advanced approaches targeting at textual affect recognition performed at the sentence-level are described in [1,3,4]. The lexical, grammatical approach introduced by Mulder et al. [4] focused on the propagation of affect towards an object. Boucouvalas [1] developed the Text-to-Emotion Engine based on word tagging and analysis of sentences. An approach for understanding the underlying semantics of language using large-scale real-world commonsense knowledge was proposed by Liu et al. [3], who incorporated the created affect sensing engine into an affectively responsive email composer called EmpathyBuddy.

Peris et al. [7] argues that online chats may stimulate rather than inhibit social relations, and chat users seem to find a media for rich, intense, and interesting experiences. The motivation behind our research is to enrich social interactivity and emotional expressiveness of real-time messaging, where a machine is used as a communication channel connecting people and transmitting human emotions. Here, a key issue is to provide the automation of multiple expressive means so that the user does not have to worry about visual self-presentation as in standard Instant Messaging (IM) systems, but can focus on the textual content of the conversation. While constructing our Affect Analysis Model we took into account crucial aspects of informal online conversation such as its specific style and evolving language [5].

The remainder of the paper is structured as follows. In Section 2 we shortly introduce the developed Affect Analysis Model. We describe the developed IM application integrated with the Affect Analysis Model and analyse the results of a user study in Section 3 and Section 4, respectively. In Section 5 we conclude the paper.

2 Rule-Based Approach to Affect Sensing from Text

We proposed a rule-based approach to affect sensing from text at a sentence-level (details are given in [6]). The algorithm for analysis of affect in text consists of five stages: (i) symbolic cue analysis, (ii) syntactical structure analysis, (iii) word-level analysis, (iv) phrase-level analysis, and (v) sentence-level analysis. The salient features of this algorithm are: (1) analysis of nine emotions and five communicative functions on the level of individual sentences; (2) the ability to handle the evolving language of online communications; (3) foundation in affect database; (4) vector representation of affective features of words, phrases, clauses and sentences; (5) consideration of syntactic relations in a sentence; (6) analysis of negation, modality, and conditionality; (7) consideration of relations between clauses in compound, complex, or complex-compound sentences; and (8) emotion intensity estimation.

An empirical evaluation of the Affect Analysis Model algorithm [6] showed promising results regarding its capability to accurately classify affective information in text from an existing corpus of informal online communication. In a study based on blog entries, the system result agreed with at least two out of three human annotators in 70% of the cases.

3 Instant Messaging Application Integrated with the Affect Analysis Model

The AffectIM, an Instant Messaging system with emotional intelligence, was developed as a web-based application running in the Internet browser. Within our research project, we could design only two avatars, one male and one female. So the graphical representative is automatically selected by the system according to the user's sex.

The main window of AffectIM system while online conversation is shown in Fig. 1. From the list of friends displayed in the left frame, the user selects the person (available online), whom he or she wishes to communicate with. The central frame allows user to type and to send the messages. It displays the conversation flow in three modes: plain, transcribed, and with emotions. Further, it displays emotional avatars (own – to the left of conversation field, and friend's – to the right). Two buttons located under the avatar animation refer to the visualization of emotion distribution (either in a color bar or pie graph) and emotion dynamics (line graph). Since the language of online communication is constantly evolving, AffectIM also provides the functionality to add new abbreviations, acronyms, and emoticons to the Affect database (see two buttons located to the left from the input text field).

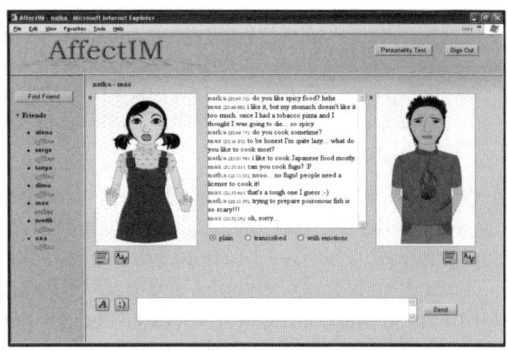

Fig. 1. AffectIM interface

4 User Study of the AffectIM System

The purpose of the user study was to evaluate "richness of experience" and "affective intelligence" of our AffectIM system. We hypothesized that user experience and effectiveness of the communication of emotions may benefit from introduction of automatic emotion recognition function and emotionally expressive avatars to IM application, as opposed to manual selection of emotional behavior or uninformed, random display of affect. Our hypotheses are tested by considering the following dimensions regarding users' experience: (1) Interactivity. (2) Involvement (engagement). (3) Sense of copresence. (4) Enjoyment. (5) Affective intelligence. (6) Overall satisfaction. In addition to these main criteria, we asked participants to give us feedback on some general questions.

4.1 Experimental Design, Subjects and Procedure

The experiment was designed as a within-subjects experiment in pairs. In particular, we compared three AffectIM interfaces using different configuration conditions.

For the user study, we prepared three versions of the system:

1. Automatic (A-condition). In this interface, affect sensing from text is performed based on the developed Affect Analysis Model, and the recognized emotions are conveyed by the avatar expressions.
2. Manual (M-condition). During this condition, no automatic emotion recognition from text is performed; however, users may select emotion (and its intensity) to be shown by avatars using "select pop-up menus".
3. Random (R-condition). Here, the avatars show a 'quasi-random' reaction. First, we process each sentence using the Affect Analysis Model, and then we apply

two rules: (1) if the output is emotional, we run two functions that randomly select the emotion out of nine available emotions and its intensity, correspondingly; (2) for the case of "neutral" output, we set the function that generates "neutral" emotion with the probability of 60% or "random" emotion with the probability of 40%.

It is important to note that in each of three interfaces five communicative functions are automatically recognized and shown by avatars. In other words, the occurrence of communicative behavior is not varied across the three experimental conditions.

Twenty university students and staff (10 males, 10 females) took part in our study. All of them were computer literate, and 19 persons had prior experience with computer based chat or Instant Messaging system.

Each pair of participants was composed by male and female subjects. Before the IM session, all participants were given instructions and their AffectIM IDs and passwords. Each pair of participants was asked to have online conversations through three interfaces given in random order. After each interface condition, users filled the corresponding page of the questionnaire in and commented on their experience.

After the participants completed the IM communication about the three topics and corresponding questionnaire, they were asked to answer some general questions about their experience with the IM system.

4.2 Analysis of Results

The average duration of sessions on each interface was 10.1 minutes (minimum 8 and maximum 12.5 minutes), excluding the time needed to fill out the questionnaires.

The 11 questions on main criteria were answered based on 7-item agreement Likert scale. Since our study involved each subject being measured under each of three conditions, we analyzed data using statistical method ANOVA (two-factor ANOVA without replications with chosen significance level $p < 0.05$).

The **interactivity** was measured using statement *"The system was interactive"*. Subjects tended to consider the condition, which allowed them to manipulate the expressed emotion manually, as most interactive. However, ANOVA resulted in no significant difference in interactivity among three interfaces.

The **involvement** was evaluated using two questionnaire items: *"I felt it was important for my conversation partner that I responded after each his/her statement"* and *"I was awaiting the replies of my conversation partner with true interest"*. The ANOVA results showed that the reported involvement of all three systems does not differ significantly, showing that the level of engagement was almost the same.

The following two questionnaire items covering the aspects of space and togetherness are intended for evaluation of **sense of copresence**, or social presence: *"I felt if I were communicating with my conversation partner in the shared virtual space"*, *"The system gave me the sense that the physical gap between us was narrowed"*. The statistic ANOVA results for the first questionnaire item indicated the significance of the difference in sense of copresence felt in A-condition and R-condition (p (R-A) < 0.05), and showed that the A-condition gave stronger feeling of communication in the shared virtual space than the R-condition. No significant difference among three interfaces was reported on the second statement.

The level of **enjoyment** was evaluated using the statement "*I enjoyed the communication using this IM system*". The high levels of enjoyment were reported during A-condition and M-condition. However, ANOVA resulted in no significant differences among all three IM interfaces.

To evaluate **affective intelligence**, four statements (three – directly related to the system and one – indirectly related) were proposed to subjects in questionnaire: "*The system was successful at conveying my feelings*", "*The system was successful at conveying my partner's feelings*", "*The emotional behavior of the avatars was appropriate*", and "*I understood the emotions of my communication partner*". Fig. 2 shows the bar graphs of means of questionnaire results for these statements.

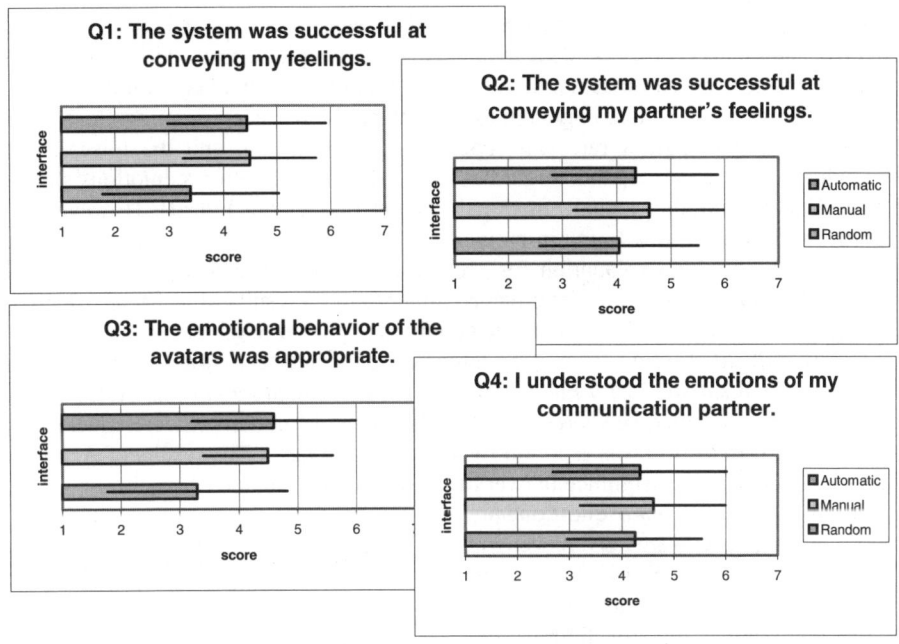

Fig. 2. Questionnaire results on affective intelligence (Q1, Q2, Q3, Q4)

As seen from the graph bar for Q1 (Fig. 2), the systems in A-condition and M-condition (with small prevalence of mean results in M-condition) were both more successful at conveying own feelings than the system in R-condition. Since M-condition is considered as a "gold standard" in communicating person's emotions, and ANOVA showed no significant difference between M-condition and A-condition, we might say that automatic emotion recognition system performed well enough to bring high affective intelligence to IM application. As was expected, significant differences were found between R-condition and M-condition (p (R-M) < 0.05), and between R-condition and A-condition (p (R-A) < 0.01).

While evaluating successfulness of the interfaces at conveying conversation partner's feelings (see Q2 in Fig. 2), the highest rate was given by subjects to M-condition, and the lowest – to R-condition. However, ANOVA for this criterion resulted in no significant difference among all interfaces. One user's comment regarding the emotional reactions of the partner's avatar was: "I concentrated too much on the reactions of my avatar and not enough on that of my partner. Reading and thinking about the answer took away the concentration on the avatar".

Interesting results were observed for the evaluation of appropriateness of emotional behavior of avatars. As seen from the graph (Q3 in Fig. 2) and statistical data of ANOVA, results for A-condition and M-condition significantly prevailed those for R-condition (p (R-A) < 0.01; and p (R-M) < 0.01). Users' comments confirmed that during R-condition subjects sometimes couldn't understand why the avatars did not correspond to their words and reacted in "wrong" ways. Although A-condition was rated a little bit higher than M-condition, no significant difference was detected between these interfaces.

The statement "*I understood the emotions of my communication partner*" measured affective intelligence of the system indirectly, since people used to derive emotional content from text based on semantic information and their empathetic abilities. Emotional expressions of avatars may help to understand the partner's emotion clearer. As was expected, the highest rate was reported in M-condition, and the lowest – in R-condition, where participants might be confused, since sometimes emotions shown by the avatar contradict actual emotional content (see Q4 in Fig. 2). However, no significant difference was found in partner's emotion comprehension among all three interfaces. A possible explanation for such results might be that a person typically relies on his/her own affective intelligence rather than on results of artificial affective intelligence. That is why the mean for R-condition appeared relatively high.

The **overall satisfaction** from using three AffectIM interfaces was evaluated using statement "*I am satisfied with the experience of communicating via this system*". Regarding the results, average scores for A-condition and M-condition were equal (4.6), whereas less satisfaction was reported for R-condition (4.25). The results of ANOVA showed no significant difference in overall satisfaction among interfaces.

In addition to the main questionnaire items, participants were given general questions. Subjects were asked to associate nine emotion states with nine avatar expressions shown on still figures. Female avatar was shown to male subjects, while male avatar was shown to female subjects. The percentages of reported correct associations within males and females are shown in Fig. 3. As seen from the graph, all 10 female subjects correctly associated 'anger', 'joy', and 'shame' emotions, while all 10 male subjects completely agreed only on 'joy' emotion.

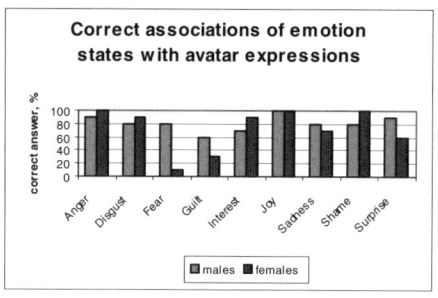

Fig. 3. Questionnaire results on emotions associated with avatar expressions

The detected pairs of most often confused emotions are 'fear' – 'surprise' and 'guilt' – 'sadness'; and less often confused emotions are 'guilt' – 'fear' and 'sadness' – 'fear'. Some participants confused emotions in 'interest' – 'joy' and 'surprise' – 'guilt' pairs. These results suggest that during the experiment some participants faced the difficulty with correct interpretations of emotional behavior of avatars.

To the question *"While online, do you use emoticons or abbreviations?"*, 19 subjects answered positively. We observed all automatically recorded dialogs, and found that to some degree the majority of participants used abbreviated language.

The participants' comments and the results of answers to the question *"To what degree do you think is necessary to look at a graphical representation of the other communicating person?"* suggest that there are two types of IM users: (1) some are open to new features of IM, and find animated graphical representation of a person helpful in understanding the partner's emotions and giving some sense of physical presence; (2) others tend to concentrate their attention on content, and prefer small emotional symbolic cues, like emoticons, to avatar expressions.

Participants also were asked to indicate whether manual selections of emotion state and intensity were helpful or not during M-condition. Only 30% of males and 60% of females answered positively. The result of answers to the question *"How often did you use this function, when you wanted?"* is represented as a bar graph in Fig. 4. As seen from these data, female subjects used emotion selection function more ardently than male subjects.

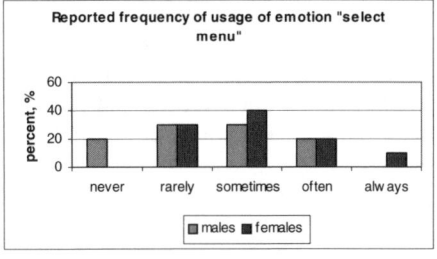

Fig. 4. Questionnaire results on frequency of usage of emotion "select menu"

The user opinions regarding the emotion "select menu" aspect were very diverse. Some users criticized the type of pop-up menu, commenting that it was difficult to use and it took long time to select. For more convenience, they proposed to replace pop-up menus by icons and spread them out. One of the subjects complained that emotion select menu disturbed the flow of the chat. Another reported problem is that since there is no preview of what the emotion expression looks like, it is unclear whether it matches the user's intention. Some subjects felt that basic emotions are too general and are not sufficient to convey emotion in many cases. Also, they suggested providing the possibility of showing more different or even mixed emotions (some state between sadness and joy). However, we think that displaying mixed emotional expressions would add more confusion and misinterpretation to the conversations.

Some subjects underlined positive aspects of manual selection of emotion states. They found this function helpful, because (1) it offered the possibility to visually express feelings and better understand them, (2) it allowed preventing inappropriate emotional reaction of avatar, and (3) guaranteed accuracy of communicated emotion. We can conclude that for sensitive conversation users would prefer manual control to avoid system mistakes that could sometimes harm the conversation.

5 Discussion and Conclusions

We implemented a web-based IM application, AffectIM, and endowed it with emotional intelligence by integrating our Affect Analysis Model. The user study conducted on AffectIM showed that the IM system with automatic emotion recognition function was successful at conveying users' emotional states during communication online, thus enriching expressivity and social interactivity of online communications. From the experiment we learned that the IM application might benefit from an integration of automatic emotion sensing with manual control of emotional behavior of avatars in one interface, which will allow users to select between two modes depending on type and sensitivity of conversation.

While analyzing the recorded conversations from our study, we detected some misspelled emotion-related words: "feiled" instead of "failed"; "promissing" instead of "promising", etc. In our future work, we plan to add correction of misspelled words to the system. Moreover, we aim to study cultural differences in perceiving and expressing emotions, and to integrate a text-to-speech engine with emotional intonations into the developed IM application.

Acknowledgments. We acknowledge and thank Dr. Ulrich Apel and Dr. Boris Brandherm for their efforts and help in organization of the user study on AffectIM. We wish also to express our gratitude to all the participants of the experiment.

References

1. Boucouvalas, A.C.: Real Time Text-to-Emotion Engine for Expressive Internet Communications. In: Being There: Concepts, Effects and Measurement of User Presence in Synthetic Environments, pp. 306–318. IOS Press, Amsterdam (2003)
2. Kim, S.-M., Hovy, E.: Automatic Detection of Opinion Bearing Words and Sentences. In: Proceedings of IJCNLP 2005 (2005)
3. Liu, H., Lieberman, H., Selker, T.: A Model of Textual Affect Sensing using Real-World Knowledge. In: Proceedings of IUI 2003, pp. 125–132 (2003)
4. Mulder, M., Nijholt, A., den Uyl, M., Terpstra, P.: A Lexical Grammatical Implementation of Affect. In: Proceedings of the 7th International Conference on Text, Speech and Dialogue, pp. 171–178. Springer, Berlin (2004)
5. Neviarouskaya, A., Prendinger, H., Ishizuka, M.: Analysis of Affect Expressed through the Evolving Language of Online Communication. In: Proceedings of IUI 2007, pp. 278–281. ACM Press, New York (2007)
6. Neviarouskaya, A., Prendinger, H., Ishizuka, M.: Textual Affect Sensing for Sociable and Expressive Online Communication. In: Paiva, A.C.R., Prada, R., Picard, R.W. (eds.) ACII 2007. LNCS, vol. 4738, pp. 220–231. Springer, Heidelberg (2007)
7. Peris, R., Gimeno, M.A., Pinazo, D., et al.: Online Chat Rooms: Virtual Spaces of Interaction for Socially Oriented People. CyberPsychology and Behavior 5(1), 43–51 (2002)
8. Strapparava, C., Valitutti, A., Stock, O.: Dances with Words. In: Proceedings of IJCAI 2007, Hyderabad, India, pp. 1719–1724 (2007)
9. Subasic, P., Huettner, A.: Affect Analysis of Text Using Fuzzy Semantic Typing. IEEE Transactions on Fuzzy Systems 9(4), 483–496 (2001)
10. Turney, P.D.: Thumbs Up or Thumbs Down? Semantic Orientation Applied to Unsupervised Classification of Reviews. In: Proceedings of ACL 2002, USA (2002)
11. Wilson, T., Wiebe, J., Hoffmann, P.: Recognizing Contextual Polarity in Phrase-level Sentiment Analysis. In: Proceedings of HLT/EMNLP 2005, Vancouver, Canada (2005)

Expressions of Empathy in ECAs

Radoslaw Niewiadomski[1], Magalie Ochs[2], and Catherine Pelachaud[1,3]

[1] IUT de Montreuil, Université Paris 8, France
niewiadomski@iut.univ-paris8.fr
[2] Université Paris 6, France
magalie.ochs@lip6.fr
[3] INRIA Paris-Rocquencourt, France
catherine.pelachaud@inria.fr

Abstract. Recent research has shown that empathic virtual agents enable to improve human-machine interaction. Virtual agent's expressions of empathy are generally fixed intuitively and are not evaluated. In this paper, we propose a novel approach for the expressions of empathy using complex facial expressions like superposition and masking. An evaluation study have been conducted in order to identify the most appropriate way to express empathy. According to the evaluation results people find more suitable facial expressions that contain elements of emotion of empathy. In particular, complex facial expressions seem to be a good approach to express empathy.

Keywords: ECA, empathy, facial expressions.

1 Introduction

Recent research has shown that virtual agents expressing empathic emotions toward users have the potentiality to enhance human-machine interaction [1,2,3]. *Empathy* is commonly defined as the capacity to "put your-self in someone else's shoes to understand her emotions" [4]. Several empathic virtual agents are already been developed. They express empathy in different ways. For instance, in [5], the virtual agent uses specific sentences (such as "I am sorry that you seem to feel a bit bad about that question"). In [6], the facial expression is used by supposing that empathic facial expressions of emotion are similar to expressions of felt emotion of the same type. Finally, virtual agent's expressions of empathy are generally fixed intuitively and are not evaluated.

In this paper, we evaluate our model that enables an ECA to display empathic emotions. Compared to other expressive empathic agents (e.g. [6,7]) our agent uses two types of facial expressions *simple* and *complex* ones. By simple facial expressions we intend spontaneous facial displays of emotional states (which can be described by one-word label) e.g. display of anger or contempt. The term *complex facial expressions* [8,9] describes expressions that are the combinations of several *simple* facial displays (e.g. superposition of two emotions) or that are modified voluntarily by the displayer (e.g. masking of one emotion by another one). Aiming at finding the appropriate facial expression of an empathic ECA we examine both types of expressions in empathic situations.

In section 2 we describe our architecture that computes the agents's empathic emotion and the expression that it will display, while Section 3 reports the results of the evaluation study we conducted.

2 Architecture of Empathic Agent

The process of generation of facial expressions in our empathic agent is the following. It takes as input a user's sentence. The dialog engine parses it and computes an answer for the agent. After the evaluation of what the user said, the Emotion Module computes which emotions are triggered for the agent. Finally the Facial Expressions Module computes the expression to be displayed by the agent called Greta [10]. The architecture is composed of three main modules:

1. The **Dialog Engine** is based on the JSA (Jade Semantics Agent) framework [11]. The JSA framework[1] enables one to implement BDI-like dialog agent. Such agents are able to interpret the received message and to respond to it automatically. The dialog engine contains a representation of goals and beliefs of the agent and those of its interlocutor. Depending on the received messages and on the agent's responses, the dialog engine updates the goals and beliefs.
2. Based on the goals and beliefs of both interlocutors: the agent and its interlocutor the **Emotion Module** computes the agent's emotional state [6]. Two kinds of emotions are computed: *egocentric emotions* and *empathic emotions*. The former ones correspond to the emotions of the agent, given its own goals (excepted altruist goals) and beliefs. The empathic emotions are elicited according to the goals and beliefs of the agent's interlocutor. For instance, an agent may be happy for its interlocutor because one of the interlocutor's goals was achieved (empathic emotion) and sad because one of its own goals just failed (egocentric emotion). The egocentric and empathic emotions are determined based on a *model of emotion elicitation* in which the conditions of emotion elicitation are described in terms of beliefs and goals [6].
3. The **Facial Expressions Module** computes the resulting facial expression that can either be simple or complex. The simple expressions are defined manually [13]. Complex facial expressions are generated basing on a face partitioning approach and on Paul Ekman's studies [13,14,15,16]. Each facial expression is defined by a set of eight facial areas. Expression is a composition of these facial areas, each of which can display signs of emotion. For complex facial expressions, different emotions can be expressed on different areas of the face; e.g., in sadness masked by happiness, sadness is shown on the eyebrows area while happiness is displayed on the mouth area (for details see [8,9,17]).

---

[1] The JSA framework is open-source [12].

3 Evaluation

The main aim of this evaluation study is to verify how empathy can be expressed by an ECA. Our hypothesis is that subjects would evaluate as more appropriate the agent which expresses empathic emotions than egocentric emotion. Moreover, we want to check which type of expression: *simple* expression of empathic emotion or *complex* facial expression of *egocentric* and *empathic* emotions is more suitable. We are also interested whether these results depend on the pair of emotions used.

3.1 Scenario Set-Up

Our evaluation study consists in presenting to subjects a set of scenarios and a set of animations of facial expressions of our agent. The subjects have to choose the facial expressions which are the most appropriate to the situations described in the scenarios.

Emotions. For the purpose of the experiment we wrote (in French) nine very short stories (scenarios SC1 - SC9) that were displayed to the subjects during the experiment. Each of them describes the events or situations which are intended to elicit unambiguous emotional states. In the experiment we focused on three emotional states. Our intention was to use emotions whose expressions are very different one from another so they could not be confused by the subjects. At the same time we wanted to evaluate the appropriateness of expressions of negative and positive empathic emotions. For that purpose we chose one positive emotion: *joy* and two very different negative emotions: *anger* and *sadness*.

Scenarios. The first three scenarios (SC1 - SC3) served only to show some facial expressions of Greta to subjects that probably had not seen any expressive ECA previously. These scenarios were not considered in the final result. The evaluation part is constituted of the remaining six scenarios (SC4 - SC9). They are short dialogues between two agents: Greta and Lucie. Greta and Lucie are presented to the subjects as close friends. In each scenario they meet and tell each other about events that have happened to them recently. Each scenario concerns a different pair of emotions. Thus (at least at the beginning of the scenario) emotions elicited in Greta are different from the emotions elicited in Lucie. However, the scenarios were designed to encompass situations where the empathic emotions (can) occur. For example scenario SC4 concerns the emotions of joy (Greta) and sadness (Lucie). In this scenario the following dialog is presented to the participants of the experiment:

Greta et Lucie se rencontrent dans un parc (Greta and Lucie meet in a park).
Greta : Salut Lucie ! je reviens de l'Université : j'ai eu l'examen ! Et toi ? (Hi Lucie! I just come back from the University: I pass the exam! and you?)
Lucie : Salut Greta ! Oh, et bien moi je viens d'apprendre que je ne l'ai pas eu... (Hi Greta! I have not passed the exam...)
Greta: oh!

The six scenarios SC4 - SC9 correspond to all combinations of pairs of the three emotions: anger, joy, and sadness. For each scenario SC_i there is a complementary scenario SC_j in which emotions of the two protagonists (Greta and Lucie) are exchanged.

Facial expressions. The output of the Emotion Module is composed of two labels of emotional states that correspond to the *egocentric emotion* and the *empathic emotion*. In order to evaluate the facial expressions of our agent we use different types of facial expressions. First of all the simple facial expressions of both emotions are used. We use also all plausible combinations of these emotions. Two types of complex facial expressions concern two emotional states at once: superposition and masking. Thus:

- egocentric emotion and empathic emotion can be superposed,
- egocentric emotion can be masked by a fake expression of empathic emotion,
- empathic emotion can be masked by a fake expression of egocentric emotion.

The last combination is impossible in the context of our evaluation study (as the scenario is built egocentric emotion cannot be fake). For this reason we use in this study only the first two complex expressions. As a consequence, for each scenario SC4 - SC9 we generated four animations each of which shows Greta displaying one facial expression. They correspond to:

- EXP1 - the **simple expression** of emotion E_i elicited by the Greta's event (e.g. expression of joy in SC4),
- EXP2 - the **simple expression** of empathic emotion E_j of Greta, *i.e.* the emotion that corresponds to the emotion elicited by the Lucie's event (e.g. expression of sadness in SC4),
- EXP3 - the expression of **superposition** of emotions E_i and E_j (e.g. superposition of joy and sadness in SC4),
- EXP4 - the expression of **masking** of E_i by E_j, (e.g. joy masked by sadness in SC4).

The expressions of the type EXP3 and EXP4 were generated with the algorithm presented in Section 2.

3.2 Procedure

In our evaluation study we asked participants to choose the appropriate facial expressions of Greta for each scenario. To have access to a greater number of participants, we set up our experiment as a web application. One experiment session consists in going through 9 different web pages. Each of them presents one scenario and four videos. After reading the scenario on one web page, subjects can play the videos (see Figure 1). Participants view the videos in the order they wish. They can re-view animations at their convenience. In the six scenarios SC4 - SC9 they have to order the animations from the most appropriate to the less appropriate one. Each animation as well as the scenarios SC_i are displayed in a random order. The participation in the experiment was anonymous.

Expressions of Empathy in ECAs 41

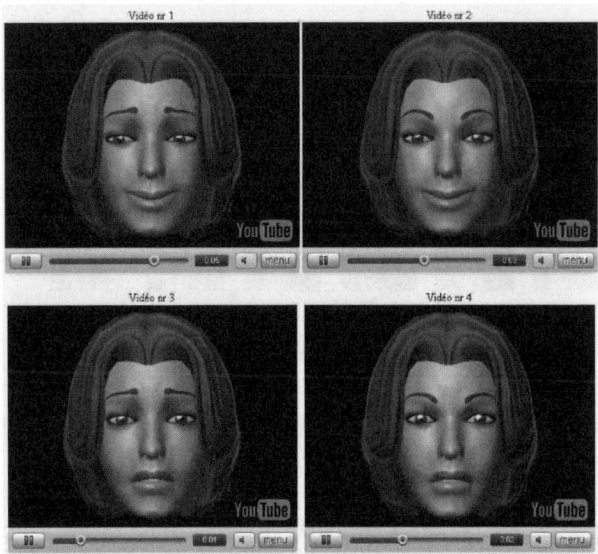

Fig. 1. The four animations displayed to the subjects (scenario SC 4)

24 persons have participated in the experiment. 44% of the participants were between 20 and 29 years old and 36% were between 30 and 39 years old. 11 participants were women, the other 13 - men.

3.3 Results

We analyzed 24 sets of answers for the scenarios SC4 - SC9. Two different criteria were applied. In the first criterion, we analyzed only the frequency of the choice as the most adequate facial expression for each type of expression. After that, in the second criterion, we considered complete subjects' answers, *i.e.* the order of animations. The results of both criteria are presented in Figure 2.

In the criterion "the most adequate facial expression" the expressions of the type EXP2 (*i.e.* simple expression of empathic emotion) was most often chosen

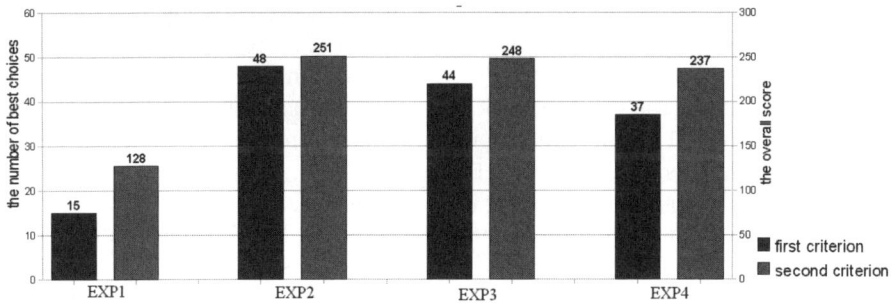

Fig. 2. The results for scenarios SC4-SC9 using both criteria

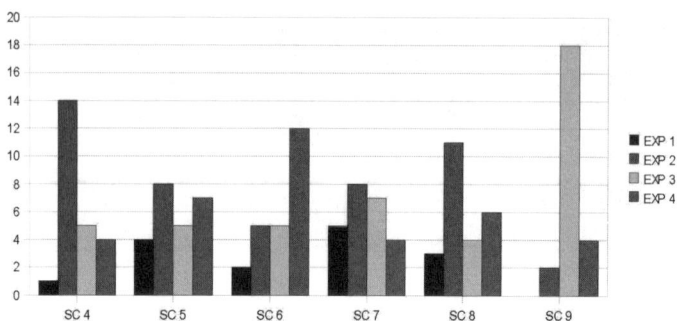

Fig. 3. The number of the most suitable expressions for each scenario

as the most appropriate display (33%). The expression of the type EXP3 (superposition) was chosen nearly as often as EXP2 (31%). The expression of masking (EXP4) got 26%. The simple expression of egocentric emotional state (EXP1) was less often evaluated as the most appropriate one (10%).

The Friedman statistical test revealed significant difference between expressions (p<.005). More particularly plan comparison test showed significant difference between EXP1 and the other expressions: EXP2 (p<.005), EXP3 (p<.005), and EXP4 (p<.005). The difference between other pairs were not significant (EXP2 and EXP3 (p >.05), EXP2 and EXP4 (p>.05), EXP3 and EXP4(p>.05)).

The difference in the evaluation of facial expressions between scenarios is significative (p<.05) (see Figure 3). For SC4, SC5, SC7, and SC8 the simple expression of empathic emotion (EXP2) was most often chosen. The facial expression of superposition of egocentric and empathic expressions (EXP3) was the most often chosen for the scenario SC9, while masking of an egocentric expression by an empathic expression (EXP4) was the most often chosen for the scenario SC6. There is no scenario for which the simple expression of egocentric emotion (EXP1) predominates.

We also compare the perception of negative and positive empathic emotions. In the scenarios SC4 and SC6 the egocentric emotion of Greta is positive (i.e. joy) while Lucie's emotion is negative (sadness and anger respectively). For SC5 and SC7 the egocentric emotion of Greta is negative and Lucie's one is positive. By summing up the results for SC4 and SC6 and the results for SC5 and SC7 one can see that subjects found the expression of the egocentric emotion more adequate if this emotion is negative (9 cases vs. 3). On the contrary, for positive egocentric emotion, the expression of empathic emotion (19 vs. 16) or masking expression (16 vs. 11) is more appropriate. Wilcoxon matched pairs test showed that the difference is significant (p<.05).

In the second criterion we consider the order of the animations (from the most adequate to the least adequate) given by the subjects. For this purpose we introduced an index i; i is set to 3 for the most adequate expression, $i = 2$ for a second place, and so on. We sum up the values of i for each type of facial expression (EXP1 - EXP4) of SC4 - SC9. The results (see Figure 2) are similar to the results obtained for the first criterion "the most adequate solution". The Friedman test showed the significant difference between the four types of expressions (p<.005).

3.4 Discussion

The aim of our experiment was to evaluate the empathic expressions for a conversational agent. In both analyzed criteria the expressions that contained the elements of empathic emotion (*i.e.* EXP2 - EXP4) were found much more suitable than the simple expression of egocentric emotion (EXP1). Indeed, the expression of the type EXP1 was not prevailing for any scenario. Thus we can say that people expect the agent displays signs of empathy in empathic situation. Moreover, analyzing the results of each scenario separately we found that subjects considered more often the expression of negative empathic expression to be the most adequate when the egocentric state is positive than in the opposite case, *i.e.* expression of positive empathic emotion in the case of negative egocentric state. It means that the expression of negative egocentric emotion can be more justified than the expression of egocentric positive emotion in case of contradictory emotions.

We observed also that complex expressions were often evaluated as the most suitable ones. Specially superposition (EXP3) of two emotions, *i.e.* the added expressions of both egocentric and empathic emotions, was considered the most adequate nearly as often as the simple expression of empathic emotion (EXP2). On the other hand we observed differences in evaluation of facial expressions between scenarios. Thus more works appears to be needed to establish if/when complex facial expressions are an improvement. Perhaps they are particularly suitable to express empathy in certain situations (for example if both agents talk about similar arguments) or to express particular pairs of emotional states.

4 Conclusions

In this paper we presented the architecture and the evaluation study of an agent that expresses empathy. The innovation of our approach consists in using to express empathy simple expressions of emotions as well as expressions of superposition and masking. As we expected people find more suitable facial expressions that contain elements of the emotion of empathy. In particular, complex facial expressions seem to be an appropriate mean to express empathy.

In the future, we aim to continue our research on empathic expressions of ECAs. We wish to study the differences between the role of simple empathic expression and various types (i.e. masking and superposition) of complex expressions. Various social contexts will also be considered. In the present experiment we evaluated only one type of interpersonal relations between the agents, namely between friends. We need to extend our findings to other interpersonal relationships.

Acknowledgement. Part of this research is supported by the EU FP6 Integrated Project CALLAS IP-CALLAS IST-034800.

References

1. Prendinger, H., Mori, J., Ishizuka, M.: Using human physiology to evaluate subtle expressivity of a virtual quizmaster in a mathematical game. International Journal of Human-Computer Studies 62, 231–245 (2005)
2. Partala, T., Surakka, V.: The effects of affective interventions in human-computer interaction. Interacting with computers 16, 295–309 (2004)
3. Picard, R., Liu, K.: Relative Subjective Count and Assessment of Interruptive Technologies Applied to Mobile Monitoring of Stress. International Journal of Human-Computer Studies 65, 375–396 (2007)
4. Pacherie, E.: L'empathie et ses degrés. In: L'empathie. Odile Jacob, 149–181 (2004)
5. Prendinger, H., Ishizuka, M.: The empathic companion: A character-based interface that addresses users' affective states. International Journal of Applied Artificial Intelligence 19, 285–297 (2005)
6. Ochs, M., Pelachaud, C., Sadek, D.: An empathic virtual dialog agent to improve human-machine interaction. In: Autonomous Agent and Multi-Agent Systems (AAMAS) (2008)
7. Reilly, S.: Believable Social and Emotional Agents. PhD thesis, Carnegie Mellon University (1996)
8. Niewiadomski, R.: A model of complex facial expressions in interpersonal relations for animated agents. PhD thesis, University of Perugia (2007)
9. Niewiadomski, R., Pelachaud, C.: Model of facial expressions management for an embodied conversational agent. In: Paiva, A., Prada, R., Picard, R. (eds.) ACII 2007. LNCS, vol. 4738, pp. 12–23. Springer, Heidelberg (2007)
10. Bevacqua, E., Mancini, M., Niewiadomski, R., Pelachaud, C.: An expressive ECA showing complex emotions. In: Proceedings of the AISB Annual Convention, Newcastle, UK, 208–216 (2007)
11. Louis, V., Martinez, T.: JADE Semantics Framework. In: Developing Multi-agent Systems with Jade, pp. 225–246. John Wiley and Sons Inc., Chichester (2007)
12. JADE (2001), http://jade.tilab.com/
13. Ekman, P.: Unmasking the Face. A guide to recognizing emotions from facial clues. Prentice-Hall, Inc., Englewood Cliffs (1975)
14. Ekman, P., Friesen, W.V.: The repertoire of nonverbal behavior's: Categories, origins, usage and coding. Semiotica 1, 49–98 (1969)
15. Ekman, P.: Darwin, deception, and facial expression. Ann. N.Y. Acad. Sci. 1000, 205–221 (2003)
16. Ekman, P.: The Face Revealed. Weidenfeld & Nicolson, London (2003)
17. Niewiadomski, R., Pelachaud, C.: Fuzzy similarity of facial expressions of embodied agents. In: Pelachaud, C., Martin, J.C., André, E., Chollet, G., Karpouzis, K., Pelé, D. (eds.) IVA 2007. LNCS (LNAI), vol. 4722, pp. 86–98. Springer, Heidelberg (2007)

Archetype-Driven Character Dialogue Generation for Interactive Narrative

Jonathan P. Rowe, Eun Young Ha, and James C. Lester

Department of Computer Science
North Carolina State University
Raleigh, NC 27695, USA
{jprowe,eha,lester}@ncsu.edu

Abstract. Recent years have seen a growing interest in creating virtual agents to populate the cast of characters for interactive narrative. A key challenge posed by interactive characters for narrative environments is devising expressive dialogue generators. To be effective, character dialogue generators must be able to simultaneously take into account multiple sources of information that bear on dialogue, including character attributes, plot development, and communicative goals. Building on the narrative theory of character archetypes, we propose an archetype-driven character dialogue generator that uses a probabilistic unification framework to generate dialogue motivated by character personality and narrative history to achieve communicative goals. The generator's behavior is illustrated with character dialogue generation in a narrative-centered learning environment, CRYSTAL ISLAND.

Keywords: Agents in narrative.

1 Introduction

Devising robust, believable virtual agents is a central problem in interactive narrative. Because characters are instrumental in defining and advancing plots, as well as in creating compelling experiences for audiences [1], creating virtual agents that play the roles of these characters is a central issue in interactive narrative generation. Recent years have seen great strides in virtual agents for interactive narrative in education [2], [3], training [4], [5], [6], and entertainment [7], [8], [9]. This work has largely centered on creating adaptive, narrative-centered interactions that afford significant degrees of user control and autonomy, with the promise of coherent and engaging experiences that satisfy the myriad criteria of narrative utility [10], [11].

Character dialogue is often the driving force in moving a plot forward. Characters drive narrative by fostering empathy and conflict within a story. Life-like characters are revealed to the audience through a combination of dialogue and events within the story world. To support believable character-character and character-player interactions, a computational model of character dialogue generation for interactive narrative must satisfy three requirements. First, it must generate character-appropriate dialogue. Because each character's dialogue must "follow clearly and validly from the character that uses it" [1], dialogue must be appropriate for character personalities and their associated motivations, preferences, and constraints. Second, it must consider the narrative context in which the dialogue will be delivered. As it generates dialogue, it must take into account narrative history to respond to the possible interactions initiated by the user and other characters. Third, it must perform robustly: it must be able to cope with combinations of goals, character personality attributes, and narrative history that cannot be anticipated prior to runtime.

To address these requirements, we propose an archetype-driven character dialogue generator. The dialogue generator builds on *character archetypes* [12], which are narrative-theoretic blueprints of well-established sets of traits for a particular character and role, such as fears, goals, motivations, and personality characteristics. Character archetypes exist outside of any one particular narrative scenario; they are adopted because of their power to define consistent sets of character traits that are both familiar and believable to audiences. The archetype-driven model of character dialogue generation employs probabilistic unification grammars that enable it to simultaneously consider multiple sources of information (character archetypes, narrative history, and communicative goals) to dynamically generate character-appropriate dialogue that achieves specific communicative goals for specific plot contexts. The model of dialogue generation has been implemented in a character dialogue generator for characters inhabiting CRYSTAL ISLAND, a narrative-centered learning environment for the domain of middle school microbiology.

This paper is structured as follows. Section 2 discusses related work in virtual agents for narrative environments and dialogue generation, which is followed by an introduction to the CRYSTAL ISLAND narrative-centered learning environment in Section 3. Section 4 describes the archetype-driven model of character dialogue generation; it presents an interactive narrative architecture that houses the character dialogue generator, the archetype-based character representation, and the probabilistic unification formalism used by the character dialogue generator. The dialogue generator's behavior is illustrated with a scenario from CRYSTAL ISLAND in Section 5. Concluding remarks and directions for future work follow in Section 6.

2 Related Work

A significant body of work has investigated intelligent virtual characters for interactive narrative environments. Among the earliest and most influential work is that by Carnegie Mellon's Oz group [7], which explored a range of issues in creating believable virtual characters for interactive drama, including extensions to their behavior specification language to support natural language generation [13]. The interactive drama Façade proposed a beat-based structuring of narrative content and character behaviors to create dramatic, adaptive vignettes [9]. Façade made significant advancements in character believability and player control, although dialogue was not dynamically generated.

Beyond incorporating rich emotional models for virtual characters [2], [4], work on character representation has informed the approach to character dialogue generation introduced here. Rizzo et al. [14] propose a goal-based model of character personality and social behavior for use in a virtual environment. One of the objectives of this approach was to ensure that "agents endowed with different personalities [could] perform the same high-level behaviors in different ways" [14]. The approach proposed here shares this objective. Mosher and Magerko [15] introduce character representations that use personality templates and "stereotypes," an approach paralleling the use of character archetypes here. The abstract character representations that they use are psychologically motivated and emphasize psychological traits whereas the proposed archetype representation has literary underpinnings and focuses on character preferences and goals.

Work on character dialogue generation in interactive narrative has explored a number of representational choices. Cavazza and Charles [8] propose a unified representation for narrative and communicative acts that takes advantage of a hierarchical task network (HTN) representation for narrative planning in the I-Storytelling environment. In contrast, THESPIAN uses a decision-theoretic approach to reasoning about character goals to control socially normative dialogue behavior [6]. THESPIAN's approach accommodates probabilistic preferences and weights. Other notable work on character dialogue includes T2D, which maps Rhetorical Structure Theory structures to DialogueNet structures to generate robust and extensible character-character dialogue from a monological text [16].

The natural language generation community has investigated generation techniques that consider the speaker's personality. Walker et al. [17] demonstrate that linguistic style is a significant influence on communication, and that listeners infer a speaker's personality from the speaker's linguistic style. Their Linguistic Style Improvisation (LSI) theory analyzes the speaker's choices of semantic content, syntactic form, and acoustic features according to inter-personal relationships. While LSI focuses on social dimensions of linguistic style, another line of research has explored personality-informed variations in computer-generated dialogue. Isard et al. [18] represent linguistic personality with n-gram language models. Their language models were trained on a corpus that is labeled with the dimensions of big-five personality models. Mairesse and Walker [19] map personality-dependent linguistic traits to their dialogue generator's parameters to control various aspects of dialogue generation, such as content selection and structure, syntactic template selection, aggregation, pragmatics, and lexical selection. While both lines of work represent significant advances, they do not consider interactive narrative factors bearing on character dialogue generation.

3 CRYSTAL ISLAND Narrative-Centered Learning Environment

The archetype-driven model of character dialogue generation is being investigated in CRYSTAL ISLAND (Figure 1), a narrative-centered learning environment for the domain of microbiology for middle school students [20]. CRYSTAL ISLAND features a science mystery set on a recently discovered volcanic island. Students play the role of the protagonist, Alex, who is attempting to discover the identity and source of an unidentified infectious disease plaguing a newly established research station. The story opens by introducing the student to the island and members of the research team for which the protagonist's father serves as the lead scientist. Several of the team's members have fallen gravely ill, including Alex's father. Tensions have run high on the island, and one of the team members, Ford, suddenly accuses another, Quentin, of having poisoned the other researchers. It is the student's task to discover the outbreak's cause and source, and either incriminate or exonerate Quentin.

The virtual world of CRYSTAL ISLAND, the semi-autonomous characters that inhabit it, and the user interface were implemented with Valve Software's Source™ engine, the 3D game platform for Half-Life 2. The Source engine currently provides much of the low-level (reactive) character behavior control. Narrative generation, pedagogical planning, and character behavior management are the subject of ongoing work.

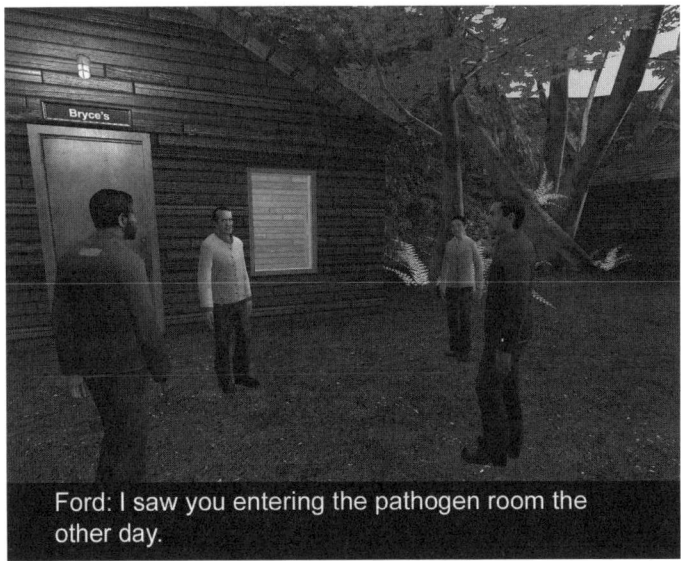

Fig. 1. CRYSTAL ISLAND narrative-centered learning environment

4 Archetype-Driven Character Dialogue Generation

Figure 2 shows an overview of the components of an interactive narrative architecture. Prior to run time, interactive narrative authors use the Character Composer's authoring facilities to instantiate characters from archetypes to populate the story world. The resultant character compositions furnish the raw materials for the Narrative Generator, which considers character attributes and the Narrative History as it dynamically constructs narrative plans. The Character Dialogue Generator is passed communicative goals by the Narrative Generator. Communicative goals consist of character objectives achievable by performing certain speech acts. To achieve communicative goals, the Character Dialogue Generator considers character compositions and the Narrative History as it generates the dialogue. The Character Behavior Controller coordinates character dialogue and action in the world model.

The four components that drive character dialogue are the Character Composer, the Narrative Generator, the Narrative History, and the Character Dialogue Generator:

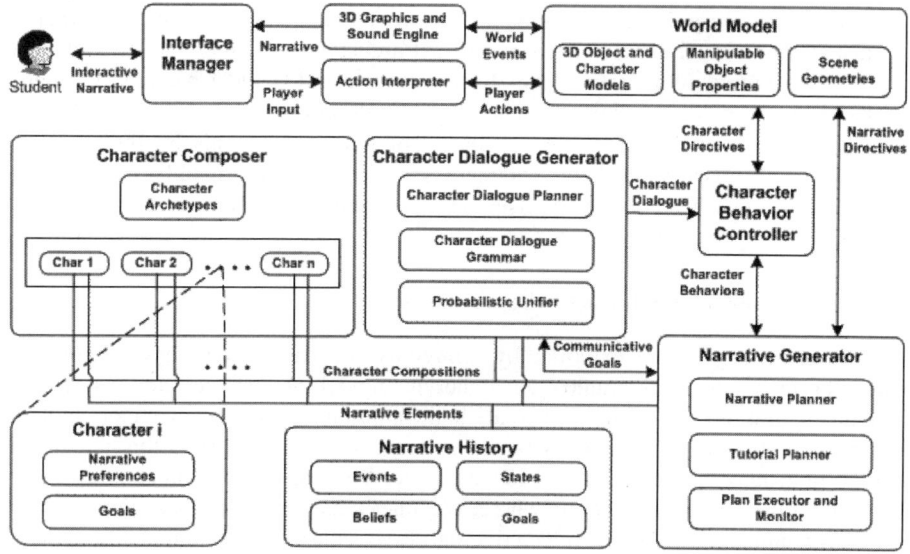

Fig. 2. Interactive narrative architecture

- **Character Composer:** A semi-automated module that manages story-independent and story-dependent representations for defining virtual characters, the character composer includes a library of character archetypes that define abstract character traits and attributes. In essence, the library is a cast of canonical character descriptions. Authors interact with the Character Composer to instantiate abstract archetypes into fully defined, story-dependent characters that will inhabit the interactive narrative environment. During execution, the instantiated characters contribute to the dialogue generation processes. Character representations are discussed in detail below (Section 4.1).
- **Narrative Generator:** Chief among the functionalities of interactive narrative is the capacity to adapt plot direction and character responses to a player's actions in the environment. The Narrative Generator's execution and monitoring facilities serve as the interface between narrative planning and the world model with which players interact. The Plan Executor reconciles narrative plans and character definitions with the current narrative state to devise communicative goals and behaviors for the narrative's characters. These goals are emitted as behavioral directives to the Character Behavior Controller or as communicative goals to the Character Dialogue Generator. As these goals are pursued within the world model, the Narrative Generator monitors goal status in order to update the Narrative History or invoke narrative re-planning. For narrative-centered learning environments, the narrative generator also includes a tutorial planner.
- **Narrative History:** The Narrative History maintains a set of both authored and generated narrative events, facts, character beliefs, and goals comprising the story-to-date. The initial state for the Narrative History encodes the narrative back-story predating any plot progress that a student has observed or participated in. As students progress through the narrative, additional elements are added to the

Narrative History. Narrative elements inform (1) the Character Composer when it instantiates virtual characters from abstract archetypes, (2) the Narrative Planner by contributing to narrative plan formulation, and (3) the Character Dialogue Generator to guide the dialogue generation search process.

- **Character Dialogue Generator:** Given communicative goals by the narrative planner, the character dialogue generator's activities are informed by character archetypes and narrative history. The character dialogue generator is discussed in detail below (Section 4.2).

4.1 Archetype-Based Character Representation

To capture the regularities exhibited by commonly occurring categories of characters, character representations employ an archetype model. Archetypes are a powerful structuring tool used by authors to create believable but distinctive characters in their stories [12]. The prevalence of character archetypes can be observed throughout literature and other narrative media. For example, Shakespeare's Hamlet is a classic example of the *Recluse* archetype, whereas Hamlet's mother Gertrude exemplifies the *Matriarch* archetype [12]. For interactive narrative, constructing computational models of character archetypes offers the potential to create more portable, story-independent character specifications than traditional story-dependent approaches. Much as they do for authors, character archetypes offer interactive narrative attractive interactional properties such as character identifiability and believability.

To support the representations required by the Character Dialogue Generator's probabilistic unification grammar framework (as described in Section 4.2), character archetypes are defined using a weighted preference schema. Character archetypes are currently heuristically defined, and are composed of the following features:

- **Narrative Preferences:** For a given archetype, narrative preferences map normalized probabilities to subsets of the elements contained in the Narrative History. These mappings quantify the importance of narrative elements, which consist of narrative events, facts, and beliefs, to a particular character archetype. In this paper, events are occurrences that modify the narrative state in some manner, facts are true assertions about characters or the world state that do not explicitly reference events, and beliefs encompass events and facts whose actual truth values are not certain. Two primary types of narrative preferences are used here: categorical preferences and abstraction preferences. *Categorical preferences* distinguish among different types of narrative elements, quantifying whether a character would value concrete facts and events from the story, or de-emphasize them in favor of beliefs. *Abstraction preferences* are story-independent descriptions used to distinguish subsets of story-specific narrative elements that are important to a particular archetype. Because archetypes are generally story-independent, it is necessary to define narrative abstractions that can label and encompass elements from a particular story. For example, characters that embody the *Abuser* archetype (Table 1) are often the target of some ego-damaging event, catalyzing them to seek revenge in response [12]. To encode this regularity, an interactive narrative "author" can create a narrative abstraction called *Revenge-Catalyst* prior to runtime. This abstraction can be specified to aggregate a subset of narrative elements once a story is authored and entered into the Narrative History. Because these elements

are important to *Abuser* characters, its archetype specification will associate a high weight with the *Revenge-Catalyst* narrative abstraction and emphasize the associated elements during runtime.
- **Goals:** Character goals are explicit objectives pursued by a character, and are assigned normalized probabilities in a manner similar to Narrative Preferences. Character archetypes can also value or de-value goals independent of a story. For example, the *Traitor* archetype typically seeks to advance his or her social or professional status. This goal is independent of any story in which the *Traitor* participates. After characters are instantiated, their goals inform the Narrative Generator's narrative planning.

The approach described offers two principle benefits. First, the use of a probabilistic, preference-driven formalism provides a degree of flexibility in character's decision-making processes that supports robust performance in character dialogue generation. Second, the use of archetypes permits the partial definition of story-independent characters. These definitions can subsequently be used to instantiate story-specific characters that automatically adopt the features defined in an associated archetype. Archetypes introduce the opportunity to partially automate the character creation process once a sufficient library of character archetypes has been encoded. The current choice of features for encoding archetypes, and the generality of our character representations, are the subjects of ongoing investigation.

To illustrate, consider two character archetypes derived from [12], *Traitor*, and *Abuser* (Table 1). The *Traitor* is an archetype that typifies a cool, rational and organized antagonistic character whose priorities lie with their job or their own professional advancement. These characters often see business and life as a game to be won, and avoid chaotic or strongly emotional situations. Examples of this archetype from literature and film include *Wall Street*'s Gordon Gekko, and *A Christmas Carol*'s Ebenezer Scrooge. In its corresponding computational representation, one can observe this archetype's preference for objective, rational information through its high weighting of facts and events. Additionally, the choice of goals pertinent to the *Traitor* archetype emphasizes professional objectives such as the success of the research expedition (e.g. *See-Organization-Succeed*) and career advancement (e.g. *Advance-Status*), rather than personal or emotional concerns. The *Traitor* also values the *Incriminate-Evidence* abstract narrative element, weighting incriminating information that may eventually be useful for competitive advantage.

In contrast, the *Abuser* archetype typifies antagonistic characters that are strongly guided by their emotions, often exhibiting intensity, passion, and aggression. This archetype strongly values the ways that others view him, and it may carry a delicate ego. Examples of the *Abuser* include Shakespeare's Othello and *The Great Gatsby*'s Tom Buchanan. We encode this archetype by more strongly weighting beliefs over events and facts to de-emphasize the rational, objective side of its personality. The archetype's tendency toward aggression and ego is characterized by its high preference for the *Embarassment-Sequence* abstract narrative element, a sequence referencing an ego-damaging event, as well as a heavy weight placed on the related *Avenge-Transgression* goal.

Table 1. Example partial definitions of the *Traitor* and *Abuser* character archetypes

Archetype	*Traitor*		*Abuser*	
Narrative Preferences:				
Element Categories	Events	0.5	Events	0.3
	Facts	0.4	Facts	0.2
	Beliefs	0.1	Beliefs	0.5
Element Abstractions	Embarass-Sequence	0.2	Embarass-Sequence	0.5
	Conflict-Catalyst	0.3	Conflict-Catalyst	0.3
	Incriminate-Evidence	0.5	Incriminate-Evidence	0.2
Goals:	Advance-Status	0.4	Advance-Status	0.2
	See-Org-Succeed	0.3	See-Org-Succeed	0.1
	Bring-Order	0.3	Avenge-Transgression	0.7

4.2 Probabilistic Unification Grammars for Character Dialogue Generation

The principle objective of the Character Dialogue Generator is to create natural language utterances that satisfy a communicative goal provided by the narrative generator, that are appropriate for the specified character archetype, and that take the narrative history into account as the syntax and semantics of the dialogue are planned. Turn-taking conversational behavior is handled separately by the Narrative Generator, and is beyond the scope of this paper. The Character Dialogue Generator considers three sources of information: communicative goals from the Narrative Generator, archetype information from the Character Composer, and narrative elements from the Narrative History. First, the communicative goal specifies the speech act that should be performed at a given dialogue turn. Second, the character archetype provides generator preferences over the types of topics reflected by the archetype. Third, the narrative history provides a repository of current (and previous) topics.

In addition, the Character Dialogue Generator considers three additional sources of information and defines them as parts of the dialogue generation grammar. These are the preferences of communicative goals both over topics and over syntactic templates, and the preferences of character archetypes over syntactic templates. Syntactic characteristics of templates are represented using three features associated with template strings: *mood* (indicative, interrogative), *modality* (explicit, implicit), and *viewpoint* (1st singular, 1st plural, 3rd, other). Explicit vs. implicit modality determines whether the topic is explicitly expressed in the utterance or not. These features have been employed to express different linguistic traits representing either different types of characters [19] or different social relationships between characters [8]. Preferences over alternate values for each feature are represented using normalized probabilities, as in the case of topic preferences.

Given these six sources of information, the Character Dialogue Generator accomplishes its objective by decomposing the generation task into two separate search processes, semantic planning and syntactic planning (Figure 3). *Semantic planning* is the process of searching for a topic (a narrative element) within the Narrative History that best satisfies the constraints imposed by the given preferences both of a character archetype and of a communicative goal. Similarly, *syntactic planning* is a search for a syntactic template to realize the selected topic that best satisfies the preferences of both the character archetype and of the communicative goal. Formulated as a search problem

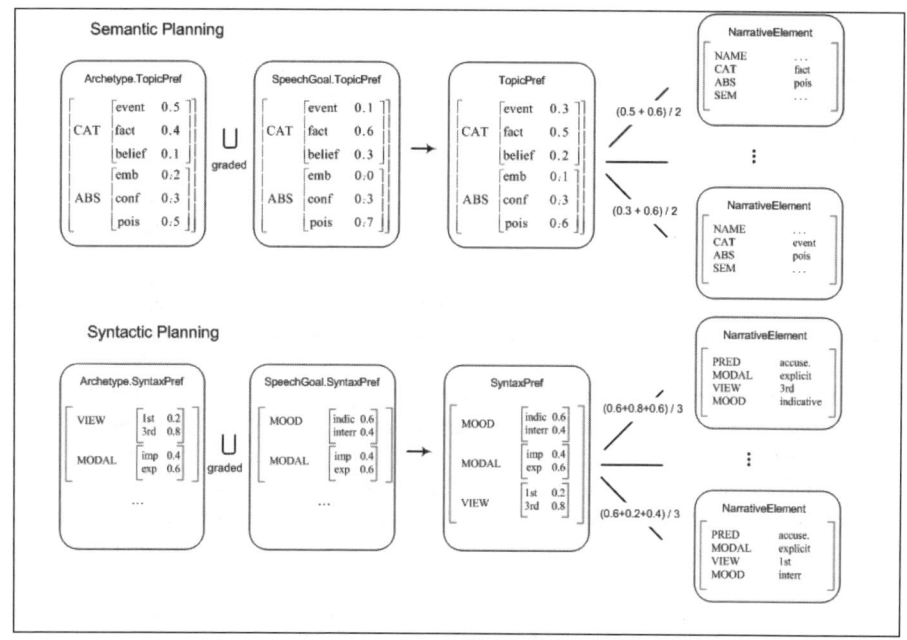

Fig. 3. Dialogue generation via probabilistic unification

with constraints, the task of dialogue generation now confronts two challenges: (1) to effectively combine constraints from different sources, and (2) to robustly handle potential conflicts in the preferences between different sources.

To address these challenges, the character dialogue generator employs a probabilistic unification grammar framework. While probabilistic unification grammars have been applied to natural language processing tasks such as parsing [21], [22], [23] and speech recognition [24], the approach proposed here is among the first attempt to apply the framework to the task of dialogue generation. Probabilistic unification grammars offer three advantages over alternative approaches: they provide (1) a unified formalism to combine two different search tasks, (2) an effective method for merging different sources of information into a single structure, and (3) a principled mechanism for robustly handling potential conflicts in values between different sources of information.

In probabilistic unification grammars, each terminal and non-terminal symbol is represented as a feature structure. A feature structure is a set of feature-value pairs in which values are either atomic symbols or nested feature structures (Figure 4). We adopt the technique of *graded unification* [21] that supports the unification of two features with different values by adjusting probabilities associated with each value. In this approach, each atomic-valued feature is associated with a probability and the probabilities of atomic values associated with a feature are normalized to sum to 1.0. When two atomic features are unified, the unification operator collects the disjoint set of all values present in the two arguments in the result and assigns a new probability to each value by computing the average of the probabilities associated with the value

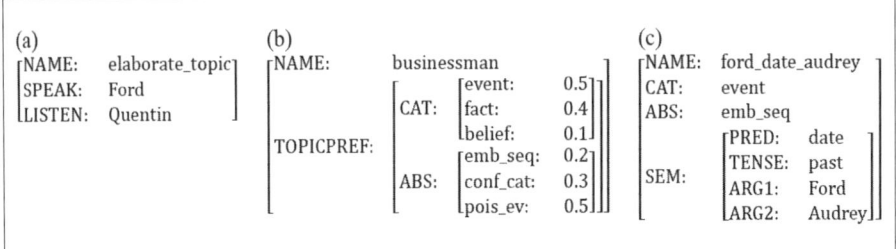

Fig. 4. Feature structure for (a) communicative goal, (b) character archetype, and (c) narrative element. In these examples, the probabilities of values equaling 1 are omitted for clarity.

in the two arguments. When a value is present only in one argument, it is assumed that the probability of the value in the other argument is zero. It is through this technique that the Character Dialogue Generator exhibits robustness with respect to potential conflicts in the values from different information sources.

Consider an illustrative example where the topic preferences of a character archetype (*Archetype.TopicPref*) and a communicative goal (*SpeechGoal.TopicPref*), are unified. Suppose that the feature *CAT* (topic category such as event, fact, and belief), common to both information sources, associates probabilities for *event* and *belief* of 0.6 and 0.4, respectively, in *Archetype.TopicPref*. The corresponding feature in *SpeechGoal.TopicPref* has only one value, *event,* with its associated probability 1.0.[1] The *graded unification* technique unifies these features into a new set with updated *event* and *belief* values. The probability of *event* is adjusted to 0.8, which is the average of its associated probability in each argument (0.6 in *Archetype.TopicPref* and 1.0 in *SpeechGoal.TopicPref*). The value *belief* is not present in *SpeechGoal.TopicPref*, so its associated probability in *SpeechGoal.TopicPref* is assumed to be 0.0 and its probability in the resulting set is adjusted to 0.2 (the average of 0.4 and 0.0).

Suppose that the Character Dialogue Generator is creating a dialogue for a character whose archetype is *Abuser* and a new communicative goal, *ElaborateTopic*, is posted. During semantic planning, the Dialogue Generator first computes the overall topic preference for the given dialogue turn by unifying the two features, *CAT* and *ABS* (topic abstraction such as Revenge-Catalyst, Embarrassment-Sequence), of the topic preference of the *Abuser* archetype and the topic preference of the communicative goal *ElaborateTopic* using the graded unification technique. The resulting topic preference contains new probabilities for each of the *CAT* and *ABS* features.

The combined topic preference is then used to search for a narrative element in the Narrative History by unifying the two features, *CAT* and *ABS*, of the combined topic preference with the corresponding *CAT* and *ABS* features of each narrative element in the Narrative History. Because more than one narrative element could be unified with the topic preference, the matching score of a narrative element is computed as the

---

[1] The current implementation of the Character Dialogue Generator uses information sources with manually authored probabilities. While the current approach demonstrates the feasibility of the probabilistic unification formalism for runtime performance, manually authoring probabilities would be prohibitively expensive. A promising direction for future work is learning probability distribution functions from observations of players' interactions with interactive narrative generators.

average probability of each feature in the topic preference that was unified with the features of the narrative element. The narrative element that yields the highest matching score is selected as the topic for the given dialogue turn.

Syntactic planning is performed in a similar manner. First, syntax preferences are computed by unifying the three features: *MOOD* (mood), *MODAL* (modality), and *VIEW* (viewpoint) of the syntax preference of the *Abuser* archetype and the syntax preference of the *ElaborateTopic* communicative goal. The combined syntax preference is matched with each syntactic template by unifying the three features. Among the templates matched with the combined syntax preference, the highest scoring one is selected as the template to realize the topic selected during semantic planning. Finally, the resulting dialogue utterance is generated by filling the slots within the selected template with the corresponding sub-features of the *SEM* (semantics) feature in the selected narrative element which represents the semantics of the corresponding narrative element.

5 An Illustrative Scenario

To illustrate the behavior of the archetype-driven approach to character dialogue generation, consider a scene in CRYSTAL ISLAND, where one of the characters, Ford, is accusing another character, Quentin, of poisoning other team members. The effects of archetype on a dialogue can be seen by holding constant the communicative goal and the narrative setting and then varying the character archetype. Table 2 shows sample dialogues created by the Character Dialogue Generator with two different settings of archetypes for Ford, the *Abuser* and the *Traitor*.

Table 2. Generated dialogue

Speaker (Communicative Goal)	Character Archetype	
	Abuser	*Traitor*
Ford (*AssertTopic*)	We all know you are doing this, Quentin.	It is evident who's doing this, Quentin.
Quentin	What are you talking about, Ford?	
Ford (*ReassertTopic*)	Don't deny you poisoned Bryce and Teresa.	Did you think nobody could figure out you poisoned Bryce and Teresa?
Quentin	You are accusing me of making Bryce and Teresa sick?	
Ford (*ElaborateTopic*)	You put salmonella in Bryce's and Teresa's foods.	I saw you entering the pathogen room the other day.
Player	I don't believe you, Ford.	
Ford (*Rebut*)	Quentin poisoned Bryce, Alex.	Think about it rationally, Alex.

The generated dialogue shows how a character's archetype drives the generation decisions in selecting topics and syntactic templates. In the dialogue generated for the *ElaborateTopic* communicative goal, the *Abuser* version of Ford selects his belief that Quentin contaminated the research team's food with salmonella to elaborate upon the current topic (accusing Quentin of poisoning) whereas the *Traitor* version selects an event in which he observed Quentin entering the pathogen room. This reflects the difference between the two archetypes: the *Abuser* makes passionate, unfounded accusations while the *Traitor* offers well-informed, controlled evidence to support his accusation. The results also show the difference in syntactic tendencies between the two archetypes. An *Abuser* character tends to be fiery and aggressive whereas the *Traitor* acts coolly and rationally. The *Abuser* selects templates with explicit modality for of all his dialogue turns, while the *Traitor* selects the explicit modality only twice (*ReassertTopic*, *ElaborateTopic*) out of four total turns.

6 Conclusions and Future Work

Interactive narratives can benefit significantly from the implementation of robust character dialogue generators. Character dialogue is one of the primary means through which characters are revealed to an audience, but hand-authoring dialogues can become prohibitive as the space of possible dialogue interactions grows large. To address this problem, character dialogue generators for interactive narratives must meet several requirements. They must generate dialogues that are appropriate for characters' traits, such as personalities, motivations, and preferences; they must consider narrative context and history as they formulate dialogue; and they must be able to robustly handle the large number of possible character-character and character-player interactions that may result in dialogue.

The model of dialogue generation introduced here uses probabilistic unification grammars to robustly create natural language character dialogue for interactive narrative. It is based on a computational representation of character archetypes that encodes story-independent character specifications. Character archetypes can be instantiated to compose characters whose sets of preferences over elements from the narrative history are utilized to generate character-appropriate dialogue that is situated within the current narrative context. The generated dialogues use preference information encoded within character archetype representations and yield character-specific variations in the dialogue.

The work to date has two principle limitations. First, it has focused on a few selected narrative contexts, an initial but not comprehensive archetype representation, and a small set of dialogues. Second, follow-up work needs to be undertaken to empirically investigate the fluidity and the "narrative appropriateness" of generated dialogues.

Several directions for future work appear promising. First, given the potentially rich inter-character relations that occur in compelling narratives, the range of information encoded by archetypes can be extended to represent character relationships and other features. The current feature set is preliminary in nature, and was largely chosen as a mechanism in service of the dialogue generation model discussed. Second, the current work focuses on natural language generation. In the future, it will be important to introduce character-appropriate prosodic markers in conjunction with text-to-speech technologies for speech generation. Third, it will be interesting to investigate techniques for coordinating archetype-driven character dialogue and characters' behaviors, with an emphasis on gesture and gaze. Finally, a particularly promising direction for future work is conducting extensive human subject evaluations. It is expected that these will yield important findings on the effects of character-driven dialogue on players, as well as insights into the set of features used to encode character archetypes and parameterize dialogues. Furthermore, we expect that human subject evaluations will furnish corpora for learning the probability distributions for the archetype representations and the unification grammars.

Acknowledgements

The authors would like to thank Bradford Mott and Scott McQuiggan for their work developing earlier versions of CRYSTAL ISLAND. We also wish to thank the other members of the IntelliMedia Center for Intelligent Systems at North Carolina State University for useful discussions and support. We are grateful to Omer Sturlovich and Pavel Turzo for use of their 3D model libraries, and Valve Software for access to the Source™ engine and SDK. This research was supported by the National Science Foundation under Grants REC-0632450 and IIS-0757535. Any opinions, findings, and conclusions or recommendations expressed in this material are those of the authors and do not necessarily reflect the views of the National Science Foundation.

References

1. Egri, L.: The Art of Dramatic Writing: Its Basis in the Creative Interpretation of Human Motives. Simon & Schuster, New York (1960)
2. Aylett, R., Figueiredo, R., Silva, A., Dias, J., Paiva, A.: Making It Up as You Go Along - Improvising Stories for Pedagogical Purposes. In: Gratch, J., Young, M., Aylett, R.S., Ballin, D., Olivier, P. (eds.) IVA 2006. LNCS (LNAI), vol. 4133, pp. 304–315. Springer, Heidelberg (2006)
3. Marsella, S., Johnson, W.L., LaBore, C.: Interactive Pedagogical Drama for Health Interventions. In: Proc. of AI in Education (AIED), pp. 341–348 (2003)
4. Gratch, J., Marsella, S.: A Domain-independent framework for modeling emotion. Journal of Cognitive Systems Research 5, 269–306 (2004)
5. Riedl, M.O., Stern, A.: Believable Agents and Intelligent Story Adaptation for Interactive Storytelling. In: Proc. Tech. for Interactive Digital Storytelling and Entertainment, pp. 1–12 (2006)
6. Si, M., Marsella, S.C., Pynadath, D.V.: Thespian: Modeling Socially Normative Behavior in a Decision-Theoretic Framework. In: Gratch, J., Young, M., Aylett, R.S., Ballin, D., Olivier, P. (eds.) IVA 2006. LNCS (LNAI), vol. 4133, pp. 369–382. Springer, Heidelberg (2006)
7. Bates, J.: Virtual Reality, Art, and Entertainment. PRESENCE: Teleoperators and Virtual Environments 1, 133–138 (1992)
8. Cavazza, M., Charles, F.: Dialogue Generation in Character-based Interactive Storytelling. In: Proc., A.I. and Interactive Digital Entertainment (AIIDE), pp. 21–26 (2005)

9. Mateas, M., Stern, A.: Structuring Content in the Façade Interactive Drama Architecture. In: Proc. AI and Interactive Digital Entertainment (AIIDE), pp. 93–98 (2005)
10. Roberts, D., Isbell, C.: Desiderata for Managers of Interactive Experiences: A Survey of Recent Advances in Drama Management. In: Proc. AAMAS Wkshp on ABSHLE (2007)
11. Rowe, J., McQuiggan, S., Lester, J.: Narrative Presence in Intelligent Learning Environments. In: Proc. AAAI Fall Sym. on Intell. Narrative Tech., pp. 126–133 (2007)
12. Schmidt, V.L.: 45 Master Characters: Mythic Models for Creating Original Characters. Writer's Digest Books: Cincinnati, OH (2001)
13. Loyall, A.B., Bates, J.: Personality-Rich Believable Agents That Use Language. In: Proc. Autonomous Agents, First International Conference, pp. 106–113 (1997)
14. Rizzo, P., Veloso, M., Miceli, M., Cesta, A.: Goal-based personalities and social behaviors in believable agents. Applied Artificial Intelligence 13, 239–271 (1999)
15. Mosher, B., Magerko, B.: Personality Templates and Social Hierarchies Using Stereotypes. In: Proc. Tech. for Interactive Digital Storytelling and Entertainment, pp. 207–218 (2006)
16. Piwek, P., Hernault, H., Prendinger, H., Ishizuka, M.: T2D: Generating Dialogues Between Virtual Agents Automatically from Text. In: Pélachaud, C., Martin, J.-C., André, E., Chollet, G., Karpouzis, K., Pelé, D. (eds.) IVA 2007. LNCS (LNAI), vol. 4722, pp. 161–174. Springer, Heidelberg (2007)
17. Walker, M., Cahn, J., Whittaker, S.: Improvising Linguistic Style: Social and Affective Bases for Agent Personality. In: Proc. Auton. Agents, 1st Int. Conf., pp. 96–105 (1997)
18. Isard, A., Brockmann, C., Oberlander, J.: Individuality and Alignment in Generated Dialogues. In: Proc. Natural Lang. Generation, 4th Int., pp. 25–32 (2006)
19. Mairesse, F., Walker, M.: PERSONAGE: Personality Generation for Dialogue. In: Proc. 45th Annual Meeting of the Assoc. for Comp. Ling (ACL), pp. 496–503 (2007)
20. Mott, B., Lester, J.: U-Director: A Decision-Theoretic Narrative Planning Architecture for Storytelling Environments. In: Proc. AAMAS, pp. 977–984 (2006)
21. Kim, A.: Graded Unification: A Framework for Interactive Processing. In: Proc. 32nd Annual Meeting of the Association for Computational Linguistics, pp. 313–315 (1994)
22. Abney, S.: Stochastic Attribute-Value Grammars. Comp. Ling. 23, 597–618 (1997)
23. Johnson, M.: Learning and parsing stochastic unification-based grammars. In: Proc. Learning Theory and Kernel Machines (COLT / Kernel), pp. 671–683 (2003)
24. Hemphil, C., Picone, J.: Speech Recognition in a Unification Grammar Framework. In: Proc. ICASSP, pp. 723–726 (1989)

Towards a Narrative Mind: The Creation of Coherent Life Stories for Believable Virtual Agents

Wan Ching Ho and Kerstin Dautenhahn

Adaptive Systems Research Group, School of Computer Science
University of Hertfordshire, Hatfield, Hertfordshire, AL10 9AB, UK
{w.c.ho, k.dautenhahn}@herts.ac.uk

Abstract. This paper describes an approach to create coherent life stories for Intelligent Virtual Agents (IVAs) in order to achieve long-term believability. We integrate a computational autobiographic memory, which allows agents to remember significant past experiences and reconstruct their life stories from these experiences, into an emotion-driven planning architecture. Starting from the literature review on episodic memory modelling and narrative agents, we discuss design considerations for believable agents which interact with users repeatedly and over a long period of time. In the main body of the paper we present the narrative structure of human life stories. Based on this, we incorporate three essential discourse units and other characteristics into the design of the autobiographic memory structure. We outline part of the implementation of this memory architecture and describe the plan for evaluating the architecture in long-term user studies.

1 Introduction

In pursuing the ultimate goal of creating agents that give an "illusion of life" [1], Intelligent Virtual Agents (IVAs) researchers in the past decade have applied different mechanisms to enhance agents' believability. In addition to the efforts in advancing the computer graphics and language engines; from the perspective of cognitive architecture modelling, human personality and emotion models (e.g. BDI [2],OCC [3] and PSI [4, 5]) have brought fruitful outcomes into this research direction. As a result, nowadays a large variety of software applications in education, entertainment and other areas are populated with believable agents for increasing the level of user engagement. So, what can still be missing?

We argue that the answer can be *long-term believability*. When considering the IVA's role as an one-off problem solving tool or a short-term interactive target, it is probably not necessary for the agent to inform human users what it remembers and learned from the past. There are circumstances, however, that require users to *repeatedly interact with the same agent*. For example, this includes a long-term companion agent which assists the user in various tasks on a day-to-day basis, or a conversational narrative agent which tells stories about itself to the user. Both types of agents are aiming at long-term and repeated interactions with a user.

Thus, they shall require at least an extra cognitive component which commonly does not seem to exist in IVA applications – a long-term *autobiographic memory*. The reason for integrating this memory is obvious: A companion agent shall learn from the user through their interaction histories; similarly, a narrative agent shall not duplicate its story in order not to bore the user.

One may then argue that, in order to generate different stories, a narrative agent can select different bits of story content from a database with some predefined rules. This approach may be systematically feasible and the stories, with the high-level control on the narrative structure (e.g. defined in [6] and [7]), can make sense to us. Alternatively, when using intelligent characters and substantial involvement from users, emergent narrative [8, 9] creates flexible structures for real-time and interactive narrative with some constraints to the richness of the story contents. Both approaches, however, do not address the issue of long-term believability – while planning to unfold a story, the user's previous interactions and the character's own experiences in the story so far have not been addressed.

Focusing on the general aim in creating IVAs that appear believable and acceptable to humans, one way to facilitate this aim is allowing human users to have a better understanding of the behaviour expressed by agents. This direction, as indicated earlier by Sengers [10], requires agents to behave in a *coherent*[1] manner so that the change from one behavior to another can be more understandable to human observers. In this paper one of our main goals is to model a human long-term memory computationally, thus we can attempt developing a sense of coherent "self" for IVAs.

In human beings, coherence is provided in the life stories of human beings [11] – since we communicate that sense of self and negotiate it with others through our life stories. Thus, we claim that the inclusion of it will be beneficial to believable agents, particularly in a narrative story-telling environment.

The rest of this paper is organised as follows: in the Related Work section, we first introduce the background literature in narrative storytelling agents, episodic memory modelling and our previous work in computational autobiographic memory. Next, we illustrate the implementation plan for long-term autobiographic memory with some preliminary results. We describe the on-going evaluation for the validation of our research hypothesis, namely that autobiographic memory can help agents improving their believability and user engagement through producing coherent life stories from the past experiences stored in their autobiographic memory. The Conclusion and Future Work sections conclude the paper.

2 Related Work

In Cognitive Science and Artificial Intelligence, the main stream of research in modelling human long-term memory and narrative tends to be associated with sense-making and problem-solving activities. "The human phenomenon" of

---

[1] Sengers discussed a general problem of behavioral incoherence in IVA research – often human observers cannot understand *why* agents behave as they do; this phenomenon was described as 'schizophrenia' [10].

rich and meaningful narrative, which can be found in all activities that involve the representation of events in time [12], has somewhat been neglected. In this section we introduce research in cognitive memory modelling and we present fundamental ideas of human/agent life stories as narrative. The design of long-term autobiographic memory for IVAs in the next section is influenced by and inspired from this literature.

2.1 Modelling Long-Term Memory

To embodied computational agents that are capable of learning and adapting themselves to the environment, memory and representation are essential to be integrated in the architecture. Therefore, modelling a human-like long-term memory has always fascinated many AI researchers – and led to various memory models contributing significantly to the understanding of human cognition. For instance, Script, from Schank and Abelson [13], captures two important aspects of human memory in the perspective of developmental psychology: 1) it represents everyday events and activities, and 2) it has social-cultural components. Early models like Frame [14] and Script, however, represent knowledge in abstract static structures with little or no relation to "organic" developmental processes, or emotion.

In recent years, the use of temporal sequences of episodic events, in both robotic and virtual agents research, is a growing area. For example, by collecting relevant events stored in episodic memory, an explorative robot is able to reduce its state-estimate computation in the tasks of localising and building a cognitive map in a partially observable office environment [15]. Also, long-term episodic memory with attributing emotions may help a virtual robot to predict rewards from human users, thus facilitating human-robot interactions in a simple Peekaboo communication task [16].

Mirza et al. [17, 18] uses the concept of *interaction histories*, defined as 'the temporally extended, dynamically constructed and reconstructed, individual sensori-motor history of an agent situated and acting in its environment including the social environment'. This work is strongly inspired by dynamical systems approaches to memory and sensori-motor coordination. The approach does not lend itself naturally to be applied to virtual characters and believable virtual agents, since the memory content is not represented symbolically and it is thus not straightforward how to visualise and, more importantly, communicate it to human users.

The current research trend towards modelling a complete human episodic memory (e.g. episodic memory in Soar [19] and a generic episodic memory module [20]) establishes a common structure that consists of context, contents and outcomes/evaluation for agents to remember past experiences. These models were created to focus on the following three different aspects:

1. Accuracy – how relevant situations can be retrieved from the memory.
2. Scalability – how to accommodate a large number of episodes and not decrease significantly the performance of the system.
3. Efficiency – how to optimise the storage and recall of memory contents

It is undoubted that these aspects are important for intelligent systems that are able to remember past events and anticipate future events. However, remembering past events and retrieving them in a timely manner, we argue, is not the only crucial part in creating IVAs interacting with human users – faster retrieval times cannot increase dramatically agents' believability. Likewise, highly accurate memory retrieval is not necessary in many situations involving believable agents that are meant to possess human-like memory: Human memory is not 'perfect' and can sometimes retrieve false memories or simply forget.

Brom et al. attempted to create a full episodic memory which stores more or less everything happening around the agent for the purpose of storytelling [21]. The authors claimed that the modelled episodic memory can answer specific questions from human users in real time regarding the agent's personal histories. Nevertheless, with the story scenario which was used in their paper, this memory was able to allow an agent to describe past actions in time but it failed to provide dramatic elements which facilitate an *interesting* piece of narrative. This may be due to the fact that the motivational and emotional states of the agents have not been fully integrated to determine the *importance* of a situation in an episode.

2.2 Narrative Agents and Autobiographic Memory

Bruner discussed in detail the narrative construction of human reality [22], in particular the social reality and the important role of stories in communication and social interaction. Following Bruner, Schank and Abelson suggested three cognitive features supported by the role of stories in individual and social understanding processes [23]:

1. Human knowledge is based on stories constructed around past experiences.
2. New experiences are interpreted in terms of old stories.
3. The content of story memories depends on whether and how they are told to others and these reconstituted memories form the basis of the individual's remembered self.

Based on these theoretical foundations, Dautenhahn proposed and discussed five different types of storytelling agents [24]:

1. **Type 0:** An agent that has a *grandfather model* of memory [23] – telling the same stories over and over again.
2. **Type I:** An agent that has a great variety of stories stored in its story-base. However each time the agent selects a single story randomly and tells it in exactly the same way as stored – not being situated to the conversational context.
3. **Type II:** A narrative agent that selects the story to tell from its large story-base based on the current context of interaction. However, this agent does not *listen* to its interaction partner; thus it does not understand and incorporate other's stories in its own.
4. **Type III:** A believable narrative agent that is able to interpret the meaning and content of the story and finds in its own story-base the most similar story which is then adapted in order to produce an appropriate response.

5. **Type IV:** An *autobiographic agent* that can tell the most rich and interesting stories. This agent, like a living human being, links a story which is being told with its current sensory data, embodiment (e.g. using non-verbal behaviours), past experiences and personality.

The term *autobiographic agents* was first defined in[25, page 5] as "agents which are embodied and situated in a particular environment (including other agents), and which dynamically reconstruct their individual history (autobiography) during their lifetimes.". This individual history can help autobiographic agents to develop individualised social relationships and to communicate with others, which are characteristics of social intelligence. The idea of modelling autobiographic memory in agents was inspired from the human psychological phenomena regarding the assimilation of memory to autobiography and sharing it with others – which were suggested as a psychological approach to narrative:

- "...this social function of (autobiographic) memory underlines all of our storytelling, history-making narrative activities, and ultimately all of our accumulated knowledge systems." [26, page 12]
- Narrative is one of the most effective means of creating, maintaining and communicating the self [11].

A simple, but nevertheless first technical implementation of an autobiographic agent, using a Webots (Cyberbotics) simulation environment, was presented by Dautenhahn and Coles [27]. Here, an agent acted on the basis of its memories of past experiences and replayed a sequence of actions that previously led to a goal. Extending this work, Ho et al. developed and experimentally evaluated different types of computational memory architectures for Artificial Life autobiographic agents, e.g. in [28]. In attempting to solve the problem of behavioural incoherence from a user's perspective, we developed an *Observer Interface* that shows the dynamically changing contents of an agent's memory as well as stories received from other agents [29]. Thus, this interface was an attempt to allow the user to see coherence in the agents' behaviour, i.e. allowing them to understand *why* particular actions were taken. Thus, in this work the agents' behaviours can be understood as intentional, narratively structured and temporally grounded. Furthermore, the communication of experience can be seen to rely on emergent mixed narrative reconstructions combining the experiences of several agents, cf. details in [29].

3 Narrative Autobiographic Memory for IVAs

In light of the preceding review and discussion, we seek an approach to develop IVAs that, with a long-term autobiographic memory, capture in particular the social aspects of the agents' life stories in order to increase their believability from a user's perspective. People perceive and interpret the (social) world in terms of stories. Thus, we expect that users interacting with IVAs that possess, express, and act upon their life stories will be better able to interact with the agents and understand the nature of the agents' actions and stories that the IVAs try to express. Specifically, we expect that a user's degree of familiarity with an IVA correlates with his knowledge of its life stories.

In this section we start with the presentation of the memory design and certain criteria that are necessary to be considered regarding the memory structure and encoding mechanisms. Next we show the initial implementation and some samples of preliminary results. At the end we describe our evaluation plan for the validation of agents' life stories reconstructed from their long-term autobiographic memory.

3.1 Design and On-Going Implementation

Before starting the technical design of a computational memory that allows IVAs to encode sensory data for their life story, a conceptual definition of life story is crucial for the task. Linde [11, page 21] stated that "A life story consists of all the stories and associated discourse units, such as explanations and chronicles and the connections between them, told by an individual during the course of his or her lifetime that satisfy the following two criteria:

1. The stories and associated discourse units contained the life story have as their primary evaluation a point about the speaker, not a general point about the way the world is.
2. The stories and associated discourse units have extended reportability; that is, they are tellable and are told and retold over the course of a long period of time[2]."

To adapt this definition and allow IVAs creating their believable and coherent life story, our design focuses on three essential story-associated discourse units: *explanations*, *event chronology* and *evaluation*; plus the characteristic of *reportability*.

Context and Agent Architecture. The application which we use to test the IVAs' long-term autobiographic memory is called FearNot!. It is an anti-bullying software developed by partners in the eCIRCUS[3] project. FearNot! provides eight-to-eleven year old users with the opportunity to visit a virtual school environment complete with characters representing most significant roles in bullying locations and scenarios that are commonplace in real-life bullying incidents. Users interact in the environment by watching narrative episodes unfolding as the result of the action executed by autonomous characters (the left screenshot in Figure 1). After each episode, the user can interact with one of the characters (the right screenshot in Figure 1), e.g. to give him/her suggestions on how to cope with the problem, and then see the result of the given advice on a subsequent episode.

---

[2] Later in her book, Linde extended this definition by including the fact that the content of a life story, as a product of cultural context, to include certain facts and exclude others. Due to the space constraint in this paper, the cultural element in agents' life story is discussed in our on-going research work in [30].

[3] http://www.e-circus.org/

Fig. 1. Sample screenshots of FearNot! software: (Left) a bullying episode. (Right) user interaction interface.

The FAtiMA architecture (see details of implementation in [31]) was used as the foundation for integrating the long-term autobiographic memory. It is an agent architecture where emotions and personality take a central role in influencing behaviour. The concept of emotion used steams from OCC cognitive theory of emotions, where emotions are defined as valenced (good or bad) reactions to events. The assessment of this relationship between events and the character's emotions is called the appraisal process.

FAtiMA agents are able to perceive events/actions happening concurrently in the environment, e.g. the bystander character can perceive both events that the victim character is mocked by the bully while he/she is running away and crying. Nevertheless, agents record these events in a chronological sequence based on the time of the received messages from the framework. Moreover, to smooth the narrative storytelling and to achieve the educational goals of FearNot!, characters' goal is authored in a way that allows the storyline in each bullying episode to be consistent and easily understood by the child user who plays the game. Therefore, multiple agents may act within the same timeframe, but it is not our focus while developing FearNot! as a character-based narrative system.

Memory Structure. As inspired by the narrative structure for life story for humans, the storage structure representing each event in the computational autobiographic memory consists of three main components: **Abstract**, **Narrative** and **Evaluation**. To support the description of each component, Figure 2 shows an illustrative example of an event stored in one of the character's memories (the character is named John).

We aim to incorporate the important features extracted from Linde's definition of life story into the design of the memory storage structure. First, the *chronological order of events* is addressed by the inclusion of the **Time** in the autobiographic memory (see Figure 2). Recording the time and the chronological order of events is crucial for agents' life stories simply because users can make inferences about the causality and thereby meaning of events as experienced by the agent. The **Time** here records three different types of data:

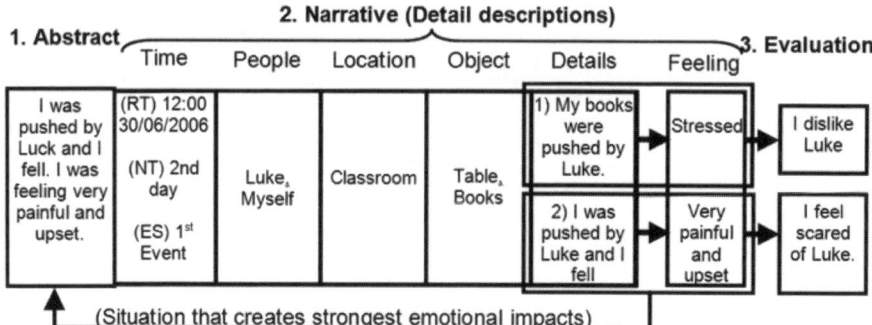

Fig. 2. A sample event constructed under an episode where a character (Luke) hits another one (John). The narrative component contains detailed descriptions of the event – when and where the event occurred, and objects and people (with their actions) involved in the event. The abstract of an event stores a summary of the event containing the most significant actions and feelings from the narrative component. Finally, evaluation corresponds to the agent's psychological evaluation of the effects of remembered actions according to the agent's social relations.

- Real time (RT) is the time that represents the real world; it will be used for calculating the time difference when the software is used over a long term with a user.
- Narrative time (NT) indicates the virtual time when an episode takes place in the whole story. It will be particularly useful during the software evaluation session.
- Event sequence (ES) simply shows the order of the event in time, e.g. the first or second event.

Story *explanation* and *evaluation* are similar discourse units in agents' life stories. They refer to **Abstract** and **Evaluation** of an event in the memory structure respectively. **Abstract** is represented by the actions which, among others in the same event, were associated with the highest emotional intensities. The example in Figure 2 example shows that the **Abstract** contains the most significant action to John in the whole event: Being pushed by Luke and falling which caused him to feel very bad. In agents' life stories, **Abstract** serves as an interactive function which often summarises the narrative or provides a "hint" to reveal the rest of the story (i.e. details stored in **Narrative**). The explanation is thus embedded in the design – in a human life story, it starts with a statement of some proposition, and then follows it with a sequence of statements of reasons showing why the proposition should be believed [11].

Based on the context of FearNot – an educational anti-bullying software, an event's **Evaluation** indicates the effects of remembered actions according to the agent's social relations, i.e., this essentially shows the action's consequence(s) in terms of inter-personal relationships remembered in an agent's memory. This evaluation in the agents' life stories is socially and interactionally important since it shows their "personalised" view, not a general point about the way the world is. It also plays a critical role in establishing the coherence of a life story, because an agent can trace back to many similar past experiences, from which the same result of evaluation was derived. Retrieving these experiences from autobiographic memory and putting them together in the agent's life story reconstructs the kind of "self" that is presented by the agent.

While retrieving events from agent's autobiographic memory, depending on the context, a set of search keys is used to look for similar events. For examples, the victim character can search for all past experiences associated with a specific person (e.g. the bully), happened in the playground, and with negative emotions. Here the content manipulations, which group similar past events together and then select representative contents for story generation, are similar to *Event Reconstruction and Event Filtering and Ranking* processes respectively from our previous research [29]. Furthermore, we expect that it would be more believable to have some random factors influencing the process of story creation, like humans telling their life story with slightly random content, structure and length. Therefore, an algorithm for constructing the remembered stories randomly is applied here to pick-up contents from some **Narrative** fields (see Figure 2). Nevertheless, being consistent with the basic structure of a narrative is very important since we want to retain the meaning of each story from an individual agent's perspective.

Last but not least, the characteristic of *reportability* is shown in the way that actions are remembered in an agent's autobiographic memory: The significance of an action to the agent is determined by the emotion attributed to that action from the appraisal process[4]. To be reportable an event/action must be unusual, unexpected, or abnormal; neutral actions/events that did not create much emotional impact to the agent are probably not "worth telling" [32] and will not be remembered as part of its life story. In our previous work an *event reconstruction process* in the agent's memory retrieval can generate short summaries covering the most significant actions in an event, derived from the autobiographic memory [33]. Thus, the agent can respond with narratives when interacting with users. We expect that similar mechanisms can be applied to create an agent's life story which contains events taking place under a specific context in different periods of the agent's life time. Therefore, stories generated from the event reconstruction process are context-dependent – the context thus serves as a matching key for selecting relevant events for the event reconstruction.

3.2 Preliminary Results

Based on our previous work [33] which incorporated the psychological view that emotions can arise in response to both internal and external events, agents embedded with long-term autobiographic memory are now able to encode events

---

[4] In the current implementation of FAtiMA, when an external event or action is perceived, a set of predefined emotional reaction rules is used by an individual agent in the appraisal process to create the meaning of the event. These rules, which include emotional threshold, importance of goals, desirability of events, etc., represent the agent's standards and attitudes towards generic events.

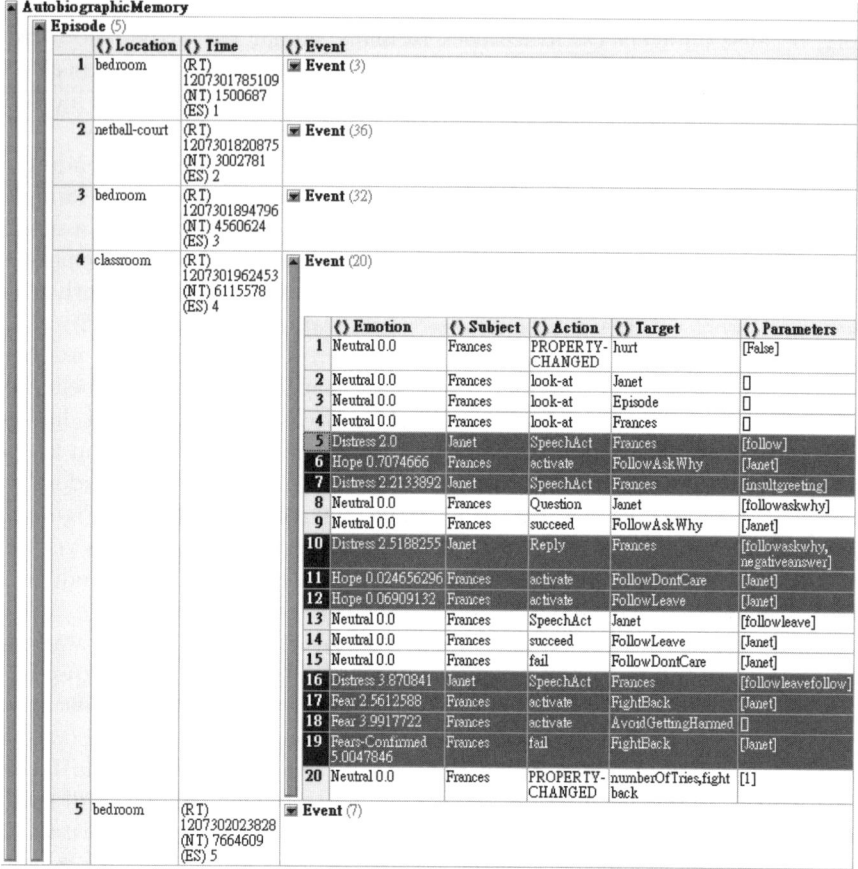

Fig. 3. Contents of an agent (a female victim character in FearNot!) showing Frances' long-term autobiographic memory in a life story structure. The highlighted part shows significant actions that created different types of emotional impact to the agent in the event. The view in this figure was generated by Altova XMLSpy®.

that happen during their life time. In the same context of using FearNot! software, Figure 3 illustrates, in XML format, part of the memory contents that an agent encoded in its memory.

The implementation is still on-going at the current stage. Both **Abstract** and **Evaluation** will be created in each event as planned in the design. A language engine will be used to convert the memory contents into agent's life story. Examples of an agent's event summary generated from our previous work [33] are as follows:

– *"During this morning. Luke made fun of me on the street and then Luke continued teasing me. I was sad. Can you believe it. I considered to fight Luke back and then I was planning to stay safe. I was feeling scared. Then, I failed to fight Luke back."*

– "This morning, I looked at Luke on the street and I hate Luke really, Luke attacked me and then I cried. Did you see that, Luke told me to get out of the way. I was feeling very desperate, I asked Luke to stop bullying me."

3.3 Evaluation Plan

Undoubtedly it would be inappropriate to compare real life stories from humans with our agents' synthetic ones, because a human life story is extremely dynamic, discontinuous and with rich social discourse units that conveys "who we are and how we got to be that way". Nevertheless, in our future evaluation with eight-to-eleven year old child users, we eagerly aim to examine the story structure and the level of coherence of life stories which will be generated from the long-term autobiographic memory. Through questionnaire and story-rewriting exercises, we hope to find out respectively 1) what length, quality and contents of the life story users would expect from our agents; and 2) how similar the agents' life story can be to users' written summary based on their observation on the narrative unfolded in FearNot!. To answer these questions, we have recently carried out an evaluation event including school children interacting with several FearNot! episodes, using an initial implementation of the architecture described in this article. We are currently analysing the results that will be presented in future publications.

Ultimately, in future user studies supported by the fully implemented architecture, we expect to find out whether the long-term believability of narrative agents can be increased through the integrated autobiographic memory.

4 Conclusion and Future Work

In this paper we discussed the issue of long-term believability for IVAs. We addressed this issue by proposing to integrate a long-term autobiographic memory for the creation of agents' coherent life stories as their narrative "self". We took inspiration from human life stories and illustrated an on-going implementation for agents' storytelling memory structure, which essentially incorporates human narrative discourse units and the reconstructive nature of human autobiographic memory. Finally, through the creation of their own life stories, we expect that this memory can help agents to 1) express their internal subjective sense of "self" and 2) show continuity of the "self" – autobiographic agents that learn and adapt themselves in the environment.

In the future we will first focus on finishing the architecture implementation and then establish the link between the long-term autobiographic memory and a language engine[5] for making agents' life story readable to human users. Furthermore, we will investigate making an agent's story more recognisable as an

---

[5] In the previous implementation FAtiMA sends the contents of a character's short summary generated from autobiographic memory to the language engine as speech acts. The language engine then makes use of the semantic parser for the generation of natural language used in the summary. See [34] for detail.

interesting life story, particularly taking into account that the coherence is a cooperative achievement of the 'story-teller' (the agent) and the 'audience' (the user). This indicates that agents must construct each narrative with an appropriate evaluation, plus giving the audience sufficient cues to understand this evaluation.

Acknowledgments

Authors would like to thank members of the eCIRCUS technical team to create the FearNot! software framework as a testbed for the work described in this paper, and particularly João Dias from INESC-ID for developing the FAtiMA architecture. This work was partially supported by European Community (EC) and is currently funded by the eCIRCUS project IST-4-027656-STP. The authors are solely responsible for the content of this publication. It does not represent the opinion of the EC, and the EC is not responsible for any use that might be made of data appearing therein.

References

[1] Bates, J.: The role of emotion in believable agents. Communication of the ACM 37(7), 122–125 (1994)
[2] Bratman, M.E.: Intention, Plans, and Practical Reason. Harvard University Press, Cambridge (1987)
[3] Ortony, A., Clore, G.L., Collins, A.: The cognitive structure of emotions. Cambridge University Press, Cambridge (1988)
[4] Dörner, D.: The mathematics of emotions. In: Dörner, D., Detje, F. (eds.) Proceedings of the Fifth International Conference on Cognitive Modeling, Bamberg, Germany, April 10-12, 2003, pp. 75–79 (2003)
[5] Dörner, D., Gerdes, J., Mayer, M., Misra, S.: A simulation of cognitive and emotional effects of overcrowding. In: Proceedings of the 7th International Conference on Cognitive Modeling, Trieste, Italy (April 5-8, 2006)
[6] Barthes, R.: Introduction to the structural analysis of narrative. In: Image Music Text (1966), pp. 79–124. Hill and Wang, New York (1977)
[7] Goguen, J.: The structure of narrative (2001),
http://www-cse.ucsd.edu/classes/sp01/cse171/story.html
[8] Aylett, R.: Narrative in virtual environments: Towards emergent narrative. In: Proc. Narrative Intelligence. AAAI Fall Symposium, pp. 83–86. AAAI Press, Menlo Park (1999); Technical Report FS-99-01
[9] Aylett, R., Figueiredo, R., Louchart, S., Dias, J., Paiva, A.: Making it up as you go along: improvising stories for pedagogical purposes. In: Gratch, J., Young, M., Aylett, R.S., Ballin, D., Olivier, P. (eds.) IVA 2006. LNCS (LNAI), vol. 4133, pp. 307–315. Springer, Heidelberg (2006)
[10] Sengers, P.: Narrative and schizophrenia in artificial agents. In: Mateas, M., Sengers, P. (eds.) Narrative Intelligence, pp. 259–278. John Benjamins, Amsterdam (2003)
[11] Linde, C.: Life Stories: The Creation of Coherence. Oxford University Press, Oxford (1993)

[12] Abbott, H.P.: The cambridge introduction to narrative. Cambridge University Press, Cambridge (2002)
[13] Schank, R.C., Abelson, R.P.: Scripts, Plans, Goals and Understanding: An Inquiry into Human Knowledge Structures. Lawrence Erlbaum Assicuates, Hillsdale (1977)
[14] Minsky, M.: A framework for representing knowledge. In: Winston, P.H. (ed.) The Psychology of Computer Vision, pp. 211–277. McGraw-Hill, New York (1975)
[15] Endo, Y.: Anticipatory robot control for a partially observable environment using episodic memories. Technical Report GIT-IC-07-03, Georgia Tech Mobile Robot Lab (2007)
[16] Ogino, M., Ooide, T., Watanabe, A., Asada, M.: Acquiring peekaboo communication: Early communication model based on reward prediction. In: Proceedings of IEEE International Conference in Development and Learning (ICDL) 2007, London, UK (July 2007)
[17] Mirza, N.A., Nehaniv, C.L., Dautenhahn, D., te Boekhorst, R.: Interaction histories: From experience to action and back again. In: Proceedings of the 5th IEEE International Conference on Development and Learning (ICDL 2006), Bloomington, IN, USA (2006) ISBN 0-9786456-0-X
[18] Mirza, N.A., Nehaniv, C.L., Dautenhahn, D., te Boekhorst, R.: Grounded sensorimotor interaction histories in an information theoretic metric space for robot ontogeny. Adaptive Behavior 15(2), 167–187 (2007)
[19] Nuxoll, A., Laird, J.E.: A cognitive model of episodic memory integrated with a general cognitive architecture. In: Proceedings of the Sixth International Conference on Cognitive Modeling, pp. 220–225. Lawrence Erlbaum, Mahwah (2004)
[20] Tecuci, D., Porter, B.: A generic memory module for events. In: Proceedings to the 20th Florida Artificial Intelligence Research Society Conference (FLAIRS 20), Key West, FL (2007)
[21] Brom, C., Pesková, K., Lukavskýz, J.: What does your actor remember? towards characters with a full episodic memory. In: Cavazza, M., Donikian, S. (eds.) ICVS-VirtStory 2007. LNCS, vol 4871, pp. 89–101. Springer, Heidelberg (2007)
[22] Bruner, J.: The narrative construction of reality. Critical Inquiry 18(1), 1–21 (1991)
[23] Schank, R.C., Abelson, R.P.: Knowledge and memory: The real story. In: Wyer, R.S. (ed.) Knowledge and Memory: The Real Story, pp. 1–85. Lawrence Erlbaum Associates, Mahwah (1995)
[24] Dautenhahn, K.: Story-telling in virtual environments. In: Working Notes Intelligent Virtual Environments, Workshop at the 13th biannual European Conference on Artificial Intelligence (ECAI 1998), Brighton Centre, Brighton, August 23-28 (1998)
[25] Dautenhahn, K.: Embodiment in animals and artifacts. In: AAAI FS Embodied Cognition and Action, pp. 27–32. AAAI Press, Menlo Park (1996); Technical report FS-96-02
[26] Nelson, K.: The psychological and social origins of autobiographical memory. Psychological Science 4, 7–14 (1993)
[27] Dautenhahn, K., Coles, S.: Narrative intelligence from the bottom up: A computational framework for the study of story-telling in autonomous agents. Artificial Societies and Social Simulation (2000) (January 31, 2001)
[28] Ho, W.C., Dautenhahn, K., Nehaniv, C.L.: A study of episodic memory-based learning and narrative structure for autobiographic agents. In: Proceedings of Adaptation in Artificial and Biological Systems, AISB 2006 conference, vol. 3, pp. 26–29 (2006)

[29] Ho, W.C., Dautenhahn, K., Nehaniv, C.L.: Computational memory architectures for autobiographic agents interacting in a complex virtual environment: A working model. Connection Science 20(1), 21–65 (2008)
[30] Ho, W.C., Dautenhahn, K., Lim, M., Enz, S., Zoll, C., Watson, S.: Towards learning 'self' and emotional knowledge in social and cultural human-agent interactions. International Journal of Agent Technologies and Systems 1 (to appear, 2009)
[31] Dias, J., Paiva, A.: Feeling and reasoning: a computational model for emotional agents. In: Bento, C., Cardoso, A., Dias, G. (eds.) EPIA 2005. LNCS (LNAI), vol. 3808, pp. 127–140. Springer, Heidelberg (2005)
[32] Bruner, J.: Actual Minds, Possible Worlds. Harvard University Press, Cambridge (1986)
[33] Ho, W.C., Dias, J., Figueiredo, R., Paiva, A.: Agents that remember can tell stories: Integrating autobiographic memory into emotional agents. In: Autonomous Agents and Multiagent Systems (AAMAS), pp. 35–37. ACM Press, New York (2007)
[34] Dias, J., Ho, W.C., Vogt, T., Beeckman, N., Paiva, A., Andre, E.: I know what i did last summer: Autobiographic memory in synthetic characters. In: Paiva, A.C.R., Prada, R., Picard, R.W. (eds.) ACII 2007. LNCS, vol. 4738, pp. 606–617. Springer, Heidelberg (2007)

Emergent Narrative as a Novel Framework for Massively Collaborative Authoring

Michael Kriegel and Ruth Aylett

School of Mathematical and Computer Sciences,
Heriot Watt University,
Edinburgh, EH14 4AS, Scotland
{michael,ruth}@macs.hw.ac.uk

Abstract. An emergent narrative is a narrative that is dynamically created through the interactions of autonomous intelligent virtual agents and the user. Authoring in such a system means programming characters rather than defining plot and can be a technically and conceptually challenging task. We are currently implementing a tool that helps the author in this task by training the characters through demonstration of example story lines (rehearsals), rather then explicit programming. In this paper we argue that this tool is best used by a group of authors, each providing an example story and that in order to achieve true emergence, collective authoring is required. We compare the rehearsal based authoring method of our authoring tool with other collaborative authoring efforts and underline why both the storytelling medium "emergent narrative" and our particular approach to authoring are better suited for massively collaborative authoring.

1 Introduction

An increasingly popular branch of AI research concerns itself with interactive narratives - computer based storytelling systems that dynamically respond to the audience and adapt the plot during its presentation. Unfortunately, so far no existing system has yet been able to fully realize these keen ambitions and building such a system remains an open research problem. One of the main reasons for this situation is that the creation of interactive stories poses a big challenge for an author. There is both a lack of theoretic approaches to authoring interactive stories and concrete authoring tools, which results in the interactive narrative medium being very inaccessible to authors. In previous work [1, 2] we have proposed the design of an authoring tool, which is currently being developed that might provide a solution to this problem by allowing the author to provide content in the form of linear stories and thus working within a well studied story creation framework (that of the traditional linear story). Our assumption is that with an increasing number of linear stories the authors provide, the complexity of the story world and thus the quality of an interactive story within this story world increases. In this paper we will argue in favour of a collaborative authoring approach using this tool.

2 Authoring Emergent Narratives

Emergent Narrative is a term that was first coined by Aylett[3] and refers to a form of interactive storytelling, where the narrative is built bottom-up from interactions of characters. Like in any other emergent system relatively simple local decisions lead to complex behaviour[4], in our example a narrative. Despite that, human authors are still needed, but their role is significantly different from authors of other forms of narrative. In an emergent narrative based system an author specifies a virtual world, characters, their goals, motivations, actions and emotions rather than defining specific plot segments. A possible visualization of an emergent narrative can be seen in Figure 1. Within this framework, characters can be implemented as autonomous intelligent virtual agents. Authoring in this kind of environment poses two challenges that are described in the following sections.

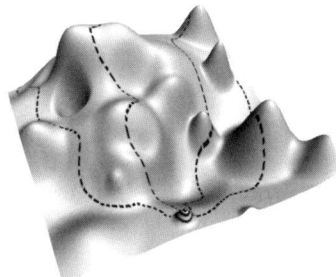

Fig. 1. A conceptual visualisation of an emergent narrative as a 3 dimensional landscape. A particular story that is experienced equals a specific path through the landscape. In this illustration several possible story paths are shown that all initially start from the same point but diverge due to decisions of the user.

2.1 Unpredictability: The Conceptual Authoring Challenge

Not only are emergent narratives non-linear but they are also unpredictable at authoring time. The very nature of emergent narrative requires the author to "let go" of specific story lines altogether and to focus on creating the elements from which the story will emerge. The authors are not even intended to predict what stories will emerge, they are merely setting the boundaries of the story world. And yet our experience during the authoring process for the educational drama *FearNot!*[5], which was an early prototype of emergent narrative technology showed that instead of letting the story emerge naturally, authors tend to iterate between modifying story elements (character settings) and simulation, until the simulation results in the desired story lines. This approach to authoring is of course very tedious and frustrating and furthermore suppresses any emergence, as such defeating the purpose of emergent narrative.

2.2 Knowledge Representation: The Technical Authoring Challenge

The other challenge is the form of authored content. In 1991, Eileen Cornell Way stated[6]: "There is a basic although not often articulated assumption in

AI that any system which is able to behave intelligently must consist, in part, of symbolic structures that in some way *represent* the knowledge and beliefs necessary for that behaviour." This still holds true today, despite the progress that non-symbolic knowledge representation methods like neural networks have made since then. For intelligent behaviour, at a high enough level to result in something like a narrative, symbolic knowledge encoding is still absolutely vital. Existing interactive storytelling systems differ in the specifics of knowledge representation, but they all face the same problem: a symbolic description of a world is always just a model and there are many different ways to model a world. Consequently it is the author's responsibility to ensure that the model is consistent and matches the need of the application. In *FearNot!*, a system based on autonomous agents with a continuous planner, most of the authored knowledge can be found inside the action and goal descriptions of the characters planning domains. According to Simpson *et al.*, knowledge representation in AI planning, i.e. the process of representing planning domains for a particular task is *"as important a research topic as the algorithmic aspects of abstract planning engines"* [7]. In the planning context that we are concerned with, problems that the knowledge engineer/author has to face are for example finding the right level of granularity (i.e. distinguish between actions and goals) and finding the right level of abstraction/generality for describing actions and goals.

3 Rehearsal Based Authoring

We are currently implementing an emergent narrative authoring system based on the *FAtiMA!* agent architecture (originally developed as part of the *FearNot!* software) that will facilitate the authoring process and help authors avoiding some of the aforementioned problems. The main requirement for the software is to make authoring as accessible and user friendly as possible. We intend to achieve this goal by equipping the software with a video game like interface, so that the process of authoring becomes more user friendly and fun. The system learns new goals and action descriptions from observing example storylines(rehearsals). An author provides the system with those rehearsals by simply playing, controlling the characters, very much like a puppeteer would control puppets. This kind of indirect authoring of planning domains has been drawn from the idea of planning operator induction as described in[8]. We have chosen this approach to allow authors to continue thinking in terms of linear stories. In order for a rehearsal to be efficient it of course has to include some new events or at least a new order of events that add something to what was already covered by previous rehearsals. With the previously described features, our work shares many similarities with other research on learning by demonstration (e.g. [9, 10]). We have however added a mixed initiative planning feature that distinguishes our authoring method from most related work. Mixed initiative planning refers to a planning software that is supervised and assisted by a human being[11]. In our case it means that the characters' or puppets' minds are activated during the authoring process. Although they are controlled by the author, they still plan their own actions. Authors can decide to let the characters perform the actions they have planned or let them perform another action they see more fitting in the situation. In those cases author and character can also engage in a dialogue, in which the author motivates the orders given to the character. The software can use this additional feedback for further refinement of the planning domain.

Adding the mixed initiative mode should provide two core benefits: First, it will provide authors with immediate feedback on a character's authored personality so far and thus make it easier for them to "debug" a character and correct parts of its personality. Second, especially after multiple rehearsals authors will be relieved from the burden of giving the characters repeated instructions. With every rehearsal a character will become more active and autonomous and authors can focus on the input of new knowledge rather then repeating knowledge the character already has. Summarising, the main benefits of our rehearsal based authoring method are a user-friendly game-like interface, the fact that it allows authors to think linearly, i.e. top-down while creating bottom-up emergence and the built-in feedback mechanism through mixed initiative planning. We will not focus further on these issues in this paper, as they are already discussed in more detail in [1] and [2].

4 Massively Collaborative Authoring

User generated content is a buzz-word of the internet industry and the motor of the Web 2.0. Online communities like *MySpace* or *Facebook*, virtual worlds like *Second Life* or the on-line encyclopaedia *Wikipedia* are all well-known examples of how to successfully leverage internet users as content providers. We envisage that a similar collective authoring process could be applied to the creation of emergent narratives. Technically, this would be possible using the rehearsal based authoring mechanism described in the previous section. Different people can provide rehearsals at different times, so every author feeds a little bit into the system.

4.1 Other Collaborative Authoring Projects

Our work bears a certain resemblance to other collaborative authoring projects. Common-sense knowledge bases / ontologies like Cyc^1 or *OpenMind Common Sense*[12] are also trying to learn symbolic representations of knowledge through collaborative input of many users, however not within a storytelling context as in our work. We only need to capture the amount of common sense knowledge that is necessary for the agent to act believably in the given story context. We do not want to create real intelligence, the illusion of intelligence for a narrative purpose is enough. *Jabberwacky*[2] is a chat-bot that learns from millions of conversations with internet users. Every user that comes to the *Jabberwacky* website to chat is an indirect author and increases the chat-bot's repertoire

---

[1] http://www.cyc.com
[2] http://www.jabberwacky.com

of utterances. A quite similar approach can be found in *The Restaurant*[13], a research experiment in using gameplay data aquired in a multiplayer game to author the AI for a single player game. The idea of "user generated content" also slowly finds its way into the commercial video game industry. The most notable example for this is *Spore*, the new game of *Sims* creator Will Wright, which through highly procedural AI and graphics allows players to create very unique content. The game automatically streams this content to other players to populate their game world. Again, players are used as authors. Finally, it is worth mentioning *a million penguins*[3], an attempt to collaboratively create a novel on the Web through a wiki. While there is no official academic publication about the lessons learned form this experiment yet, several blog posts[4] summarize the resulting narrative as incoherent and chaotic. Most participants enjoyed the project as an interesting experiment with the conclusion that the literary form of a novel and the collaborative approach of a wiki are too different to be combined.

4.2 Advantages of Collaboration

The collaboration of many users on the web can be seen as an emergent phenomenon itself, so it shares an important feature (emergence) with emergent narratives. Mapping the emergent processes of the web onto the creation of emergent narrative systems thus seems like a sensible step. As we have pointed out, an emergent narrative resembles a whole landscape full of possible stories. In order to create that landscape, a high quantity as well as quality of input is required - an enormous task for a single individual. As the previous literature review has shown, the internet community is both capable of providing vast amounts of data (quantity) and also possesses a great collective creativity (quality) that was ultimately too much to fit in a single novel and resulted in chaos in the case of *a million penguins*. However, the medium of emergent narrative is different. While collaboration in *a million penguins* meant extending the storyline further in length[5], in an emergent narrative it means making the story world richer. Another author just adds more possibilities to what might happen when someone plays through an emergent narrative. Figure 2 illustrates this contrast.

While the collaboration in an emergent narrative results in a reshaping and refinement of the story landscape, in a linear narrative the only way forward is to add to the already existing storyline. Doing that can result in undesired jumps in the story's coherence (as seen in the Figure after the 3rd author's input) and thus a disconnected narrative.

---

[3] http://www.amillionpenguins.com
[4] e.g. http://www.futureofthebook.org/blog/archives/2007/02/
a_million_penguins_a_wikinovel.html
[5] ...or creating branches in a linear medium, where they are not desired. In *a million penguins* quite a few side novels split of the main plot line when authors did not agree with the development of the storyline.

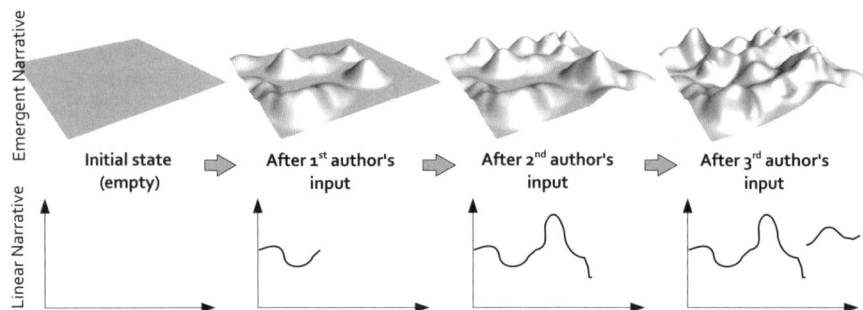

Fig. 2. Conceptual diagram comparing collaborative authoring of an emergent and a linear narrative

4.3 Avoiding Incoherence

One could argue that the same incoherence can also occur in an emergent narrative (e.g. characters acting inconsistently). How can we make sure that several authors do not contradict each other or in other words, how do we keep the story landscape smooth? The potential answer lies within the previously described rehearsal based authoring and its mixed initiative feature. Since the characters are "alive" during the authoring process and give constant feedback on their so far authored personality, a second author cannot easily create a personality for a character that completely contradicts the one given to it by previous authors. Doing that will require the author to justify their decisions to the character and will as a result override the personality that was given to the character by previous authors. This might partly invalidate the previous authors' work but at least it will ensure a smooth story landscape. However, the mixed initiative feature will hopefully help authors to fit in their rehearsal plots nicely with previous rehearsals, i.e. the author will let himself be guided by the characters and vice versa.

4.4 Ignorance Is Useful

There is another advantage of collaboration. Everything authors have done before will inevitably influence their continuing authoring. They cannot just simply forget what they have rehearsed before and will base their future rehearsal plots on that (consciously or subconsciously). If there is only one single author that provides all the rehearsals this might thus decrease the resulting emergence. If the rehearsals are however distributed to many different authors, none of them will know the exact rehearsal story-lines that their predecessors have rehearsed. The only connection between their work that they are aware of comes through the mixed initiative feature, i.e. the character's feedback during authoring, but this feedback is limited to only the current plot situation. In a collaborative authoring situation every single author is to a certain extent unaware of the work of the other authors and that unawareness provides a very useful prerequisite for creativity. This is in fact a general truth for all emergent systems, as Steven

Johnson states in his book *Emergence*[4], when talking about the requirements of emergent behaviour (using the example of an ant colony):

> **Ignorance is useful:** ... Having individual agents capable of directly assessing the overall state of the system can be a real liability in swarm logic, for the same reason that you don't want one of the neurons in your brain to suddenly become sentient."

Exactly the same applies to emergent narratives (a character-centric system, with individual agents limited to their own world view) and as explained above to the emergent collaborative authoring process. It is hard for a single brain to design all the elements of an emergent system, as that brain will always try to predict what emerges from the system and thus suppress true emergence. Only multiple brains can achieve that task.

5 Conclusion

In this paper we have described a rehearsal based approach to emergent narrative authoring and argued for a collaborative authoring process using this method. Emergent narrative seems to be a narrative medium that is well suited for collaboration and the rehearsal based approach that lets each author perform a number of rehearsal scales up much better than the more obvious solution of assigning one author to each character. Until a truly emergent narrative (that includes unpredictable things happening) has been built, it will remain unclear, whether this is a desirable and enjoyable experience, but in order to build one, collaboration seems inevitable.

5.1 Future Work

In the last months we have started the implementation of the rehearsal based authoring tool described in this paper. After finishing the software development, we plan to perform a small-scale user study to verify the claims that have been made in this paper. For this experiment, we will provide authors with a small set of characters and a limited story domain and then investigate the effects of collaboration. Provided this yields a satisfying result, a long term goal would be a large scale study. To prepare that however, practicalities such as motivating a large number of authors to participate, preventing spamming and vandalism and the question of whether human editors should be involved need to be addressed.

Acknowledgements

This paper is supported by the eCIRCUS (Contract no. IST-4-027656-STP) project carried out with the provision of the European Community in the Framework VI Programme. The authors are solely responsible for the content of this publication. It does not represent the opinion of the European Community, which is not responsible for any use that might be made of data appearing therein.

References

[1] Kriegel, M., Aylett, R.: A mixed initiative authoring environment for emergent narrative planning domains. In: Proceedings of the AISB Annual Convention, pp. 453–456 (2007)
[2] Kriegel, M., Aylett, R., Dias, J., Paiva, A.: An authoring tool for an emergent narrative storytelling system. In: Papers from the AAAI Fall Symposium on Intelligent Narrative Technologies, Technical Report FS-07-05, pp. 55–62 (2007)
[3] Aylett, R.: Narrative in virtual environments: Towards emergent narrative. In: AAAI Press Fall Symposium, Technical report FS-99-01, pp. 83–86 (1999)
[4] Johnson, S.: Emergence. Penguin Books (2001)
[5] Aylett, R., Dias, J., Paiva, A.: An affectively driven planner for synthetic characters. In: International Conference on Automated Planning and Scheduling (ICAPS), pp. 2–10. AAAI Press, Menlo Park (2006)
[6] Way, E.C.: Knowlege Representation and Metaphor. Kluwer Academic Publishers, Dordrecht (1991)
[7] Simpson, R.M., McCluskey, T.L., Zhao, W., Aylett, R., Doniat, C.: An integrated graphical tool to support knowledge engineering in ai planning. In: European Conference on Planning, Toledo, Spain (2001)
[8] McCluskey, T.L., Richardson, N.E., Simpson, R.M.: An interactive method for inducing operator descriptions. In: Proceedings of the 6th International Conference on AI Planning and Scheduling (AIPS 2002), Toulouse, France (2002)
[9] Cypher, A. (ed.): Watch What I Do: Programming by Demonstration. MIT Press, Cambridge (1993)
[10] Lockerd, A.: Thought streams: Simulating commonsense, web.media.mit.edu/~alockerd/research.html
[11] Burstein, M.H., McDermott, D.V.: Issues in the development of human-computer mixed-initiative planning systems. In: Gorayska, B., Mey, J.L. (eds.) Cognitive Technology: In Search of a Humane Interface. Elsever Science B.V. (1996)
[12] Singh, P.: The open mind common sense project. Online Article at KurzweilAI.net (2002) (retrieved August 6, 2007), http://www.kurzweilai.net/meme/frame.html?main=/articles/art0371.html?
[13] Orkin, J., Roy, D.: The restaurant game: Learning social behavior and language from thousands of players online. Journal Of Game Development 3(1), 39–60 (2007)

Towards Real-Time Authoring of Believable Agents in Interactive Narrative

Martin van Velsen

Institute for Creative Technologies,
13274 Fiji Way, Marina del Rey, California 90292 USA,
vvelsen@ict.usc.edu

Abstract. In this paper we present an authoring tool called Narratoria[1] that allows non-technical experts in the field of digital entertainment to create interactive narratives with 3D graphics and multimedia. Narratoria allows experts in digital entertainment to participate in the generation of story-based military training applications. Users of the tools can create story-arcs, screenplays, pedagogical goals and AI models using a single software application. Using commercial game engines, which provide direct visual output in a real-time feedback-loop, authors can view the final product as they edit.

Keywords: Authoring, Interactive Narrative, Believable Agents, Simulations, Games, Training, Machinima.

1 Introduction

The software presented in this document seeks to provide sufficient authoring capabilities to allow users to create training applications using interactive narrative without the need of programmers. Traditional film has come a long way from the days of black and white movies accompanied by live piano playing. Video games have progressed even faster but are still catching up with the narrative possibilities of film. Furthermore, those who have traditionally created story based content for film do not yet have the ability to intuitively create content for new media forms such as games and digital interactive training systems. New approaches in both interactive drama and video-game technologies pave the way for exploration of what could be called interactive films [17], but much work remains to be done to bring these capabilities to those who benefit most from these developments. Narratoria attempts to provide an integrated intuitive authoring solution by giving those users who are experts in story telling in traditional media easy to use drag and drop tools for compositing complex dramatic scenes using game technologies.

2 Authoring Interactive Narrative

In our first authoring effort, the Institute for Creative Technologies' (ICT) Leaders project [6] used a commercial game engine (Unreal Tournament™) to immerse the

---

[1] Narratoria is Copyright 2008, University of Southern California, All Rights Reserved.

user in an interactive decision making game. The Leaders simulation uses cinematic cut-scenes linked together through pre-defined decision points in a story graph. Decisions are elicited from users through a Naïve Baysean natural language classification interface, which triggers the system to progress the story to a different decision point through the display of a rendered scene in the simulation (game). In our Army oriented training application, the production started with the collection of anecdotes from Army Captains who recently returned from the field. Anecdotes were analyzed by pedagogical experts and turned into plot points or decision points. Our screenplay writers linked plot points into a story graph, where each full branch represent a story arc and nodes represent plot or decision points. Story arcs ensure a coherent and engaging experience, whereas the plot points ensure that users have to navigate decisions. Screenplay writers turned the story graph into a concrete movie script that maps decisions into engaging story arcs. Traditional scripts are linear, to be played all the way through one scene after another. The writers clustered scenes into what we call story molecules [9] ensuring scenes could lead to multiple outcomes and therefore other scenes. Each molecule then becomes self contained in that it can be played by itself but could also be combined with other molecules. This modular script in the form of a story graph was handed over to a director and our art director. A director matched the script to storyboards, whereas the art director created an inventory of characters and animations needed to express the story. Finally a team of animators in collaboration with a director composed the final game from all the available digital content stored in libraries. Collectively the authoring modules form a workflow for interactive Narrative as used by the Leaders project. Other workflows can be created by re-sequencing the existing modules or by adding new modules.

3 Authoring Believable Worlds: A Question of Language

A major hurdle in creating an authoring tool for film based interactive narrative lies in the fact that multiple experts are needed. Each of these experts might use a different expression language specific to a field. An expression language could be thought of the as way experts express the models most salient to their field. For example, software developers write code that operates on spatial X, Y and Z coordinates, whereas film directors work in terms of medium and long camera shots, etc. The software tool presented in this paper hides from the user the fact that each expert works in his or her own language. It does this by integrating a number of editing tools, such as script editors, story graph editors and timeline editors. Using this approach we ensure that the learning curve for users not intimately familiar with authoring software is gradual. Especially those interfaces such as storyboards [11], which are traditionally a paper and pencil tasks, are excellent candidates to be added to the production workflow as digital incarnations. Behind the scenes the authoring application translates the interaction into shared software structures. For example screenplays can be written using the Narratoria movie script editor module, whereas film editors use the timeline editor module. The only link between narrative information and 3D visualization is made through anchors or markers placed in the virtual environment. They are used to place and move actors or cameras, props, etc. All other control over the world is established through code interpreting the intentions of a director who uses this to give hints to the game engine. Further character performance is achieved by code that blends animations from a library in different ways to create new gestures. Blending is accessible from the timeline module where the final visuals are composed. The game engine then acts as a digital stage where directors interact with virtual actors. By allowing software to interpret the intentions of the creative users, those users can express themselves more freely because they do not feel burdened by the demands of the computational knowledge needed to built interactive worlds. Feedback is obtained from observing the output of the visualization drivers as well as timing feedback in the timeline module. Since events in the simulation may not play when expected, the timeline gives indications as to when an action started and how long it played. Synchronization in the simulation is similar to the 'rising-edge' approach [4] where the system attempts to honor the starting times of actions.

3.1 Authoring Believable Agents

Computer controlled software agents act out the interactive script. These digital actors are under semi-autonomous control of the simulation and take performance hints from a director who plans their movements, gestures and dialog. Since the digital actors, much like their real counterparts, act from marker to marker, a bit of freedom exists for the characters to choose how to get from A to B. Actors climb hills or descent stairs to reach their destination marker and beforehand it is not known how much time this will take or what obstacles might be encountered. Additional code attempts to ensure that actions play out as they should, taking into account other events that might interfere. Another tool, assisting directors in interacting with virtual actors, is the facility of game engines to blend animations. New animations can be derived using this approach, reducing the need for large libraries of gestures and movements. Blending is accessible from the timeline module and is achieved by overlapping multiple animation actions in time.

3.2 The Cinematic Authoring Process

New projects start with a screenplay. A screenplay can be written in software such as Finaldraft™ or composed directly within the Narratoria editor. Scripts are internally segmented into scenes either by extraction or through annotation by authors. Authors depend on scenes since they represent the longest uninterrupted sequence of events and therefore form a consistent reference point throughout the authoring process. Scene data can now be accessed in the timeline editor and the story graph editor simultaneously. Those working on the story graph map out plot points and link them with story arcs using drag and drop methods. Interactions and transitions are both depicted using scenes from the script and authoring the story is achieved by assigning the appropriate scenes to either arcs or interacts that are contained in the nodes. Authors map scenes, depicting interactions with virtual actors, to classes of text input that will be typed by users playing the game. Interact scenes can be set to also trigger arcs, which means for that specific user input maps to a decision and moves the plot to the next node. While the script is being revised and the story structure is further developed, editors familiar with non-linear film editing use the timeline to refine the cinematic content of the scenes. Some of the information comes from the script, such as dialogs, subtitles and camera shots. Other details are added manually, for instance: transitions, animations, and other aspects not found in screenplays. Editors add these visualization instructions as actions on timeline tracks. Each of the actions can be timed relative to the start of the scene and stretched to change the playback speed. Story editors try out interactions in the game and timeline editors play sequences to see how they appear on the screen. Finally the dataset compiles into a single XML file ready to be used by the game.

4 Technical Discussion

Narratoria consists of two major components: an authoring application and a visualization driver. The authoring application manages authoring modules and is written in C++ using the Qt interface development toolkit. Authoring modules load during startup and provide dedicated authoring capabilities for one particular part of a production pipeline. For example, one module manages the character bible and another presents a timeline canvas. On the visualization side we provide a driver that integrates into a number of different commercial game engines. So far we have tested this driver with Unreal Tournament™, Gamebryo™ and TVML™. Scene data can be streamed on the spot from the timeline editor to the visualization drivers by an author pressing 'play' for either a track, a scene or the entire screenplay.

4.1 Authoring Interoperability

With multiple authoring modules operating on the same information and with a number of displays interpreting visual data, the single most important aspect of the Narratoria architecture is its data design. Concepts from film and theatre form the basis of a global ontology, embedded in extendable core document classes [10]. Core concepts range from camera models to definitions of time and time fragments (beats) or even learning objectives for more recent interactive training applications. Heterogeneous software components can act on data that is altered or even managed by other components through an open data framework. The following characteristics contribute to the open nature of the document definitions:

- Using introspection, each part and sub part of a Narratoria data set can be extracted, identified and used by itself. For example, the XML instance of a character can be used independently from the XML containing a character bible.
- Every single part of the data set can be individually streamed into an external format such as XML, a method frequently used by similar efforts [14]. Recursive streaming allows hierarchies of data types (such as an entire document) to be streamed to an output module or visualization driver.
- Every individual Narratoria data type can be instantiated from a data stream. Given a well formed fragment of XML, that contains a valid Narratoria data type, a new instance of that type can be generated by the core system.
- The Narratoria data referencing design, described below, allows new data types to be added at runtime.

When deriving new classes, using the document and data types provided, developers are required to provide streaming facilities for novel data types. Developers are assisted in this by a collection of tools designed for this purpose. For those situations where new data types need to extend or reference other data types, three mechanisms provide means to interoperate and design new modules:

- Reference by derivation, where new capabilities are added by building on existing Narratoria classes (documents and data types). The newly created classes only need to provide a small amount of information such as a class name for introspection.
- Reference by linking, in this case new data is added to existing classes at runtime in the form of anonymous class variables. This would be analogous to adding new attributes to an object at runtime.
- Reference by association and the least intrusive of relating two pieces of data is the ability to freely associate data instances. In this case a third object, which we call the insulator object is placed in between two objects and forms the actual reference.

Using insulator objects allow users to have full control over the creation and dissolution of data references. For example, we could imagine a situation where a developer created a new document class containing a list of speech acts. Some of these speech acts might be specific to instances of characters from the character bible. In this case some character objects are connected to speech act objects by insulator objects and stored in the project document, which is designed to manage all instances of insulator objects. When the link between a speech act instance and virtual character is no longer valid the insulator object is deleted. Each insulator object contains methods for triggering a message to the user, indicating a potentially important link is about to be broken (for example a user deletes a speech act) and the action can be either allowed or dismissed.

4.2 Heterogeneous Collaboration

We described how specialized editing modules can be placed in a workflow, a workflow that can be adjusted depending on the tasks and domain. Users working in the workflow can collaborate on the same data set by dividing the data over a number of different users. We make use of the open nature of the data model for splitting up documents and allowing users to work on individual fragments. Those fragments are completely independent units, which can be integrated back into one large data set using facilities provide from within the authoring tool. For example, during the ICT Leaders project, six animators would work on scenes in the timeline editor in their own copy of the software. Finished scenes were integrated back into the final data set by our art director and were manually checked by a director who would ensure that the separate pieces conformed to the overall vision. Finer grained extraction of data allowed animators to work on individual tracks within a scene for those scenes where many characters interact and act simultaneously. In those cases scenes assemble from individual track documents into scene documents, which are in turn combined into the final product.

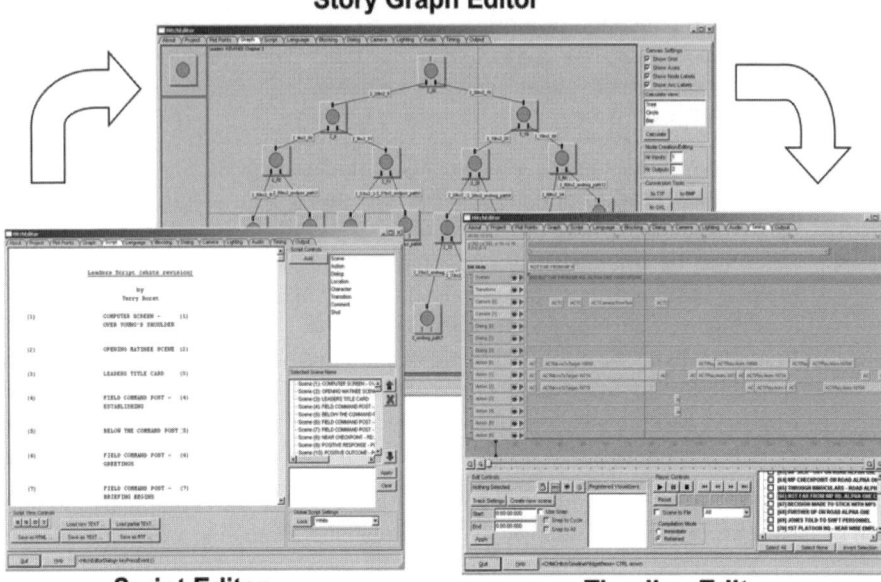

Fig. 1. Authoring Interoperability

5 Narratoria in Production, Preliminary Results

The Narratoria authoring tools have been used in numerous projects at the ICT. In the ICT Leaders project, before the introduction of authoring tools, animations were called by custom code in the game. Programmers moved the virtual actors around and had them speak lines using pre-recorded audio. Additional camera control code added means to place and move cameras. With this approach the screenplay took six months to implement. After introducing Narratoria into the pipeline the production time was reduced to three months and handed over to six animators, one art director and one cinematic director. During the last three months of production, no programmers were required to create any of the narrative content. Using the extendibility of Narratoria we recently added a pedagogical authoring module for those games that use an automated tutor. For example, in authoring the ICT's ELECT BiLAT game [8], authors use drag and drop techniques to associated learning objectives with game content that models a user's interactions. An automated tutor then uses this information to track a student's performance. Narratoria has also been applied to ICT's Spatial Cognition project, where it generated varied scenarios for virtual characters tasked with navigating complex virtual environments [12].

6 Related Work

Similar combinations of tools exist, which take on the separate areas of screenplay writing, movie-clip sequencing and animation creation [1][13], but none integrate the tools into a configurable workflow. Some more recent applications originating in the research community do allow traditional media types to be combined with novel abstract action sequences using XML documents but these systems either do not provide an intuitive mechanism for artists to access the materials [3][4][5] or they do not allow non-traditional actions such as character animation or camera actions to be used [2][7][16].

7 Future Work

With a framework in place that is targeted towards cinema style training applications we need to address a number of factors. First of all the software needs to be adapted to different forms of interactive narrative. Second, better collaboration mechanisms are needed to support users remotely located working on data simultaneously. Finally, further research could uncover how much artists will let the simulation interpret their direction and what additional acting behaviors can be installed within the virtual actors.

Acknowledgments. The author would like to thank Roland Lesterlin and Kurosh Valanejad for supporting the development of the tools.

This research was sponsored by the U.S. Army Research, Development, and Engineering Command (RDECOM). Statements and opinions expressed herein do not necessarily reflect positions or policies of the U.S. Government; no endorsement should be inferred.

References

1. Baecker, R., Rosenthal, A.J., Friedlander, N., Smith, E., Cohen, A.: A multimedia system for authoring motion pictures. In: Proceedings of the Fourth ACM international Conference on Multimedia (1996)
2. Bailey, B.P., Konstan, J.A., Carlis, J.V.: DEMAIS: designing multimedia applications with interactive storyboards. In: Proceedings of the Ninth ACM international Conference on Multimedia (2001)
3. Bulterman, D.C., Hardman, L.: Structured multimedia authoring. ACM Transactions on Multimedia Computing, Communications and Applications (2005)
4. Drapeau, G.D.: Synchronization in the MAEstro multimedia authoring environment. In: Proceedings of the First ACM international Conference on Multimedia (1993)
5. Gebhard, P., Kipp, M., Klesen, M., Rist, T.: Authoring scenes for adaptive, interactive performances. In: Proceedings of the Second international Joint Conference on Autonomous Agents and Multiagent Systems (2003)
6. Gordon, A., van Lent, M., van Velsen, M., Carpenter, M., Jhala, A.: Branching Storylines in Virtual Reality Environments for Leadership Development. In: Proceedings of the Innovative Applications of Artificial Intelligence Conference (IAAI) (2004)
7. Harada, K., Tanaka, E., Ogawa, R., Hara, Y.: Anecdote: a multimedia storyboarding system with seamless authoring support. In: Proceedings of the Fourth ACM international Conference on Multimedia (1996)

8. Hill Jr., R.W., Belanich, J., Lane, H., Core, M., Dixon, M., Forbell, E., Kim, J., Hart, J.: Pedagogically Structured Game-Based Training: Development of the ELECT BiLAT Simulation. In: Proceedings of the 25th Army Science Conference (2006)
9. Iuppa, N., Borst, B.: Story and Simulations for Serious Games. Focal Press, Oxford (2006)
10. Ivanovic, M., Budimac, Z.: A framework for developing multimedia authoring systems: Systems, Man, and Cybernetics. In: IEEE International Conference, October 14-17, 1996, vol. 2, pp. 1333–1338 (1996)
11. Jhala, A., Rawls, C., Munilla, S., Young, R.M.: Longboard: A Sketch Based Intelligent Storyboarding Tool for Creating Machinima. In: Proceedings of the Florida AI Research Society Conference (2008)
12. Kim, Y., van Velsen, M., Hill, R.: Modeling Dynamic Perceptual Attention in Complex Virtual Environments. In: Proceedings of the conference on Behavior Representation in Modeling and Simulation (BRIMS) (2005)
13. Robertson, J., Good, J.: Children's narrative development through computer game authoring. In: Proceeding of the 2004 Conference on interaction Design and Children: Building A Community (2004)
14. Robertson, J., Good, J.: Adventure Author: An Authoring Tool for 3D Virtual Reality Story Construction. In: the Proceedings of the AIED 2005 Workshop on Narrative Learning Environments, pp. 63–69 (2005)
15. Sung, M., Lee, D.: A collaborative authoring system for multimedia presentation. In: IEEE International Conference on Communications, June 20-24, 2004, vol. 3, pp. 1396–1400 (2004)
16. Ueda, H., Miyatake, T., Yoshizawa, S.: IMPACT: an interactive natural-motion-picture dedicated multimedia authoring system. In: Proceedings of the SIGCHI Conference on Human Factors in Computing Systems: Reaching Through Technology (1991)
17. Young, R.M., Mark, R.: Towards an Architecture for Intelligent Control of Narrative in Interactive Virtual Worlds. In: Proceedings of the International Conference on Intelligent User Interfaces (2003)

Male Bodily Responses during an Interaction with a Virtual Woman

Xueni Pan[1], Marco Gillies[1], and Mel Slater[1,2]

[1] University College London, London, UK
s.pan@cs.ucl.ac.uk
[2] ICREA-Universitat Politécnica de Catalunya, Spain

Abstract. This work presents the analysis of the body movement of male participants, while talking with a life-size virtual woman in a virtual social encounter within a CAVE-like system. We consider independent and explanatory variables including whether the participant is the centre of attention in the scenario, whether the participant is shy or confident, and his relationship status. We also examine whether this interaction between the participant and the virtual character changes as the conversation progresses. The results show that the participants tend to have different hand movements, head movements, and posture depending on these conditions. This research therefore provides strong evidence for using body movement as a systematic method to assess the responses of people within a virtual environment, especially when the participant interacts with a virtual character. These results also point the way towards the application of this technology to the treatment of social phobic males.

1 Introduction

When a virtual character smiles at you, will you smile back, and why? In this research we consider whether people respond to virtual characters as if they were real, in the context of an interaction within an immersive virtual environment (VE). Our motivation is to assess the extent to which such environments can be used in place of physically real environments for studying issues such as social phobia [7,10] and paranoia [3], and also treatment programmes for such social conditions. In previous work we have relied on questionnaire responses and physiological responses [7]. Here we consider the gestural and postural responses of male participants to an approach by an attractive and friendly virtual woman in a CAVE-like system.

In this paper, we discuss a new methodology where we examine participants' behavioural responses as a direct measure of the extent to which they respond as if the virtual encounter were real. Non-verbal behaviour (facial expression and body movements) reflects human beings' automatic responses. However, unlike verbal responses, body movement is an easily observable, gross overall indicator of a person's state. It therefore could offer us an additional method for assessing the realism of people's responses within a virtual environment. Moreover, while interacting with a virtual character, the participants' body movements reflect not only the states of the participants themselves but also how much are they engaged with the ongoing interaction with the virtual character.

Section 2 introduces the importance of using body movement as a measurement to VE. Section 3 describes the methodology. Results are presented in Section 4 and discussion in Section 5. The conclusions are given in Section 6.

2 Related Work

A common framework for measuring VE is "presence" which may be defined as the extent to which participants act and respond as if what they experience in the virtual reality were real [8]. This tendency of acting realistically towards virtually generated sense data distinguishes VE from all other media, such as films and books leading to a range beneficial applications from virtual psychotherapy to training. Therefore, instead of using only subjective responses as obtained from questionnaires and interviews in evaluating how people behave and respond within a VE, our work emphasises the importance of participants' measurable responses.

Most previous research on presence has used questionnaire and postexperiment interviews as the measurement instrument [13,16]. However, the participant can only complete the questionnaire and interviews after the experiment, therefore only reflecting what is in their memory. Moreover, as subjective responses from the participant, questionnaires also have serious methodological problems in this context [12]. Therefore questionnaires are best seen as supplements to behavioural and physiological data rather than the central means for assessment.

Physiological data, such as heart rate, heart rate variability and electrodermal activity provide excellent evidence of participant's physical reaction in real time (for example [9]). However, this is limited to a person's autonomic nervous system responses rather than higher level behavioural responses.

Here in addition we consider that the inclusion of actual bodily behaviour in our repertoire of possible factors could add significantly to our understanding people's responses, since they are clear and visible cues to which other people respond in turn. Several studies have used the behaviour of the participant in VE to evaluate presence [11,14]. One of these studies observed the participants actual responses to "danger", i.e. whether they ducked when virtual objects flew towards them [11]. More importantly, the interaction between human participants and virtual characters is crucial in many VE applications. In particular in therapeutic and training applications, where the involvement and behavioural response are the key factors, how the participant behaviour in the interaction is even more important than their reported "feeling" towards the virtual character. Few studies have examined the participants' interactive behaviour in VEs. Bailenson et al. observed the behaviour of participants towards a virtual character and found that they have showed greater hesitation in approaching the character which was with more human like movement [2]. Vinayagamoorthy et al.

suggested that participants tended to adopt a socially-acceptable spatial behaviour with virtual characters in immersive system [15]. Krämer assessed the verbal responses towards different interface and found that, when confronted with a virtual agent, the participants adopt more natural speech [6].

The paper examines the interpersonal behaviour of the participant while interacting with a virtual character in a virtual social encounter. Many important interpersonal communication factors were considered, such as: anxiety, domination, flirtation, affiliation, and avoidance.

3 Methodology

The work we present here is part of a larger experimental study carried out in an immersive (Cave) system, where shy and confident males interacted with a forward virtual woman, in a virtual bar [7]. In this paper we concentrate on male participants' bodily responses towards a virtual female in a virtual social encounter. Twenty-four participants were invited and attended the experiment. They were recorded with a camera from behind during the experiment and the recording were annotated afterwards by a body movement expert from UCLIC (UCL Interaction Centre) who otherwise had no involvement in the experiment and had no knowledge of the purpose of our research.

3.1 Body Movement Annotations

Compared to using motion capture for body movement annotation, manually annotating body movement through videos data has some own advantages. First, the results generated manually are more semanticly meaningful; secondly, the observation procedure is unobstrusive so that more spontaneous behaviours could be captured [4]. However, in our work, due to restrictions in the immersive projection system and the experimental setup, only video data taken from behind the participants was available for the annotation. As shown in Fig. 1, the video data was obtained with limited lighting condition and restricted view point. Therefore instead of using a standard annotation scheme we have formed our own body movement annotations which serves our purpose. Here we consider conversational behaviours which are related with anxiety, domination, flirtation, affiliation, and avoidance. We decomposed the body movement annotations into 3 categories: hand movement, head movement, and posture movement.

We included nodding, head cocking, head shaking, looking around, and looking down in the head movement analysis. Increased nodding and head cocking (see Fig.1(b)) shows affiliation and higher involvement, whilst looking around (Fig.1(d)) and looking down(Fig.1(e)) indicates lower involvement [1].

For hand movement we included: hands on hips, head-touching, hands in pockets, hands behind back, hands in front, hands making conversational gesture, and arm crossing. We are particularly interested in hands on hips (see Fig.1(f), 1(g), 1(h), and 1(i)), which is a dominant behaviour used by males while courting [1]. Also we considered head-touching, which is related to self-consciousness

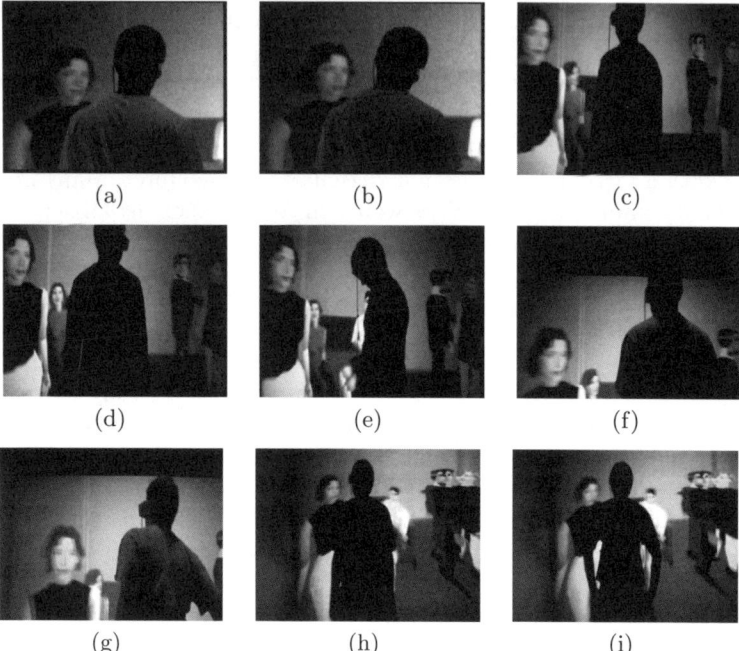

Fig. 1. (a) and (b) are the same participant where in (a) his head is straight and in (b) there is head-cocking; (c), (d) and (e) shows a participant with normal head position, looking around, and looking down. (f) and (h) show two participants' original positions, (g) and (i) show the same two participants with their hands on hips.

within uncomfortable social situations [5]. Certain types of head-touching can also be interpreted as preening which is associated with courting; however, such behaviour is more common in women than men [1].

For posture movement, we look at posture shifting, shrugging, wiggling/ swaying, and shifting weights. Similarly to head-touching, shifting weight reflects self-consciousness related to uncomfortable social situation [5].

3.2 Independent and Explanatory Variables

There are four independent and explanatory variables considered. The first two factors are *shy/confident* and *observed/not observed* [7]. Additionally, because of the particular interaction generated by this experiment, another explanatory variable that might influence behaviour is the participant's reported relationship status. Therefore we consider the factor *single/involved* as one of the explanatory variables. During the experiment the virtual female asked the participant if he is involved or not with someone. The answers to this question were taken to determine their relationship status. Among the 24 participants, 12 of them reported themselves to be single, 12 of them were involved (we did not verify this with them afterwards). The last explanatory variable considered in this study is the changing level of intimacy as the conversation progressed. As shown in Table 1, the whole interaction is segmented into 3 different periods where the level of intimacy changes:

Approach, *Mundane*, and *Intimate*. With this segmentation we are able to monitor participants' behavioural changes as the conversation progresses.

Table 1. Sequence of Events and Virtual female Questions in the Virtual Encounter. Event here refers to a triggered utterance. It is segmented into 3 different period as the intimacy level increases.

Approach: the virtual female initiates the conversation.	
1	Experiment starts and the Virtual female stares at the participant
2	Virtual female stars approaching the participant
3	"Hi, It looks like we are the only people alone here,right"?
4	"My name is Christina."
5	"So, what are you doing for a living?"
6	"I'm an air hostess; I just arrived in London yesterday. Where do you live?"
7	"I don't know London very well, but actually, I am thinking about moving here, what do you think?"
Mundane: "everyday life" conversation.	
8	"But I heard it rains all the time here, is that true?"
9	"Do you like it here?"
10	"I've noticed that people dressed very well around here. By the way, that shirt looks great on you. How much was it?"
11	"Ah, I really want to find a pair of trousers, something like these (Looking down) for my brother. Where did you get these?"
Intimate: the conversation became more personal and intimate	
12	"So, do you know anyone here?"
13	"I feel a bit shy about talking with other people, do you mind if I talk with you for a bit longer?"
14	The virtual character approaches to an intimate distance
15	"If you don't mind me saying, I think you look very nice."
16	"I was wondering actually, are you single, or involved with someone at this time?"
17	"Maybe we should meet up."

3.3 Assessing Body Movements

On the assessment form a matrix is given defined by the 3 main periods as rows, and the columns defined by different body annotations. Each element of this matrix is the number of occurrences of a particular behaviour annotation at a certain period. The assessment form was given to the body movement expert who filled in it for each participant by screening the video data from the experiment; 10 hours were needed to conclude this task and she was paid for this work.

4 Results

Consider any particular action such as 'head touching'. We were interested in whether there were any systematic variations of this response with the independent and explanatory variables of the experiment. The null hypothesis is that 'head touching' occurs at random through time. Under this null hypothesis the distribution the number of head-touches should for each individual follow a Poission distribution. Therefore we use the Poisson log-linear model as the appropriate model for analysis of variance of the response variable on the independent and explanatory variables.

In Table 2 we show the results of a series of such log-linear regressions. In each case the Poisson model fits well within the bounds of the traditional 5% significance level, and we show the significant explanatory variables, whether their association with the response is positive or negative, and the corresponding significance level. These results show that:

- For head movement, participants who were *observed* by other virtual characters tended to look around more than participants who were *not observed*, and in the *Mundane* period participants tended to look around less than in the *Approach* period; *shy* participants tended to look down less than *confident* participants; in the *Mundane* period, participants tended to nod more than in the *Approach* period, and *observed* participants tended to nod less than participants who were *not observed*. In both *Mundane* and *Intimate* periods, participants tended to do more nodding and cocking than in the *Approach* period.
- For hand movement, participants who were *observed* tended to head-touch less, and *single* participants tended to head-touch more.
- For posture, *observed* participants tended to shift weights less than participants who were *not observed*.

Table 2. Variations of Response with Independent and Explanatory Variables

Categorise	Response Variable	Explanatory Variables	Association	Significance Level
Head Movement	Looking around	Observed	+	0.00
		Mundane	–	0.05
	Looking down	Shy	–	0.01
	Nodding	Mundane	+	0.03
		Observed	–	0.06
	Nodding + Cocking	Mundane	+	0.00
		Intimate	+	0.05
Hand Movement	Head-touching	Observed	–	0.00
		Single	+	0.01
Posture	Shifting weights	Observed	–	0.00

5 Discussion

The results suggested that *confident* participants tended to look down more during the conversation. This might indicate that confident participants paid less attention to the virtual female, and were less involved with the interaction.

The results also show that participants who were *observed* by other virtual characters looked around more, and nodded, head-touched, and shifted weights less. The fact that they looked around more when being observed fits our expectation because it coincides with people's normal social behaviour of looking around when being observed by others. Less nodding, head-touching, and shifting weights furthermore suggest that participants who were observed may have been distracted and therefore paid less attention to the virtual female. However, in our previous analysis with only the questionnaire data and physiologically data, no difference was found between participants who were observed and those who were not.

Moreover, the result indicates that participants who are *single* head-touched more, which could be explained as greater attention being paid to the virtual female with more involvement in the conversation. This is also a factor that failed to stands out with other measurements in our previous work.

Finally, compared to the *Approach* period, there is more nodding and head cocking in *Mundane* and *Intimate* periods. Also in *Mundane* period the participants looked around less. All these results suggest that the participants became more involved in the interaction as the conversation progressed, which coincides with our physiological results.

6 Conclusion and Future Work

The results support the notion that participants tended to act towards the virtual character with appropriate interpersonal behaviour. Body movement is a gross overall indicator of a person's state, and is relatively easily observable. It therefore could offer us additional methods to measure the responses of people in VEs, especially when the participant interacts with a virtual character. In our previous report [7] we evaluated the reactions of participants and the results showed that the participants tended to respond to the situation at the subjective and physiological level as if it were real. In the previous study we have also evaluated their verbal responses to assess their behaviour, yet one can argue that verbal responses can be playfully delivered by the participants without being serious. This new evaluation of bodily responses, however, underlines the findings of our previous study, since it is unlikely that people deliberately and consciously choose their bodily responses - they are an automatic action. Another contribution of this work is that a bodily annotation system which focuses on conversational behaviours was proposed for annotating body movement through video data with poor view and lighting conditions. In further research, the analysis of facial expressions, voice and other automatic human responses is being considered. These results further emphasise that virtual reality technology can be used in the treatment of social phobia.

Acknowledgments

This work is funded through the EPSRC Empathic Avatar project EP/D505542/1. We would like to thank Andrea Kleinsmith from UCLIC for her professional help in the annotation of body movement.

References

1. Argyle, M.: Bodily Communication, 2nd edn. Methuen & Co Ltd, second edition (1988)
2. Bailenson, J.N., Swinth, K.R., Hoyt, C.L., Persky, S., Dimov, A., Blascovich, J.: The independent and interactive effects of embodied agent appearance and behavior on self-report, cognitive, and behavioral markers of copresence in immersive virtual environments. In: Presence (2005)
3. Freeman, D., Pugh, K., Antley, A., Slater, M., Bebbington, P., Gittins, M.: A virtual reality study of paranoid thinking in the general population. pp. 627–633 (2008)
4. Kipp, M., Neff, M., Albrecht, I.: An annotation scheme for conversational gestures: how to economically capture timing and form. Language Resources and Evaluation, December
5. Knapp, M.L.: Nonverbal communication in human interaction. Holt, Rinehart and Winston, New York (1978)
6. Krämer, N.: Social communicative effects of a virtual program guide. In: Panayiotopoulos, T., Gratch, J., Aylett, R.S., Ballin, D., Olivier, P., Rist, T. (eds.) IVA 2005. LNCS (LNAI), vol. 3661. Springer, Heidelberg (2005)
7. Pan, X., Slater, M.: A preliminary study of shy males interacting with a virtual female. In: Presence: The 10th Annual International Workshop on Presence (2007)
8. Sanchez-Vives, M.V., Slater, M.: From presence to consciousness through virtual reality. Nature Reviews Neuroscience 6(4), 332–339 (2005)
9. Slater, M., Antley, A., Davison, A., Swapp, D., Guger, C., Barker, C.: A virtual reprise of the stanley milgram obedience experiments. PLoS ONE 1(1), 39 (2006)
10. Slater, M., Pertaub, D.P., Barker, C., Clark, D.M.: An experimental study on fear of public speaking using a virtual environment 9(5), 627–633 (2006)
11. Slater, M., Usoh, M.: Representations systems, perceptual position, and presence in immersive virtual environments. In: Presence: Teleoperators and Virtual Environments (1993)
12. Slater, M.: How colourful was your day?: Why questionnaires cannot assess presence in virtual environments. Presence: Teleoperators and Virtual Environments 13(4), 484–493 (2004)
13. Slater, M., Sadagic, A., Usoh, M., Schroeder, R.: Small group behaviour in a virtual and real environment: A comparative study. Presence: Teleoperators and Virtual Environments 9(1), 37–51 (2000)
14. Uno, S., Slater, M.: The sensitivity of presence to collision response. In: Virtual Reality Annual International Symposium (1997)
15. Vinayagamoorthy, V., Brogni, A., Steed, A., Slater, M.: The role of posture in the communication of affect in immersive virtual environments. In: The 2nd ACM SIGGRAPH International Conference on Virtual Reality Continuum and Its Applications (2006)
16. Witmer, B., Singer, M.: Measuring presence in virtual environments: a presence questionnaire. Presence 7(3), 225–240 (1998)

A Virtual Agent for a Cooking Navigation System Using Augmented Reality

Kenzaburo Miyawaki and Mutsuo Sano

Faculty of Information Science and Technology
Osaka Institute of Technology
Osaka, Japan
miyawaki@is.oit.ac.jp

Abstract. In this paper, we propose a virtual agent for a cooking navigation system which use the ubiquitous sensors. The cooking navigation system can recognize the progress of cooking and can show appropriate contents suitable to the situation. Most cooking navigation systems show only a movie and a text. However, the recognition system of cooking is applicable to realize a virtual agent which can perform helpful actions suitable to the users' behavior. We implemented the virtual agent, and demonstrated the cooking navigation system including the agent.

1 Introduction

Context-aware systems using ubiquitous sensors are attracting more and more attention. Based on such a situation, we are studying the support systems for human life in the ubiquitous environment. We make much of the development of cooking navigation systems particularly, because cooking is a significant task in everyday life. A cooking navigation system can recognize the progress of cooking, and can offer appropriate helpful contents for the situation of cooking. We have developed some systems [1]. These systems could recognize a state of cooking, and could show an appropriate instruction by a movie and a text. In addition to those functions, it will be possible to realize an intelligent agent which can perform useful actions for users by using the state recognition. Such an agent can make a cooking task more smooth and more enjoyable. The rest of this paper is organized as follows. Section 2 introduces the related works. Section 3 describes the system overview. Section 4 shows the mechanism of state recognition of cooking. Section 5 explains our virtual agent, and section 6 shows a demonstration of the agent. Finally, Section 7 concludes this paper.

2 Related Works

Some recipe presentation systems have been already developed [2][3]. The systems use a movie and a text mainly, and should be controlled manually. The agent using voice synthesis merely reads instructions [2]. Nakauchi et al. [4] have developed the cooking support system using a robot. Their system can predict a users' next action of cooking. The cooking tasks are, however, very simple (e.g., making a cup of coffee). Fadil et al. [5] have investigated the cooking navigation system using mixed reality, and proposed the interactive cooking support system, but they did not mention agents and system interfaces in detail.

3 System Overview

Fig. 1 shows the overview of our system. The system is composed of a task controller and an interaction controller. The task controller is a recognition system of cooking state. It receives the events detected by sensors, and recognizes a state of cooking. The interaction controller uses the recognition result to show the appropriate contents and to make a virtual agent perform the action suitable to the situation.

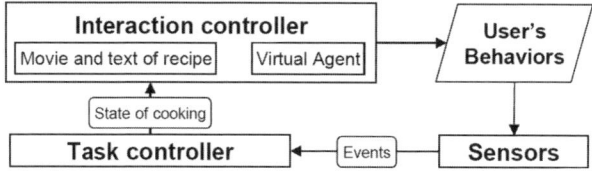

Fig. 1. System overview

4 State Recognition of Cooking

A state transition model of cooking is very important in order to recognize a state of cooking. The task controller recognizes the cooking progress by the model and sensors. Cooking is composed of various operations (cut, mix, boil, etc) in processes of cooking ingredient, and the operations require many preparing actions (e.g., moving something). The cooking operations and the preparing actions change a state of objects. These state changes are detectable with sensors. For example, cameras, thermometers or location sensors, such as RFIDs, are useful. A model, which is described by the state changes of objects, makes the progress of cooking recognizable. We define some terms as follows for explanation of the model.

1. **Cooking object**
   Cooking ingredients, seasonings, cooking utensils and kitchen equipment are cooking objects.
2. **State of a cooking object**
   This term means a cooking object location, such as kitchen worktop, cooking stove, and shelves. States of kitchen equipment are also included. (e.g., power level of cooking stove.)
3. **Event**
   An event is a state change of cooking objects. That is detected by sensors.

Our model is described with the elements defined above.

4.1 Sensing Environment for the State Recognition

Fig. 2-(a) shows our kitchen. The cameras are used to recognize an object location with the AR ToolKit Plus [6] library. Fig. 2-(b) shows the AR ToolKit Plus markers pasted on the cooking utensils. The system can detect the actions of preparation such as putting something by these markers. In the case of mixing bowls, two markers, a location marker and a status marker, are pasted on a bowl. The status marker is used to detect whether a bowl contains the cooking ingredient. In addition to the cameras, micro switches are embedded in the knife holder (Fig. 2-(a)). These switches are used to detect the events of replacing/taking the kitchen knife.

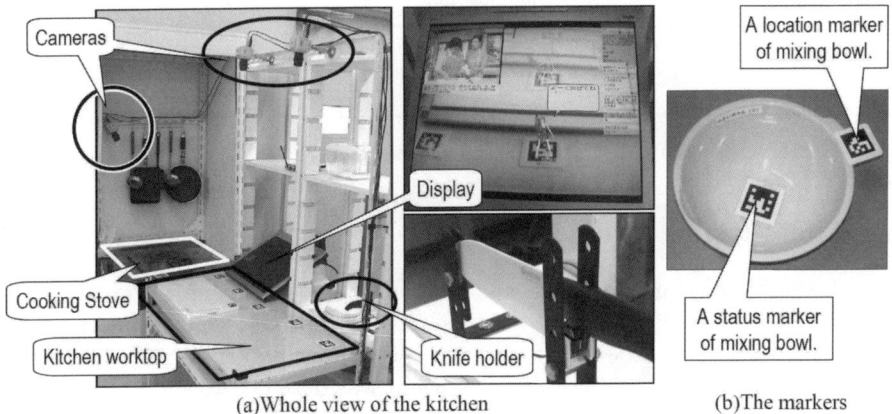

(a)Whole view of the kitchen   (b)The markers

Fig. 2. Whole view of the kitchen

4.2 State and Transition Model of Cooking

Recipes generally contain some steps such as following examples.

Example of recipe, "Japanese omelet"
S_1. Beat eggs in a bowl.
S_2. Stir sugar, Japanese sweet sake, salt, soy source and cooking liquor, into the bowl.
S_3. Put a frying pan on the cooking stove, and heat cooking oil.
S_4. Pour the egg to the frying pan.
S_5. $\cdots$

All steps must be converted to state and transition models in order to recognize a state of cooking. In this paper, the state machine diagram of UML is used for the description of the models (Fig 3). It is significant that all of the triggers must correspond to the events detected by sensors. The task controller inputs the events to the state transition models and recognizes the progress of cooking.

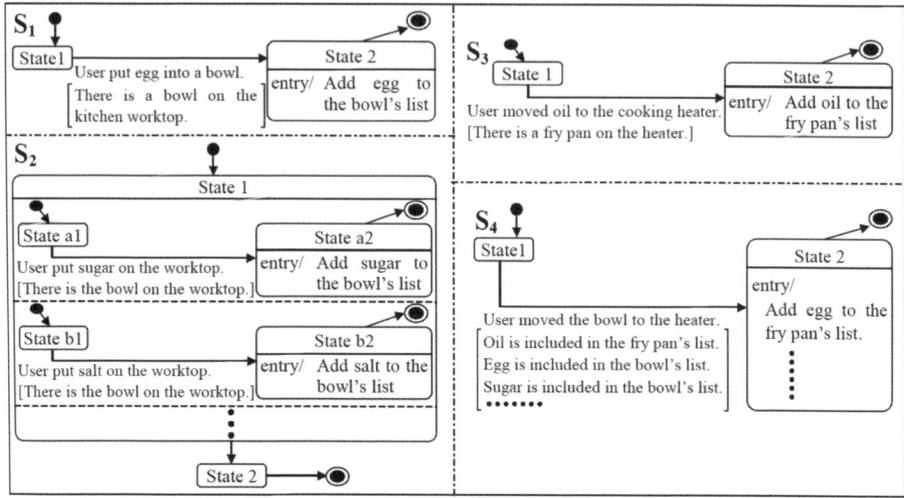

Fig. 3. Examples of state machine diagrams

5 The Virtual Agent

Fig. 4 shows our virtual agent and its actions. This agent is rendered by Direct3D [7]. The agent controller receives recognition results from the task controller, and makes the agent perform the actions which correspond to the category of a current cooking step. The agent was designed to synchronize with users' actions for friendship with users and to cheer users up for the pleasure of cooking. Along that policy, three actions were designed. The actions were one cheering action and two similar actions to users' cooking operation. The actions of cutting and stirring are performed during users' cooking operations which correspond to that. The cheering action is performed during heating operations, such as fry or boil. The agent performs the actions and gives users the messages to warn or to cheer. The agent controller changes the pitch of the agent actions depending on the intensity of the user's movement in order to avoid monotonousness. We use the frame differential of images to estimate the users' movement. it is calculated as follows [8].

$$M = \frac{\sum_{i=1}^{m}\sum_{j=1}^{n}(I_{t(i,j)} - I_{t-1(i,j)})}{m \times n} \qquad (1)$$

M denotes average intensity of movement. I denotes brightness of pixels, and m and n denote width and height of images. The images are shot from top of kitchen worktop and cooking stove. Depending on the result, the pitch of the agent actions changes.

5.1 A Wearable Camera and Markers for the Agent

Fig. 5 shows a wearable camera for the system. The camera is fixed on a user's chest, and the camera angle is regulated to shoot images which are similar to the

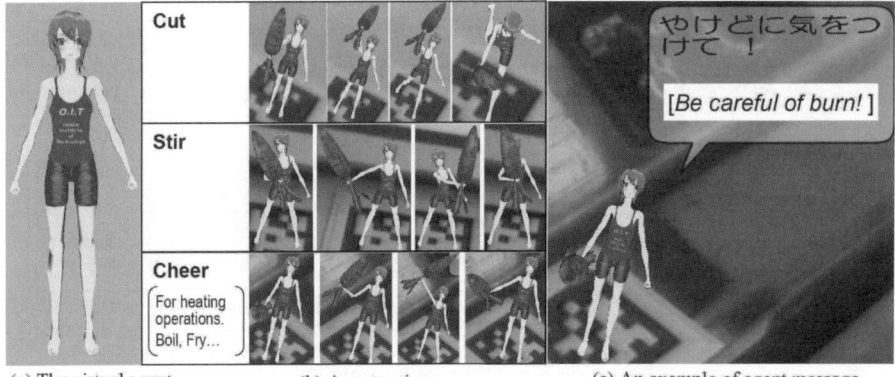

(a) The virtual agent (b) Agent actions (c) An example of agent message

Fig. 4. The virtual agent

view from users' eye. Fig. 6 shows the markers for the agent. The markers were put on the kitchen worktop and around the cooking stove. The chest mounted camera shoots the cooking operations of a user, and the markers inside of the user's sight are detected. The virtual agent appears on the detected marker.

Fig. 5. The wearable camera

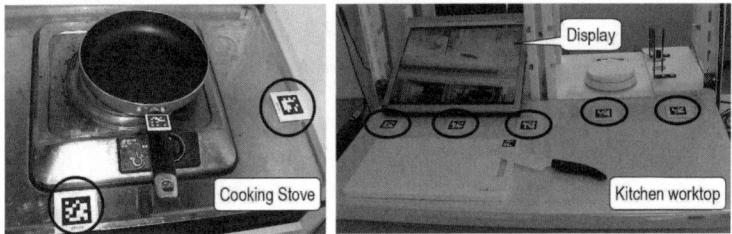

Fig. 6. Markers for the virtual agent

5.2 The User Interface

Fig. 7 is a screenshot of the system. The top left block is a movie of a recipe. The right side shows a list of cooking steps. Fig. 7-(b) explains the detail of a node of

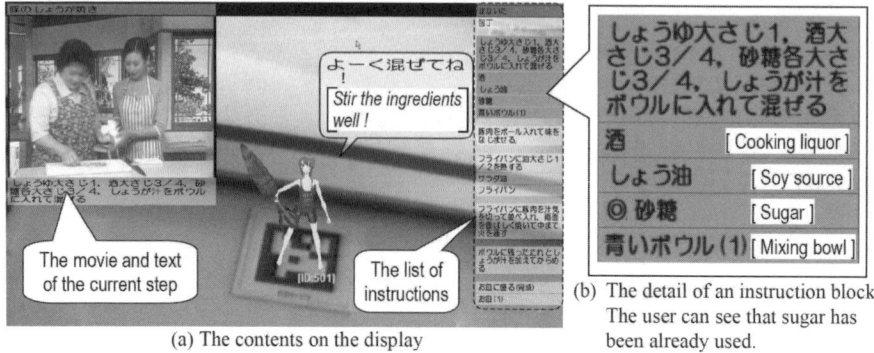

(a) The contents on the display

(b) The detail of an instruction block. The user can see that sugar has been already used.

Fig. 7. The user interface

the list. One node consists of an instruction text and a list of the objects which are required for the step. The list of the objects changes its state depending on the state machine diagrams. This list is useful to prevent user from failure, such as forgetting to add seasonings or adding same seasonings twice, of cooking. The background is the image captured by the chest mounted camera. The LCD was located on the kitchen worktop and tilted to make the situation that users can see easily both the display and his/her hands (Fig. 5, Fig. 6).

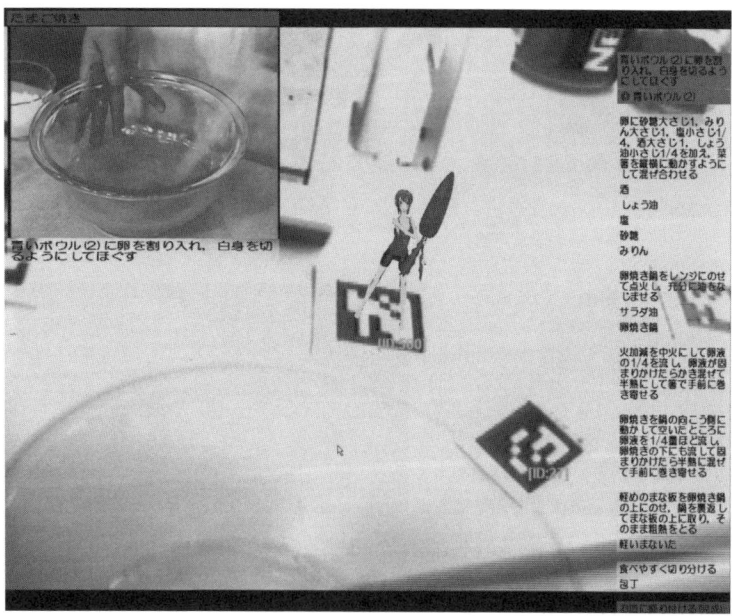

Fig. 8. Example image of the demonstrations

6 Demonstration

We selected three cooking menus and demonstrated the system. The menus were Japanese omelet, Pork fry with ginger and Mashed tofu mixed with leek and kelp. The user cooked the menus one by one. Fig. 8 shows an example image of the demonstrations. This is a scene of stirring eggs for Japanese omelet. The virtual agent synchronized with the user's action, and it started to stir. The agent speeded up its action when the user stirred eggs quickly. We confirmed the correct control of the agent. Based on the result, we plan to investigate users' evaluation of the virtual agent and to develop more various actions along the evaluation.

7 Conclusion and Future Works

We proposed a virtual agent for cooking navigation systems. A state recognition of cooking was necessary to control the agent. The agent could perform appropriate actions by the recognition result, and the system could control the pitch of agent actions dynamically depending on the action speed of users. We demonstrated the cooking navigation system and the virtual agent, and confirmed that the system could control the agent correctly. We suppose that the synchronization can generate the entrainment between users and the agent. In the future studies, we plan to evaluate that possibility and to design agent actions for more delightful cooking.

References

1. Kenzaburo, M., Mutsuo, S., Satoshi, N., Katsuo, I.: Modeling of Concurrent Cooking for Cooking Navigation System with Ubiquitous Sensor Network. In: Proc. of the Image Electronics and Visual Computing Workshop 2007, 4C-1 (2007)
2. NINTENDO. Shaberu! DS Oryouri-Navi (in Japanese),
   http://www.nintendo.co.jp/ds/a4vj/
3. Hamada, R., Okabe, J., Ide, I.: Cooking Navi: Assistant for Daily Cooking in Kitchen. In: Proc. of 13th ACM Intl. Multimedia Conf., pp. 371–374 (2005)
4. Yasushi, N., Tsukasa, F., Katsunori, N., Takashi, M.: Intelligent Kitchen: Cooking Support by LCD and Mobile Robot with IC-Labeled Objects. In: Proc. of the 2005 IEEE/RSJ International Conference on Intelligent Robotics and Systems, pp. 2464–2469 (2005)
5. Fadil, Y., Mega, S., Horie, A., Uehara, K.: Mixed Reality Cooking Support System Using Content-Free Recipe Selection. In: Proc. of the IEEE International Symposium on Multimedia, pp. 845–850 (2006)
6. AR ToolKit Plus Home page,
   http://studierstube.icg.tu-graz.ac.at/handheld_ar/artoolkitplus.php
7. Microsoft DirectX SDK,
   http://msdn.microsoft.com/en-us/directx/default.aspx
8. Nagasaka, A., Tanaka, Y.: Automatic Video Indexing and Full - Video Search for Object Appearances. Trans. of Information Processing Society of Japan 33(4), 543–550 (1992)

Social Perception and Steering for Online Avatars

Claudio Pedica[1,2] and Hannes Vilhjálmsson[1]

[1] Center for Analisys and Design of Intelligent Agents, Reykjavik University, Iceland
[2] Camerino University, Italy

Abstract. This paper presents work on a new platform for producing realistic group conversation dynamics in shared virtual environments. An avatar, representing users, should perceive the surrounding social environment just as humans would, and use the perceptual information for driving low level reactive behaviors. Unconscious reactions serve as evidence of life, and can also signal social availability and spatial awareness to others. These behaviors get lost when avatar locomotion requires explicit user control. For automating such behaviors we propose a steering layer in the avatars that manages a set of prioritized behaviors executed at different frequencies, which can be activated or deactivated and combined together. This approach gives us enough flexibility to model the group dynamics of social interactions as a set of social norms that activate relevant steering behaviors. A basic set of behaviors is described for conversations, some of which generate a social force field that makes the formation of conversation groups fluidly adapt to external and internal noise, through avatar repositioning and reorientations. The resulting social group behavior appears relatively robust, but perhaps more importantly, it starts to bring a new sense of relevance and continuity to the virtual bodies that often get separated from the ongoing conversation in the chat window.

1 Introduction

Massively Multiplayer Online Role Playing Games (MMORPGs) are a rapidly growing form of mass entertainment delivered over the Internet in the form of live game worlds that persist and evolve over time. Players connect to these worlds using client software that renders the world from their perspective as they move about and meet fellow players. The community is the cornerstone of these games. Therefore, any effort spent on supporting communication and social interaction between players has to be considered valuable for the application. When games wish to use avatars to represent players in environments where they can virtually meet face-to-face, all the animated behaviors that normally support and exhibit social interaction become important. Since players cannot be tasked with micro-management of behavior, the avatars themselves have to exhibit a certain level of social intelligence [1] [2]. The purpose of this avatar AI is in fact twofold: to give players helpful cues about the social situation and to ensure that the whole scene appears believable and consistent with the level of game world realism.

This paper presents ongoing work, which is a collaborative effort between a major MMORPG developer and a research center in the field of embodied conversational agents that specializes in multi-modal behavior generation based on social and cognitive models. The work presented here is one piece of the project which deals with modeling the relatively low-level motion dynamics in conversational interaction, which despite several ongoing efforts, is largely an unsolved issue.

Many approaches propose interesting solutions for generating the stream of actions that an agent, or in our case, an automated avatar needs to perform in order to believably simulate the positional and orientational movements of a human engaged in a conversation. Each action usually triggers some motor function directly at locomotion level in order to animate the agent. The sequence of actions usually needs to pass through an intermediate layer in order to achieve the desired fluidity of movements and reactions. This extra level between action generation and locomotion, is responsible for smoothing the agent's overall behavior by applying steering forces directly to the underling physical model. Therefore a steering layer provides a suitable solution for filling the gap between two consecutive actions, generating a net continuous fluid behavior. This approach is particular well suited for modeling unconscious reactions and motion dynamics in conversational interactions.

A conversation defines a positional and orientational relationship of its participants. This arrangement has been described by Kendon [3] as an instance of an F-formation system. Moreover, since a conversation is not a fully rigid formation, external events from the environment or individual behaviour of single participants may produce fluctuations inside this system that lead to compensational rearrangement that avatars can automate without requiring input from their human users. In fact, this is the kind of reactive behaviour which is ill suited for explicit control [1].

2 Related Work

2.1 Automating Avatar Control

In most commercial avatar-based systems, the nonverbal expression of communicative intent and social behavior relies on explicit user input. For example, in both Second Life and World of Warcraft, user can make their avatars emote by entering special emote commands into the chat window. This approach is fine for deliberate acts, but as was argued in [4], requiring the users to think about how to coordinate their virtual body every time they communicate or enter a social situation places on them the burden of too much micromanagement. When people walk through a room full of people, they are not used to thinking explicitly about their leg movements, body orientation, gaze direction, posture or gesture, because these are things that typically happen spontaneously without much conscious effort [3]. Some of these behaviors are continuous and would require very frequent input from the user to maintain, which may be difficult, especially when the user is engaged in other input activities such as typing a chat message. In the same way that avatars automatically animate walk cycles so that users won't have to worry about where to place their virtual feet, avatars should also provide the basic nonverbal foundation for socialization.

Automating the generation of communicative nonverbal behaviors in avatars was first proposed in BodyChat where avatars were not just waiting for their own users to issue behavior commands, but were also reacting to events in the online world according to preprogrammed rules based on a model of human face-to-face behavior [4]. The focus was on gaze cues associated with establishing, maintaining and ending conversations.

A study showed that the automation did not make the users feel any less in control over their social interactions, compared to using menu driven avatars. In fact, they reported they felt even more in control, suggesting that the automated avatars were providing some level of support [1]. The Spark system took this approach further by incorporating the BEAT engine [5] to automate a range of discourse related co-verbal cues in addition to cues for multi-party interaction management, and was able to demonstrate significant benefits over standard chat interaction in online group collaboration [2]. The Demeanor system [6] blends user control at several different levels of specification with autonomous reactive behavior to generate avatar posture based on affinity between conversation partners.

However, both Spark and Demeanor assume that the user will bring their avatar to the right location and orient correctly for engaging other users in conversation. Interestingly, even though users of online social environments like Second Life appear sensitive to proximity by choosing certain initial distances from each other, they rarely move when approached or interacted with, but rely instead on the chat channel for engagement [7]. Since locomotion and positioning is not being naturally integrated into the interaction when relying on explicit control, it is worth exploring its automation as well. Some have suggested that once an avatar engages another in a conversation, a fixed circular formation should be assumed [8]. This is a simplification, as the circle we often see is merely an emergent property of a complex space negotiation process, and therefore the reliance on a fixed structure could prevent the avatars from arranging themselves in more organic and natural ways. Therefore we decided on an approach that combines a higher level organizational structure (that of conversations) with low level steering behaviors, affected by a social force field inspired by work on dynamic group behavior in embodied agents such as [9]. We focus on continuous social avatar positioning in this current work, while keeping in mind that other layers of behavior control introduced in previous work will need to be added for fully supporting the social interaction process.

2.2 Dynamic Group Behavior

Numerous works have been published in the area of dynamic group behavior. Most of them concern simulating crowds of people or formations of animals, such as flocks of birds or schools of fish. These kinds of global collective phenomena have been modeled with different approaches but the most interesting and successful of them define the group dynamic as an emergent behavior. In this direction, there are basically two main approaches:

- A particle-based system approach, where particles are animated in real time by the applications of forces.
- An agent-based systems approach, in which each agent is managed in real time by rules of behavior.

The main difference between these two approaches is the order of magnitude of the simulated group behavior. Particle-based systems are well suited for modeling global collective phenomena (such as group displacement and collective events) where the number of individuals is huge. They are all quasi-similar objects and the overall focus is on the global collective behavior. Couzin et al. [10] use an interesting approach to model schools of fish. They define three proximity zones for each particle, where each zone can exert a force on the particle's constant velocity. These forces have different priorities and each of them models a particular behavior. Heigas et al. [11] use a physically-based particle model to simulate emergent human crowd behavior. They use a model of fluid dynamics that incorporates two elementary repulsive forces to simulate jamming and flowing behavior. Treuille et al. [12] present a model for crowd dynamic continuously driven by potential fields. The integration of global navigation planning and local collision avoidance into one framework produces very good video results. Furthermore, their model seems to integrate well with several agent-based models promoting interesting future integrations. Particle-based systems present good performance and scalability but they put more emphasis on the large-scale collective movement and do not model the micro-dynamic of interpersonal interactions.

On the other side, agent-based systems are more suitable for modeling small scale group dynamics. Thalmann et al. [13] use complex finite automata to determine the behavior of actors. These automata represent intelligent autonomous behaviors defined by a set of clever rules. The purpose of the model is still to simulate human crowds but this time the introduction of structured behavior of groups and individuals is remarkable. A hierarchical model describes the behavior of each part, but still the rules of social interaction such as conversation are not taken into account. The approach of Reynolds [14] to modeling emergent collective phenomena with an agent-based system, consist of adding a steering motor force to the agent-particle. Despite its simplicity, the idea of an agent-particle with a force-driven behavior leads to the modeling of many complex collective behaviors such as crowd path following, leader following, queuing at a doorway and flocking. Although there isn't any specific reference to conversation dynamics, the elegance and the simplicity of the solution was of great inspiration for the current work. Very comprehensive is also the work of Shao et al. [15], which present fully autonomous pedestrian interacting in a virtually reconstructed Pennsylvania Station. Interesting for the purposes of our work is how they use perceptual information to drive low level reactive behaviors in a social environments. Still motion dynamics for conversations and social interactions are taken into account.

Moving to a more fine-grained social behavior, Rehm et al. [16] recognize the value of social proxemics and formation theories. They use them to inform their models of dynamic distance and orientation between pairs of humanoid agents based on their interpersonal relationship. The focus of their work is modeling how the relationship changes over time based on simulated personality and a theory driven classification of small group interactions. Their underlying model provides an interesting representation of the interpersonal relationships in focused social interactions and how they influence distances and orientations in small groups of two participants. While interpersonal relationships are necessary for simulating full simulation of social group dynamics, they are not sufficient, as is evident from Kendon's work [3].

The state-of-the-art work on dynamic movement and positioning of social agents, which inspired the current research, is the work of Jan et al. [9] who propose to drive behavior by a social force field. At any point in time, an agent is motivated to move toward a certain position and this motivation is modeled as a force which depends on the position of the other participants. Each social force is activated by certain rules and then they are used to compute a destination point for the agent. While the approach looks promising, the main problem is that motivational forces applied to agent orientation are not taken into consideration.

3 Approach

3.1 Underlying Social Theory

To define a formal model that approximates the physical group dynamics of a real conversation, we start from the basic concepts of social theory. Like some of the more interesting previous work, our primary inspiration has been Kendon's [3] research that defines a conversation as an instance of an F-formation system. Individuals in a conversation tend to arrange in a way that gives all of them equal, direct and exclusive access to a common space. This positional and orientational arrangement is called an F-formation and the set of the behavioural relationships among the participants defines a behavioural system called the F-formation system. From the definition of the F-formation comes an explanation of the usual circular arrangement of people in conversation with more than two participants: it is simply the best way to give everybody equal access to a common focused space. At the moment, we are not dealing with interpersonal relationships, emotions, affiliation and other important social aspects of group dynamics because they are extensively addressed in many other works. In our approach we start from the fundamental concepts of the F-formation theory in order to cover those aspects of social group dynamics still open and not taken into account by the state-of-the-art avatar environments. From the simple fact that participants in a focused social interaction share a common space with equal access rights to it, comes a series of compensatory movements that have to be simulated before one can hope to completely model social group dynamics.

The common focused space, called the o-space (innermost white circle in Fig. 1, left), is necessary for the social interaction. All the participants perform a series of behaviours that aim to create and defend the o-space. The set of these behaviours defines a system of positional and orientational reactionary behaviours for maintaining an equilibrium in the system. Typically the o-space is sustained through the appropriate orientation of the lower body.

Any external or internal perturbation to the system leads to a series of compensational movements to again reach a stable formation. From this point of view, conversation formation are more similar to bubbles than rigid circles. However, the shape of the formation strongly depends on the number of people involved, the interpersonal relationship among them and from environmental constraints (e.g. the position of furniture inside a room). In the present work, we start by concentrating on simulating the dynamics of conversations in an open space setting, assuming they are strangers with equal social status. Thus we are first dealing with something that tends to proximate a circular formation, but other settings will follow.

Kendon explains a connection between the F-formation system and Goffman's concept of a frame [17]. A frame is a set of rules and social norms that all the participants

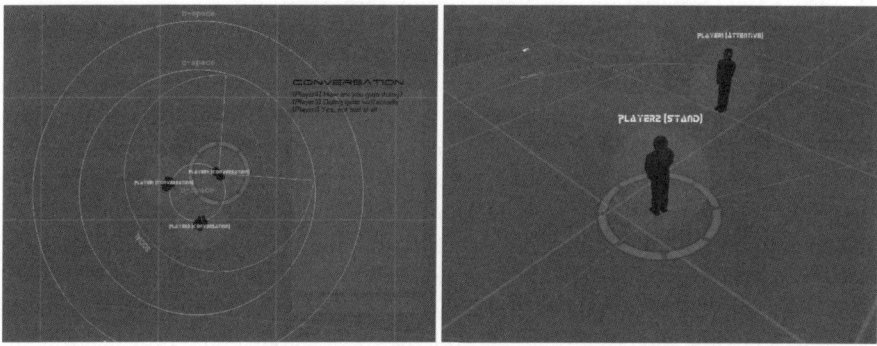

Fig. 1. The first picture is a top-down view of a conversation; notice the participants social spaces and convesersation domain with its o-, c- and b-space. The second picture shows an avatar focused and attentive to another; notice how the head is oriented toward its target and how the beam gets narrow to represent an avatar focused gaze.

in an interaction silently accept. The frame comes from the experience of the individual and states what behaviours are meaningful and what conduct is expected in that particular face-to-face interaction. The process of frame-attunement is tightly linked to the F-formation system. By actively maintaining a formation, participants inform each other that they share the same frame for the situation. This further reinforces the definition of an F-formation as a system and moreover describes the system as a unit of behaviour at the interactional level of organization, not at the individual level. Thus, an F-formation system, and consequently a conversation, is a shared social unit.

Since an F-formation system defines a positional relationship among the participants, it is quite natural to take into account the work of Hall and his Proxemics Theory [18]. Basically, in his work Hall claims the existence of four important distances between people as they interact. From shorter to longer, these are: intimate, personal, social and public distances (colored lines in Fig. 1). Proxemics can be seen as an interaction parameter in the sense that for a particular interaction a distance within a specific range is respected. Usually, normal conversations between acquaintances take place in the social area of the participants, while the personal area is reserved for interaction between close friends. In the present work, we are first looking at acquaintances carrying on normal, daily conversations, but plan to look at other relationships as well.

As part of their autonomous behaviour, avatars must exhibit proper reactions to the positional and orientational fluctuation inside conversation groups and larger changes within their various proximity areas. This is true, even if the avatars are otherwise under user control. In fact, the avatars cannot trust their users to always exhibit proper distances between them in order to preserve the believability of the interaction. In our automated approach, this complex social dynamic is defined as emergent avatar behaviour where a set of rules based on proximity perceptual data, defines a social force field that guides the active steering behaviours of each individual. Furthermore, the idea of conversation as a frame, and thus a social unit, leads to an architecture where we need to treat each social interaction as an important entity in the world.

3.2 The Social Simulation and Visualization Platform

For conducting this research, we felt we needed a special virtual environment that supported and highlighted the type of information and behavior we wanted to develop. We created a tool called CADIA Populus that combines full online multi-player capability with clear visual annotation of the social environment in terms of the theoretical models being used to inform avatar behavior (See Fig.1). The avatars themselves are human 'clay' figures with articulated necks, drawn to scale in the environment for accurately reflecting distances. Users can drive their avatars around and strike up conversations with each other using the built-in text chat. A single user can generate any number of avatars and switch between them at will, which is perfect for manipulating the social situation. A flexible and powerful framework for avatar automation is provided with the tool, part of which is inspired by the OpenSteer[1] model which allows steering behaviors to be activated and deactivated. Access to the social environment, through what we call a social perception interface, can be used to trigger behaviors. Examples of social perception include how many individuals are within a certain proximity range (intimate, personal, social and public) or whether the avatar has stepped within the domain of an active conversation according to Kendons F-Formation model. Notice that the perception of social and public space has a blind cone behind the avatar, of respectively 90 and 150 degrees. Both perceptual visualization and the set of active behaviors can be shown inside the environment. The tool is written in Python and uses the Panda 3D[2] game development library from CMU and Disney.

3.3 Social Situations and Activation of Steering Behaviors

In order to let an avatar generate the proper reactive behavior in every social situation, we activate a set of steering behaviors each of which represents a rule underlying the social frame that the avatar is participating in. For example, as soon as an avatar joins a conversation (i.e. enters a new social frame) it will start to keep personal distance, social equality, group cohesion, common attention and domain awareness. Each of them are behaviors in the steering layer and they can generate motivational forces for repositioning or reorienting the avatar. The activation of this set of behaviors produces the reactive dynamics expected from a group of people that has accepted the same social frame, i.e. share the same social norms. In our platform, a steering behavior usually implements activation/deactivation rules, generates motivational force or calls other steering behaviors. Instead of computing steering forces, like in Reynolds [14], our behaviors compute motivational forces such as a desired velocities or a desired direction for a particular avatar's body part. Each behavior has a priority, a weight and a frequency. Priority and weight are used during combination of forces. In particular, we linearly combine (using weights) the desired velocities generated by the active highest priority behaviors and choose the desired direction generated by the highest priority behavior. Some behaviors define in their body particular activation rules that depend on perceptual data. For example, the behavior for keeping personal distance will generate a repulsion force only if an individual steps into the avatar's personal space.

---

[1] http://opensteer.sourceforge.net/
[2] http://www.panda3d.org

In order to provide an example of how this behavior activation works, we are going to succinctly describe how an avatar joins a conversation. As soon as an individual gets close enough to an ongoing conversation, it steps inside the conversation domain. If it keeps moving closer to the conversation, the individual receives an *associated* status. An associated person will be allowed to join a conversation if certain requirements are met. Since our conversations are only modeled as open at the moment, the requirements are very relaxed. In fact is sufficient to have the body oriented toward the o-space of the conversation (intersection of transactional segments) and not move too much (i.e. not be running somewhere). Once an avatar is allowed to join, it is considered inside the conversation social frame, and therefore it is necessary to activate the proper set of rules listed above in order to adapt the agent's behavior to the ongoing social situation and smoothly blend in.

3.4 Conversation Force Field

As mentioned earlier, as soon as an avatar enters a new conversation, a particular set of steering behaviors is activated and run. Some of them generate the low level reactive behaviors of the avatar, which produce repositions and reorientations for keeping the formation's equilibrium. Our purpose is to simulate the compensational movements of the formation when the system is subjected to some fluctuations and for doing so we define a field of forces that drives the avatar movements. This force field is produced and updated by the following steering behaviors: KeepPersonalDistance, KeepConversationEquality and KeepConversationCohesion. Each of these behaviors has its own priority and generates motivational forces at a given frequency using perceptual information. KeepPersonalDistance prevents avatars from stepping inside the personal space of someone else generating a repulsion force. The activation rule generates a repulsion only when someone steps into the avatar's personal space. KeepConversationEquality forces an avatar to keep a shared group space of constant size, generating an attraction or a repulsion force toward a specific point. An orientational force toward the group is also generated. This behavior produces motivational forces only if someone is inside the avatar's social space. KeepConversationCohesion prevents an avatar from getting isolated from a group and keeps participants close enough to each other, by generating an attraction force when someone is inside the avatar's public space. Furthermore, an orientational force is generated towards those participants, that for some reason, are going far away from the conversation. The cohesion forces have the highest priority over all the other forces, while repulsion and equality forces are blended together. What follows is a more detailed discussion on how to compute these forces for the social force field of a conversation.

To maintain a minimal distance between individuals we calculate a *repulsion force* (Fig. 2) as follows.

Let N_p be the number of individuals inside the personal area of the avatar, $r \in \Re^3$ the position of the avatar and $r_i \in \Re^3$ the position individual i inside the personal area and Δ_p the personal distance:

$$F_{repulsion} = -(\Delta_p - d_{min})^2 \frac{R}{\|R\|} \qquad (1)$$

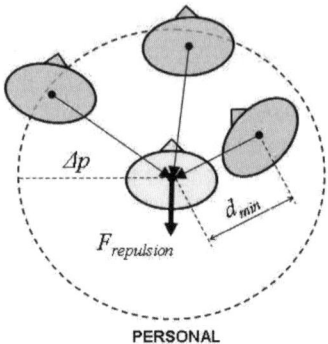

Fig. 2. Repulsion force diagram

where $R = \sum_{i}^{N_p} (r_i - r)$ and d_{min} is the distance of the closest individual.

To sustain the o-space with the individual inside the social area, we calculate an *equality force* (Fig. 3) as follow. This equality force acts like a repulsion or attraction force, depending on the difference between the distance of the avatar from the center of the group in its social area and the mean of every group member's distance from the same center.

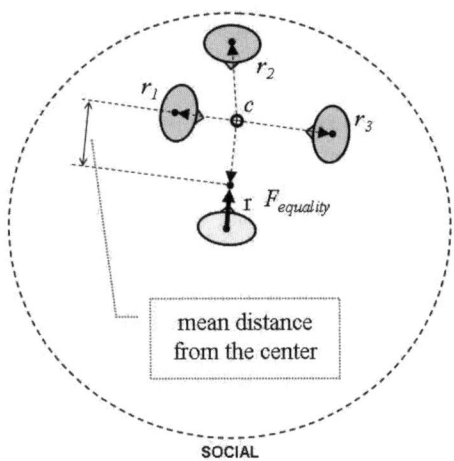

Fig. 3. Equality force diagram

Let N_s be the number of individuals in the avatar's social area, and m the mean distance of the members from the group center. We can calculate the centroid, mean distance and equality force and equality orientation as follow:

$$c = \tfrac{1}{N_s+1} \left(r + \sum_{i}^{N_s} r_i \right) \qquad (2)$$

$$m = \tfrac{\|r-c\|}{N_s+1} \sum_{i}^{N_s} \|r_i - c\| \qquad (3)$$

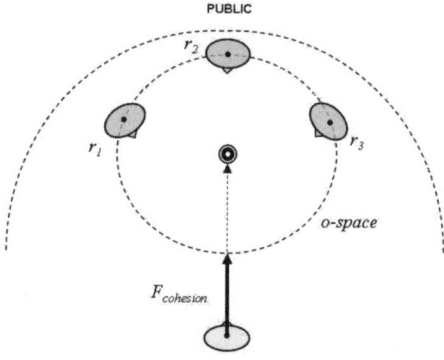

Fig. 4. Cohesion force diagram

$$F_{equality} = \left(1 - \frac{m}{|c-r|}\right)(c-r) \quad (4)$$

$$D_{equality} = \sum_{i}^{N_s} (r_i - r) \quad (5)$$

To avoid being isolated, individuals are attracted toward the o-space by means of a *cohesion force* (Fig. 4).

Let N_a be the number of individuals in the avatar's public area, $o \in \Re^3$ the center of the conversation and s the radius of the o-space. Then we can calculate the cohesion force and cohesion orientation as follow:

$$\alpha = \frac{N_a}{(N_s + 1)} \quad (6)$$

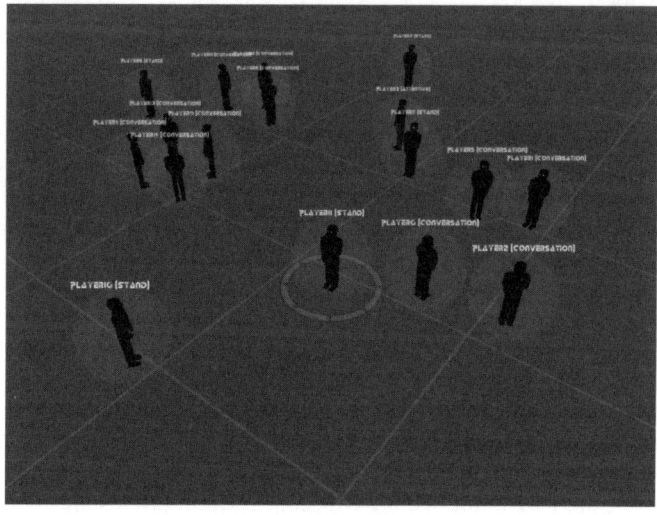

Fig. 5. An open setting with several ongoing conversations

$$F_{cohesion} = \alpha \left(1 - \frac{s}{\|o - r\|}\right)(o - r) \tag{7}$$

$$D_{cohesion} = \sum_{i}^{N_a} (r_i - r) \tag{8}$$

the scaling factor α for the cohesion force is used to reduce the magnitude of this force if the avatar is surrounded by individuals in its social area. The cohesion force is intended to be stronger for those participants far away and isolated from the conversation.

At each simulation time step, using these forces we have a force field that drives the avatar's repositioning and reorienting behavior. Notice that the conversation force field is activated only when the conversation is not in a steady formation and a rearrangement is necessary. Therefore, we are going to perform forces calculation only when strictly necessary. In fact, KeepConversationEquality and KeepConversationCohesion have a further activation/deactivation rule that runs the behaviors only if strong movements in the conversation is perceived and the formation is not well formed anymore.

3.5 Conversation Unit

In our social force field model, some forces are calculated using parameters (e.g. the center of the o-space or its radius) provided by a *conversation unit*, a small separate module shared by all the participants that represent the conversation as a social unit. The rationale for a high-level social interaction unit comes from the necessity of dealing with group behavior, where an individual is aligned with the same behavioral frame of the other participants but still maintain its own individuality. This is a first step towards simulating the aforementioned frame attunement process. The next step is to keep a conversation frame inside each avatar's mind in addition to the shared conversation unit in the world. The former will maintain all of the avatar's own beliefs about the conversation and its participants and, moreover, will activate context-specific reactive behavior. The latter will uniquely represent the conversation as a world object, with its own position, dimensions and generally visible information about its members. This way it will be possible to let an agent perceive a conversation through its perception system and then, recognizing that conversation as a specific social situation (eg. a frame in its mind), activate the proper reactive behaviors in order to produce a continuous dynamic. Another advantage of this approach is that the conversation unit will be a suitable entry point in case will be necessary to impose scripted behavior on a group of agents for artistic purposes.

4 Conclusions and Future Work

While formal evaluation of the effectiveness of the approach for multi-player games has not been performed yet, the visual results (available in video form[3]) and informal interaction tests support the direction of the chosen approach. One of the most immediate impacts on the user is the feeling that conversation groups are robust elements

---

[3] http://www.ru.is/faculty/hannes/movies/CADIAPopulus/iva2008.html

in the environment, that organically maintain sensible formations as players bring their avatars in and out of contact with other players. Additionally, social perception and reactive maneuvering seems to give the avatars a heightened level of realism and a greater relevance to the ongoing social interaction, potentially avoiding the dreaded disconnect between the visual environment and the chat box.

An obvious current limitation that we plan to address is that the locomotion model we currently apply is a simple point mass that, while good in its simplicity, is really far from legged locomotion. We will investigate better techniques for combining steering forces, avoiding jitter and achieving real-time performance with truly large crowds of agents. We will also continue to refine the force field model but, most of all we want to further investigate the benefits of explicitly representing conversation frames. We believe we have the right underlying model here to simulate a range of different social situations within the same complex environment.

Finally, as we move up from the lower levels of purely reactive behaviors we continue to add higher levels of intelligent control, incorporating some of the behaviors that we have ready from previous work, while also adding new ones that address social scenarios that the specific MMORPG setting calls for. The user interface will also need to grow and adapt to handle the increased complexity without risking the seamless user experience. This is something we believe automation makes easier, but that is an empirical question we look forward to answering.

Acknowledgments

We are very grateful to the other project team members, in particular Marta Larusdottir, Dr. Kristinn R. Thorisson and Eng. Eric Nivel, for numerous discussions and suggestions. We would also like to thank Torfi Olafsson and the rest of the team at CCP Games for their productive collaboration. Special thanks go to Dr. Adam Kendon for his precious material on face-to-face interaction and valuable personal communication. This work is supported by the Humanoid Agents in Social Game Environments Grant of Excellence from The Icelandic Research Fund.

References

1. Cassell, J., Vilhjalmsson, H.: Fully embodied conversational avatars: Making communicative behaviors autonomous. Autonomous Agents and Multi-Agent Systems 2(1), 45–64 (1999)
2. Vilhjalmsson, H.: Animating conversation in online games. In: Rauterberg, M. (ed.) ICEC 2004. LNCS, vol. 3166, pp. 139–150. Springer, Heidelberg (2004)
3. Kendon, A.: Conducting Interaction: Patterns of behavior in focused encounters. Cambridge University Press, New York (1990); Main Area (nonverbal behavior)
4. Vilhjalmsson, H., Cassell, J.: Bodychat: Autonomous communicative behaviors in avatars. In: Autonomous Agents, pp. 269–276. ACM Press, New York (1998)
5. Cassell, J., Vilhjalmsson, H., Bickmore, T.: Beat: the behavior expression animation toolkit. In: SIGGRAPH 2001, August 12-17, 2001, pp. 477–486. ACM Press, New York (2001)
6. Gillies, M., Ballin, D.: Integrating autonomous behavior and user control for believable agents. In: Autonomous Agents and Multi-Agent Systems, July 19-23, 2004, pp. 336–343. ACM Press, New York (2004)

7. Friedman, D., Steed, A., Slater, M.: Spatial social behavior in second life. In: Pélachaud, C., Martin, J.-C., André, E., Chollet, G., Karpouzis, K., Pelé, D. (eds.) IVA 2007. LNCS (LNAI), vol. 4722, pp. 252–263. Springer, Berlin (2007)
8. Salem, B., Earle, N.: Designing a non-verbal language for expressive avatars. In: Collaborative Virtual Environments, pp. 93–101. ACM Press, New York (2000)
9. Jan, D., Traum, D.: Dynamic movement and positioning of embodied agents in multiparty conversation. In: Proc. of the ACL Workshop on Embodied Language Processing (June, 2007) 59–66 (2007)
10. Couzin, I., Krause, J., James, R., Ruzton, G., Franks, N.: Collective memory and spatial sorting in animal groups. Journal of Theoretical Biology (218), 1–11 (2002)
11. Heigeas, L., Luciani, A., Thollot, J., Castagne, N.: A physically-based particle model of emergent crowd behaviors. In: Proc. of GraphiCon., September 5-10 (2003)
12. Treuille, A., Cooper, S., Popovic, Z.: Continuum crowds. In: SIGGRAPH 2006 Papers, pp. 1160–1168. ACM, New York (2006)
13. Musse, S.R., Thalmann, D.: Hierarchical model for real time simulation of virtual human crowds. IEEE Transactions on Visualization and Computer Graphics 7(2), 152–164 (2001)
14. Reynolds, C.W.: Steering behaviors for autonomous characters. In: Proc. of the Game Developers Conference, San Francisco, CA, pp. 763–782. Miller Freeman Game Group, New York (1999)
15. Shao, W., Terzopoulos, D.: Autonomous pedestrians. In: Proc. of the ACM SIGGRAPH Symposium on Computer Animation, July 29-31, 2005. ACM Publishing, New York (2005)
16. Rehm, M., Andre, E., Nisch, M.: Let's come together - social navigation behaviors of virtual and real humans. In: Maybury, M., Stock, O., Wahlster, W. (eds.) INTETAIN 2005. LNCS (LNAI), vol. 3814, pp. 124–133. Springer, Heidelberg (2005)
17. Goffman, E.: Frame Analyses: An Essay on the Organization of Experience. Harvard University Press, Cambridge (1974)
18. Hall, E.T.: The Hidden Dimension. Doubleday, New York (1966)
19. Brown, B., Bell, M.: Cscw at play: 'there' as a collaborative virtual environment, November 6-10, 2004, vol. 6, pp. 350–359. ACM Press, New York (2004)
20. Gillies, M., Dodgson, N.A.: Behaviourally rich actions for user-controlled characters. Computers & Graphics, 28(6), 945–954 (2004)
21. Shi, J., Smith, T.J., Graniere, J., Badler, N.: Smart avatars in jackmoo. In: Virtual Reality, March 1999, IEEE Computer Society Press, Los Alamitos (1999)

Multi-party, Multi-issue, Multi-strategy Negotiation for Multi-modal Virtual Agents

David Traum[1], Stacy C. Marsella[2], Jonathan Gratch[1], Jina Lee[2], and Arno Hartholt[1]

[1] Institute for Creative Technologies, University of Southern California
Marina del Rey, CA, USA
traum@ict.usc.edu

[2] Information Sciences Institute, University of Southern California
Marina del Rey, CA, USA

Abstract. We present a model of negotiation for virtual agents that extends previous work to be more human-like and applicable to a broader range of situations, including more than two negotiators with different goals, and negotiating over multiple options. The agents can dynamically change their negotiating strategies based on the current values of several parameters and factors that can be updated in the course of the negotiation. We have implemented this model and done preliminary evaluation within a prototype training system and a three-party negotiation with two virtual humans and one human.

1 Introduction

In the most general case negotiation can include trade-offs between multiple issues, can involve multiple parties, each with their own agendas, and can be negotiated through multiple modalities, including speech and face to face bodily communication. Moreover, the parties involved need not maintain a constant position, but can dynamically vary their goals, strategies to achieve those goals, and agenda for carrying out those strategies as the negotiation proceeds.

In this paper, we describe work on pushing the frontier of what can be accomplished by virtual agents in negotiation with humans and other virtual agents. We extend previous work in several directions. In [1] we presented a model of team negotiation involving multiparty dialogue. This model allowed virtual humans to engage in multimodal negotiation over multiple options and discussion among multiple agents as to which options were best for satisfying the shared goals. However, it did not allow for more general negotiation, including a range of different utility valuations and relationships among the agents, including adversarial and more neutral as well as team members. Several factors were taken into account, including the roles of agents, the previous dialogue history, and the utility calculations, but there was only a single fixed negotiation algorithm mapping the value of these factors to a negotiation move.

In [2], we extended this model to handle other kinds of relationships, including adversarial negotiation. Agents could assess their own view of utilities of actions as well as utilities of a negotiating partner. A model of trust was created, using factors of credibility, solidarity, and familiarity. Agents had a choice of strategies to select, depending on factors including utility and controlability. However, this model was also limited in a number of ways. First, it only handled negotiation of whether or not to select a single action, rather than allowing a broader set of possible decisions. Strategies were always with respect to this single action. There were only two parties involved in the negotiation, the agent and one other (e.g., a human trainee).

In this paper we describe work that takes the next step toward a fully general and human-like model of negotiation. We combine the strengths of both previous models, as well as some further extensions. We allow negotiation over simultaneous courses of action. The trust model is extended to refer to specific individuals, rather than a general trust level. Strategies are made specific to each possible issue of negotiation, and one can consider different strategies for each issue. Moreover, we have expanded the set of strategies that the agents may choose from.

The primary purpose of the negotiation model is to enable the virtual humans to act as role players in a training environment, in which a human trainee can practice different styles and tactics of negotiation and analyze the results. Things are set up so that the trainee must generally balance three different goals in order to be successful at more difficult negotiations:

Solve problems - The most basic matter is figuring out a mutually acceptable solution, just based on the utilities for all the participants. All things being equal, people will act rationally and agree to proposals that are in their own interest. The trainees must be able to go beyond their initial starting points and see how to make a solution attractive to others, e.g. by offering additional resources, committing to important actions, and removing obstacles. The trainee must also consider alternative plans that might lead to a win-win or compromise situation that is an adequate even if not optimal solution.

Gain Trust - Generally, all things are *not* equal. The trainee must also work on an interpersonal level to develop and maintain the trust of other participants. With our model this involves working at three aspects:

> **Familiarity.** The trainees must show that they know how to behave appropriately in this situation, for this culture, including polite pleasantries, and adhering to norms of topic management.
>
> **Credibility.** The trainees must be truthful and say things that are believable, and also stand by their word and follow through on promises.
>
> **Solidarity.** The trainees must show that there is some alignment in goals between themselves and agents – that they want some of the same things.

Manage Interaction - It is also important for the trainees to properly manage the interaction. By properly setting the agenda and controlling the topic progression they can lead to more successful results (assuming they are solving problems and gaining trust). They still must be reactive to the concerns that the agents express and not be too heavy-handed and unilateral. On the other hand, if they lose control of the agenda, the agents may agree on an undesirable plan or refuse to consider other options.

This model has been implemented in our virtual humans and in our current test scenario controls the behavior of two different virtual humans in a three-party negotiation (with a human user) in a prototype negotiation-training application. The virtual humans recognize speech and a limited set of gestures and body postures, and produce speech and gestures. The architecture of the whole system is described in [3]; in this paper we focus on the cognitive aspects of the negotiation model and multimodal realization of negotiation strategies.

In section 2, we describe the extensions to the multiparty dialogue model. In section 3, we describe the negotiation strategies and how they typically affect behavior. In Section 4 we give more details on how the strategies are implemented, including relevant factors to consider, selection criteria, and how strategies are realized. In section 5 we give some examples of how the dialogue model and strategies are used to influence the behavior of virtual humans in negotiation with a person. Finally, in section 6 we conclude with a discussion of related and future work.

2 Multi-modal Multi-party Dialogue Model

The negotiation is carried out in the context of a multi-party meeting with multiple individuals involved in a (virtual) face to face setting. The agents obey the norms of conversation, including deciding who or what to look at, how to orient their bodies, which posture to adopt, when to speak or listen, and what to say. As outlined in [4,1,5], the dialogue model uses the information-state approach to dialogue management [6], with multiple layers of interaction. Each layer consists of information state components and dialogue acts that change values of the components. Decisions of listening, processing utterances, and speaking are made asynchronously, and the agents have the capacity to both respond to communications from other human and virtual agents, and to initiate communication based on their internal state and decisions. There are specific representations of each conversation the agent is aware of, with its conversational state. We have extended previous work with a tighter coupling between the dialogue modelling, emotion modelling, and non-verbal expression. In the rest of this section we briefly describe some of these extensions, in particular, the gaze and listener reactions model, and the use of motivations for tailoring output, and tracking of focus and strategy.

The gaze model [7] has been extended to include different styles of gaze depending on the reason for the gaze. There is also much more non-verbal feedback during the listening and processing of utterances of others, depending on whether the agent agrees or disagrees with what is said and trusts or does not trust the speaker. Specifically, the listener's dialogue model informs its non-verbal behavior generation process [8] whether the speaker is agreeing or disagreeing with a prior speaker and whether the listener itself agrees or disagrees with that stance. If the listener agrees, it may nod while the other speaks. On the other hand, if it disagrees, the non-verbal behavior generator will select other behaviors, such as lowering its head and frowning (lower the brow) or pulling back its head and lowering its eyebrows (inner and outer brow are raised). The particular behaviors chosen depend on the cultural, physical, and personal features of the specific agent. For example, elderly listeners may nod more slowly or different agents may use more idiosyncratic behaviors.

Output motivations are also used to guide the generation of both verbal and non-verbal negotiation behavior, depending on not only the main message to be expressed, but also the reason for saying it and the issue and negotiation strategy that motivate that reason [9]. This allows the agent's text and behavior generation components to subtly tailor the output, e.g. distinguishing between pointing out a plan flaw to defeat a proposal vs bringing it up as an issue to be addressed and overcome.

Agents actively compute motivations for strategies of all of the options under consideration. The current topic of conversation is tracked based on understanding of the content of utterances and the dialogue history. References to the topic or any constituent actions and states contribute to selecting or maintaining the current topic. The agents dynamically activate the strategy for the current topic from among the motivations for all issues under discussion.

3 Multi-party Negotiation Strategies

In previous work [2], our negotiation strategies were based on orientations to the negotiation [10,11]. In a multi-party situation, these orientations are not so straightforward, as one must distinguish the attitudes about the negotiated items, the individual participants, the whole group, and subgroups. Thus one may wish to avoid the whole negotiation, or just one issue. One may wish to avoid the whole group interaction, or just one participant. One may feel distributive or integrative with individuals, the whole group, or subgroups (coalitions). One may simultaneously be integrative toward some while distributive towards others, and wanting to avoid yet others.

We defer these issues for the time being, and focus on specific strategies that take some of these orientations into account but focus on concrete objectives rather than the orientations that lead to them. Strategies have applicability conditions, tactics to carry out the strategies, and behaviors (verbal and non-verbal) to communicate the external impressions that are appropriate for that strategy. We describe these features and the strategy informally in this section and discuss the formal and implementational details in the next. We have so far implemented the following negotiation strategies:

Find Issue. This strategy is appropriate in the case where there is a negotiation meeting currently occurring, but there is no issue that is a current topic of negotiation. The possible moves include requesting a topic of negotiation from another agent (human or virtual), proposing a topic, or proposing constraints on topic selection. In addition to the kinds of gestures associated with request and proposal moves, this strategy might be signaled non-verbally by a more open body posture. Open postures place no physical barriers between conversants, for example, no crossing of arms or legs, indicating a willingness to participate.

Avoid. This strategy is appropriate in the case when there is no topical issue or the focused issue is undesirable but seen as avoidable. The moves include talking off-topic, e.g. small talk, trying to leave the meeting or the topic, or switching the topic to another issue. Avoid can be signaled non-verbally by a more closed, negative and defensive posture. For example, crossed arms while standing is a sign of defensiveness and protection (e.g., [12,13]) in many cultures.

Attack. This strategy is appropriate in the case where the topic is seen as not avoidable and having negative utility, with little potential for improving the utility. It is an assumed bad outcome. The moves include stating flaws in the issue under discussion - negative outcomes that are likely, or pre-conditions that are not met, attempts to propose alternative, better issues, and ad-hominem comments about the advocates of the issue. Attack can be signalled non-verbally by an open but more aggressive, dominant posture. For example, arms akimbo (on hips) while standing is a sign of disliking, dominance and even anger (e.g., [14,13]).

Negotiate. This strategy is appropriate in the case where it is not clear what the outcome of adopting the issue will be – there is a potential for either negative, neutral or positive results, depending on how the plan is carried out, and whether all individuals involved will do their parts. Here, the agent is not necessarily for or against the issue, but willing to consider whether it can be made to work or not. Moves in this strategy include stating flaws, as for the attack strategy, but also proposals of solutions to fix the flaws, and bargains that would give up some utility on some aspects while gaining utility on others. Conditional commitments, contingent on the commitment of others and fixing of flaws are also appropriate. Because of the potential outcome is unclear, we currently associate a mixed non-verbal signal with this strategy, for example one hand on hip.

Advocate. This strategy is appropriate when one has good reason to believe that the outcome of the issue will have positive utility. The moves involved include proposing plans to bring about the outcome, proposing solutions or ameliorations to flaws that have been introduced, and offering commitment to the issue or its component parts. Because of the potential outcome is positive, we signal this strategy with an open, relaxed posture.

Success. This strategy involves the follow-through of a successful mutual commitment to an issue - it may involve formalizing remaining details of how to carry it out, as well as friendly disengagement from the meeting. Because a positive outcome has been achieved, we currently associate an open, relaxed posture with this strategy.

Failure. This strategy follows from the commitment against a course of action. It involves disengagement from the issue and possibly the meeting, seeing the issue as settled. The agent may have either positive or negative emotions associated with the failure and the non-verbal behavior may need to vary accordingly.

4 Implementing the Strategies

In this section we describe in more detail how the strategies described in Section 3 are implemented. In 4.1, we describe the factors that are used to decide which strategies to adopt. We then describe these factors in more detail in 4.2 and 4.3. We describe how strategies are selected in 4.4 and how they are performed in 4.5.

4.1 Factors in Strategy Selection

There are several factors that are examined when deciding which strategy to chose:

Topic. Foremost is the question of which issue is the topic of the current conversation. If there is no topic, then only the find-issue and avoid strategies are applicable. If there is a current topic, then appraisals of plans related to this issue will be the source of further decisions.

Control. This is the agent's estimated ability to control the discussion about the topic under discussion. Control is a pre-requisite for successful avoidance.

Utility. This is the agent's calculation of how good the outcome will be if the issue is carried out, using current assumptions about plans and likelihoods of effects holding and commitments being carried out. More details on how these are calculated are given in Section 4.2. An agent who thinks the an issue has positive utility will generally be an advocate. There is also a consideration of absolute utility (positive or negative) vs. relative utility compared to other options.

Potential. This is the agent's estimation of how good the utility can get, assuming that everyone will "do the right thing". For issues with negative utility, the potential is the principal factor in deciding whether to attack or negotiate.

Trust. How much does the agent trust the other agents in the negotiation? With low trust an agent will not want to continue and commitments will not increase the estimated probability that actions will succeed. With high trust, other agents can be believed.

Commitment. This involves whether participants have committed themselves for or against issues. Mutual commitment is generally a pre-requisite of the success strategy, while negative commitment is a result of the failure strategy. There are also commitments to actions that support one or another of the issues, which can lead to different predictions of utility and potential of that issue.

4.2 Multi-issue Utilities

The ability of our agents to negotiate with humans and other agents stems from their understanding of the goals of each party, the actions that can achieve or thwart those goals, and the commitments and preferences agents have towards competing courses of action. To provide this understanding, our agents use domain-independent reasoning algorithms operating over a general partial-order plan representation: see [15,1]. Plans provide a concise representation of the causal relationship between actions and agents' goals, including causal links and causal threats between the plans of different agents. The representation includes decision theoretic information to represent the perceived utility of different goals and their likelihood of satisfaction. Finally, the representation includes a simplified theory of mind, allowing agents to represent and reason about the beliefs, intentions and preferences of other agents.

A key aspect of multi-party negotiation involves discussion of alternative ways to achieve goals. To support such negotiation, agents reason about alternative, mutually exclusive courses of action (plans) for achieving goals, and incorporate a general decision-theoretic method for evaluating the relative strengths and weaknesses of different alternatives. Strengths of a plan include states of positive utility that would result from the plan's execution, weighted by their probability of attainment. Weaknesses include resulting negative utility states and basic flaws that might block the plan's execution. For example, a plan may contain unsatisfied preconditions that would require (negotiated) help from other agents to satisfy. It may also contain causal threats, as when the expected actions of another agent might block the plan's successful execution.

Agents also reason about the *potential* strengths of a plan, meaning the expected utility of beneficial effects assuming that any potential flaws are successfully resolved. Consider, for example, the situation where Bob has to borrow Mary's car in order to buy groceries. The likelihood that this plan succeeds depends on the likelihood of Mary performing the "lend car" action. Initially, Bob may have some *a priori* probability that Mary will lend the car. If Mary verbally commits to lending the car, this probability is likely to increase, although this depends on Mary's perceived trustworthiness as a negotiation partner. The expected benefit of the plan includes the current estimate that Mary will perform this action, accounting for any stated commitments and the current measure of trust Bob has for Mary, whereas the potential benefit calculation assumes Mary will help with probability 1.0.

The (potential) strengths and weaknesses of a plan serve as talking points for the negotiation and criteria for moving between negotiation stances. Strengths are points that should be emphasized when advocating a certain course of action whereas weaknesses are objections that can be raised. The relative magnitude of strengths and weaknesses of a course of action inform its strategy for negotiation. A plan with more severe weaknesses than strengths, and no potential for improvement should be avoided or fought against. A plan with some positive potential might merit negotiation.

4.3 Models of Other Agents

As well as calculating ones own beliefs, goals, intentions, and computations of the expected utilities of various actions and outcomes, the agents also engage in (limited) reasoning about the mental states of others. They track the beliefs of others (which might be positive, negative, or unknown), intentions to act, and utilities of others. These contribute to the estimation of the likelihood of other agents to act in particular ways, and thus the estimated utility of a course of action. Trust of each other agent is also computed as described in Section 1.

There are also models of the interactional structure between the agents. Part of this is the dialogue model, discussed in section 2. The dialogue model tracks the topic, as discussed above. Control is calculated using a heuristic that an agent has control over (avoiding) a topic if it has not been referred to more times by other agents than a threshold amount. Commitments are also calculated on the basis of dialogue utterances: if an agent makes a (grounded) assertion or promise, this leads to a social commitment.

4.4 Choosing Strategies

Table 1 shows the applicability conditions for choosing among the strategies. In general, only a subset of the factors are relevant for any given strategy. Also, there is some

Table 1. Choosing Negotiation Strategies based on Factors

	topic	control	utility	potential	trust	commitment
find-issue	-				some	
avoid		+	-		some	
attack	+	-	-	-	some	
negotiate	+		-	+	some	
advocate	+		+		some	
success	+				moderate	mutual
failure	+				very low	negative

overlap in the set of applicable strategies. Our initial algorithm chooses deterministically, preferring first to find the topic, and then decide (based on the utility, potential utility, and control) whether to avoid, attack, negotiate, or advocate. Once commitments have been established the agents follow the success or failure strategy for that topic.

4.5 Dialogue Realizations of Strategies

Once the strategy has been chosen, the agent will have the option of selecting from a number of moves that go with the strategy. These moves are in competition in the agent's decision space with other kinds of actions. These include dialogue actions, such as giving grounding feedback and addressing questions, as well as non-dialogue actions such as emotion reasoning and acting in the virtual world.

For **Find-issue**, the two main actions are requesting a topic from the meeting initiator, and proposing a topic. The initiative parameter for the agent determines this choice. With no initiative, the agent will not bring up the topic at all. With a medium level, the agent will ask for the topic. If the agent has high initiative and control, it will introduce a high utility topic.

In the **Avoid** strategy, the possible move types are:

- change topic to high utility issue
- talk about non-issues (ad hominem, small-talk)
- disengage from meeting

The agent will prefer to change the topic if there is a good one, otherwise will try non-issue talk and if that does not work, but there is still some control, will try to leave.

For the Attack strategy, the agent will choose either ad-hominem attacks, e.g., blaming the topic-initiator for the problems, or pointing out flaws with this issue, that have been identified as described in section 4.2. Flaws include pre-conditions that are not likely to be met, negative outcomes, and lack of necessary commitments from participating agents. The agents also compare the plans unfavorably with higher-utility options. No possible solutions are presented to the flaws.

In the Negotiate strategy, the same flaws are used, but as well as stating the problems, the agents may also choose to propose solutions.

In the Advocate strategy, agents will talk about the high-utility outcomes, and will also address any mentioned flaws. They will also offer and solicit commitments.

The negotiation is considered successful when all participants make a positive commitment towards an issue. The agents will make a negative commitment when entering the failure strategy. Once commitments have been made to an issue, the agents will attempt to disengage from the meeting and move on to other tasks on their action agenda.

5 The SASO-EN Three-Party Negotiation Domain

Our current test scenario is an expansion of that used in [1]. This scenario involves a negotiation about the possible re-location of a medical clinic in an Iraqi village. As well as the virtual Doctor Perez and a human trainee playing the role of a US Army Captain, there is a local village elder, al-Hassan, who is involved. The doctor's main objective is to treat patients. The elder's main objective is to support his village. The captain's main

Fig. 1. SASO-EN Negotiation in the Cafe: Dr Perez (left) looking at Elder al-Hassan

objective is to move the clinic out of the marketplace, which is considered an unsafe area. Figure 1 shows the doctor and elder in the midst of a negotiation, from the perspective of the trainee. There are three main issues under discussion, corresponding to different options for and plans to accomplish the location of the clinic:

– whether to move the clinic near to the US Base (the captain's preferred option, unsuitable for the elder)
–whether to keep the clinic in the marketplace (the preferred option of both the elder and the doctor, though initially with negative utility, unsuitable for the captain)
–whether to move the clinic to an old hospital location in the center of the village (no one's preferred option because of the large amount of work needed to make it viable, but with potential for positive utility).

The bulk of the authoring for the cognition is done using a central ontology [16], for constructing the task model resources, intrinsic utility, plans, and language semantics. Additional work includes the creation of the external visage and behaviors of the characters. As mentioned previously, the agents have characteristic postures corresponding to their negotiation strategies. Table 2(a) shows the mapping for the doctor (a westerner), while Table 2(b) shows the posture for the Iraqi elder. Knowing these mappings we can guess that in Figure 1, the doctor is employing the negotiate strategy to the

Table 2. Mapping of Strategies to Postures for different cultural types

(a) Western Doctor.

Strategy	Posture
Find Issue	Hands at Side
Avoid	Crossed Arms
Attack	Hands on Hips (Akimbo)
Negotiate	Left Hand on Hip
Advocate	Hands at side
Success	Hands at side
Failure	Arms Crossed in Front

(b) Middle-Eastern Elder.

Strategy	Posture
Find Issue	Hands at Side
Avoid	Hold Wrist in Front Low
Attack	Hold Wrist behind Back
Negotiate	Hold Wrist in Front High
Advocate	Hands at side
Success	Hands at side
Failure	Hands at side

current topic, while the elder is employing the attack strategy. We can also guess that the elder is the current turn holder, because the doctor is looking at the elder, while the elder looks at the captain (represented as the camera position).

1 C Hello gentlemen.
2 D Hello captain.
3 E Hello captain.
4 C I have orders to move this clinic to a camp near the US base.
5 D we need to help the victims of this conflict you started
6 C I understand, but it is imperative that we move the clinic out of this area.
7 D do you see that girl over there her mother was killed by American gunfire today
8 C It is not safe here
9 D look at these people they are injured because of your operations
10 C i have my orders to move you to the camp
11 D Elder i think staying at the market would be best
12 E we have many matters to attend to
13 C i understand
14 E we must stop this killing insanity
15 E captain you would do better to protect the town
16 C we cannot protect you here
17 E i must refuse
18 D i would have to refuse this decision

Fig. 2. Unsuccessful negotiation dialogue between C, a captain (human trainee), D, a doctor (virtual human), and E, a village elder (virtual human)

Figures 2 and 3 show examples of negotiations between virtual characters employing these models and a human negotiator playing the role of the captain. For ease of reading we have linearized the dialogue into turns, though in practice the agents often interrupt each other and the human. Figure 2 shows an unsuccessful negotiation – the captain makes several errors here including lack of sensitivity to the concerns the agents are bringing up, no search for a win-win compromise, and lack of maintenance of trust. In particular, the captain loses a lot of trust by showing no solidarity with the agents and insisting on a plan they do not like. The agents are trying to avoid this option, starting with line 5, and also express other preferences and concerns, with lines 11 and 15.

1 C Hello Doctor Perez.
2 D Hello captain.
3 E Hello captain.
4 C Thank you for meeting me.
5 E How may I help you?
6 C I have orders to move this clinic to a camp near the US base.
7 E We have many matters to attend to.
8 C I understand, but it is imperative that we move the clinic out of this area.
9 E This town needs a clinic.
10 D We can't take sides.
11 C Would you be willing to move downtown?
12 E We would need to improve water access in the downtown area, captain.
13 C We can dig a well for you.
14 D Captain, we need medical supplies in order to run the clinic downtown.
15 C We can deliver medical supplies downtown, Doctor.
16 E We need to address the lack of power downtown.
17 C We can provide you with power generators.
18 E Very well captain, I agree to have the clinic downtown.
19 E Doctor, I think you should run the clinic downtown.
20 D Elder, the clinic downtown should be in an acceptable condition before we move.
21 E I can renovate the downtown clinic, Doctor.
22 D OK, I agree to run the clinic downtown, captain.
23 C Excellent.
24 D I must go now.
25 E I must attend to other matters.
26 C Goodbye.
26 D Goodbye.
26 E Farewell, sir.

Fig. 3. Successful negotiation dialogue between C, a captain (human trainee), D, a doctor (virtual human), and E, a village elder (virtual human)

Eventually the trust is so low that the agents lose patience with the captain and break off the negotiation.

In Figure 3, we see a more successful negotiation. Here the captain pays more attention to both building trust, and compromising and addressing the concerns of the others. In line 5, the elder politely looks for the topic of the meeting. When the captain proposed this topic in line 6, the elder tries to avoid this topic, not wanting the clinic to be moved away from the town. When the Captain persists in line 8, both the doctor and elder choose attack strategies, pointing out problems with the proposed plan - lack of a clinic in the town for the elder, and the loss of neutrality that proximity to the US base would bring for the doctor. The captain proposed a new solution in 11. This plan has potential for both agents. In 12 the elder shows the negotiate strategy, not just pointing out a problem with the plan, but suggesting an avenue for improvement. This suggestion is taken up by the captain in 13 and satisfactorily addressed. The doctor has his own issues with this plan though, as illustrated in 14, and dealt with in 15. The elder continues with another issue in 16, and after the captain deals with this in 17, the plan actually has positive utility to the elder, causing him to agree to the plan in 18 and become an advocate, as shown in 19, where he in turn tries to convince the doctor to adopt this plan as well. The doctor has a remaining issue as shown in 20. When the elder satisfactorily addresses this issue in 21, the doctor is also ready to accept this plan, and the negotiation is successfully concluded with a resolution to move the clinic to the old hospital downtown, which will be supplied by the captain and renovated by the elder, in return for improved water and power provided by the captain.

5.1 Evaluation

Evaluation of a negotiation model for virtual humans such as the one presented in this paper is a very challenging process. The most important questions are:

- Does it lead the virtual human to negotiate in a manner similar to real humans?
  - Does it make the same decisions humans tend to make in those situations?
  - Does it realize the decisions in the same manners?
  - Does it show the breadth and diversity of behaviors that humans show?
- Can virtual humans using the model help people become better negotiators?

Unfortunately these are generally not easy questions to answer. Those related to human-like behavior are binary distinctions that are hard to turn into scaled metrics that can show progress before final completion. Also for many aspects, it requires the full virtual human performance rather than an isolated component. In this case it can be hard to attribute specific degree of success or failure to an isolated set of components, and it may become unclear whether the problem lies e.g., with the negotiation reasoning or with speech recognition, language understanding or generation.

We have started work in this area by having people try to negotiate with Doctor Perez and Elder al Hassan. Our preliminary results show that people are able to achieve similar rates of successful interaction as with our previous system at a similar level of development, but with a richer multi-party experience. More work is needed, however, especially in building a bigger corpus of training examples for the natural language understanding component, in order to increase the performance of the topic reference components.

6 Limitations, Related and Future Work

While our negotiation model significantly extends the generality and expressiveness of previous negotiation models for virtual humans, it is still far from the general case that we aspire to. First, while the agents may consider several different issues, they still can't consider arbitrary deals, and thus their ability to initiate and respond to novel bargains is very limited. Also, more strategies are needed to cover cases of interactional as well as transactional goals. Further factors such as power, status, interpersonal distance, and autonomy need to be taken into account. We also need to develop meta-strategies that take into account the (assumed) current strategies of other agents and the desired strategy in order to be able to manipulate the negotiation (or react to being manipulated). We would also like to improve the topic management algorithms, including experimenting with domains with more items to negotiate, and in which multiple options can be simultaneously compared and considered. In addition, there is still much work to be done in tying together the negotiation strategies, emotion and non-verbal behavior.

Although our negotiation model is ambitious in breadth – integrating multi-party dialogue, emotion, and nonverbal communication – other research has addressed aspects of this problem in more detail and these suggest obvious improvements to our work. Creating convincing embodied conversational agents is an ongoing challenge and several related projects are advancing the state-of-the-art in multi-modal speech generation and expressive character behavior (e.g. [17,18,19,20]). Other work has explored the cognitive aspects of negotiation, especially the challenge of identifying win-win deals that recognize and incorporate the potentially different beliefs and preferences of all negotiation partners. For example, [21] propose a general multi-issue negotiation approach that finds higher value agreements than human negotiators, though assumes perfect information about all party's preferences. [22] addresses negotiations where the other party's preferences are unknown, illustrating how this information can be inferred through a series of offers and counter-offers. Several psychological studies have also explored how emotion influences bargaining behavior and and illustrated how negotiation partners often deploy emotion displays strategically to influence outcomes. For example, [23] demonstrate that displays of anger by one partner will tend to elicit larger concessions unless the recipient of this display feels powerful, in which case anger tends to backfire. Although our approach incorporates a general model of emotion, it does not address such strategic displays and incorporating such findings could enhance the training value of our approach.

Acknowledgments

We would like to thank the rest of the Virtual Human team at USC, as well as many others for interesting discussions on the role of cooperation and non-cooperative dialogue. This work was sponsored by the U.S. Army Research, Development, and Engineering Command (RDECOM), and the content does not necessarily reflect the position or the policy of the Government, and no official endorsement should be inferred.

References

1. Traum, D., Rickel, J., Marsella, S., Gratch, J.: Negotiation over tasks in hybrid human-agent teams for simulation-based training. In: Proceedings of AAMAS 2003: Second International Joint Conference on Autonomous Agents and Multi-Agent Systems, July 2003, pp. 441–448 (2003)
2. Traum, D., Swartout, W., Marsella, S., Gratch, J.: Fight, flight, or negotiate: Believable strategies for conversing under crisis. In: Panayiotopoulos, T., Gratch, J., Aylett, R.S., Ballin, D., Olivier, P., Rist, T. (eds.) IVA 2005. LNCS (LNAI), vol. 3661, pp. 52–64. Springer, Heidelberg (2005)
3. Kenny, P., Hartholt, A., Gratch, J., Swartout, W., Traum, D., Marsella, S., Piepol, D.: Building interactive virtual humans for training environments. In: Proceedings of I/ITSEC (2007)
4. Traum, D.R.: Ideas on multi-layer dialogue management for multi-party, multi-conversation, multi-modal communication. In: Theune, M., Nijholt, A., Hondorp, H. (eds.) CLIN, Rodopi. Language and Computers - Studies in Practical Linguistics, vol. 45, pp. 1–7 (2001)

5. Traum, D., Swartout, W., Gratch, J., Marsella, S.: A virtual human dialogue model for non-team interaction. In: Dybkjaer, L., Minker, W. (eds.) Recent Trends in Discourse and Dialogue. Springer, Heidelberg (2008)
6. Traum, D., Larsson, S.: The information state approach to dialogue management. In: van Kuppevelt, J., Smith, R. (eds.) Current and New Directions in Discourse and Dialogue, pp. 325–353. Kluwer, Dordrecht (2003)
7. Lee, J., Marsella, S., Traum, D.R., Gratch, J., Lance, B.: The Rickel Gaze Model: A window on the mind of a Virtual Human. In: Pélachaud, C., Martin, J.-C., André, E., Chollet, G., Karpouzis, K., Pelé, D. (eds.) IVA 2007. LNCS (LNAI), vol. 4722, pp. 296–303. Springer, Heidelberg (2007)
8. Lee, J., Marsella, S.: Nonverbal behavior generator for embodied conversational agents. In: Gratch, J., Young, M., Aylett, R., Ballin, D., Olivier, P. (eds.) IVA 2006. LNCS (LNAI), vol. 4133, pp. 243–255. Springer, Heidelberg (2006)
9. Lee, J., DeVault, D., Marsella, S., Traum, D.: Thoughts on FML: Behavior generation in the virtual human communication architecture. In: Proceedings of FML 2008, The First Functional Markup Language Workshop at AAMAS (2008)
10. Walton, R.E., Mckersie, R.B.: A behavioral theory of labor negotiations: An analysis of a social interaction system. McGraw-Hill, New York (1965)
11. Sillars, A.L., Coletti, S.F., Parry, D., Rogers, M.A.: Coding verbal conflict tactics: Nonverbal and perceptual correlates of the avoidance-distributive- integrative distinction. Human Communication Research 9(1), 83–95 (1982)
12. Richmond, V.P., McCroskey, J.C., Payne, S.K.: Nonverbal Behavior in Interpersonal Relations, 2nd edn. Prentice-Hall, Englewood Cliffs (1991)
13. Morris, D.: Bodytalk: The Meaning of Human Gestures. Crown Publishers (1990-1991)
14. Mehrabian, A.: Significance of posture and position in the communication of attitude and status relationships. Psychological Bulletin 71, 359–372 (1969)
15. Gratch, J., Marsella, S.: A domain-independent framework for modeling emotion. Journal of Cognitive Systems Research (2004)
16. Hartholt, A., Russ, T., Traum, D., Hovy, E., Robinson, S.: A common ground for virtual humans: Using an ontology in a natural language oriented virtual human architecture. In: Language Resources and Evaluation Conference (LREC) (May 2008)
17. Mancini, M., Pelachaud, C.: Dynamic behavior qualifiers for conversational agents. In: 7th International Conference on Intelligent Virtual Agents, Paris, France. Springer, Heidelberg (2007)
18. Kopp, S., Krenn, B., Marsella, S., Marshall, A., Pelachaud, C., Pirker, H., Thorisson, K., Vilhjlmsson, H.: Towards a common framework for multimodal generation in ECAs: The behavior markup language. In: Intelligent Virtual Agents, Marina del Rey, CA (2006)
19. Cavazza, M., Lugrin, J.L., Pizzi, D., Charles, F.: Madame bovary on the holodeck: Immersive interactive storytelling. In: ACM Multimedia, Augsburg, Germany (2007)
20. André, E., Rist, T., van Mulken, S., Klesen, M.: The automated design of believable dialogues for animated presentation teams. In: Cassell, J., Sullivan, J., Prevost, S., Churchill, E. (eds.) Embodied Conversational Agents, pp. 220–255. MIT Press, Cambridge (2000)
21. Kraus, S., Hoz-Weiss, P., Wilkenfeld, J., Andersen, D.R., Pate, A.: Resolving crises through automated bilateral negotiations. Artif. Intell. 172(1), 1–18 (2008)
22. Faratin, P., Sierra, C., Jennings, N.: Using similarity criteria to make issue tradeoffs in automated negotiations. Artificial Intelligence 142(2), 205–237 (2002)
23. Dijk, E.V., Kleef, G.A.V., Steinel, W., Beest, I.G.V.: A social functional approach to emotions in bargaining: When communicating anger pays and when it backfires. Journal of Personality and Social Psychology 94, 600–614 (2008)

A Granular Architecture for Dynamic Realtime Dialogue

Kristinn R. Thórisson and Gudny Ragna Jonsdottir

Center for Analysis & Design of Intelligent Agents
and School of Computer Science
Kringlunni 1, 103 Reykjavik, Iceland
{thorisson,gudny04}@ru.is

Abstract. We present a dialogue architecture that addresses perception, planning and execution of multimodal dialogue behavior. Motivated by realtime human performance and modular architectural principles, the architecture is full-duplex ("open-mic"); prosody is continuously analyzed and used for mixed-control turntaking behaviors (reactive and deliberative) and incremental utterance production. The architecture is fine-grain and highly expandable; we are currently applying it in more complex multimodal interaction and dynamic task environments. We describe here the theoretical underpinnings behind the architecture, compare it to prior efforts, discuss the methodology and give a brief overview of its current runtime characteristics.

Keywords: Architecture, turntaking, dialogue, realtime, incremental, planning, interaction, full-duplex, granular.

1 Introduction

Many researchers have pointed out the lack of implemented systems that can manage full-duplex ("open-microphone") dialogue (cf. [1,2,3]), that is, systems that can interrupt – and be interrupted – at any point in time, in a natural manner. As pointed out by Allen et al. [3], Moore [1] and others, much of the work in the field of dialogue over the last 2-3 decades has enforced strict turntaking between the system and the user, resulting in fairly unnatural, stilted dialogue. The challenge in building such systems lies, among other things, in the complexity of integration that needs to be done: Several complex systems, each composed of several complex subsystems – and those possibly going another level down – need to be combined in such a way as to produce coordinated action in light of complex multimodal input.

In this paper we describe work on building an architecture that can address many of the rich features of realtime multimodal dialogue. As many have pointed out (cf. [4]), a proper theory of turntaking should cover varied situations ranging from debates, to lectures, negotiations, task-oriented interactions, media interviews, dramatic performances, casual chats, formal meetings, task-oriented communication on a noisy factory floor, communication between the deaf, successful communication on the telephone with no multimodal information but plenty of paraverbal information, etc. While we do not claim to be close to any such comprehensive theory or model, our approach nods in the direction of methods, techniques and architectural features that promise to take us closer to such a comprehensive state.

Among the criticisms fielded by Moore [1] towards the present state of the art in dialogue systems are the brittleness of current approaches to both recognition and synthesis and lack of holistic integrative approaches. We address these with a two-prong approach: A more granular architecture that is easier to extend and manage than alternative approaches, and a more thorough methodology for building the architecture, growing out of prior work of one of the authors on similar topics [5]. While clearly not addressing all of the topics relevant to dialogue, the architecture has already shown itself to be highly expandable, in particular supporting complex modeling of turntaking [6,7]. The system has been successfully outfitted with learning capabilities for adjusting realtime behavior to match prosodical speech patterns of interlocutors [8]. In this paper we focus on the gross architecture and describe – from a 10k-foot viewpoint – the fundamental theoretical pillars of our approach.

2 Related Work

The present work builds directly on the Ymir architecture and framework [6], and the Ymir Turn-Taking Model (YTTM) [7]. Thórisson [6] presents the goals behind the Ymir architecture in 12 main points, chief among them being full-duplex, natural multimodal interaction, with natural response times (realtime). This includes also incremental interpretation and output generation, and the requirement that generation can progress in parallel with interpretation. Of the architectures built for spoken discourse, most have left one or more of the above constraints unaddressed, often focusing on isolated integration problems in dialogue such as concept-to-speech generation with proper intonation [9] and integration of dialogue modeling and speech generation [10].

Notable prior work has aimed to identify so-called "turn-constructional units" from a proposed set of candidates that has included the sentence [4,12], syllables [5], multimodal cues [4], phoneme timing [11] and semantics [4]. We take the view that turntaking is an *emergent* property of a large set of interacting systems [13]. Therefore, we consider attempts at modeling turntaking based on such "units of construction" a futile exercise. A less obvious results is that any attempt that shies away from addressing a substantial amount of the gross features and richness of dialogue up front may be doomed, as reductionist approaches are in general bad for studying highly emergent phenomena [13,14]. Thus, any attempt at building such integrative architectures in stages must directly address the challenges of expandability and incremental construction up front.

Allen et al.'s [3] system integrates a large number of functionalities in a comprehensive architecture that, while using a push-button interface for turn exchange, performs interpretation and output generation planning incrementally, in parallel. This supports user barge-in, with supported backtracking and other sophisticated dialogue handling. The architecture goes beyond the Gandalf system [6] in content interpretation; although Gandalf could interpret gestures, body language and prosody incrementally the content of speech was done "batch-processing style" (caused by limitations in the speech recognition technology used). Similar to Thórisson [6,7], the authors argue for a separation of dialogue skills and topic skills.

Raux [2] presents a broad dialogue-capable architecture, extending ideas from Ymir [7] with more extensive dialogue management techniques. The architecture is blackboard-based like Ymir with two rather large-grain modules: An Interaction Manager, bridging between high-level and lower-level system components, controlling directly reactive behaviors (e.g. back-channel feedback), and a Dialogue Manager that plans the contributions to the conversation. The former corresponds roughly to the turntaking mechanisms in Ymir, the latter to Knowledge Bases in the Content Layer (like Ymir, the DM makes no prescriptions for the particular technology used for utterance planning). The work demonstrates the flexibility of the blackboard approach for building mixed-granularity architectures [15]. Although the resulting system is comparable to Ymir at the high level, it proposes different solutions for integrating reactive behaviors, turntaking and dialogue state, while promising to handle a wider range of dialogue styles and phenomena. However, as no comparative evaluation is provided it is difficult to assess the benefits of the alternative approach. As it is based on more coarse-grain components, we would expect it to be less expandable than the finer-grain approach we have chosen.

3 Theoretical Underpinnings

In addition to an underlying theory of turntaking in multimodal realtime dialogue, outlined in [7], our approach rests on three main theoretical pillars. The first is motivated by arguments from – or rather a critique of – the standard scientific reductionist method, the second by architectural methodology and the third by data and models from psychological studies.

We view embodied dialogue as a heterogeneous, large, densely-coupled system (HeLD), identifying dialogue in a class of *complex systems,* in the sense of Simon [14] and his concept of *near decomposability*. The main arguments for this view have been presented in [13] and [16]. HeLD systems embody and express emergent properties that have been difficult to understand without resorting to large, detailed computational models built to relatively high levels of fidelity. Without the ability to experiment with changes and modifications to the architecture at various levels of detail we cannot differentiate between a large set of models that, on paper, look like they might all work. The conclusion can only be that models of dialogue produced by a standard divide-and-conquer approach can only address a subset of a system's behaviors (and are even quite possibly doomed at the outset). This view is echoed in some recent work on dialogue architectures (cf. [1]). This, however, may seem to present an impossible difficulty; creating even *approximately* complete models of dialogue – ones that address multiple modes, prosody, realtime content generation, etc. – may seem to entail an insurmountable effort. Counterintuitively, for most complex systems, however, if we attempt to take *all* of the most significant behaviors of the system into account, the set of possible contributing underlying mechanisms will be greatly reduced [17] – quite possibly to a small finite set (while initial formulation of plausible mechanisms may be harder, because the constraints of the work are greater up front, the search is now over a manageable set of possibilities). One way to build satisfactory models of dialogue is to bring results from a number of disciplines to the table, at various levels of abstraction. It is the use of levels of abstraction that is especially important for cognitive phenomena: Use of hierarchical approaches is common in other scientific fields such as physics; for example, behind models of optics lie more detailed models of electromagnetic waves [18].

Following this view we have built our present model using a methodology that helps us create complex multi-component systems at a fairly high fidelity without losing control of the development process, the Constructionist Design Methodology (CDM) [5]. Many of the extant methodologies that have been offered in the area of agent-based simulation and modeling suffer from lack of actual use case experiences, especially for artificial intelligence projects that involve construction of single-mind systems. CDM's 9 iterative principles (semi-independent steps) have already been applied in the construction of several systems, both for robots and virtual agents (cf. [5,15,19]). Our approach has followed it fairly closely.

When dealing with HeLDs we must try to constrain the possible design space. A powerful way to do this is to build multilevel representations (cf. [17,20,21]); this may in fact be the only way to get our models right when trying to understand natural HeLDs. Notice that the thrust of the argument is not that multiple levels are "valid" or even "important", as that is a commonly accepted view in science and philosophy, but rather that to map correctly to the many ways subsystems interact in HeLDs the multiple levels are a *critical necessity*: Without simulations at fairly high levels of fidelity we cannot expect manipulations to the architecture (at various levels of detail) to produce valid results. Without this ability we cannot select from a large set of candidate approaches that, on paper, look like they might all work. We have tried to do this using data from the psychological literature to constrain the design achieved, most significantly temporal characteristics. As we are in the beginning stages of building such multi-level-of-detail models, work remains in this respect. However, the effort has already helped with the construction; Bonaiuto and Thórisson's work [22] on systems that mutually learn to perceive and produce multimodal turntaking cues is based on the turntaking mechanisms described here, built using neurally plausible models of planning which are modeled after brain research on Macaques.

4 The Architecture

The architecture is composed of two main functional clusters. First, a turntaking (TT) system that is able to support realtime turntaking [8], built such that it can interface with external systems in a disciplined way. The second cluster manages (a) continuous assessment of dialogue state, (b) the internal drive for delivering content, (c) and planning for future actions, including utterances and multimodal behavior.

Following the CDM [5] we started at the low level with several perception modules for extracting prosodical information from a person's speech (pitch, silences, speaking volume and compounds of these). Then we expanded the system with a set of control/decision modules that, based on the perceptual processing output of the perception modules, decide which turntaking context was most likely (I-Have-Turn, Other-Has-Turn, etc.). The design process is described further in [23]. Last but not least, we make extensive use of *contexts* – semi-global states that determine which modules are active when [15], allowing us to control large groups of modules in the architecture as one unit, improving management; they also allow us to better view the runtime operation of the full system and more carefully control CPU load.

We will now give a short overview of some of the key architectural components, some of which are depicted in Figure 1. *Prosody Tracker* (low-level module): Analyses the pitch, pitch derivative, speech on/off, silences and hums in a continuous manner. *Prosody Analyzer* (mid-level): Analyzes data from the PT to identify speech overlap and silences (also quantizes continuous pitch for efficiency). *Speech Recognizer* (high-level): Produces semi-continuous text from continuous audio signal [8]. *Interpreters* (high-level): A group of perception/interpretation modules that take in output from a single SR; each having its specified identification task, e.g. finding nouns, dates, fillers, etc. *Interpretation Director* (high-level): Receives input from Content Planner (see below) on what to look for at any point in time and analyzes the output of all Interpreters based on that information [8]. *Turntaking System* (TT) (low-level): The TT consists of unimodal/multimodal preceptors and deciders; its role is to maintain a coupling between the interlocutors of turn "state", or *disposition – a grouping of goals and expectations, in the form of perceptor and decider activations*. Based on it, all content-unaware decisions and actions related to the delivery and interpretation of speech acts can be coordinated. Global states (contexts) prescribe which perceptions and decisions are appropriate at any point in time, e.g. whether to expect a certain turntaking cue, whether it is relevant to generate a particular behavior on a particular cue, etc. *Dialogue Planner* (high-level): Is responsible for delivering the next "thought unit", embodied as a short segment of speech spanning roughly 1-2 second of delivery time, on average, when it is available from the CP, producing fillers and gracefully giving turn when content is not generated within a set time frame. The DP contains a motivation-to-speak, which currently is linked directly to have-something-to-say. *Content Planner* (high-level): Decides *what* to say based on inner goals and information from the ID. *Speech Synthesizer* (mid-level): Takes commands from CP and TT to start/stop speech, raise or lower speech volume. *Learner* (mid-level): Computes a decision strategy incrementally while learning on-line, and publishes it to the Other-Gives-Turn-Decider-2, which determines how long to wait until starting to speak as a silence is detected, using as an indication the prosody of the last 300 msec of the other's utterance right before the silence [8].

The Dialogue Planner and Learning module share control of turntaking, even though they might be considered to be outside of the TT system proper; typically affecting the contexts I-Want-Turn, I-Accept-Turn and I-Give-Turn. If the Content Planner has some content ready to be communicated – irrespective of what the perceptions inform the TT system that the coupled dialogue context is – the agent can signal that it wants turn via this "deliberate" route; it can also signal I-Give-Turn when content queue is empty – the TT system will not be forced to handle this as our theory suggests it should be kept content-unaware. These decisions made by the DP typically override decisions made in the reactive level.

Currently the system is implemented on three workstations, running within the Psyclone framework [15]. The distribution of modules across the three machines has been carefully tuned by hand to achieve sufficient processing and transmission speeds to support the realtime running of the system. The system has been tested both in simulated interactions with itself and in interactions with people (part of this data is published in [8]). Table 1 summarizes key operating characteristics in terms of message passing times.

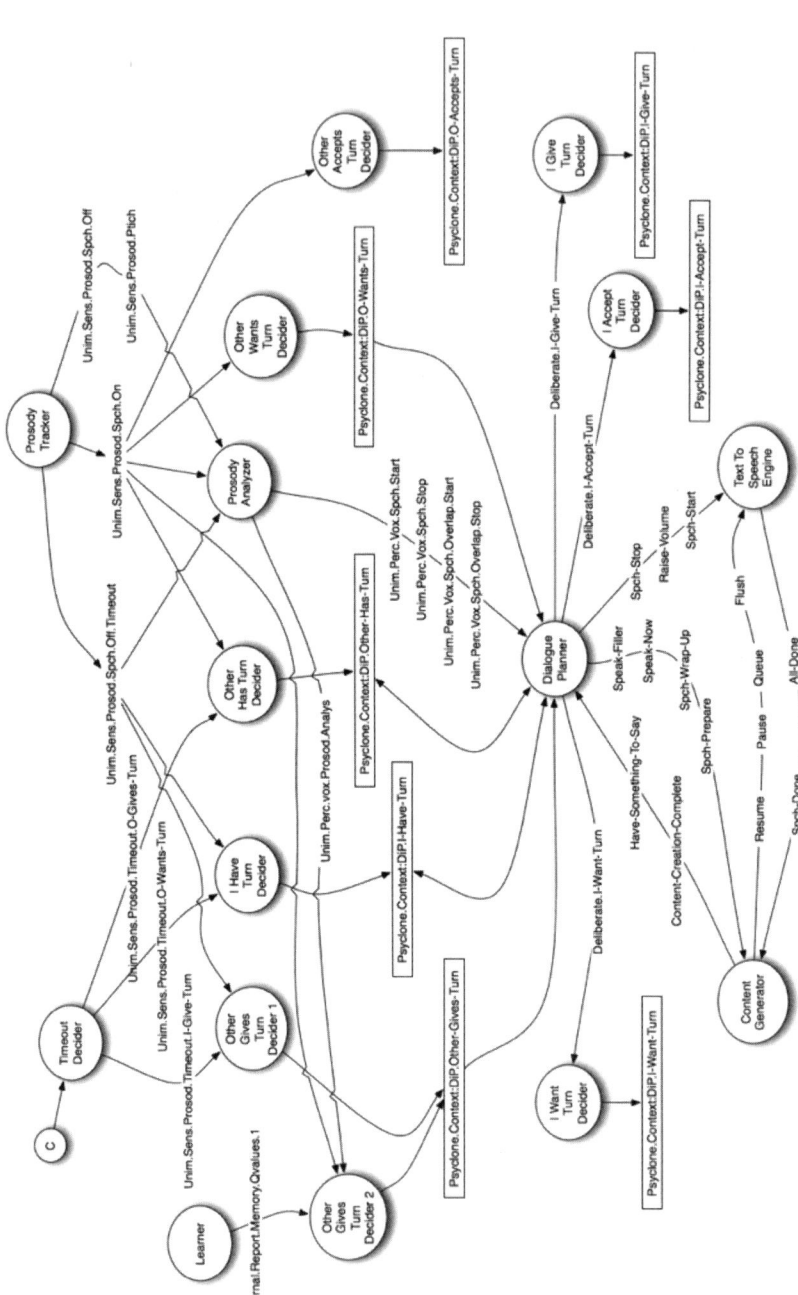

Fig. 1. Partial view of architectural components, routhing and message types. Messages are given names in the order from the most significant to least significant descriptor of the message's contents [15]

Table 1. Temporal characteristics of selected message passing at runtime. Time shows how long, in msec, each type of message takes to be transmitted between sender and receiver. (Resolution = 16 msec.)

Message Type	Receiver	Ave	Min	Max
Deliberate.I-Accept-Turn	I-accept-turn	107	97	207
Psyclone.Context:DiP.I-Accept-Turn	I-have-turn	39	15	63
Unim.Sens.Prosod.Spch.Off.Timeout	I-have-turn	4	0	16
Psyclone.Context:DiP.I-Give-Turn	Other-accepts-turn	59	4	176
Unim.Sens.Prosod.Spch.On	Other-accepts-turn	9	0	47
Unim.Sens.Prosod.Spch.On	Other-gives-turn-2	20	0	297
Internal.Report.Memory.Qvalues	Other-gives-turn-2	18	15	47
Unim.Perc.vox.Prosod.Analys	Other-gives-turn-2	82	62	469
Psyclone.Context:DiP.I-Want-Turn	Other-gives-turn-1	53	3	207
Unim.Sens.Prosod.Spch.Off.Timeout	Other-gives-turn-1	7	0	31
Unim.Sens.Prosod.Timeout.I-Give-Turn	Other-gives-turn-1	31	0	47
Psyclone.Context:DiP.Other-Accepts-Turn	Other-gives-turn-1	87	16	329
Psyclone.Context:DiP.Other-Accepts-Turn	Other-has-turn	39	0	203
Unim.Sens.Prosod.Spch.On	Other-has-turn	18	0	110
Unim.Sens.Prosod.Spch.On	Other-wants-turn	5	0	16
Unim.Sens.Prosod.Spch.Off	ProsodyAnalyzer	73	62	110
Unim.Sens.Prosod.Spch.Pitch	ProsodyAnalyzer	6	0	31
Internal.Instruct.Spch.Prepare	ContentGenerator	4	0	15
Internal.Instruct.Spch.Start	ContentGenerator	2	0	16
Output.Plan.Task.Spch.Done	ContentGenerator	4	0	16
Output.Plan.Task.Speak.Now	ContentGenerator	2	0	15
Output.Plan.Task.Spch.All-Done	ContentGenerator	0	0	0
Internal.Content-Creation-Complete	DialoguePlanner	19	0	32
Have-Something-To-Say	DialoguePlanner	25	0	47

Acknowledgments. This work was supported in part by a research grant from RANNÍS, Iceland, and by a Marie Curie European Reintegration Grant within the 6th European Community Framework Programme. The authors would like to thank Eric Nivel for his work on the *Prosodica* prosody tracker and Yngvi Björnsson for his contributions to the learning methods.

References

1. Moore, R.K.: PRESENCE: A Human-Inspired Architecture for Speech-Based Human-Machine Interaction. IEEE Transactions on Computers 56(9), 1176–1188 (2007)
2. Raux, A., Eskenazi, M.: A Multi-Layer Architecture for Semi-Synchronous Event-Driven Dialogue Management, ASRU, Japan, 514–519 (2007)
3. Allen, J., Ferguson, G., Stent, A.: An Architecture for More Realistic Conversational Systems. In: IUI, pp. 14–17. ACM Press, Santa Fe (2001)
4. O'Connell, D.C., Kowal, S., Kaltenbacher, E.: Turn-Taking: A Critical Analysis of the Research Tradition. Journal of Psycholinguistic Research 19(6), 345–373 (1990)
5. Thórisson, K.R., Benko, H., Arnold, A., Abramov, D., Maskey, S., Vaseekaran, A.: Constructionist Design Methodology for Interactive Intelligences. A.I. Magazine. American Association for Artificial Intelligence 25(4), 77–90 (2004)

6. Thórisson, K.R.: A Mind Model for Multimodal Communicative Creatures and Humanoids. International J. Appl. Artif. Intell. 13(4-5), 449–486 (1999)
7. Thórisson, K.R.: Natural Turn-Taking Needs No Manual: Computational Theory and Model, from Perception to Action. In: Granström, B., House, D., Karlsson, I. (eds.) Multimodality in Language and Speech Systems, pp. 173–207. Kluwer Academic Publishers, Dordrecht (2002)
8. Jonsdottir, G.R., Thórisson, K.R., Nivel, E.: Learning Smooth, Human-Like Turntaking in Realtime Dialogue. In: Prendinger, H., Lester, J., Ishizuka, M. (eds.) IVA 2008. LNCS (LNAI), vol. 5208, Springer, Heidelberg (2008)
9. Pan, S., McKeown, K.R.: Integrating Language Generation with Speech Synthesis in a Concept to Speech System. In: Proceedings of the ACL Workshop on Concept to Speech Generation Systems. ACL/EACL (1997)
10. Grote, B., Hagen, E., Teich, E.: Matchmaking: Dialogue Modeling and Speech Generation Meet. In: Proceedings of the 1996 International Workshop on Natural Language Generation, Herstmonceux, England, pp. 171–180 (1996)
11. Wilson, M., Wilson, T.P.: An oscillator model of the timing of turn-taking. Psychonomic Bulletin and Review 12(6), 957–968 (2005)
12. Sacks, H., Schegloff, E.A., Jefferson, G.A.: A Simplest Systematics for the Organization of Turn-Taking in Conversation. Language 50, 696–735 (1974)
13. Thórisson, K.R.: Modeling Multimodal Communication as a Complex System. In: Wachsmuth, I., Lenzen, M., Knoblich, G. (eds.) Modeling Communication with Robots and Virtual Humans, pp. 143–168. Springer, New York (2008)
14. Simon, H.A.: Can there be a science of complex systems? In: Bar-Yam, Y. (ed.) Unifying themes in complex systems: Proceedings from the International Conference on Complex Systems, pp. 4–14. Perseus Press, Cambridge (1999)
15. Thórisson, K.R., List, T., Pennock, C., DiPirro, J.: Whiteboards: Scheduling Blackboards for Semantic Routing of Messages & Streams. Proceedings of AAAI 2005, AAAI Technical Report WS-05-08, 8-15 (2005)
16. Thórisson, K.R.: Integrated A.I. Systems. Minds & Machines 17, 11–25 (2007)
17. Scwabacher, M., Gelsey, A.: Multi-Level Simulation and Numerical Optimization of Complex Engineering Designs. In: 6th AIAA/NASA/USAF Multidisciplinary Analysis & Optmization Symposium, Bellevue, WA, AIAA-96-4021 (1996)
18. Schaffner, K.F.: Reduction: the Cheshire cat problem and a return to roots. Synthese 151(3), 377–402 (2006)
19. Ng-Thow-Hing, V., List, T., Thórisson, K.R., Lim, J., Wormer, J.: Design and Evaluation of Communication Middleware in a Distributed Humanoid Robot Architecture. In: IROS 2007 Workshop Measures and Procedures for the Evaluation of Robot Architectures and Middleware, San Diego, California, 29 October - 2 November (2007)
20. Gaud, N., Gechter, F., Galland, S., Koukam, A.: Holonic Multiagent Multilevel Simulation Application to Real-time Pedestrians Simulation in Urban Environment. In: Proceedings of IJCAI 2007, pp. 1275–1280 (2007)
21. Arbib, M.A.: Levels of Modeling of Visually Guided Behavior (with peer commentary and author's response). Behavioral and Brain Sciences 10, 407–465 (1987)
22. Bonaiuto, J., Thórisson, K.R.: Towards a Neurocognitive Model of Realtime Turntaking in Face-to-Face Dialogue. In: Wachsmuth, I., Lenzen, M., Knoblich, G. (eds.) Embodied Communication in Humans and Machines. Oxford University Press, U.K (2008)
23. Thórisson, K.R., Jonsdottir, G.R., Nivel, E.: Methods for Complex Single-Mind Architecture Designs. In: Proc. AAMAS, Portugal (June 2008)

Generating Dialogues for Virtual Agents Using Nested Textual Coherence Relations

Hugo Hernault[1,3], Paul Piwek[2], Helmut Prendinger[1], and Mitsuru Ishizuka[3]

[1] National Institute of Informatics
2-1-2 Hitotsubashi, Chiyoda-ku, Tokyo 101-8430, Japan
hugo@nii.ac.jp, helmut@nii.ac.jp
[2] NLG Group, Centre for Research in Computing
The Open University, Walton Hall, Milton Keynes MK7 6AA, UK
p.piwek@open.ac.uk
[3] Graduate School of Information Science and Technology, The University of Tokyo
7-3-1 Hongo, Bunkyo-ku, Tokyo 113-8656, Japan
ishizuka@i.u-tokyo.ac.jp

Abstract. This paper describes recent advances on the Text2Dialogue system we are currently developing. Our system enables automatic transformation of monological text into a dialogue. The dialogue is then "acted out" by virtual agents, using synthetic speech and gestures. In this paper, we focus on the monologue-to-dialogue transformation, and describe how it uses textual coherence relations to map text segments to query–answer pairs between an expert and a layman agent. By creating mapping rules for a few well-selected relations, we can produce coherent dialogues with proper assignment of turns for the speakers in a majority of cases.

1 Introduction

The task of automatically generating dialogues is a crucial step for the wide dissemination of agent-based multimodal presentations. Previous research has aimed at generating dialogues using various inputs. For example, in the systems described in [1] and [2], dialogues are generated from structured data (such as product databases and story plans, respectively).

However, the ever-growing collection of textual information found on the Web makes text an input of choice given its quantity, its availability, and the diversity of its content. Early work on dialogue based on text can be found in the automatic information presenters Web2TV and Web2TalkShow [3], where the output dialogue is intended to be humorous (exaggerated and distorted).

In our approach, the Text2Dialogue system based on [4], we want to implement the following three features:

- The input is plain, un-formatted text rather than a knowledge base.
- The transformation is meaning-preserving rather than intentionally humorous.
- The output is dialogue which is specified in a format that can drive the performance of a team of computer-animated virtual agents.

The two interlocutors performing the dialogue are rendered as virtual agents, and assume the roles of expert (e.g. instructor) and layman (e.g. student). Presenting information as a dialogue is a popular means of conveying information. It can be witnessed in commercials, edutainment, and even video games.

The Multimodal Presentation Markup Language MPML3D [5], based on MPML [6], is used to control the verbal and non-verbal behavior of the agent characters. Non-verbal behavior includes body gestures, posture, and eye gaze.

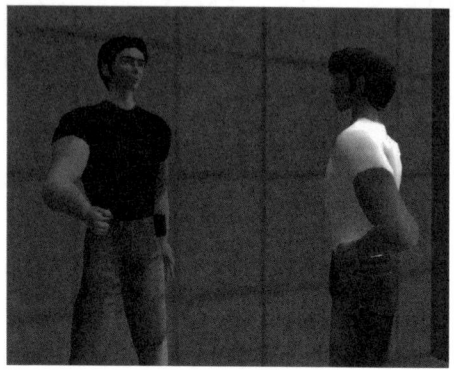

Fig. 1. Two virtual agents having a dialogue in the "Second Life" 3D online virtual world

Fig. 1 shows two MPML3D-controlled agents conversing in "Second Life", a 3D online virtual world.

In brief, there are two reasons why our approach is worth exploring. Firstly, it enables users – as content creators – to conveniently generate powerful multimodal presentations, and secondly, the system enables users – as content consumers – to "learn by observation" (vicarious learning). Studies such as those described in [7] have shown that dialogues convey information efficiently, and that one can learn effectively by watching them.

This paper presents a significant extension of previously published work [4] in that it introduces a method for generating dialogues from not just flat but also nested coherence relations in text.

The rest of the paper is organized as follows: The next section gives an overview of how the components of our system work. We describe the mechanisms used in the process of generating dialogues. Then, we address the issue of assigning turns to speakers. Finally, we briefly report on the results of a preliminary evaluation of the characteristics and quality of our generated dialogues.

2 The Text2Dialogue System

2.1 Text Analysis Component

First, our system uses a discourse analyzer to determine the coherence relations underlying the input text, i.e., the text's discourse or rhetorical structure. Compatible parsers are DiscourseAna [8] and SPADE [9]. Those parsers analyze the input text in terms of a specific theory of coherence relations known as Rhetorical Structure Theory (RST) [10]. In RST, a text is segmented into non-overlapping spans, connected by arrows that indicate the rhetorical (i.e., coherence) relations between the different spans. The endpoint of the arrow represents the "nucleus" (the more important argument of the relation) whilst the origin represents the "satellite" (containing less important information).[1]

Here is a text snippet followed by its associated RST representation:

[Mexico exported an average of 1,296,800 barrels of crude oil a day at an average of $15.31 a barrel during 1989's first eight months for a total of $4.82 billion,]^{1A} [Petroleos Mexicanos S.A. said.]^{1B} [The state petroleum monopoly said]^{1C} [sales in the period gained 15%, and $262.4 million more than originally projected at an average of $10 a barrel on an export platform of 1,250,000 barrels a day.]^{1D} (wsj_1104 from the RST Discourse Treebank)

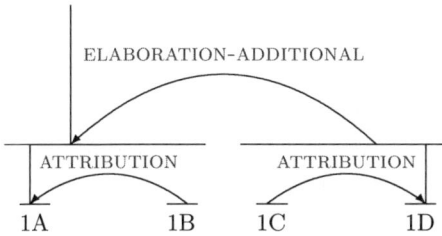

Fig. 2. RST representation of example text

2.2 Mapper Component

For each rhetorical relation, we define one or more mapping rules that determine how to generate a corresponding fragment of dialogue. For the ATTRIBUTION relation, one of the most common relations making up 13.87% of all relations in the RST Treebank [11], we created the following rule:

MAPPING RULE: **Attribution, principal rule**

ATTRIBUTION$(P, Q) \implies$
L: *What did* + *getSubject(P+Q)* +
     getMainVerbLemma(P+Q)?
E: *AddIfNotPresentIn(Q, That)* + *Q*

An example dialogue generated from the sentence ATTRIBUTION("John said", "he likes apples.") would be: "L: What did John say? E: That he likes apples."

In an RST structure, relations are nested most of the time, and we have to find systematic rules to decide how and when to perform a mapping. For instance, ATTRIBUTION takes plain text for P and Q, and therefore the mapping rule given above cannot be applied if there are other mapping rules that need to be applied to the rhetorical relations underlying P or Q. Thus, we define the notion of *relation-incompatibility* to express that certain mapping rules can not be applied at the same time:

---

[1] RST also has multi-nuclear relations, i.e., relations in which none of the arguments is more prominent than any of the other.

RELATION-INCOMPATIBILITY: We say that a relation R_1 is incompatible with a relation R_2 if, when R_1 is a child of R_2 in a RST tree, the mapping rules for R_2 cannot be applied.

There are more than 50 rhetorical relations [11]. Some of them, such as ELABORATION-ADDITIONAL and LIST are very common, whereas others (such as TOPIC-SHIFT or ANALOGY) are more specific and occur rarely. Defining mapping rules for all relations would be a time consuming task. Instead we need a way to classify and compare relations, in order to decide which are more essential for the dialogue. Therefore, we define the notion of *relation importance*:

RELATION IMPORTANCE: We use a function $Imp : \mathcal{R} \rightarrow \mathbb{N}$ that associates a natural number to all relations. Unimplemented relations are given the importance 0.

This allows us to compare relations based on their importance, and enables us to filter out non-important relations: unimplemented and secondary relations will not be mapped as they have the lowest importance. Depending on the domain of application of our system, we might want to emphasize specific relations. Setting a high importance for these relations will ensure that they are always mapped. Finally, this mechanism allows us to tune the verbosity of the system. For instance, when all relations are given the same priority, they will all be mapped to query–answer pairs.

With the notions of relation importance and relation compatibility defined, we are now in a position to describe the mapping algorithm of our system. Our dialogue mapper parses the RST tree in a top-down fashion and maps the text segments into question-answer pairs. At each recursive call, we decide to perform mapping of a child relation if

1. The child is more or as important as its parent relation, and
2. The child is not incompatible with its ancestors, and
3. The child relation is implemented (i.e., has at least one mapping rule).

Else, if no mapping is performed, the child relation's subtree will be flattened as a single line of text and spoken by the expert.

The algorithm, when applied to the tree represented in Fig. 2, returns:

a. L: What did Petroleos Mexicanos S.A. say?
b. E: That Mexico exported an average of 1,296,800 barrels
c.   of crude oil a day at an average of $15.31 a barrel
d.   during 1989's first eight months for a total of $4.82 billion.
e. E: Should I tell you more?
f. L: Yes, please.
g. L: What did the state petroleum monopoly say?
h. E: That sales in the period gained 15%, and $262.4 million
i.   more than originally projected at an average of $10 a barrel
j.   on an export platform of 1,250,000 barrels a day.

2.3 Assigning Turns to Speakers

In the previous dialogue, we can notice some discrepancies due to improper turn assignment between the two interlocutors (see the transition from line f. to g.). Indeed, when the mapping produced by a relation ends with a certain agent asking a question, the other agent is expected to produce an answer. A solution is to define an alternate-speaker rule for each relation. For instance, our alternate-speaker rule for the ATTRIBUTION relation is:

MAPPING RULE: **Attribution, alternate-speaker rule**

ATTRIBUTION$(P, Q) \Longrightarrow$
E: *RemoveIfPresentIn(Q, That) + Q*
L: *Who getMainVerbFromSentence(P+Q) + that?*
E: *getSubjectFromSentence(P+Q) +*
   generateWordForm(do, getMainVerbMorphoTagsFromSentence(P+Q))

The problematic passage in the previous dialogue is now resolved as:

f. L: Yes, please.
g. E: Sales in the period gained 15%, and $262.4 million
h.   more than originally projected at an average of $10 a barrel
i.   on an export platform of 1,250,000 barrels a day.
j. L: Who said that?
k. E: The state petroleum monopoly did.

2.4 Preliminary Evaluation

To evaluate our system's "Mapper Component", we applied it to RST-hand-annotated texts from the RST Treebank [11] and collected statistics. We implemented mapping rules for 3 of the most common relations: ELABORATION-ADDITIONAL (2 rules), ATTRIBUTION (2 rules), CIRCUMSTANCE (1 rule). We gave the same importance to all 3 relations. We ran the system on all 347 annotated newspapers articles contained in the RST Treebank, and obtained the following results:

31.8% of relations in the text are mapped. This is almost equal to the cumulated occurence rate of the 3 implemented relations, which make up 33.3% of all relations. The relatively small difference can be explained by the fact that nested relations sometimes lead to incompatibilities that prevent mapping.

The mean length of a turn in the resulting dialogues is 13.28 words (SD=32.28, median=5.0). For the layman specifically, the mean length of a turn is 2.71 words (SD=1.42, median=2.0). For the expert, the mean length of a turn is 18.88 words (SD=38.77, median=12.0). The dialogues generally have a reasonable size, and the layman's turns are short because they typically consist of sentences such as "Yes, please." or "What else?". Looking at the number of turns per dialogue, we found that the mean was 46.98 (SD=43.39, median=32.0). In summary, on average each text was rendered as a dialogue consisting of a good number of turns of a reasonable length.

In order to also evaluate the quality of the dialogues, we manually analyzed a sample of randomly selected generated dialogues. On 100 extracts of our generated dialogues, the first author evaluated their quality based on 1) whether the dialogues were syntactically sound, 2) whether they conveyed the meaning of the original text, and 3) whether they made sense when taken in context of the surrounding dialogue. The results were: 87% of our dialogues fulfilled 1), 75% fulfilled 1) and 2), and 60% satisfied 1), 2), and 3).

3 Conclusions

We have presented the latest advances on our Text2Dialogue system. Using abstract mapping rules and an algorithm based on relation incompatibility and relation importance, we can automatically generate dialogues from text. While the first version of the system [4] was confined to non-nested relations, the version described in this paper can also handle nested rhetorical structures in text. Our algorithm deals with incompatible nested relations by using a preference ordering on relations to decide which relation to realize. The problem of properly assigning turns to speakers has been addressed using alternate-speaker rules for each relation. Our evaluation of the latest prototype indicates that the proposed algorithm results in reasonably organized dialogue for a majority of the test inputs.

In the future, we will be working on (1) implementing more mapping rules for frequently occurring relations, (2) improving the syntax of the generated dialogues, and (3) investigating how to improve the coherence of the generated dialogues by considering context.

Acknowledgements

The first author was supported by an International Internship Grant from NII under a Memorandum of Understanding with the Institut National Polytechnique de Toulouse, and a "Strategic Project" Grant from NII.

References

1. André, E., Rist, T., van Mulken, S., Klesen, M., Baldes, S.: The automated design of believable dialogues for animated presentation teams. In: Embodied Conversational Agents, pp. 220–255. MIT Press, Cambridge (2000)
2. Cavazza, M., Charles, F.: Dialogue generation in character-based interactive storytelling. In: Proceedings of the AAAI First Annual Artificial Intelligence and Interactive Digital Entertainment Conference, Marina Del Rey, California, USA (2005)
3. Nadamoto, A., Tanaka, K.: Complementing your TV-viewing by web content automatically-transformed into TV-program-type content. In: Procs. 13th Annual ACM Intl. Conf. on Multimedia, pp. 41–50. ACM Press, New York (2005)

4. Piwek, P., Hernault, H., Prendinger, H., Ishizuka, M.: T2D: Generating dialogues between virtual agents automatically from text. In: Pélachaud, C., Martin, J.-C., André, E., Chollet, G., Karpouzis, K., Pelé, D. (eds.) IVA 2007. LNCS (LNAI), vol. 4722, pp. 161–174. Springer, Heidelberg (2007)
5. Nischt, M., Prendinger, H., André, E., Ishizuka, M.: MPML3D: A reactive framework for the Multimodal Presentation Markup Language. In: Gratch, J., Young, M., Aylett, R.S., Ballin, D., Olivier, P. (eds.) IVA 2006. LNCS (LNAI), vol. 4133, pp. 218–229. Springer, Heidelberg (2006)
6. Prendinger, H., Descamps, S., Ishizuka, M.: MPML: A markup language for controlling the behavior of life-like characters. Journal of Visual Languages and Computing 15(2), 183–203 (2004)
7. Craig, S., Gholson, B., Ventura, M., Graesser, A.: the Tutoring Research Group: Overhearing dialogues and monologues in virtual tutoring sessions. Intl. Journal of Artificial Intelligence in Education 11, 242–253 (2000)
8. Le, H.T., Abeysinghe, G.: A study to improve the efficiency of a discourse parsing system. In: Gelbukh, A. (ed.) CICLing 2003. LNCS, vol. 2588, pp. 101–114. Springer, Heidelberg (2003)
9. Soricut, R., Marcu, D.: Sentence level discourse parsing using syntactic and lexical information. In: Procs. HLT/NAACL 2003, Edmonton, Canada (2003)
10. Mann, W.C., Thompson, S.A.: Rhetorical structure theory: Toward a functional theory of text organization. Text 8(3), 243–281 (1988)
11. Carlson, L., Marcu, D.: Discourse tagging reference manual. Technical Report ISI-TR-545, ISI (September 2001)

ns and Dialogue in a Lifestyle Agent

Cameron Smith[1], Marc Cavazza[1], Daniel Charlton[1], Li Zhang[1],
Markku Turunen[2], and Jaakko Hakulinen[2]

[1] School of Computing, University of Teesside, Middlesbrough, United Kingdom
{c.g.smith,m.o.cavazza,d.charlton,l.zhang}@tees.ac.uk
[2] Department of Computer Sciences, Speech-based and Pervasive Interaction Group,
University of Tampere, Finland
{Markku.Turunen,Jaakko.Hakulinen}@cs.uta.fi

Abstract. In this paper, we describe an Embodied Conversational Agent advising users to promote a healthier lifestyle. This embodied agent provides advice on everyday user activities, in order to promote a healthy lifestyle. It operates by generating user activity models (similar to decompositional task models), using a Hierarchical Task Network (HTN) planner. These activity models are refined through various cycles of planning and dialogue, during which the agent suggests possible activities to the user, and the user expresses her preferences in return. A first prototype has been fully implemented (as a spoken dialogue system) and tested with 20 subjects. Early results show a high level of task completion despite the word error rate, and further potential for improvement.

Keywords: Multimodal interaction with intelligent virtual agents, Embodied Cognitive Modelling, Conversational and non-verbal behavior.

1 Introduction

Intelligent Virtual Agents are a privileged application for the integration of conversational systems and reasoning technologies. This became especially true since the development of a new generation of knowledge-intensive applications, such as collaborative problem solving, advisory systems, tutoring systems, etc. More recently, there has been a growing interest in assistive systems and health advisors providing guidance on daily activities or well-being.

In this paper, we describe the first integrated prototype of the Health and Fitness Companion, an Embodied Conversational Agent serving as a lifestyle assistant, which helps the user in adapting her daily activities to follow a healthier lifestyle. In order to fulfil this role, this agent integrates cognitive and communication abilities.

The system presents itself as an embodied conversational agent based on the Nabaztag™ ubiquitous computing device[1], which is a wireless rabbit, originally designed as a domestic interface to the Internet. Such a system needs to support flexible dialogue, while being able to reason on user activities and possess appropriate knowledge in relevant domains (daily activities, food and nutrition, basic exercise

---

[1] www.nabaztag.com

physiology) while building detailed activity models. One central idea for such a Companion is to blend into the user's everyday life without being either too directive or too intrusive. Our proposal to achieve this is to let the Companion act as an embodied agent at home, so that the user will only engage in conversation when planning her daily activities, not when actually carrying them out. In a similar fashion, the object of the dialogues between the user and the Companion should be the activities themselves, not any direct goals that would be perceived as the Companion's.

2 Reasoning on User Activities with HTN Planning

Since our lifestyle agent (the Health and Fitness Companion, henceforth HFC) should assist the user in planning her daily activities, an appropriate way to reason on the user's tasks is to be able to generate them using a task-based planning system.

We use a HTN Planner to generate an activity model for the user, in the form of a solution AND/OR graph, which constitutes a decomposition of typical user activities. The Planner uses standard decomposition algorithms [8] for HTN with several extensions to the traditional HTN representations [12] as well as the attachment of semantic features to these methods. Semantic features can represent domain application constraints (weather conditions, financial cost, time available …) and are handled via semantic evaluation rules.

In previous applications which have integrated planning and dialogue, user interaction was most often centred on collaborative planning, or on real-time response to an evolving situation that required planning adaptation or re-planning. Here, Planning is used to establish a model of user daily activities, with an emphasis on lifestyle, which manifests itself through system suggestions when planning user activities ("why don't you cycle to work?"). This overall strategy has been preferred over explicit advice ("you should exercise more") or negotiation ("you can eat takeaway food if you go to the gym twice this week").

There are thus two different dialogue phases. One establishes the user activity model and is similar to joint plan construction (Sample Dialogue in Fig. 1). The other consists in reporting about the activities that actually took place during the day, and is using the explicit activity model for assessment.

System: Good Morning! Do you have anything special planned for today?
User: No, not really. I'm just going to work.
System: Alright. Why don't you cycle to work today?
User: No, it is raining so I'd rather take the train.
System: Okay, so you'll take the train to work?
User: Yes, thank you.

Fig. 1. Nabaztag^{TM} with Sample Dialogue

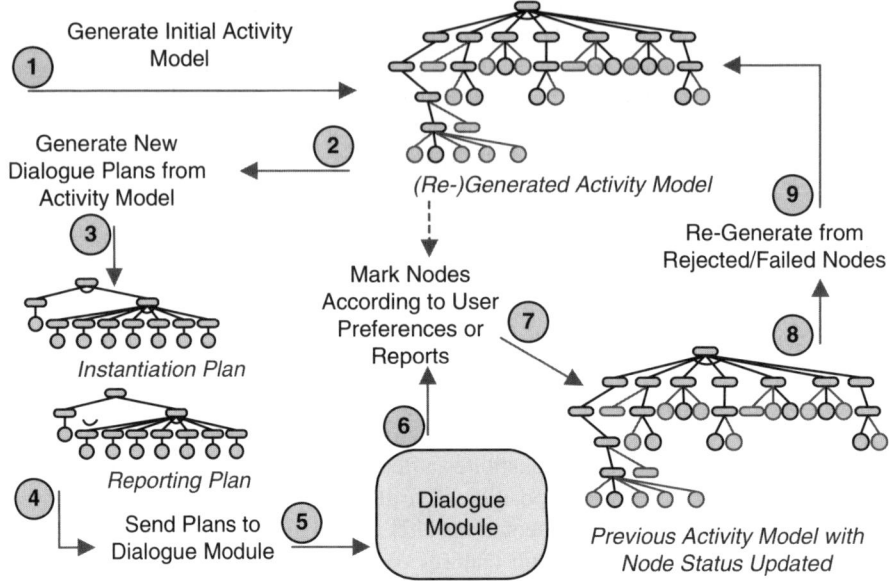

Fig. 2. Activity Model Update Cycle

The system operates through a global interaction cycle, which integrates dialogue and planning for the construction of the activity model (Fig. 2). The initial step consists in generating an initial activity model for the user, based on default knowledge, as well as any persistent information on user preferences acquired from previous interactions. The default knowledge consists in the decomposition of a typical working day comprising: transportation to the office, the office day as seen through the prism of food intake, post-work leisure activity which includes both sports and sedentary activities (e.g. watching TV), and dinner options. It is embedded in the definition of HTN methods.

3 Ad Hoc Dialogue Control Generation

Throughout this interaction cycle, each time the Planner generates a candidate activity model, it also generates a corresponding ad hoc dialogue controller which will be used to enquire about user preferences in relation to the planned activities. This ad hoc dialogue controller also takes the form of an AND/OR graph, and is produced after the generation of a complete activity tree.

There are several advantages to working with an explicit task decomposition as a dialogue controller, rather than fully interleaving activity planning and dialogue. The most obvious one is that we can use a standard HTN algorithm rather than a real-time version. Others include the ability to jump to different parts of the AND/OR graph as a function of the user utterance, providing more dialogue flexibility.

4 Refining and Updating the Activity Model through Dialogue

The dialogue controller manages agent's questions, which are initiated by the agent but can have an open formulation, such as "do you have any plans for tonight?". In return, the semantics of user utterances update the planning state directly.

User utterances, whether they answer a system question or express a preference, are used to update the activity model. Whenever the user's reply is incompatible with the current activity model (because it rejects the current proposal, or expresses a preference which is not part of the current task decomposition), this will trigger a re-planning step, re-generating the activity model so as to retain those tasks which have been accepted and producing alternatives compatible with the user's preference.

Re-planning takes place regularly throughout an interaction cycle. Fig. 3 describes an example of re-planning, corresponding to the dialogue of Fig. 1. The system, using its default knowledge, has generated an activity model for short distance transportation which involved cycling (the cycling node having the highest heuristic value). This task is tagged as being weather-sensitive as the planning methods take into account such external conditions.

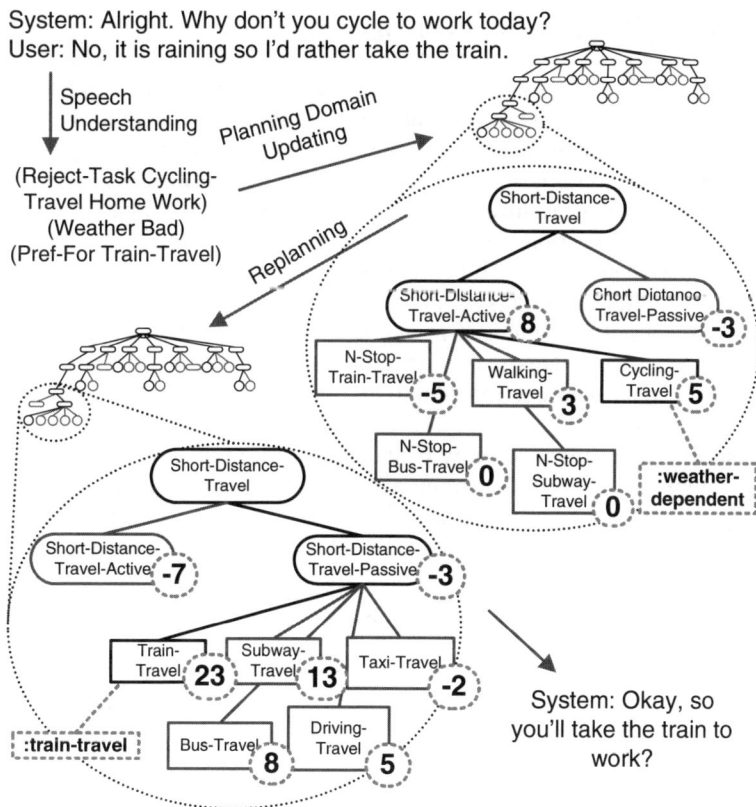

Fig. 3. Generating a new activity model by re-planning from user input

When the system suggests cycling as a means of transportation, the user rejects it invoking the weather ("it is raining") and providing an alternative option (train travel). This additional information leads to re-planning and the generation of a new activity model (Fig. 3, bottom). The explicit preference for the train leads to the selection of train travel in the new activity model. It can be noted that, had the user only mentioned poor weather, this would still have ruled out the options on the left side of the tree which all involve, in part, walking in the rain.

Because this application is intended for everyday activities, with a certain routine in mind, the major emphasis is on minimising dialogue effort, as previously stated. This means ruling out complex negotiations [15], for instance re-discussing previously accepted tasks at a later stage. Another important feature is flexibility: it is possible to mention activities in variable order, not bound by their temporal ordering during the day, as well as skipping certain topics for further reference.

While negotiation is not explicitly taking place, it is however perfectly possible that during the user's day, the agreed actions will be abandoned for other options, such as the user choosing a good meal instead of the agreed light lunch. This will be detected during the reporting dialogue and may yield an opportunity to re-plan ulterior activities based on the assessment performed.

5 Integrating Planning and Dialogue

The integration of Planning and Dialogue is at the heart of establishing a cognitive ability for the Companion. One traditional mechanism for the integration of dialogue with collaborative planning (or joint task description) consists in the explicit specification of communication primitives which bridge the gap between dialogue and (collaborative) planning. For instance, Allen et al. [1] have mapped linguistic expressions to corresponding modifications to a plan under elaboration, and Traum et al. [15] have defined "negotiation primitives". It is important to note that negotiation primitives can represent both the transaction taking place in the course of dialogue and the interaction unit in collaborative planning. One example of the latter case has been described by Kim and Gratch [10]. This traditional integration approach is the convergence of speech acts and plan modification primitives which we could refer to as being inspired by a "pragmatic" layer.

We wish to explore the possibility of integration at a semantic level, by mapping semantic features extracted from user utterances to predicates in the planning domain, leaving the planner to deduce the consequences of new information as part of the re-planning process itself. This extension of the HTN approach with semantic categories offers a new possibility for dialogue integration, based precisely on a semantic mapping of utterance contents to semantic categories attached to HTN methods. This in turn relieves the dialogue system from accurate speech acts recognition and makes it possible to: i) process implicit rejections and alternative preferences, ii) become more robust in terms of recognition.

6 Results and Discussion

A first end-to-end implementation has been completed using a dialogue architecture previously developed by some of the authors [17]. The planning domain includes

16 axioms, 111 methods (enhanced with 42 semantic categories and 113 semantic rules), and 49 operators. The speech recognition / NLP component uses a vocabulary of 1383 words and a total of 917 CFG grammar rules. At this stage, the system's vocabulary only contains a limited number of entries for food types, covering current situations, meaning that vocabulary coverage is balanced across the four relevant domains that are part of the activity model: transportation, physical activity, leisure and food.

The objective of this first evaluation is: i) to test the system stability and performance and, at the same time, ii) to assess the concept of the Health and Fitness Companion, by observing whether end-users can spontaneously interact with it. These evaluations involved 20 subjects, who were briefly introduced to the concept of the HFC. They were explained the scenario using slides illustrating the topics the system "knew about" (using images of activities and food types). At no stage were they shown examples of possible utterances or dialogues. This created realistic albeit quite challenging conditions for this evaluation.

There are several dimensions to be considered for the evaluation of this sort of conversational agent. Task completion is a traditional measure of task-based dialogue systems; however, in the present case, the task at hand is more complex than typical tasks for dialogue systems, such as completing a transaction. When the task itself becomes more complex, its success can no longer be assessed by a binary mode (success versus failure, as in information dialogues) and one should use instead at least a graded score such as the 5-scale proposed by Frokjaer et al. [7].

Paek [13] has advocated comparative analysis based on a quantitative relation between task completion and Word Error Rate (WER). For more complex tasks (also involving planning) Traum et al. [16] have introduced a new methodology based on response appropriateness which supports a finer analysis and has the additional advantage of remaining applicable for high WER. Since the dialogue's objective is to instantiate an activity model, we have taken as a measure of task completion the percentage of correct instantiation of an activity model corresponding to the scenario.

Table 1. Evaluation Results

Phase	Average WER	Average CER	Average Task Model Completion	Average Utterance Length
Planning	42%	24%	80%	4.2
Reporting	44%	24%	95%	3.3

Our results are summarised in Table 1. The level of WER is easily explained by the realistic experimental conditions and the absence of any user training with the system (nor any speaker adaptation). It can still be regarded as compatible with the state-of-the-art and is certainly consistent with data reported from similar systems, such as [11] [16]. Task model instantiation was higher for reporting dialogues, which can be explained by the fact that these tend to be simpler with shorter responses (as seen by the lower average length for user utterances).

7 Relationship to Previous Work

De Rosis et al. [6] have previously described an Embodied Conversational Agent advising on eating disorders. Another medical advisory system has been described by Allen et al. [2], which tended to be mostly user-driven dialogue systems. Assistive systems are a particular case in which Planning or scheduling systems reason on user's activities, rather than on a common task. Several systems have used explicit models of activities [9], for instance as predefined plans ("straw plans" in TRIPS [3]). Similar integration of Planning and dialogue have been described in previous architectures such as Ravenclaw ([5]) and COLLAGEN ([14]).

Finally, as far as Planning is concerned, we can observe significant variability in the use of planning as a reasoning and representation mechanism. For instance, Bickmore and Sidner [4] represent the informational task, while Traum et al. [15] represent the real-world task, and so do Allen et al. [1] [3]. We have opted here to represent the real-world task from the strict perspective of the user.

8 Conclusions

We have described the concept of a Health and Fitness Companion and the specific modes of interaction associated with it. We have produced a first end-to-end implementation, which has enabled us to test the system under realistic conditions. The level of task completion obtained with untrained users suggests a good potential for acceptance of the HFC. The analysis of system performance, in particular for its spoken dialogue component, suggests possible directions for performance improvement. Other possibilities have yet to be explored in terms of dynamic dialogue strategies, shifting dialogue control from the user to the system. Further work should be dedicated to determining the level of scalability required to implement a consistent application in a given domain, so that usability testing can proceed with the confidence that its results are not obscured by technical performance issues.

Acknowledgements

This work was funded in part by the Companions project (www.companions-project.org) sponsored by the European Commission as part of the Information Society Technologies (IST) programme under EC grant number IST-FP6-034434.

References

1. Allen, J., Chambers, N., Ferguson, G., Galescu, L., Jung, H., Swift, M., Taysom, W.: PLOW: A Collaborative Task Learning Agent. In: AAAI 2007 Proceedings, Vancouver, British Columbia, pp. 1514–1519 (2007)
2. Allen, J., Ferguson, G., Blaylock, N., Byron, D., Chambers, N., Dzikovska, M., Galescu, L., Swift, M.: Chester: Towards a Personal Medication Advisor. Journal of Biomedical Informatics 39(5), 500–513 (2006)

3. Allen, J., Ferguson, G., Swift, M., Stent, A., Stoness, S., Galescu, L., Chambers, N., Campana, E., Aist, G.: Two Diverse Systems Built using Generic Components for Spoken Dialogue (Recent Progress on TRIPS). In: ACL 2005 Proceedings of the Interactive Poster and Demonstration Sessions, pp. 85–88. Ann Arbor Michigan (2005)
4. Bickmore, T.W., Sidner, C.L.: Towards Plan-Based Health Behavior Change Counseling Systems. In: Proceedings of AAAI Spring Symposium on Argumentation for Consumers of Healthcare, Stanford, CA, pp. 14–18 (2006)
5. Bohus, D., Rudnicky, A.I.: RavenClaw: Dialog Management using Hierarchical Task Decomposition and an Expectation Agenda. In: Proceedings of Eurospeech 2003, pp. 597–600 (2003)
6. De Rosis, F., Novielli, N., Carofiglio, V., Cavalluzzi, A., De Carolis, B.: User modeling and adaptation in health promotion dialogs with an animated character. Journal of Biomedical Informatics 39(5), 514–531 (2006)
7. Frokjaer, E., Hertzum, M., Hornbaek, K.: Measuring usability: Are effectiveness, efficiency, and satisfaction really correlated. In: Proceedings of CHI 2000 Conference, pp. 345–352. ACM Press, New York (2000)
8. Ghallab, M., Nau, D., Traverso, P.: Automated Planning: Theory and Practice. Morgan Kaufmann, San Francisco (2004)
9. Gruenstein, A., Cavedon, L.: Using an activity model to address issues in task-oriented dialogue interaction over extended periods. In: AAAI Spring Symposium on Interaction between Humans and Autonomous Systems over Extended Operation, March 2004, AAAI Press, Menlo Park (2004)
10. Kim, H.-S., Gratch, J.: A planner-independent collaborative planning assistant. In: AAMAS 2004 Proceedings, pp. 764–771 (2004)
11. Leuski, A., Kennedy, B., Patel, R., Traum, D.: Asking Questions to Limited Domain Virtual Characters: How Good Does Speech Recognition have to be? In: 25th Army Science Conference, Orlando, Florida (2006)
12. Nau, D.S., Smith, S.J.J., Erol, K.: Control Strategies in HTN Planning: Theory versus Practice. In: AAAI 1998/IAAI 1998 Proceedings, pp. 1127–1133 (1998)
13. Paek, T.: Empirical methods for evaluating dialog systems. In: ACL 2001 Workshop on Evaluation Methodologies for Language and Dialogue Systems 2001, pp. 3–10 (2001)
14. Rich, C., Sidner, C.L.: COLLAGEN: A Collaboration Manager for Software Interface Agents. User Modeling and User-Adapted Interaction 8(3-4) (1998)
15. Traum, D., Rickel, J., Gratch, J., Marsella, S.: Negotiation over Tasks in Hybrid Human-Agent Teams for Simulation-Based Training. In: AAMAS 2003 Proceedings, Melbourne, Australia, July 2003, pp. 441–448 (2003)
16. Traum, D.R., Robinson, S., Stephan, J.: Evaluation of multi-party virtual reality dialogue interaction. In: LREC 2004 Proceedings, pp. 1699–1702 (2004)
17. Turunen, M., Salonen, E.-P., Hartikainen, M., Hakulinen, J.: Robust and Adaptive Architecture for Multilingual Spoken Dialogue Systems. In: Proceedings of ICSLP 2004 (2004)

Audio Analysis of Human/Virtual-Human Interaction

Harold Rodriguez[1], Diane Beck[1], David Lind[2], and Benjamin Lok[1]

[1] University of Florida, Gainesville FL 32607, USA
drharold@ufl.edu, lok@cise.ufl.edu, beck@cop.ufl.edu
http://www.cise.ufl.edu/research/vegroup/
[2] Medical College of Georgia, Augusta GA 30912, USA
dlind@mail.mcg.edu

Abstract. The audio of the spoken dialogue between a human and a virtual human (VH) is analyzed to explore the impact of H-VH interaction. The goal is to determine if conversing with a VH can elicit detectable and systematic vocal changes. To study this topic, we examined the H-VH scenario of pharmacy students speaking with immersive VHs playing the role of patients. The audio analysis focused on the students' reaction to scripted empathetic challenges designed to generate an unrehearsed affective response from the human. The responses were analyzed with software developed to analyze vocal patterns during H-VH conversations. The analysis discovered vocal changes that were consistent across participants groups and correlated with known H-H conversation patterns.

Keywords: Conversational agents, speech visualization, digital signal processing.

1 Introduction

Recently, significant work has been applied towards providing opportunities to practice human-human interactions through interactions with a virtual human (VH). These VHs play the role of conversational agents and have been used in military [1], medical [2], and entertainment [3] scenarios to train soldiers, detect empathy, and practice communication skills. In these systems, there has been a consistent effort to increase the communication cues that are tracked by the VH simulation to increase interaction fidelity.

In this paper, we focus on the information contained within the H-VH audio stream. The goal of this work is to determine if the audio stream of a person speaking with a VH contains prosodic information that can 1) quantify the impact of the VH experience and 2) be used to improve the VH experience. That is, can VHs elicit detectable and systematic vocal changes in humans? Using a digital signal processing approach, critical moments in a H-VH conversation are analyzed to measure the human response to VHs.

First, this work presents a signal processing approach to H-VH audio analysis. The approach focuses on the application of frequency-spectrum audio analysis and visualization techniques. It describes the most applicable metrics to quantifying the user's tone while conversing with a VH. Second, a user study designed to generate an unrehearsed affective response from the human is used to investigate the efficacy of the approach. Finally, this paper presents directions on applying the detectable vocal changes as an input into VH simulations, as signs of user attention and as a valuable teaching tool to help poor students remediate their communication skills.

Related Work. Significant work on vocal analysis and emotion detection has been conducted by J. Bachorowski, K. Scherer, R. Cowie, and others. Studies have shown the value of extracting speech or acoustic *parameters* [4,5], features that describe a person's speech, as opposed to focusing on lexical and grammatical measures. In fact, these acoustic parameters are language-independent [6,7] and convey subtle information to the listener. For example, lowered tone and a softer voice may indicate sadness without having to verbally express this sentiment. Hence, some preliminary emotion detection has been attempted using these attributes, based on prosodic training algorithms [8], Bayesian inference [9], and promotion/demotion schemes [10].

It is useful to set aside strict emotion detection and instead focus on understanding the basic acoustic properties that provide the best heuristics for researchers in this field. There is only a weak link between speech streams and their signature traits because a good "Rosetta stone" to decode the traits has yet to be developed [11]. However, a select few have shown to be significant: fundamental frequency (F0), speech rate, and amplitude [12]. These three main measures, a few novel ones, and some from [13] are explored within a H-VH context.

2 A Signal Processing Approach to H-VH Audio Analysis

Overview. Spoken dialogue, natural language, and embodied conversational agents systems (among many others) all aim to derive higher level meaning from the content of the words presented by the user. This work aims to supplement the content information by quantifying global and local *frequency* and *temporal* metrics of the user's audio.

In this section, we will describe an approach for analyzing user speech as well as software that was developed to support the unique requirements of H-VH interaction analysis. In our user studies (Sect. 3), each student had a conversation with a VH for about 13 minutes. The audio was extracted from WMV video file recordings of the students (32 kHz, 16 bit, 32 kbps) and the audio was loaded into CFMetrix, an in-house developed audio analysis and visualization program that processes the data. Coupled with user interaction, the system highlights significant moments in the audio stream.

Metrics. In processing the audio signal of a H-VH interaction, there are dozens of potential metrics to compute and analyze. However, metrics which have been shown to have a large impact on H-H conversations [4,5,12] were implemented and grouped by domain (frequency or temporal).

Frequency metrics focus on the frequency domain characterization of the user's audio stream. While the primary frequency metric is the F0 (fundamental frequency) of the user's audio, other frequency metrics can be derived:

- *F0 Mean* - the average fundamental frequency (main pitch) of the signal over some time. F0 is obtained by performing an FFT of the audio stream and finding the first major peak in the signal (typically around 120 Hz for men and 220 Hz for women).
- *F0 Variance* - the statistical dispersion of the fundamental frequency over the support time. Defines amount of inflection in an utterance.
- *Frequency Range* - the largest range of frequencies above a given power threshold. Provides the size of most speech energy.
- *Frequency Histogram* - a running sum of waveform amplitudes over all available frequencies. Conveniently exaggerates the locations of user speech and ambient noise.

Temporal metrics focus on the "loudness" of speech and speech timing. In addition to the audio's signal amplitude, RMS amplitude, power mean, and power variance, the following calculations are useful:

- *Power Level* - $10 * log_{10}(Power/10pW)$. Relatable loudness (units in dB).
- *Power Range* - RMS amplitude of a superposition of FFTs over all time. Typical loudness of a user.
- *Speaking Contour* - represented as a graph. When someone is speaking, a rising slope is shown, but as soon as the voice ceases, it drastically declines.
- *Signal-to-noise ratio (SNR)* - defined as $10*log10(Power/Power_{min})$, where $Power_{min}$ is a user-defined minimum power level. Global SNR provides information on the viability of ambient noise filtering, whereas local SNR justifies the inclusion or exclusion of an utterance for analysis.
- *Rate of Speech* - divides the number of words (number of times the power of the spectrum is above $Power_{min}$) by the sound duration (in minutes).

Filters and Visualizations. The audio stream of a human talking to a VH has unique properties that assist in analysis. Specifically, in most H-VH interactions 1) usually only one conversationalist is speaking, 2) audio recordings of the user are separable since they have a high SNR, and 3) the user is speaking into a speech recognition system (or is under that assumption) and thus tends to enunciate more than usual.

Nevertheless, several averaging, normalization, and bandpass filters were created and used to aid signal "clean-up". The visualizations provide instant feedback on the effect of these adjustable filters. Furthermore, the researcher can interact with visualizations to identify the sound level under which no meaningful signal lies. This made for a very easy, intuitive, and effective model for facilitating metric calculations.

Metrics (e.g. F0) were rendered in both 2D and 3D modes (Fig. 1). In the 2D visualization, frequency vs. amplitude was visualized in real-time. Multiple audio streams can be easily compared using such a visualization: A "difference

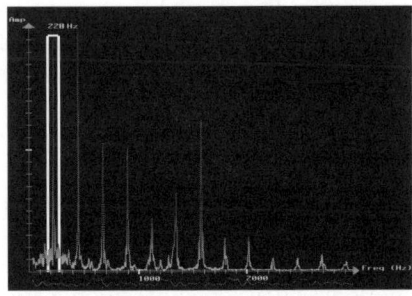 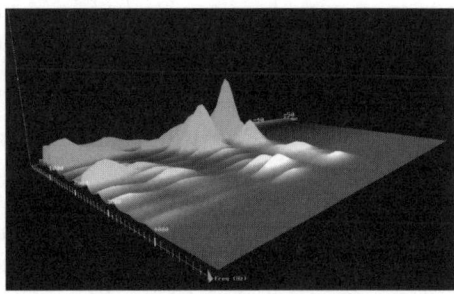

Fig. 1. Left: A white box whose width is proportional to the F0 Variance is centered on the F0 of a user saying "Uhh." **Right:** 3D FFT visualization of a sound stream (*horiz., vert.*) over time (*depth*) with moving average filtering enabled (for smoother heightmap).

spectrum" is generated by analyzing two streams at once. The differences in their FFT spectra, as well as differences in metrics, can be color-coded and displayed simultaneously. Thus, 2D frequency visualizations enable an efficient method to visually explore audial metrics and the impact of filter parameters.

The 3DFFT visualization presents the Fourier transform of the sound stream over an extended time (e.g. 10 seconds), allowing a user to not only identify frequency-related metrics at a certain time (2D) but trends and metric interactions over time. Consequently, this was one method used to identify metrics to explore in the user study.

3 User Study

Motivation. Given a system that can analyze the audio from a H-VH interaction, we examined 27 audio streams of pharmacists talking to a VH playing the role of a patient. The study had the following three goals:

1. Verify the metrics most applicable to processing a conversation with a VH.
2. Identify if audio analysis could characterize a H-VH interaction.
3. Understand the subconscious, affective impact of speaking with a VH.

To magnify the empathetic response, the audio analysis focused on "empathetic challenges". An empathetic challenge is VH-initiated dialogue meant to generate an unrehearsed affective response from the human (e.g. "I'm scared. Could this be cancer?"). Empathetic challenges were chosen as they tend to elicit the strongest, most genuine responses in previous studies [2] and were good indicators of user involvement [14].

Test-Bed Platform: The Interpersonal Simulator (IPS). The Interpersonal Simulator (IPS) is a VH experience used to train interpersonal skills [15]. The IPS combines VR tracking, natural speech processing, and immersive high-fidelity VHs to produce an engaging VH experience. It is constantly evolving through extensive user testing with over 412 medical, nursing, physician assistant, and pharmacy students and professionals. The virtual experience has been compared to H-H interaction [14], demonstrated construct validity for the evaluation of interpersonal skills [16], and applied to scenarios involving empathy [2,17], sexual history [18], and breast cancer [19].

Application: Pharmacy Patient Interview. Interpersonal skills have been shown to improve patient care and reduce litigation costs, making them critical for health professionals [20,21]. For this study, the researchers collaborated with a pharmacy college to train practicing pharmacists with H-VH interactions. Instead of training with human actors pretending to have a disease, the pharmacists interacted with a VH. The scenario was developed based on a standard pharmaceutical peptic ulcer disease case found in [22]. The VH simulates a 35-year old Caucasian male patient complaining about increasing pain in his abdominal region. A special response occurred during the scenario denoted as an *empathetic moment*.

About seven minutes into the interview, the VH relates to the participant that "my dad died of cancer"and asks, "Could this be cancer?"This moment was designed to evoke sadness and an empathetic response from the participant. Pharmacists are trained to handle this situation delicately and professionally, expressing empathy and providing reassurance that the medical team will do everything they can to find out what is wrong.

Thirty-nine pharmacy students (12 men, 27 women) participated in the study. Of these, audio for 8 men and 19 women was recorded (n=27). The goal was for the pharmacists to 1) comfort the concerned patient and 2) identify the cause of the pain. A Wizard-of-Oz (WoZ) interface was used to prevent (poor) speech recognition performance from detrimentally influencing user tone. Audio responses were recorded using a wireless microphone attached to a baseball cap used for head-tracking (Fig. 2), and all audio was analyzed after the experiment had concluded.

Data Analysis. The audio files were analyzed at two levels of granularity, global and local. The global metrics were computed by splitting the audio file at

Fig. 2. A user interacts with the VH using natural speech

the empathetic moment ($\approx$7 minutes into the conversation). The two parts are labeled Global$_{before}$ and Global$_{after}$. For local metrics, one sentence before and one sentence after the empathetic challenge were analyzed, labeled Local$_{before}$ and Local$_{after}$. So for each users speech, each of the metrics in Sect. 2 was computed four times. The four comparisons made are:

1) Local$_{before}$ vs. Local$_{after}$ the immediate impact of the empathetic moment
2) Global$_{before}$ vs. Global$_{after}$ the impact of conversing with a VH for $\approx$7 min.
3) Global$_{before}$ vs. Local$_{after}$ the startle response of a VH challenge
4) (Local$_{after}$ - Local$_{before}$) vs. (Global$_{after}$ - Global$_{before}$) - compares the measuring time-span's effect on metrics

Results. *Vocal changes were detected and measured at the empathetic moment.* A "downward" trend in user pitch and volume arose, which could signal an empathetic response and highlight the VHs affective capabilities.

Paired two-tailed Student T-Tests were performed on the data set. As seen in Table 1, the F0 variance, range, signal amplitude, RMS amplitude, and power level of the audio was significantly different (p<0.03) after the empathetic moment.

Table 1. Local user vocal changes near the empathetic moment

Metric	Change	%	σ	P-value
F0 Mean (Hz)	-8	-4	2.02	.002
Freq. Range (Hz)	-120	-11	27.5	.022
RMS Amplitude	-266	-17	13.8	.006
Power Level (dB)	-1.39	-2	.082	.024

Gender has an impact on tonal change elicited by the VH. F0 mean and RMS amplitude were significantly lower (p<0.03) immediately after the challenge for male participants. On top of that, women also showed a significant change in frequency range and power level (Table 2, 3). F0 measurements tend to be reliable indicators of gender [23], but an affective situation such as an empathetic challenge could invalidate this assumption, as F0 is rather unstable here.

Table 2. Local vocal change for men

Metric	Change	%	σ	P-value
F0 Mean	-7	-5	.787	.018
RMS Amplitude	-424	-30	20.6	.049

Table 3. Local vocal change for women

Metric	Change	%	σ	P-value
F0 Mean	-8	-4	.589	.014
Freq. Range	-132	-11	-29.2	.024
RMS Amplitude	-199	-12	10.6	.038
Power Level	-1.1	-2	.062	.051

Analyzing the local effect of the empathetic challenge is a good indicator of the entire conversation. (Local$_{after}$ - Local$_{before}$) significantly correlated with (Global$_{after}$ vs. Global$_{before}$). P-values were high (p>.5) meaning that there is no statistical difference attributed to changing granularity when considering differences. Hence, a person's reaction to the empathetic challenge permeated across to the other side of the global arena as the user was compelled to extend the VH experience beyond that of a mere chatbot.

The trend to permeate on until the end of the interaction was realized when manually comparing the raw percentage differences: for all statistically significant metrics (and even some which were not shown to be significant), the local vs. global values were all "in step" (i.e. if one dropped, the other dropped).

4 Conclusions and Future

This paper proposed a signal analysis approach to exploring the audio of a H-VH interaction. The user's vocal metrics changed in a manner similar to those in H-H interactions [5,24], indicating a subconscious response to the VH. Hence, a strong case is made for having a system capable of tracking and responding to audio changes, thus enhancing the use of VHs as a valuable teaching tool. Particularly *real-time* analysis of the user's speech should be explored and fed as an input to VH simulations for judging attention and engagement.

Further social experiments can be conducted making vocal comparisons of the same user in various scenarios (or different users in the same scenario) to identify the inherent differences of the interaction. For example, this would enable systems to identify if business professionals speak in a different tone with VHs representing clients with different racial, gender, and weight backgrounds. Most of all, VH researchers would benefit from employing the use of metrics, methods, and correlations of these auditory research findings to reinforce and support their affection results. It is through this integration of validated metrics and findings that the VHs' effects can best be understood.

References

1. Swartout, W., et al.: Toward the holodeck: Integrating graphics, sound, character, and story. In: Proceedings of the 5th International Conference on Autonomous Agents, pp. 409–416. ACM Press, New York (2001)
2. Deladisma, A., Cohen, M., et al.: Do medical students respond empathetically to a virtual patient? American Journal of Surgery 193(2), 756–760 (2007)
3. Dow, S., et al.: Presence and engagement in an interactive drama. In: Proceedings of the SIGCHI Conference on Human factors in Computing Systems, pp. 409–416. ACM Press, New York (2007)
4. Scherer, K.R.: Vocal affect expression: A review and model for future research. Psychological Bulletin 99(2), 143–165 (1986)
5. Cowie, R., et al.: Emotion recognition in human-computer interaction. IEEE Signal Processing Magazine 1, 32–80 (2001)

6. Scherer, K.R., et al.: Emotion inferences from vocal expression correlate across languages and cultures. Journal of Cross-Cultural Psychology 32(1), 76–92 (2001)
7. Campbell, N.: Perception of affect in speech - towards an automatic processing of paralinguistic information in spoken conversation. In: ICSLP Proceedings, pp. 881–884 (2004)
8. Polzin, T.S., Waibel, A., et al.: Emotion-sensitive human-computer interfaces. In: Proceedings of the ISCA Workshop on Speech and Emotion: A Conceptual Framework for Research, pp. 201–206 (September 2000)
9. McGilloway, et al.: Approaching automatic recognition of emotion from voice: A rough benchmark. In: Proceedings of the ISCA Workshop on Speech and Emotion: A Conceptual Framework for Research, pp. 207–212 (September 2000)
10. Rodriguez, H.: Detecting emotion in the human voice: a model and software implementation using the FMOD sound API (unpublished, 2005)
11. Russell, J.A., Bachorowski, J.A., Fernandez-Dols, J.M.: Facial and vocal expressions of emotion. In: Annual Review of Psychology, pp. 329–349 (February 2003)
12. Owren, M.J., Bachorowski, J.A.: In: Measuring Emotion-Related Vocal Acoustics. Oxford University Press Series in Affective Sciences (2004)
13. Scherer, K., Banse, R.: Acoustic profiles in vocal emotion expression. Journal of Personality and Social Psychology 70(3), 614–636 (1996)
14. Raij, A., Johnsen, K., et al.: Comparing interpersonal interactions with a virtual human to those with a real human. IEEE Transactions on Visualization and Computer Graphics 13(3), 443–457 (2007)
15. Johnsen, K., Dickerson, R., et al.: Evolving an immersive medical communication skills trainer. Presence: Teleoperators and Virtual Environments 15, 33–46 (2006)
16. Johnsen, K., Raij, A., et al.: The validity of a virtual human system for interpersonal skills education. In: ACM SIGCHI (2007)
17. Cohen, M., Stevens, A., et al.: Do health professions students respond empathetically to a virtual patient? Presented at Southern Group on Educational Affairs (2006)
18. Deladisma, A., Mack, D., et al.: Virtual patients reduce anxiety and enhance learning when teaching medical student sexual-history taking skills. Association for Surgical Education 2007 Surgical Education Week 193(2), 756–760 (2007)
19. Kotranza, A., Lind, D., Pugh, C., Lok, B.: Virtual human + tangible interface = mixed reality human. an initial exploration with a virtual breast exam patient. In: IEEE Virtual Reality 2008, pp. 99–106 (2008)
20. Vincent, C., Phillips, A., Young, M.: Why do people sue doctors? a study of patients and relatives taking legal action. Obstetrical and Gynecological Survey 50, 103–105 (1995)
21. Duffy, F.D., Gordon, G.H., et al.: Assessing competence in communication and interpersonal skills: The kalamazoo ii report. Academic Medicine 79, 495–507 (2004)
22. Schwinghammer, T.L.: Pharmacotherapy Casebook: A Patient-Focused Approach. McGraw-Hill Medical, New York (2005)
23. Childers, D., Wu, K.: Gender recognition from speech. part ii: Fine analysis 90, 1841–1856 (1991)
24. Murray, I., Arnott, J.: Toward the simulation of emotion in synthetic speech: A review of the literature on human vocal emotion. J. Acoust. Soc. Amer. 93(2), 1097–1108 (1993)

Learning Smooth, Human-Like Turntaking in Realtime Dialogue

Gudny Ragna Jonsdottir, Kristinn R. Thorisson, and Eric Nivel

Center for Analysis & Design of Intelligent Agents & School of Computer Science
Reykjavik University
Ofanleiti 2, IS-103 Reykjavik, Iceland
{gudny04,thorisson,eric}@ru.is

Abstract. Giving synthetic agents human-like realtime turntaking skills is a challenging task. Attempts have been made to manually construct such skills, with systematic categorization of silences, prosody and other candidate turn-giving signals, and to use analysis of corpora to produce static decision trees for this purpose. However, for general-purpose turntaking skills which vary between individuals and cultures, a system that can learn them on-the-job would be best. We are exploring ways to use machine learning to have an agent learn proper turntaking during interaction. We have implemented a talking agent that continuously adjusts its turntaking behavior to its interlocutors based on realtime analysis of the other party's prosody. Initial results from experiments on collaborative, content-free dialogue show that, for a given subset of turntaking conditions, our modular reinforcement learning techniques allow the system to learn to take turns in an efficient, human-like manner.

Keywords: Turntaking, Machine Learning, Prosody, End-of-utterance detection.

1 Introduction

Fluid turntaking is a dialogue skill that most people handle with ease. To signal that they have finished speaking and are expecting a reply, for example, people use various multimodal behaviors including intonation and gaze [1]. Most of us pick up on such signals without problems, automatically producing information based on data from our sensory organs to infer what the other participants intend. In amicable, native circumstances conversations usually go smoothly enough for people to not even realize the degree of complexity inherent in the process responsible for dynamically deciding how each person gets to speak and for how long.

Giving synthetic agents similar skills has not been an easy task. The challenge lies not only in the integration of perception and action in sensible planning schemes but especially in the fact that these have to be coordinated while marching to a real-world clock. Efficient handling of time is one of a few key components that sets current dialogue systems clearly apart from humans; for example, speech recognition systems that have been in development for over a decade are still far from addressing the needs of realtime dynamic dialogue [2]. In spite of moderate progress in speech recognition technologies most systems still rely on silence duration as the main method for detection of end-of-utterance. However, as is well known and discussed by e.g. Edlund et al. [3], natural speech contains a lot of silences that do not indicate end-of-speech or end-of-turn, that is, silences where the speaker nonetheless does not want to be interrupted.

Although syntax, semantics and pragmatics indisputably can play a large role in the dynamics of turntaking, we have argued elsewhere that natural turntaking is partially driven by a content-free planning[1] system [4]. For this, people rely on relatively primitive signals such as multimodal coordination, prosody and facial expressions. In humans, recognition of prosodical patterns, based on the timing of speech loudness, silences and intonation, is a more light-weight process than word recognition, syntactic and semantic processing of speech [5]. This processing speed difference is even more pronounced in artificial perception, and such cues can aid in the process of recognizing turn signals in artificial dialogue systems.

J.Jr. was an agent that could interject back-channel feedback and take turns in a human-like manner without understanding the content of dialogue [6]; the subsequent Gandalf agent [7] adopted the key findings from J.Jr. in the Ymir architecture, an expandable granular architecture for cognition. We build directly on this work, introducing schemes for the automatic, realtime learning of a key turntaking decisions, that had to be built by hand in these prior systems. The system learns on-line to become better at taking turns in realtime dialogue, specifically improving its own ability to take turns correctly and quickly, with minimal speech overlap.

In the present work turntaking is viewed as a negotiation process between the parties involved and the particular observed patterns produced in the process are considered emergent [8], on many levels of detail [9], based on an interaction between many perception, decision and social processes, and their expression in a humanoid body. In our task two talking agents, each equipped with a dialogue model and dynamic speech planning capabilities, speak to each other. One of them listens to the prosody of the other (intonation and speech-on/off) and uses machine learning to best determine - as the speaker falls silent - how long to wait until starting to speak, that is, to take the turn. One way to think of this task is as a bet against time: The longer the duration of the silence the more willing we are to bet that the speaker expects us to start talking; the challenge is to reliably bet on this in realtime, during the silence, as soon as possible after the silence starts, with whatever information (prosody etc.) has been processed. In the present work we leave aside issues (perception, interpretation and actions) related to switching between different conversational topics. We also limit ourselves to detecting turn-giving indicators in deliberately-generated prosody, leaving out the topic of *turn-opportunity* detection (i.e. turn transition without prior indication from the speaker that she's giving the turn), which would e.g. be

---

[1] We use the term "planning" in the most general sense, referring to any system that makes a priori decisions about what should happen before they are put in action.

necessary for producing (human-like) interruptions. Our experiments show that our architecture and learning methodology allow the system to learn to take turns with human-like speed and reliability in about 5 minutes of interaction in which turn is successfully exchanged 7 times per minute, on average.

The rest of this paper is organized as follows. In section 2 we discuss related work in the field; we describe the system we have built in section 3, with section 4 detailing the learning mechanism. In section 5 we present the evaluation setup; sections 6 and 7 show results from experiments with the system interacting with itself humans. Section 8 is conclusion and future work.

2 Related Work

The problem of utterance segmenting has been addressed to some extent in prior work. Sato et. al [10] use a decision tree to learn when a silence signals to take turn. They annotated various features in a large corpus of human-human conversation to train and test the tree. Their results show that semantic and syntactic categories, as well as understanding, are the most important features. Their experiments have currently been limited to single domain, task oriented scenarios with annotated data. Applying this to a casual conversation scenario would inevitably increase the recognition time - as the speech recognizers vocabulary is enlarged - to the extent that content interpretation results are already obsolete for turn taking decisions by the time they are produced [2].

Traum et al. [11] and others have also addressed the problem of utterance segmenting, showing that prosodic features such as boundary tones do play a role in turntaking and Schlangen [12] has successfully used machine learning to categorize prosodic features from corpus, showing that acoustic features can be learnt. As far as we know, neither of these attempts have been applied to a real-time scenario.

Raux and Eskenazi [13] presented data from a corpus analysis of an online bus scheduling/information system, showing that a number of dialogue features, including type of speech act, can be used to improve the identification of speech endpoint, given a silence. They reported no benefits from prosody for this purpose, which is surprising given the many studies showing the opposite (cf. [1,7,11,12,14,15]). We suspect one reason could be that the pitch and intensity extraction methods they used did not work very well on the data selected for analysis. The authors tested their findings in a realtime system: Using information about dialogue structure - speech act classes, a measure of semantic completeness, and probability distribution of how long utterances go (but not prosody) - the system improved turntaking latency by as much as 50% in some cases, but significantly less in others. The Gandalf system [7] also used a measures of semantic, syntactic (and even pragmatic) completeness to determine turntaking behaviors, but data about its benefit for the turntaking per se is not available. The major lessons that can be learned from Raux and Eskenazi [raux08], echoing the work on Gandalf, is that turntaking can be improved through an integrated, coordinated use of various features in context.

Prosodic information has successfully been used to determine back-channel feedback. The Rapport Agent [14] uses gaze, posture and prosodic perception to among other things detect backchannel opportunities. The J.Jr. system, a communicative agent that could take turns in realtime dialogue with a human without understanding the content of the speech, used only prosodical information to make decisions about when to ask questions and when to interject backchannel feedback. The system was based on a finite state-machine formalism that was difficult to expand into a larger intelligent architecture [7]. Subsequent work on Gandalf [7] incorporated mechanisms from J.Jr. into the Ymir architecture, which was built as a highly expandable, modular system of preceptors, deciders and action modules; this architecture has recently been used in building an advanced vision and planning system for the Honda ASIMO robot [16].

3 System Architecture

We have built a multi-module dialogue system using the methodology described in [9,17]. The gross architecture of the system will be detailed elsewhere; here we focus on parts of the turntaking - the preceptors, deciders and action modules - needed to support learning for efficient turntaking. Following the Ymir architecture [7], our systems modules are categorized based on their functionality; perception modules, decider modules and action modules (see Figure 1). We will now describe these.

3.1 Perception

There are two main perceptors (perception modules) in the system, the Prosody-Tracker and the Prosody Analyzer. The Prosody-Tracker is a low-level perceptor whose input is raw audio signal [18]. It computes speech signal levels and compares this to a set of thresholds to determine information about speech activity, producing timestamped Speech-On and Speech-Off messages. It also analyzes the speech pitch incrementally (in steps of 16 msec) and produces pitch values, in the form of a continuous stream of pitch message updates.

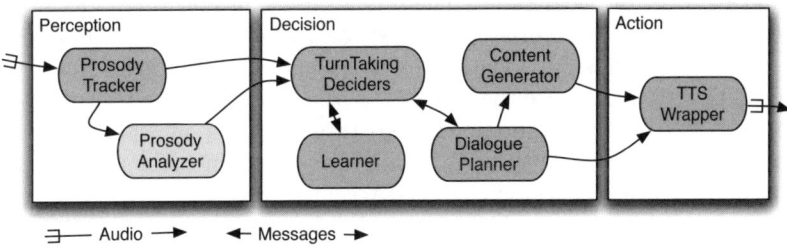

Fig. 1. System components, each component consists of one or more modules

Similar to [19], pitch is analyzed further by a Prosody-Analyzer perceptor to compute a more compact representation of the pitch pattern in a discrete state space, to support the learning: The most recent tail of speech right before a silence, currently the last 300 msec, are searched for minimum and maximum values to calculate a tail-slope value of the pitch. Slope is then split into semantic categories, currently we are using 3 categories for slope: Up, Straight and Down and 3 for relative value of pitch right before silence: Above, At, Below, as compared to the average pitch. Among the output of the Prosody-Analyzer is a symbolic representation of the particular prosody pattern identified in this tail period (see Figure 2). More features could be added into the symbolic representation, with the obvious side effect of increasing the state space. Figure 2 shows a 9 second long frame with speech periods, silences and categories. As soon as a silence is encountered (indicated by gray area) the slope of the most significant continuous pitch direction of the tail is computed.

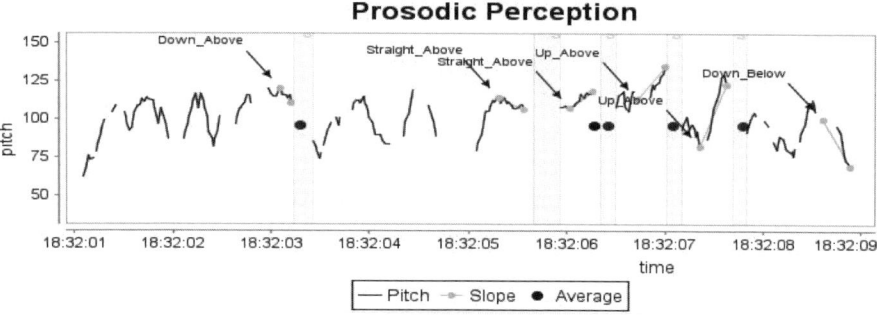

Fig. 2. A window of 9 seconds of speech, containing 6 consecutive utterances, categorized into descriptive groups. Slope is only categorized before silences.

3.2 Deciders

The dialogue state (I-have-turn, Other-has-turn etc.) is modeled with a distributed context system, implementing what can approximately be described as a distributed finite state machine. Context transition control in this system is managed by a set of deciders [9]. There is no limit on how many deciders can be active in a single system-wide context. Likewise, there is no limit to how many deciders can manage identical or non-identical transitions. Reactive deciders (IGTD,OWTD,...) are the simplest, with one decider per transition. Each contains at least one rule about when to transition, based on both temporal and other information. Transitions are made in pull manner; the Other-Accepts-Turn-Decider transits to context Others-Accepts-Turn (see Figure 3).

The Dialogue Planner and Learning modules can influence the dialogue state directly by sending context messages I-Want-Turn, I-Accept-Turn and I-Give-Turn. These decisions are under the supervisory control of the Dialogue Planner: If the Content Planner has some content ready to be communicated, the agent might want to signal that it wants turn and it may want to signal I-Give-Turn when content queue is empty (i.e. have-nothing-to-say). Decisions made by these

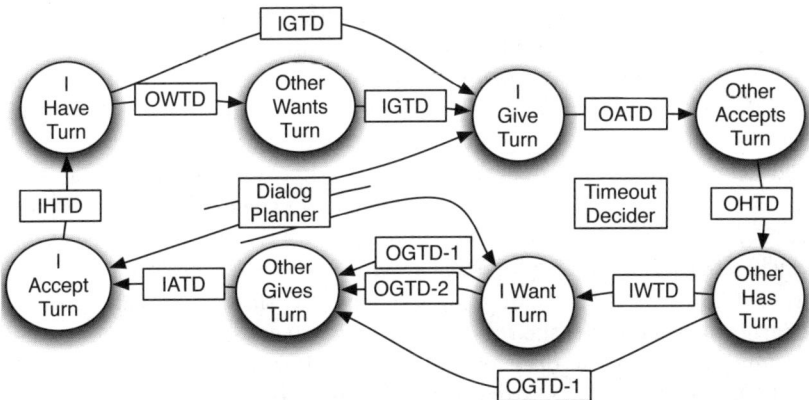

Fig. 3. The turntaking system can be viewed as a set of 8 context-states and 11 deciders. Each context-state has at least one associated decider for determining transition to it but each decider is only active in a limited set of contexts. In context-state I-Have-Turn, both I-Give-Turn-Decider (IGTD) and Other-Wants-Turn-Decider (OWTD) are active. Unlike other modules, the Dialog Planner can transition independent of the system's current context-state and override the decisions from the reactive deciders. A Timeout-Decider handles transitions if one of the negotiating context-states is being held unacceptably long but it's transitions are not included in the diagram.

modules override decisions made by other turntaking modules. The DP also manages the content delivery, that is, when to start speaking, withdraw or raise one's voice. The Content-Generator is responsible for creating utterances incrementally, in "thought chunks", typically of shorter duration than 1 second. While we are developing a dynamic content generation system based on these principles the CG simulates this activity by selecting thought units to speak from a predefined list; it signals when content is available to be communicated and when content has been delivered.

In the present system the module Other-Gives-Turn-Decider-2 (OGTD-2) uses the data produced by the Learner module to change the behavior of the system. At the point where the speaker stops speaking the challenge for the listening agent is to decide how long to wait before starting to speak. If the agent waits too long, and the speaker does not continue, there will be an unwanted silence; if he starts too soon and the speaker continues speaking, overlapping speech will result. We solve this by having OGTD-2 use information about prosody before prior silences to select an optimal wait-and-see time. This will be described in the next section.

4 The Learner

The learning mechanism is implemented as a highly isolated component in the modular architecture described above. It is based on the Actor-Critic distribution of functionality [20], where one or more actors make decisions about which actions to perform and a critic evaluates the effect of these actions on the environment; the separation between decision and action is important because in our system a decision can be made to act in the future. In the highly general and distributed mechanism we have implemented, any module in the system can take the role of an actor by sending out decisions and receiving, in return, an updated decision policy from an associated Learner module. A decision consists of a state-action pair: the action being selected and the evidence used in making that action represents the state. Each actor follows its own action selection policy, which controls how he explores his actions; various methods such as e-greedy exploration, guided exploration, or confidence value thresholds, can be used [20]. The Learner module takes the role of a critic. It consists of the learning method, reward functions, and the decision policy being learnt. A Learner monitors decisions being made in the system and calculates rewards based on the reward function, a list of decision/event pairs, and signals from the environment - in our case overlapping speech and too long silences - and publishes updated decision policy (the environment consists of the relevant modules in the system).

We use a delayed one-step Q-Learning method according to the formula:

$$Q(s,a) = Q(s,a) + \alpha[reward - Q(s,a)] \qquad (1)$$

Where $Q(s,a)$ is the learnt estimated return for picking action a in state s, and *alpha* is the learning rate. The reward functions - what events following what actions lead to what reward - need to be pre-determined in the Learner's configuration in the form of rules: A *reward* of x if *event* y succeeds at *action* z. Each decision has a lifetime in which system events can determine a reward, but reward can also be calculated in the case of an absence of an event, after the given lifetime has passed (e.g. no overlapping speech). Each time an action gets reward the return value is recalculated according to the formula above and the Learner broadcasts the new value.

In the current setup, Other-Gives-Turn-Decider-2 (OGTD-2) is an actor, in Sutton's [20] sense, that decides essentially what its name implies. This decider is only active in state I-Want-Turn. It learns an "optimal" pause duration so as not to speak on top of the other, while minimizing the lag in starting to speak. Each time a Speech-Off signal is detected, OGTD-2 receives analysis of the pitch in the last part of the utterance preceding the silence, from the Prosody-Analyzer. The prosody information is used to represent the state for the decision, a predicted safe pause duration is selected as the *action* and the Decision is posted. This pause duration determines when, in the future, the listener will start speaking/take the turn. In the case where the interlocutor starts speaking again before this pause duration has passed, two things can happen: (1) If he starts speaking before the predicted pause duration passes, the decider doesn't signal Other-Giving-Turn, essentially canceling the plan to start speaking. This leads to a better reward, since no overlapping speech occurred. (2) If he starts talking just after the pause duration has passed, after the decider signals Other-Gives-Turn, overlapping speech will likely occur, leading to negative reinforcement for this pause duration, based on the prosodic information. This learning strategy is based on the assumption that both agents want to take turns politely and efficiently. We have already begun expanding the system to be able to interrupt dynamically and deliberately - i.e. be "rude" - and the ability to switch back to being polite at any time, without destroying the learned data.

5 Evaluation Setup

We are aiming at an agent that can adapt its turntaking behavior to dialogue in a short amount of time. In this initial evaluation we focus exclusively on detecting turn-giving indicators in deliberately-generated prosody, leaving out the topic of turn-opportunity detection (i.e. turn transition without prior indication from the speaker that she's giving the turn), which would e.g. be necessary for producing (human-like) interruptions. The goal of the learning system is to learn to take turns with (ideally) no speech overlap, yet achieving the shortest possible silence duration between speaker turns. In this setup both agents always have something to say, leaving out issues such as turntaking perception and actions related to goal-directed switching between different conversational topics (think stream-of-consciousness conversations). First we describe an experiment where the system learns to take turns while interacting with itself. We then compare this to a pilot study of turntaking in human-human dialogue, using the same analysis mechanisms as in the autonomous system.

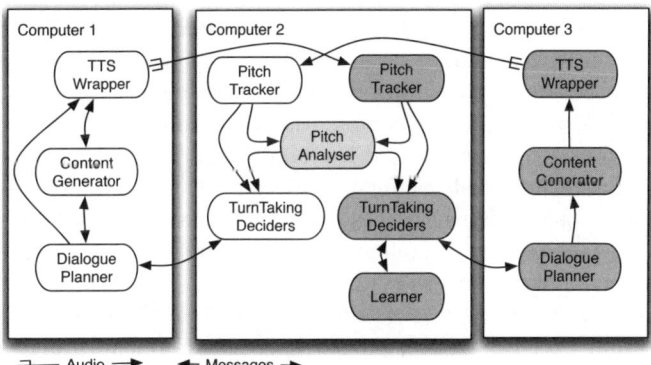

Fig. 4. Two agents distributed over a set of 3 computers. Time syncing and time drift pose a problem when measuring small time units, so all modules that we are specifically measuring time for are located on the same computer.

We have set up two instances of the system (agents) talking to each other (see Figure 4). One agent, Simon, is learning, the other, Kate, is not learning. Kate will only start speaking when she detects a 2-second pause; pause duration = 2 sec (controlled by her Other-Gives-Turn-Decider-1). Simon has the same 2 sec default behavior, but in addition he learns a variable pause duration (controlled by OGTD-2); its goal is to make this duration as small as possible, as described above. Content is produced in groups of between 1 and 5 randomly-selected sentence fragments, using the Loquendo speech synthesizer, which has been enabled

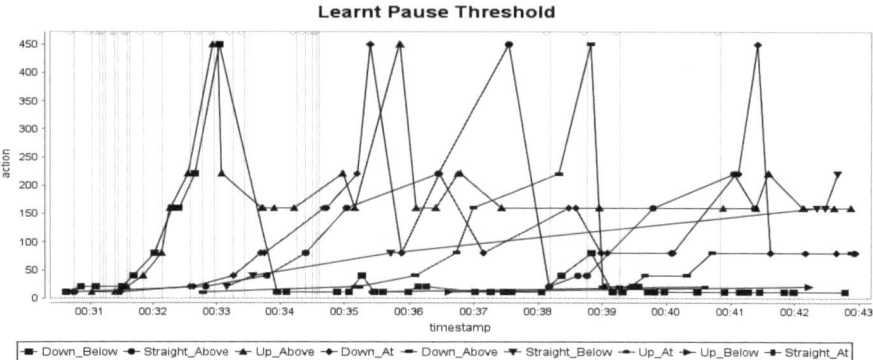

Fig. 5. The Learner's view of what is the best action (predicted "safe" pause duration) for each state, over a period of 13 minutes of learning, starting with no knowledge. Examining the actions, from smallest to largest pause duration, the system finds the optimal pause duration in as little as 3 minutes for the quickest learnt state (Down_Below), which is also the state with the shortest pause duration, indicating that this prosody pattern is a good sign of turn giving.

to start and stop synthesis at a moment's notice (around 100 msec). Loquendo uses markup to control the prosody; by adding a comma (,) to each fragment except the last one we can suppress a final fall [15] and keep the intonation up; by appending a period (.) we get a typical finall fall, e.g. "I went to my summer house," - "this weekend," - "the weather was nice," - "the whole time.". This way the intonation approximates typical spontaneous speech patterns. We selected sentences and fragment combinations to sound as natural as possible. (However, the fragments are selected randomly and assigned a role as either an intermediate fragment or last fragment.) Speech is thus produced incrementally by the combined speech synthesizer/speech planner system (speech synthesizer wrapper) as it receives each fragment, and the planner never commits to more than two fragments to the speech synthesizer at a time. As Loquendo introduces a short pause whenever there is a comma, and because of fluctuations in the transmission and execution time, pauses between fragments range from 31 to 4296 msecs, with 355 msecs being the average (see Figure 8).

5.1 Parameter Settings

Formulating the task as a reinforcement learning problem, the latest pitch tail represents the *state* and the pause duration is the *action* being selected. We have split the continuous action space into discrete logarithmic values starting with 10 msec and doubling the value up to 2 sec (the maximum pause duration where the system takes the turn by default). The action selection policy for OGTD-2 is e-greedy with 10% exploration and always selecting the shortest wait time if two or more actions share the top spot.

The reward given for decisions that do not lead to overlapping speech is the milliseconds in the selected pause; a decision to wait 100 msec from a Speech-Off signal until the I-Take-Turn decision is taken, receives a reward of -100 and -10 for deciding on a pause threshold of 10 msec. If, however, overlapping speech results from the decision, a reward of -2000 (same as waiting the maximum amount of time) is given. All rewards in the system are thus negative, resulting in unexplored actions being the best option at each time since return starts at 0.0 and once a reward has been giving the return goes down (exploration is guided towards getting faster time than the currently best action, so a pause duration larger than optimal is not explored). This can be seen in Figure 5 where each action is tried at least once before the system settles down at a tighter time range.

6 Results from System Interacting with Self

Looking at Figure 6, we clearly see that the learning agent starts by selecting pause durations that are too short and numerous overlaps are detected. The agent quickly learns the "safe" amount of time to wait before deciding turn was given, based on the prosody pattern. After 5 minutes of interaction interruptions have dropped below one per minute and occurrences after the 11th minute are solely caused by continued exploration. Each agent gets turn on average 4 times per minute (i.e. 8 successful turn exchanges). Average silence between turns is always larger than pause duration since it also includes delays resulting from processing the pitch and the time from deciding to speak until speech is delivered. These will be improved as we continue to tune the architecture's runtime performance. Interestingly, although pause duration is increased considerably for some prosody patterns, the influence on silence is minimal.

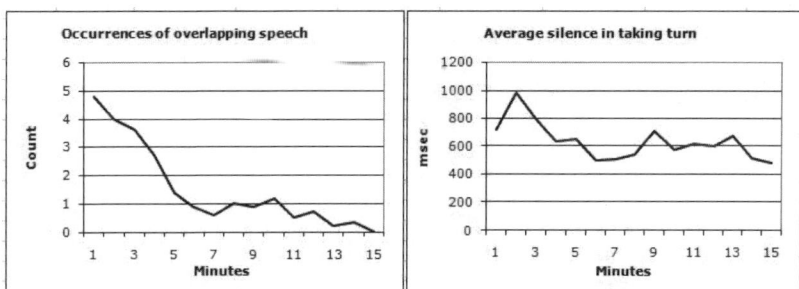

Fig. 6. When the system learns correct relations between prosody and pause duration, overlaps cease to occur (except as a function of exploring). Initial exploring of the action space is responsible for the average silence observed between turns increasing in the 2nd and the 3rd minute.

Based on our model of turntaking as a loosely-coupled system, it makes sense to compare the synchronization of "beliefs" in the agents about what turntaking context they are in (see Figure 7). At the beginning of the run there is not much consensus on who has the turn - the learning agent is constantly erroneously interrupting the other. After a few minutes of conversation the learning agent has adjusted his policy and turntaking goes smoothly.

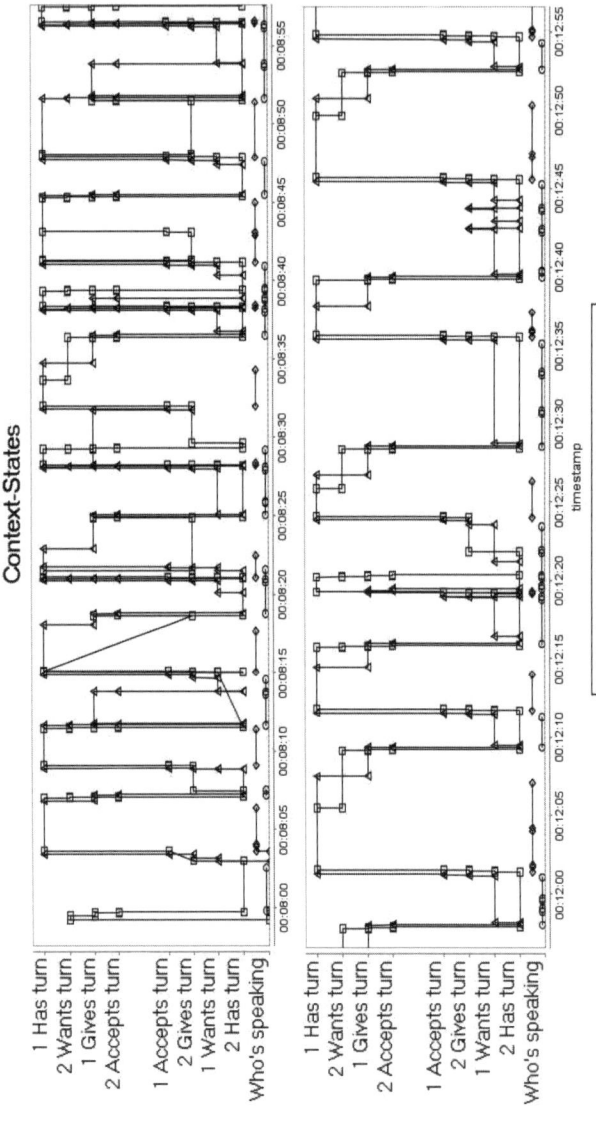

Fig. 7. Difference in dialogue context-states in the beginning of learning (upper graph) and after 5 minutes of learning (lower graph). At the bottom of each graph we plot who is speaking. The lines in each graph (1 Has Turn, 2 Wants turn, etc.) represent context-states in the two agents (see Figure 3). Each agent tries, using its perceptions, to align its current context-state with the other's. As one learns to do this, overlaps almost disappear, - a less dense graph means agent 1 has learned to couple his context-states to agent 2.

7 Comparison to Human-Human Interaction Study

We conducted a human-subject study to compare this system to. A priori we had reason to believe that people have more variable prosody generation patterns than the Loquendo speech synthesizer, with more variable silences within turn. In our human-subject study 8 telephone conversations were recorded in dual channel mode, and analyzed using the same mechanisms that the agents in our system use for deciding their turn behavior; each phone conversation was fed through our system just as if the sound was coming from two agents. The results show that silences within turn are considerably longer than expected (see Figure 8). Participants are also on average longer to take the turn than our agents, but minimum *successful* lag between turns is considerably shorter for humans (31 msec vs. 60 msec).

Length of silences in dialogue

Human-Human conversation	Average	Min	Max
Silences within turns	565	60	8468
Silences between turns	932	62	3671
Computer-Computer Coversation	Average	Min	Max
Silences within turns	355	31	4296
Silences between turns	582	172	1093

Fig. 8. Comparison of silence duration in human dialogue and between our agents. Within-turn silences: silences where the person is not giving turn; between-turn silences: successful turntaking opportunities.

For a more accurate comparison of human-human and computer-computer data, certain types of turntakings may need to be eliminated from the corpus, namely those that clearly involve a switch in topic (as these were casual conversations they include some long silences where neither party has anything to say). Another source of bias may be the fact that these dialogues were collected over Skype, which typically contains somewhat larger lag times in audio transmission than landline telephone, and certainly more lag than face-to-face conversation.

8 Conclusions and Future Work

We have built a system that uses prosody to learn optimal pause durations for taking turns, minimizing speech overlaps. The system learns this on the fly, in a full-duplex "open-mic" (dynamic interaction) scenario, and can take turns very efficiently in dialogues with itself, in human-like ways. The system uses prosodic information for finding categories of pitch that can serve a predictor of turn-giving behavior of interlocutors. As the system learns on-line it will be able to adjust to the particulars of individual speaking styles; while this remains to be tested our preliminary results indicate that this is indeed possible. At present the system is limited to a small set of turntaking circumstances where content does not play a role, assuming "friendly" conversation where both parties want to minimize overlaps in speech. Silences caused by outside interruptions - extended durations caused by searching for "the right words", and deliberate interruption techniques - are all topics for future study. The system is highly expandable, however, as it was built as part of a much larger system architecture that addresses multiple topic- and task-oriented dialogue, as well as multiple modes. In the near future we expect to expand the system to more advanced interaction types and situations, and to start using it in dynamic human-agent interactions. The learning mechanism described here will be expanded to learn not just the shortest durations but also the most efficient turntaking techniques in multimodal interaction under many different conditions. The turntaking system is architected in such a way as to allow a mixed-control relationship with outside processes. This means that we can expand it to handle situations where the goals of the dialogue may be very different from "being polite", even adversarial, as for example in on-air open-mic political debates. How easy this is remains to be seen; the main question revolves around the learning systems - how to manage learning in multiple circumstances without negatively affecting prior training.

Acknowledgments. This work was supported in part by a research grant from RANNIS, Iceland, and by a Marie Curie European Reintegration Grant within the 6th European Community Framework Programme. The authors wish to thank Yngvi Bjornsson for his contributions to the development of the reinforcement mechanisms.

References

1. Goodwin, C.: Conversational organization: Interaction between speakers and hearers. Academic Press, New York (1981)
2. Jonsdottir, G.R., Gratch, J., Fast, E., Thórisson, K.R.: Fluid semantic backchannel feedback in dialogue: Challenges and progress. In: Pélachaud, C., Martin, J.-C., André, E., Chollet, G., Karpouzis, K., Pelé, D. (eds.) IVA 2007. LNCS (LNAI), vol. 4722, pp. 154–160. Springer, Heidelberg (2007)
3. Edlund, J., Heldner, M., Gustafson, J.: Utterance segmentation and turn-taking in spoken dialogue systems (2005)
4. Thórisson, K.R.: Natural turn-taking needs no manual: Computational theory and model, from perception to action. In: Granström, B., House, D.I.K. (eds.) Multimodality in Language and Speech Systems, pp. 173–207. Kluwer Academic Publishers, Dordrecht (2002)
5. Card, S.K., Moran, T.P., Newell, A.: The model human processor: An engineering model of human performance. In: Handbook of Human Perception, vol. II. John Wiley and Sons, Chichester (1986)
6. Thórisson, K.R.: Dialogue control in social interface agents. In: INTERCHI Adjunct Proceedings, 139–140 (1993)
7. Thórisson, K.R.: Communicative Humanoids: A Computational Model of Psycho-Social Dialogue Skills. PhD thesis, Massachusetts Institute of Technology (1996)
8. Sacks, H., Schegloff, E.A., Jefferson, G.A.: A simplest systematics for the organization of turn-taking in conversation. Language 50, 696–735 (1974)

9. Thórisson, K.R.: Modeling multimodal communication as a complex system. In: Wachsmuth, I., Knoblich, G. (eds.) ZiF Research Group International Workshop. LNCS (LNAI), vol. 4930, pp. 143–168. Springer, Heidelberg (2008)
10. Sato, R., Higashinaka, R., Tamoto, M., Nakano, M., Aikawa, K.: Learning decision trees to determine turn-taking by spoken dialogue systems. In: ICSLP 2002, pp. 861–864 (2002)
11. Traum, D.R., Heeman, P.A.: Utterance units and grounding in spoken dialogue. In: Proc. ICSLP 1996., Philadelphia, PA, vol. 3, pp. 1884–1887 (1996)
12. Schlangen, D.: From reaction to prediction: Experiments with computational models of turn-taking. In: Proceedings of Interspeech 2006, Panel on Prosody of Dialogue Acts and Turn-Taking, Pittsburgh, USA (September 2006)
13. Raux, A., Eskenazi, M.: Optimizing endpointing thresholds using dialogue features in a spoken dialogue system. In: Proceedings of the 9th SIGdial Workshop on Discourse and Dialogue, Columbus, Ohio, Association for Computational Linguistics, pp. 1–10 (June 2008)
14. Gratch, J., Okhmatovskaia, A., Lamothe, F., Marsella, S., Morales, M., van der Werf, R.J., Morency, L.P.: Virtual rapport. In: Gratch, J., Young, M., Aylett, R.S., Ballin, D., Olivier, P. (eds.) IVA 2006. LNCS (LNAI), vol. 4133, pp. 14–27. Springer, Heidelberg (2006)
15. Pierrehumbert, J., Hirschberg, J.: The meaning of intonational contours in the interpretation of discourse. In: Cohen, P.R., Morgan, J., Pollack, M. (eds.) Intentions in Communication, pp. 271–311. MIT Press, Cambridge (1990)
16. Ng-Thow-Hing, V., List, T., Thórisson, K.R., Lim, J., Wormer, J.: Design and evaluation of communication middleware in a distributed humanoid robot architecture. In: Prassler, E., Nilsson, K., Shakhimardanov, A. (eds.) IEEE/RSJ Int. Conf. on Intelligent Robots and Systems (IROS 2007) Workshop on Measures and Procedures for the Evaluation of Robot Architectures and Middleware (2007)
17. Thorisson, K.R., Benko, H., Arnold, A., Abramov, D., Maskey, S., Vaseekaran, A.: Constructionist design methodology for interactive intelligences. A.I. Magazine 25(4), 77–90 (2004)
18. Nivel, E., Thórisson, K.R.: Prosodica: A realtime prosody tracker for dynamic dialogue. Technical report, Reykjavik University Department of Computer Science, Technical Report RUTR-CS08001 (2004)
19. Thórisson, K.R.: Machine perception of multimodal natural dialogue. In: McKevitt, P., Nulláin, S.Ó., Mulvihill, C. (eds.) Language, Vision & Music, pp. 97–115. John Benjamins, Amsterdam (2002)
20. Sutton, R.S., Barto, A.G.: Reinforcement Learning: An Introduction. The MIT Press, Cambridge (1998)

Predicting Listener Backchannels:
A Probabilistic Multimodal Approach

Louis-Philippe Morency[1], Iwan de Kok[2], and Jonathan Gratch[1]

[1] Institute for Creative Technologies, University of Southern California,
13274 Fiji Way, Marina del Rey CA 90292, USA
{morency,gratch}@ict.usc.edu
[2] Human Media Interaction Group, University of Twente,
P.O. Box 217, 7500AE, Enschede, The Netherlands
i.a.dekok@student.utwente.nl

Abstract. During face-to-face interactions, listeners use backchannel feedback such as head nods as a signal to the speaker that the communication is working and that they should continue speaking. Predicting these backchannel opportunities is an important milestone for building engaging and natural virtual humans. In this paper we show how sequential probabilistic models (e.g., Hidden Markov Model or Conditional Random Fields) can automatically learn from a database of human-to-human interactions to predict listener backchannels using the speaker multimodal output features (e.g., prosody, spoken words and eye gaze). The main challenges addressed in this paper are automatic selection of the relevant features and optimal feature representation for probabilistic models. For prediction of visual backchannel cues (i.e., head nods), our prediction model shows a statistically significant improvement over a previously published approach based on hand-crafted rules.

1 Introduction

Natural conversation is fluid and highly interactive. Participants seem tightly enmeshed in something like a dance, rapidly detecting and responding, not only to each other's words, but to speech prosody, gesture, gaze, posture, and facial expression movements. These "extra-linguistic" signals play a powerful role in determining the nature of a social exchange. When these signals are positive, coordinated and reciprocated, they can lead to feelings of rapport and promote beneficial outcomes in such diverse areas as negotiations and conflict resolution [1,2], psychotherapeutic effectiveness [3], improved test performance in classrooms [4] and improved quality of child care [5].

Not surprisingly, supporting such fluid interactions has become an important topic of virtual human research. Most research has focused on individual behaviors such as rapidly synthesizing the gestures and facial expressions that co-occur with speech [6,7,8,9] or real-time recognition the speech and gesture of a human speaker [10,11]. But as these techniques have matured, virtual human research has increasingly focused on dyadic factors such as the feedback a listener provides in the midst of the other participants speech [12,13]. These include recognizing and generating backchannel or jump-in points [14] turn-taking and floor control signals, postural mimicry [15] and emotional feedback [16,17]. In particular, backchannel feedback (the nods and paraverbals such as "uh-huh" and "mm-hmm" that listeners produce as some is speaking) has received considerable interest due to its pervasiveness across languages and conversational contexts and this paper addresses the problem of how to predict and generate this important class of dyadic nonverbal behavior.

Generating appropriate backchannels is a notoriously difficult problem. Listener backchannels are generated rapidly, in the midst of speech, and seem elicited by a variety of speaker verbal, prosodic and nonverbal cues. Backchannels are considered as a signal to the speaker that the communication is working and that they should continue speaking [18]. There is evidence that people can generate such feedback without necessarily attending to the content of speech [19], and this has motivated a host of approaches that generate backchannels based solely on surface features (e.g., lexical and prosodic) that are available in real-time.

This paper describes a general probabilistic framework for learning to predict and generate dyadic conversational behavior from multimodal conversational data, and applies this framework to listener backchanneling behavior. As shown in Figure 1, our approach is designed to generate real-time backchannel feedback for virtual agents. The paper provides several advances over prior art. Unlike prior approaches that use a single modality (e.g., speech), we incorporate multi-modal features (e.g., speech and gesture). We present a machine learning method that automatically selects appropriate features from multimodal data and produces sequential probabilistic models with greater predictive accuracy than prior approaches.

The following section describes previous work in backchannel generation and explains the differences between our prediction model and other predictive models. Section 3 describes the details of our prediction model including the encoding dictionary and our feature selection algorithm. Section 4 presents the way we collected the data used for training and evaluating our model as well as the methodology used to evaluate the performance of our prediction model. In Section 5 we discuss our results and conclude in Section 6.

2 Previous Work

Several researchers have developed models to predict when backchannel should happen. In general, these results are difficult to compare as they utilize different corpora and present varying evaluation metrics. In fact, we are not aware of a paper that makes a direct comparison between alternative methods.

Ward and Tsukahara [14] propose a unimodal approach where backchannels are associated with a region of low pitch lasting 110ms during speech. Models were produced manually through an analysis of English and Japanese conversational data.

Nishimura et al. [20] present a unimodal decision-tree approach for producing backchannels based on prosodic features. The system analyzes speech in 100ms

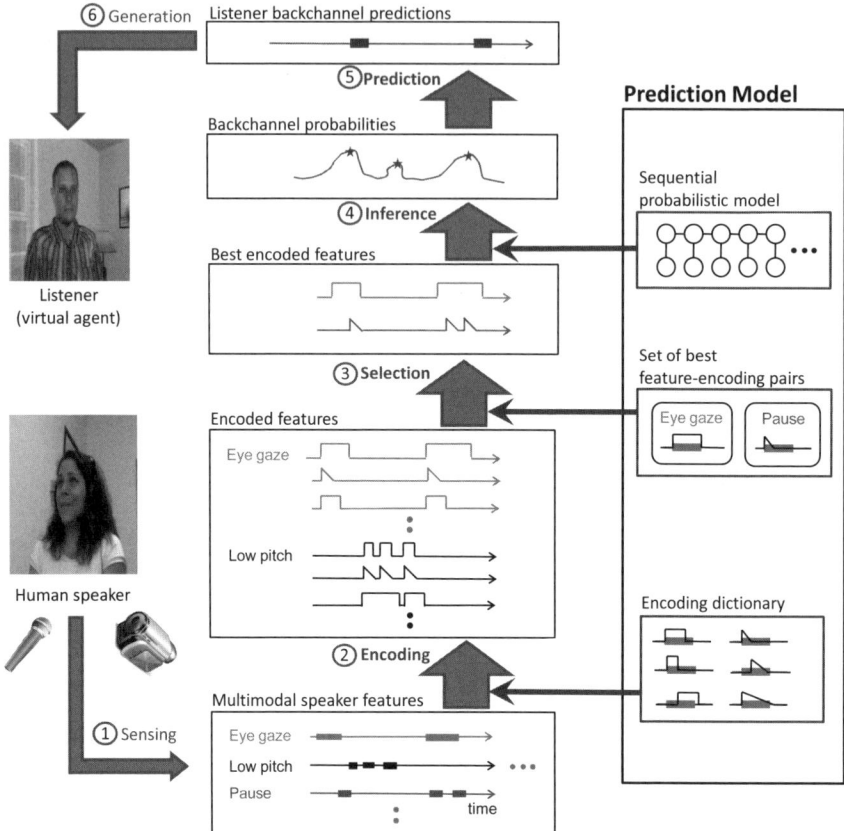

Fig. 1. Our prediction model is designed for generating in real-time backchannel feedback for a listener virtual agent. It uses speaker multimodal features such as eye gaze and prosody to make predictions. The timing of the backchannel predictions and the optimal subset of features is learned automatically using a sequential probabilistic model.

intervals and generates backchannels as well as other paralinguistic cues (e.g., turn taking) as a function of pitch and power contours. They report a subjective evaluation of the system where subjects were asked to rate the timing, naturalness and overall impression of the generated behaviors but no rigorous evaluation of predictive accuracy.

Cathcart et al. [21] propose a unimodal model based on pause duration and trigram part-of-speech frequency. The model was constructed by identifying, from the HCRC Map Task Corpus [22], trigrams ending with a backchannel. For example, the trigram most likely to predict a backchannel was (<NNS> <pau> <bc>), meaning a plural noun followed by a pause of at least 600ms. The algorithm was formally evaluated on the HCRC data set, though there was no direct comparison to other methods. As part-of-speech tagging is a challenging requirement for a real-time system, this approach is of questionable utility to the design of interactive virtual humans

Fujie et al. used Hidden Markov Models to perform head nod recognition [23]. In their paper, they combined head gesture detection with prosodic low-level features from the same person to determine strongly positive, weak positive and negative responses to yes/no type utterances.

Maatman et al. [24] present a multimodal approach where Ward and Tsukhara's prosodic algorithm is combined with a simple method of mimicking head nods. No formal evaluation of the predictive accuracy of the approach was provided but subsequent evaluations have demonstrated that generated behaviors do improve subjective feelings of rapport [25] and speech fluency [15].

No system, to date, has demonstrated how to automatically learn a predictive model of backchannel feedback from multi-modal conversational data nor have there been definitive head-to-head comparisons between alternative methods.

3 Prediction Model

The goal of our prediction model is to create real-time predictions of listener backchannel based on multimodal features from the human speaker. Our prediction model learns automatically which speaker feature is important and how they affect the timing of listener backchannel. We achieve this goal by using a machine learning approach: we train a sequential probabilistic model from a database of human-human interactions and use this trained model in a real-time backchannel generator (as depicted in Figure 1).

A sequential probabilistic model takes as input a sequence of observation features (e.g., the speaker features) and returns a sequence of probabilities (i.e., probability of listener backchannel). Two of the most popular sequential models are Hidden Markov Model (HMM) [26] and Conditional Random Field (CRF) [27]. One of the main difference between these two models is that CRF is discriminative (i.e., tries to find the best way to differentiate cases where the listener gives backchannel to cases where it does not) while HMM is generative (i.e., tries to find the best way to generalize the samples from the cases where the listener gives backchannel without looking at the cases where the listener did not give backchannel). Our prediction model is designed to work with both types of sequential probabilistic models.

Machine learning approaches like HMM and CRF are not magic. Simply downloading a Matlab toolbox from the internet and applying on your training dataset will not magically give you a prediction model (if it does, you should go purchase a lottery ticket right away!). These sequential models have constraints that you need to understand before using them:

- **Limited learning.** The more informative your features are, the better your sequential model will perform. If the input features are too noisy (e.g., direct signal from microphone), it will make it harder for the HMM or CRF to learn the important part of the signal. By pre-processing your input features

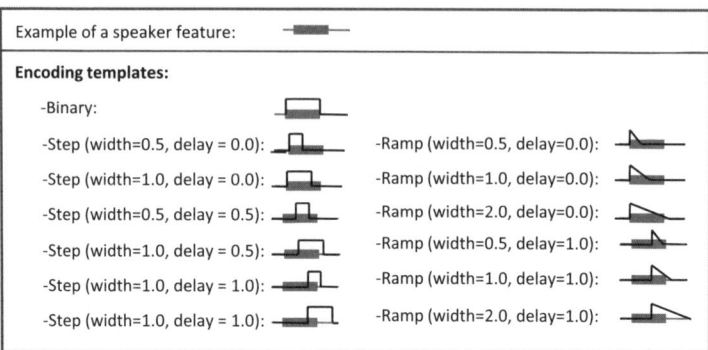

Fig. 2. Encoding dictionary. This figure shows the different encoding templates used by our prediction model. Each encoding templates were selected to model different relationships between speaker features (e.g., a pause or an intonation change) and listener backchannels. We included a delay parameter in our dictionary since listener backchannels can sometime happen later after speaker features (e.g., Ward and Tsukahara [14]). This encoding dictionary gives a more powerful set of input features to the sequential probabilistic model which improves the performance of our prediction model.

to highlight their influences on your label (e.g., listener backchannel) you improve your chance of success.
– **Over-fitting.** The more complex your model is, the more training data it needs. Every input feature that you add increases its complexity and at the same time its need for a larger training set. Since we usually have a limited set of training sequences, it is important to keep the number of input features low.

In our prediction model we directly addressed these issues by focusing on the feature representation and feature selection problems:

– **Encoding dictionary.** To address the limited learning constraint of sequential models, we suggest to use more than binary encoding to represent input features. Our encoding dictionary contains a series of encoding templates that were designed to model different relationship between a speaker feature (e.g., a speaker in not currently speaking) and listener backchannel. The encoding dictionary and its usage are described in Section 3.1.
– **Automatic feature and encoding selection.** Because of the over-fitting problem happening when too many uncorrelated features (i.e., features that do not influence listener backchannel) are used, we suggest two techniques for automatic feature and encoding selection based on co-occurence statistics and performances evaluation on a validation dataset. Our feature selection algorithms are described in Section 3.2.

The following two sections describe our encoding dictionary and feature selection algorithm. Section 3.3 describes how the probabilities output from our sequential model are used to generate backchannel.

3.1 Encoding Dictionary

The goal of the encoding dictionary is to propose a series of encoding templates that potentially capture the relationship between speaker features and listener backchannel. The Figure 2 shows the 13 encoding templates used in our experiments. These encoding templates were selected to represent a wide range of ways that a speaker feature can influence the listener backchannel. These encoding templates were also selected because they can easily be implemented in real-time since the only needed information is the start time of the speaker feature. Only the binary feature also uses the end time. In all cases, no knowledge of the future is needed.

The three main types of encoding templates are:

- **Binary encoding.** This encoding is designed for speaker features which influence on listener backchannel is constraint to the duration of the speaker feature.
- **Step function.** This encoding is a generalization of binary encoding by adding two parameters: width of the encoded feature and delay between the start of the feature and its encoded version. This encoding is useful if the feature influence on backchannel is constant but with a certain delay and duration.
- **Ramp function.** This encoding linearly decreases for a set period of time (i.e., width parameter). This encoding is useful if the feature influence on backchannel is changing over time.

It is important to note that a feature can have an *individual* influence on backchannel and/or a *joint* influence. An *individual* influence means the input feature directly influences listener backchannel. For example, a long pause can by itself trigger backchannel feedback from the listener. A *joint* influence means that more than one feature is involved in triggering the feedback. For example, saying the word "and" followed by a look back at the listener can trigger listener feedback. This also means that a feature may need to be encoded more than one way since it may have a *individual* influence as well as one or more *joint* influences.

One way to use the encoding dictionary with a small set of features is to encode each input feature with each encoding template. We tested this approach in our experiment with a set of 12 features (see Section 5) but because of the problem of over-fitting, a better approach is to select the optimal subset of input features and encoding templates. The following section describes our feature selection algorithm.

3.2 Automatic Feature Selection

We perform the feature selection based on the same concepts of *individual* and *joint* influences described in the previous section. Individual feature selection is designed to asses the individual performance of each speaker feature while the joint feature selection looks at how features can complement each other to improve performance.

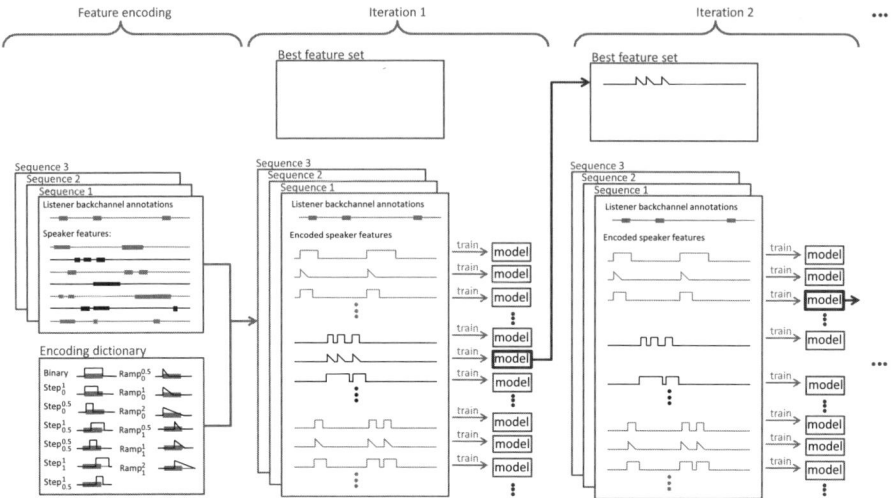

Fig. 3. Joint Feature selection. This figure illustrates the feature encoding process using our encoding dictionary as well as two iterations of our joint feature selection algorithm. The goal of joint selection is to find a subset of features that best complement each other for prediction of listener backchannel.

Individual Feature Selection. Individual feature selection is designed to do a pre-selection based on (1) the statistical co-occurence of speaker features and listener backchannel, and (2) the individual performance of each speaker feature when trained with any encoding template and evaluated on a validation set.

The first step of individual selection looks at statistics of co-occurence between backchannel instances and speaker features. The number of co-occurence is equal to the number of times a listener backchannel instance happened between the start time of the feature and up to 2 seconds after it. This threshold was selected after analysis of the average co-occurence histogram for all features. After this step the number of features is reduced to 50.

The second step is to look at the best performance an individual feature can reach when trained with any of the encoding templates in our dictionary. For each top-50 feature a sequential model is trained for encoding template and then evaluated. A ranking is made based on the best performance of each individual feature and a subset of 12 features is selected.

Joint Feature Selection. Given the subset of features that performed best when trained individually, we now build the complete set of feature hypothesis to be used by the joint feature selection process. This set represents each feature encoded with all possible encoding templates from our dictionary. The goal of joint selection is to find a subset of features that best complements each other for prediction of backchannel. Figure 3 shows the first two iterations of our algorithm.

The algorithm starts with the complete set of feature hypothesis and an empty set of *best* features. At each iteration, the best feature hypothesis is selected and added to the best feature set. For each feature hypothesis, a sequential model is trained and evaluated using the feature hypothesis and all features previously selected in the best feature set. While the first iteration of this process is really similar to the individual selection, every iteration afterward will select a feature that best complement the current best features set. Note that during the joint selection process, the same feature can be selected more than once with different encodings. The procedure stops when the performance starts decreasing.

3.3 Generating Listener Backchannel

The goal of the prediction step is to analyze the output from the sequential probabilistic model (see example in Figure 1) and make discrete decision about when backchannel should happen. The output probabilities from HMM and CRF models are smooth over time since both models have a transition model that insures no instantaneous transitions between labels. This smoothness of the output probabilities makes it possible to find distinct peaks. These peaks represent good backchannel opportunities. A peak can easily be detected in real-time since it is the point where the probability starts decreasing. For each peak we get a backchannel opportunity with associated probability.

Interestingly, Cathcart et al. [21] note that human listeners varied considerably in their backchannel behavior (some appear less expressive and pass up "backchannel opportunities") and their model produces greater precision for subjects that produced more frequent backchannels. The same observation was made by Ward and Tsukahara [14]. An important advantage of our prediction model over previous work is the fact that for each backchannel opportunity returned, we also have an associated probability. This makes it possible for our model to address the problem of expressiveness. By applying an expressiveness threshold on the backchannel opportunities, our prediction model can be used to create virtual agents with different levels of nonverbal expressiveness.

4 Experiments

For training and evaluation of our prediction model, we used a corpus of 50 human-to-human interactions. This corpus is described in Section 4.1. Section 4.2 describes the speaker features used in our experiments as well as our listener backchannel annotations. Finally Section 4.3 discusses our methodology for training the probabilistic model and evaluate it.

4.1 Data Collection

Participants (67 women, 37 men) were recruited through Craigslist.com from the greater Los Angles are and compensated $20. Of the 52 sessions, two were excluded due to recording equipment failure, resulting in 50 valid sessions.

Participants in groups of two entered the laboratory and were told they were participating in a study to evaluate communication technology. They completed a consent form and pre-experiment questionnaire eliciting demographic and dispositional information and were randomly assigned the role of listener or speaker.

The listener was asked to wait outside the room while the speaker viewed a short video clip taken from a sexual harassment awareness video by Edge Training Systems, Inc dramatizing two incidents of workplace harassment. The listener was then led back into the computer room, where the speaker was instructed to retell the stories portrayed in the clips to the listener. Elicited stories were approximately two minutes in length on average. Speakers sat approximately 8 feet apart from the listener.

Finally, the experimenter led the speaker to a separate side room. The speaker completed a post-questionnaire assessing their impressions of the interaction while the listener remained in the room and spoke to the camera what s/he had been told by the speaker. Participants were debriefed individually and dismissed.

We collected synchronized multimodal data from each participant including voice and upper-body movements. Both the speaker and listener wore a lightweight headset with microphone. Three camcorders were used to videotape the experiment: one was placed in front the speaker, one in front of the listener, and one was attached to the ceiling to record both speaker and listener.

4.2 Speaker Features and Listener Backchannels

From the video and audio recordings several features were extracted. In our experiments the speaker features were sampled at a rate of 30Hz so that visual and audio feature could easily be concatenated.

Pitch and intensity of the speech signal were automatically computed from the speaker audio recordings, and acoustic features were derived from these two measurements. The following prosodic features were used (based on [14]):

- Downslopes in pitch continuing for at least 40ms
- Regions of pitch lower than the 26th percentile continuing for at least 110ms (i.e., lowness)
- Utterances longer than 700ms
- Drop or rise in energy of speech (i.e., energy edge)
- Fast drop or rise in energy of speech (i.e., energy fast edge)
- Vowel volume (i.e., vowels are usually spoken softer)

Human coders manually annotated the narratives with several relevant features from the audio recordings. All elicited narratives were transcribed, including pauses, filled pauses (e.g. "um"), incomplete and prolonged words. These transcriptions were double-checked by a second transcriber. This provided us with the following extra lexical and prosodic features:

- All individual words (i.e., unigrams)
- Pause (i.e., no speech)
- Filled pause (e.g. "um")
- Lengthened words (e.g., "I li::ke it")
- Emphasized or slowly uttered words (e.g., "ex_a_c_tly")
- Incomplete words (e.g., "jona-")
- Words spoken with continuing intonation

- Words spoken with falling intonation (e.g., end of an utterance)
- Words spoken with rising intonation (i.e., question mark)

From the speaker video the eye gaze of the speaker was annotated on whether he/she was looking at the listener. A test on five sessions we decided not to have a second annotator go through all the sessions, since annotations were almost identical (less than 2 or 3 frames difference in segmentation). The feature we obtained from these annotations is:

- Speaker looking at the listener

Note that although some of the speaker features were manually annotated in this corpus, all of these features can be recognized automatically given the recent advances in real-time keyword spotting [28], eye gaze estimation and prosody analysis.

Finally, the listener videos were annotated for visual backchannels (i.e., head nods) by two coders. These annotations form the labels used in our prediction model for training and evaluation.

4.3 Methodology

To train our prediction model we split the 50 session into 3 sets, a training set, a validation set and a test set. This is done by doing a 10-fold testing approach. This means that 10 sessions are left out for test purposes only and the other 40 are used for training and validation. This process is repeated 5 times in order to be able to test our model on each session. Validation is done by using the holdout cross-validation strategy. In this strategy a subset of 10 sessions is left out of the training set. This process is repeated 5 times and then the best setting for our model is selected based on the performance of our model.

The performance is measured by using the F-measure. This is the weighted harmonic mean of precision and recall. Precision is the probability that predicted backchannels correspond to actual listener behavior. Recall is the probability that a backchannel produced by a listener in our test set was predicted by the model. We use the same weight for both precision and recall, so called F_1. During validation we find all the peaks in our probabilities. A backchannel is predicted correctly if a peak in our probabilities (see Section 3.3) happens during an actual listener backchannel.

As discussed in Section 3.3, the expressiveness level is the threshold on the output probabilities of our sequential probabilistic model. This level is used to generate the final backchannel opportunities. In our experiments we picked the expressiveness level which gave the best F_1 measurement on the validation set. This level is used to evaluate our prediction model in the testing phase.

For space constraint reason, all the results presented in this paper are using Conditional Random Fields [27] as sequential probabilistic model. We performed the same series of experiments with Hidden Markov Models [26] but the results were constantly lower. The hCRF library was used for training the CRF model [29]. The regularization term for the CRF model was validated with values $10^k, k = -1..3$.

Algorithm 1. Rule Based Approach of Ward and Tsukahara [14]

Upon detection of
P1: a region of pitch less than the 26th percentile pitch level and
P2: continuing for at least 100 milliseconds
P3: coming after at least 700 milliseconds of speech,
P4: providing you have not output backchannel feedback within the preceding 800 milliseconds,
P5: after 700 milliseconds wait,
you should produce backchannel feedback.

5 Results and Discussion

We compared our prediction model with the rule based approach of Ward and Tsukahara [14] since this method has been employed effectively in virtual human systems and demonstrates clear subjective and behavioral improvements for human/virtual human interaction [15]. We re-implemented their rule based approach summarized in Algorithm 1. The two main features used by this approach are *low pitch regions* and *utterances* (see Section 4.2). We also compared our model with a "random" backchannel generator as defined in [14]: randomly generate a backchannel cue every time conditions P3, P4 and P5 are true (see Algorithm 1). The frequency of the random predictions was set to 60% which provided the best performance for this predictor, although differences were small.

Table 1 shows a comparison of our prediction model with both approaches. As can be seen, our prediction model outperforms both random and the rule based approach of Ward and Tsukahara. It is important to remember that a backchannel is correctly predicted if a detection happens during an actual listener backchannel. Our goal being to objectively evaluate the performance of our prediction model, we did not allow for an extra delay before or after the actual listener backchannel. Our error criterion does not use any extra parameter (e.g., the time window for allowing delays before and/or after the actual backchannel). This stricter criterion can explain the lower performance of Ward and Tsukahara approach in Table 1 when compared with their published results which used a time window of 500ms [14]. We performed an one-tailed t-test comparing our prediction model to both random and Ward's approach over our 50 independent sessions. Our performance is significantly higher than both random and the hand-crafted rule based approaches with p-values comfortably below 0.01. The one-tailed t-test comparison between Ward's system and random shows that that difference is only marginally significant.

Our prediction model uses two types of feature selections: individual feature selection and joint feature selection (see Section 3.2 for details). It is very interesting to look at the features and encoding selected after both processes:

- *Pause* using binary encoding
- *Speaker looking at the listener* using ramp encoding with a width of 2 seconds and a 1 second delay
- *'and'* using step encoding with a width 1 second and a delay of 0.5 seconds

Table 1. Comparison of our prediction model with previously published rule-based system of Ward and Tsukahara [14]. By integrating the strengths of a machine learning approach with multimodal speaker features and automatic feature selection, our prediction model shows a statistically significant improvement over the unimodal rule-based and random approaches.

	Results			T-Test (p-value)	
	F_1	Precision	Recall	Random	Ward
Our prediction model (with feature selection)	0.2236	0.1862	0.4106	<0.0001	0.0020
Ward's rule-based approach [12]	0.1457	0.1381	0.2195	0.0571	-
Random	0.1018	0.1042	0.1250	-	-

Table 2. Compares the performance of our prediction model before and after joint feature selection(see Section 2). We can see that joint feature selection is an important part of our prediction model.

	Results			T-Test
	F_1	Precision	Recall	(p-value)
Joint and individual feature selections	0.2236	0.1862	0.4106	0.1312
Only individual features selection	0.1928	0.1407	0.5145	

– *Speaker looking at the listener* using binary encoding

The joint selection process stopped after 4 iterations, the optimal number of iterations on the validation set. Note that *Speaker looking at the listener* was selected twice with two different encodings. This reinforces the fact that having different encodings of the same feature reveals different information of a feature and is essential to getting high performance with this approach. It is also interesting to see that our prediction algorithm outperform Ward and Tsukahara without using their feature corresponding of low pitch.

In Table 2 we show that the addition joint feature selection improved performance over individual feature selection alone. In the second case the sequential model was trained with all the 12 features returned by the individual selection algorithm and every encoding templates from our dictionary. These speaker features were: pauses, energy fast edges, lowness, speaker looking at listener, "and", vowel volume, energy edge, utterances, downslope, "like", falling intonations, rising intonations.

Table 3. Compares the performance of our prediction model with and without the visual speaker feature (i.e., speaker looking at the listener). We can see that the multimodal factor is an important part of our prediction model.

	Results			T-Test
	F_1	Precision	Recall	(p-value)
Multimodal Features	0.1928	0.1407	0.5145	0.1454
Unimodal Features	0.1664	0.1398	0.3941	

In Table 3 the importance of multimodality is showed. Both of these models were trained with the same 12 features described earlier, except that the unimodal model did not include the *Speaker looking at the listener* feature. Even though we only added one visual feature between the two models, the performance of our prediction model increased by approximately 3%. This result shows that multimodal speaker features is an important concept.

6 Conclusion

In this paper we presented how sequential probabilistic models can be used to automatically learn from a database of human-to-human interactions to predict listener backchannel using the speaker multimodal output features (e.g., prosody, spoken words and eye gaze). The main challenges addressed in this paper were automatic selection of the relevant features and optimal feature representation for probabilistic models. For prediction of visual backchannel cues (i.e., head nods), our prediction model was showed a statistically significant improvement over a previously published approach based on hand-crafted rules. Although we applied the approach to generating backchannel behavior, the method is proposed as a general probabilistic framework for learning to recognize and generate meaningful multimodal behaviors from examples of face-to-face interactions including facial expressions, posture shifts, and other interactional signals. Thus, it has importance, not only as a means to improving the interactivity and expressiveness of virtual humans but as an fundamental tool for uncovering hidden patterns in human social behavior.

Acknowledgements

The authors would like to thank Nigel Ward for his valuable feedback, Marco Levasseur and David Carre for helping to build the original Matlab prototype, Brooke Stankovic, Ning Wang and Jillian Gerten. This work was sponsored by the U.S. Army Research, Development, and Engineering Command (RDECOM) and the National Science Foundation under grant # HS-0713603. The content does not necessarily reflect the position or the policy of the Government, and no official endorsement should be inferred.

References

1. Drolet, A., Morris, M.: Rapport in conflict resolution: accounting for how face-to-face contact fosters mutual cooperation in mixed-motive conflicts. Experimental Social Psychology 36, 26–50 (2000)
2. Goldberg, S.: The secrets of successful mediators. Negotiation Journal 21(3), 365–376 (2005)
3. Tsui, P., Schultz, G.: Failure of rapport: Why psychotheraputic engagement fails in the treatment of asian clients. American Journal of Orthopsychiatry 55, 561–569 (1985)

4. Fuchs, D.: Examiner familiarity effects on test performance: implications for training and practice. Topics in Early Childhood Special Education 7, 90–104 (1987)
5. Burns, M.: Rapport and relationships: The basis of child care. Journal of Child Care 2, 47–57 (1984)
6. Cassell, J., Vilhjlmsson, H., Bickmore, T.: Beat: The behavior expressive animation toolkit. In: Proceedings of the SIGGRAPH (2001)
7. Lee, J., Marsella, S.: Nonverbal behavior generator for embodied conversational agents. In: Gratch, J., Young, M., Aylett, R.S., Ballin, D., Olivier, P. (eds.) IVA 2006. LNCS (LNAI), vol. 4133, pp. 243–255. Springer, Heidelberg (2006)
8. Kipp, M., Neff, M., Kipp, K., Albrecht, I.: Toward natural gesture synthesis: Evaluating gesture units in a data-driven approach. In: Pélachaud, C., Martin, J.-C., André, E., Chollet, G., Karpouzis, K., Pelé, D. (eds.) IVA 2007. LNCS (LNAI), vol. 4722, pp. 15–28. Springer, Heidelberg (2007)
9. Thiebaux, M., Marshall, A., Marsella, S., Kallmann, M.: Smartbody: Behavior realization for embodied conversational agents. In: AAMAS (2008)
10. Morency, L.P., Sidner, C., Lee, C., Darrell, T.: Contextual recognition of head gestures. In: ICMI (October 2005)
11. Demirdjian, D., Darrell, T.: 3-d articulated pose tracking for untethered deictic reference. In: Int'l Conf. on Multimodal Interfaces (2002)
12. Heylen, D., Bevacqua, E., Tellier, M., Pelachaud, C.: Searching for prototypical facial feedback signals. In: Pélachaud, C., Martin, J.-C., André, E., Chollet, G., Karpouzis, K., Pelé, D. (eds.) IVA 2007. LNCS (LNAI), vol. 4722, pp. 147–153. Springer, Heidelberg (2007)
13. Kopp, S., Stocksmeier, T., Gibbon, D.: Incremental multimodal feedback for conversational agents. In: Pélachaud, C., Martin, J.-C., André, E., Chollet, G., Karpouzis, K., Pelé, D. (eds.) IVA 2007. LNCS (LNAI), vol. 4722, pp. 139–146. Springer, Heidelberg (2007)
14. Ward, N., Tsukahara, W.: Prosodic features which cue back-channel responses in english and japanese. Journal of Pragmatics 23, 1177–1207 (2000)
15. Gratch, J., Wang, N., Gerten, J., Fast, E.: Creating rapport with virtual agents. In: Pélachaud, C., Martin, J.-C., André, E., Chollet, G., Karpouzis, K., Pelé, D. (eds.) IVA 2007. LNCS (LNAI), vol. 4722. Springer, Heidelberg (2007)
16. Jónsdóttir, G.R., Gratch, J., Fast, E., Thórisson, K.R.: Fluid semantic backchannel feedback in dialogue: Challenges and progress. In: Pélachaud, C., Martin, J.-C., André, E., Chollet, G., Karpouzis, K., Pelé, D. (eds.) IVA 2007. LNCS (LNAI), vol. 4722. Springer, Heidelberg (2007)
17. Allwood, J.: Dimensions of Embodied Communication - towards a typology of embodied communication. In: Embodied Communication in Humans and Machines, Oxford University Press, Oxford
18. Yngve, V.: On getting a word in edgewise. In: Proceedings of the Sixth regional Meeting of the Chicago Linguistic Society (1970)
19. Bavelas, J., Coates, L., Johnson, T.: Listeners as co-narrators. Journal of Personality and Social Psychology 79(6), 941–952 (2000)
20. Nishimura, R., Kitaoka, N., Nakagawa, S.: A spoken dialog system for chat-like conversations considering response timing. In: Matoušek, V., Mautner, P. (eds.) TSD 2007. LNCS (LNAI), vol. 4629, pp. 599–606. Springer, Heidelberg (2007)
21. Cathcart, N., Carletta, J., Klein, E.: A shallow model of backchannel continuers in spoken dialogue. In: European ACL, pp. 51–58 (2003)
22. Anderson, H., Bader, M., Bard, E., Doherty, G., Garrod, S., Isard, S., Kowtko, J., McAllister, J., Miller, J., Sotillo, C., Thompson, H., Weinert, R.: The mcrc map task corpus. Language and Speech 34(4), 351–366 (1991)

23. Fujie, S., Ejiri, Y., Nakajima, K., Matsusaka, Y., Kobayashi, T.: A conversation robot using head gesture recognition as para-linguistic information. In: RO-MAN, pp. 159–164 (September 2004)
24. Maatman, M., Gratch, J., Marsella, S.: Natural behavior of a listening agent. In: Panayiotopoulos, T., Gratch, J., Aylett, R.S., Ballin, D., Olivier, P., Rist, T. (eds.) IVA 2005. LNCS (LNAI), vol. 3661. Springer, Heidelberg (2005)
25. Kang, S.H., Gratch, J., Wang, N., Watt, J.: Does the contingency of agents' nonverbal feedback affect users' social anxiety? In: AAMAS (2008)
26. Rabiner, L.R.: A tutorial on hidden Markov models and selected applications in speech recognition. Proceedings of the IEEE 77(2), 257–286 (1989)
27. Lafferty, J., McCallum, A., Pereira, F.: Conditional random fields: probabilistic models for segmenting and labelling sequence data. In: ICML (2001)
28. Igor, S., Petr, S., Pavel, M., Luk, B., Michal, F., Martin, K., Jan, C.: Comparison of keyword spotting approaches for informal continuous speech. In: MLMI (2005)
29. hCRF library, http://sourceforge.net/projects/hcrf/

IGaze: Studying Reactive Gaze Behavior in Semi-immersive Human-Avatar Interactions

Michael Kipp and Patrick Gebhard

DFKI
Embodied Agents Research Group
michael.kipp@dfki.de, patrick.gebhard@dfki.de

Abstract. We present IGaze, a semi-immersive human-avatar interaction system. Using head tracking and an illusionistic 3D effect we let users interact with a talking avatar in an application interview scenario. The avatar features reactive gaze behavior that adapts to the user position according to exchangeable gaze strategies. In user studies we showed that two gaze strategies successfully convey the intended impression of dominance/submission and that the 3D effect was positively received. We argue that IGaze is a suitable setup for exploring reactive nonverbal behavior synthesis in human-avatar interactions.

1 Introduction

While embodied agents have a wide range of applications, interactive systems with a face-to-face conversation are of particular interest [1]. However, most current HCI systems are turn-based instead of being *reactive*. In reactive systems user actions should trigger an instantaneous response on the agent side which in turn influences the user, resulting in a tightly coupled feedback loop. Such reactive behavior can only be explored with continuous user input, for instance by visually tracking the user. As theater expert Johnstone observed: "the bodies of the actors continually readjusted. As one changed position so all the others altered their postures." [2]. To simulate and study such effects in human-avatar interactions, *reactive* agents are required in an *immersive* setup. In this paper we discuss a minimalistic approach to creating immersiveness and implementing reactive gaze behavior that instantaneously adapts to the user's current position.

Gaze is a powerful interaction modality with many functions like signaling attention, regulating turn-taking or deictic reference [3]. Gaze also serves as an indicator for mood, personality and status. The latter has been explored by social scientists, semioticists and theater professionals alike [2,4]. Because of its communicative importance gaze is highly relevant for embodied virtual agents [5,6], in robotics [7] and human-computer interaction (e.g. COGAIN[1]).

STEVE was one the first immersive human-avatar interaction systems [8]. Users were instructed by a 3D-situated virtual tutor who displayed a number of gaze behaviors, including continuous gaze following and gaze aversion [9]. However, no empirical studies on the impact on personality/status were reported.

---
[1] http://www.cogain.org

Moreover, the used VR goggles had the possible risk of *VR sickness*. Heylen et al. [5] investigated gaze behavior of a cartoon-style talking head. Results showed that users found the functionally optimized gaze strategy easiest to use, they found the character more friendly and completed the task in less time. Fukayama et al. [10] showed that varying the gaze pattern in terms of amount, mean duration and target points has a significant impact on impression formation. Bente et al. [11] proposed a system for investigating social gaze and found that prolonged gaze led to better evaluation of the interlocutor, a finding that explains a part of our results. Poggi et al. [6] created a formalism for generating gaze using a meaning-signal mapping. They leave open the question how to *react* to the user's continually changing position. Lee et al. [12] created the *Eyes alive* system where pupil movement (saccades) was generated using statistical models of real people. Their data-driven approach outperformed random gaze. The system is complementary to ours which neglects saccade movement.

Except for [8], most systems have a 2D view of the agent with the user at a fixed position that was not tracked. Immersive systems like STEVE run the risk of VR sickness. IGaze intends to study and apply tightly coupled interactions between user behavior and agent behavior in a semi-immersive setup. While empirical studies traditionally compare the usage of embodied agents to more traditional interfaces [13,14], more recent studies try to specifically validate the effects of *particular behaviors* [15,16]. In this paper, we empirically validate the effect of specific gaze behaviors.

2 The IGaze System

The IGaze system is an immersive human-avatar interaction system for studying reactive nonverbal behavior. Immersiveness is established by two factors: an illusionistic 3D effect makes the user feel like s/he is moving in 3D and a continuous gaze adjustment that makes the avatar follow the user with the head.

The setup consists of a 42" display and an IR camera behind the screen, pointing at the user. The user wears glasses with 2 infrared LEDs attached. For the IR camera we use Nintendo's Wii remote (1024x768 resolution). In the current modular architecture, the input module computes a hypothetical head position from the detected IR lights, based on J. Lee's *WiiDesktopVR* software[2]. Head position values are used by the behavior control system (a) to position the virtual camera at user's location oriented toward the avatar and (b) to orient the avatar's head (e.g. always looking at the user). The animation controller handles facial viseme animation and the combination of procedural animation (head rotation) with keyframe animation (breathing). We use Horde3D[3], developed by N. Schulz, for rendering. The 3D character (Figure 1) has a 40-joints skeleton and 4 viseme morph targets. Speech is synthesized using OpenMARY[4].

---

[2] http://www.cs.cmu.edu/ johnny/projects/wii
[3] http://www.horde3d.org
[4] http://mary.dfki.de

Fig. 1. The illusionistic 3D effect consists of adapting the virtual camera according to the user's head position (distance, height, sideways position)

The 3D illusion is created by positioning the camera according to the position of the user in front of the screen. The user can "look into" the room by moving his/her head (see Figure 1). To avoid jumps in the camera movement due to misreadings or flare, we smooth incoming values v_{in} using a change factor η to obtain the new value $v_{new} = (1-\eta)\ v_{old} + \eta\ v_{in}$. Especially the z-value (distance), estimated from the distance of the two IR blobs, is quite unstable at distances of $> 1m$; we therefore applied stronger smoothing to the distance z-update ($\eta = .1$) than to the x/y-update ($\eta = .6$).

2.1 Gaze Strategies

We implemented 3 gaze strategies: the *Mona Lisa* strategy (= continuous gaze following), dominant and submissive strategy. The gaze aversion behavior was different for dominant and submissive. Strategies were modeled using timed finite state automata depicted in Fig. 2.

The **Mona Lisa Strategy (ML+, ML-)** consists of following the user's position with the eyes all the time. Two variants should check on the impact of the 3D effect: With the 3D effect switched on (ML+) the avatar looks at the position of the virtual camera. When switched off (ML-) it looks at the hypothesized position in the real world (usually *not* the camera). The Mona Lisa gaze is related to Poggi's *magnetic eyes*, hypothesized to mean dominance [4]. It also fits *stare* (request for attention), *look in the face* and *look straight into someone's eyes* (expression of dominance, defy), or even *cold anger*.

Dominant Strategy (Dom): High status, according to Johnstone, is gained by outstaring your interlocutor [2]. Moreover, he observed that if A breaks eye contact and does not look back, A is higher. Also, according to the Visual Dominance Ratio (VDR) measure, the higher status person gazes roughly the same amount while listening and while speaking, whereas the lower status person spends more time gazing while listening [17]. Our dominant strategy consists of maintaining eye contact while speaking and randomly changing from gazing to averting while listening. More precisely, the avatar establishes and holds eye contact when speaking, and after speaking, immediately looks away. When listening, the avatar establishes eye contact after 0–3 sec., then holds it for 4.5–7.5

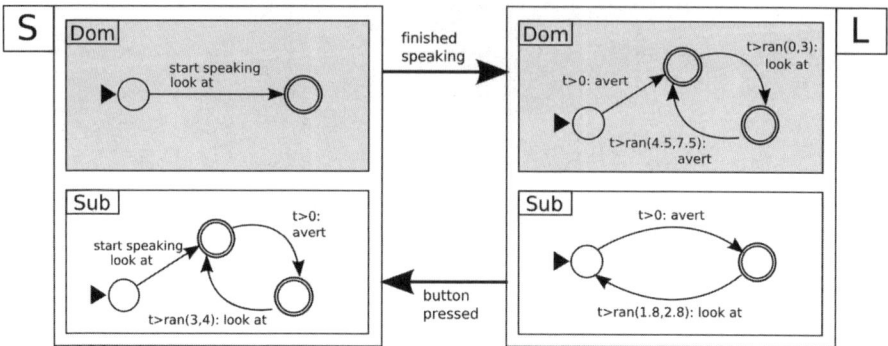

Fig. 2. Timed automata were used to model the gaze behaviors of dominant (upper states) and submissive (lower). S/L refer to speaking/listening modes

sec. before looking away. The dominant avert behavior consists of a movement 12° *away* from the user and 5° *upward* from the default up-down angle.

Submissive Strategy (Sub): Low status, according to Johnstone, means being outstared by your interlocutor. Moreover, if A breaks eye contact and looks back, A is lower. Our submissive strategy makes the avatar only look briefly every now and then and immediately avert the gaze again. In the submissive strategy, the avatar establishes eye contact when starting to talk but averts his gaze immediately after eye contact. His gaze remains averted for 3–4 sec. He then establishes eye contact again and looks away immediately. During listening, the pattern is the same with the difference that the avatar holds eye contact for 1.8–2.8 sec. The submissive avert behavior consists of a movement *away* from the user (5° while speaking, 8° while listening) and 15° *downward*.

3 Experiment

In our experiment subjects played applicants in a virtual application interview. 14 subjects (aged 21–36, 5 female, 9 male, German native speakers) participated. Subjects interacted with the avatar (interviewer) in a private cabin and wore a headset with microphone to make them believe that speech input is understood. The avatar was remote controlled by the experimenter (wizard of oz) who triggered the utterances. We had the following *hypotheses* regarding the outcome of our experiment: **(H1)** The 3D effect is not uncomfortable, **(H2)** the 3D effect helps people to immerse, **(H3)** dominant gaze behavior is perceived as dominant, **(H4)** submissive gaze behavior is perceived as submissive.

3.1 Pilot Study

In a pilot study we asked the 10 subjects to take the application interview as seriously as possible and to answer truthfully. Many subjects displayed a high degree of stress similar to a real application setting. This had three negative side-effects: (1) the subjects hardly moved, thus not noticing the 3D effect, (2) they were so focused on their answers that little attention was given to the avatar's behavior, and (3) the avatar was judged by the content of the interview questions (subjects found the avatar getting "too personal" or sometimes being "more relaxed"). When we found no effects in the analysis we modified the design in various ways: (a) we demonstrated the 3D effect prior to the interview, (b) we asked subjects to pay attention to the avatar's gaze behavior, (c) we told subjects not to take the interview too seriously (e.g. invent answers), (d) we changed the answer scale from 5 points to 7 points because only few subjects had used the extreme points.

3.2 Main Study

Procedure. The subjects were told to participate in an experiment about a "virtual application interview training". They should act as if in an application interview for an academic position. However, they were asked to pay attention to the avatar's gaze and not take the application answers themselves too seriously. Moreover, we demonstrated the 3D effect before the interview.

Table 1. The four conditions of our experiment

condition	gaze behavior	3D effect
ML-	continuous "Mona Lisa" gaze following	inactive
ML+	continuous "Mona Lisa" gaze following	active
Dom	dominant gaze behavior	active
Sub	submissive gaze behavior	active

During one session the avatar asked 32 interview questions. Each question was 2–3 sentences long to give room for avatar head movement. The subject had to answer each question and after 4 questions the screen was turned blank and the subject filled in a paper/pencil *in-session questionnaire* to rate the past experience. The subject's saying "ready" triggered the next 4-question session. Thus, we had 8 4-question sessions. In each session the condition was changed: ML-, ML+, Dom or Sub (Table 1). The order of conditions was random and balanced across subjects. They virtual character performed gaze behavior both while speaking and listening. In order for the system to know that an answer was finished, a human operator had to press a button when the subject finished his/her answer. The whole interaction lasted between 15–25 minutes, and was followed by a post-questionnaire.

Each *in-session questionnaire* (paper and pencil) asked for 5 ratings on a 7-point scale[5]. The subject was asked whether s/he found the avatar (1) likable, (2) dominant, (3) extrovert, (4) natural, and (5) how stressed the subject him/herself felt. In the *post-questionnaire* we first asked the subjects to describe (free form) any differences between the session's segments. We then explained the 3D effect to the subject and asked in 3 questions (7-point scale) whether

---

[5] Extreme values were labeled *not at all* and *very much*, middle value was labeled *neutral*.

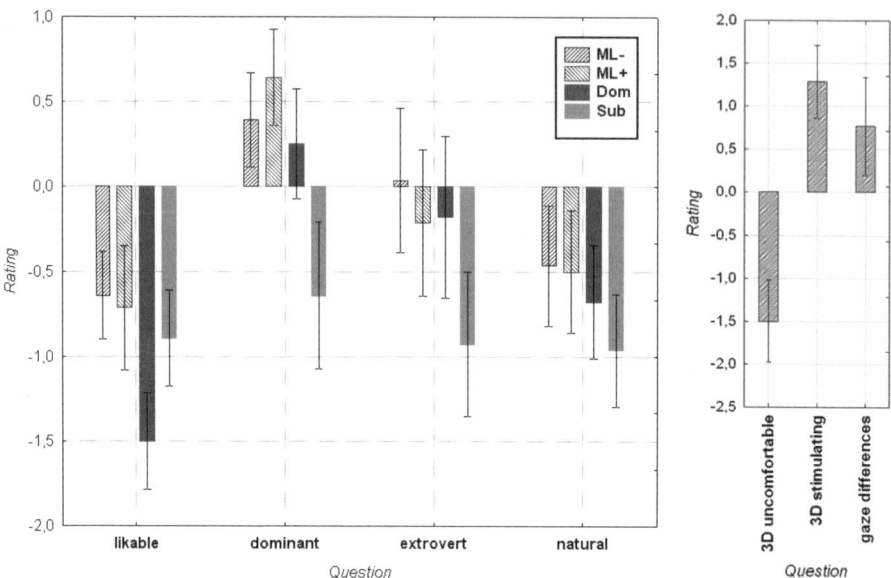

Fig. 3. (a) Left figure shows mean values and standard error of the 4 questions vs. 4 conditions. (b) Right figure shows mean and std. err. for 3 debriefing questions.

(1) the 3D effect was uncomfortable, (2) the 3D effect was enjoyable, and (3) whether any differences in gaze behavior were discernible.

Results. Fig. 3 (left) shows the mean values of our four questions (likable, dominant, extrovert, natural) over the four conditions (ML-, ML+, Dom, Sub). We first checked whether and how the conditions differed with regard to the questions using ANOVA which yielded significant main effects for condition (F(3,39)=3.70, $p < .05$), question (F(3,39)=3.60, $p < .05$) and condition-question interaction (F(9,117)=2.59, $p < .01$), therefore the four conditions had different answer patterns.

We computed ANOVAs for each question to find out whether specific questions differed with respect to condition, using the Fisher LSD test over all condition pairs to single out the exact differences. For question *likable*, we found a main effect (F(3,39)=3.12, $p < .05$) in the ANOVA with significant difference between ML- vs. Dom ($p < .01$) and ML+ vs. Dom ($p < .05$) using Fisher LSD. Question *dominant* had a main effect (F(3,39)=4.01, $p < .05$). Significant differences were ML- vs. Sub ($p < .05$), ML+ vs. Sub ($p < .01$) and Dom vs. Sub ($p < .05$). We found similar results for question *extrovert*: a main effect (F(3,39)=3.56, $p < .05$) and differences between ML- vs. Sub ($p < .01$), ML+ vs. Sub ($p < .05$) and Dom vs. Sub ($p < .05$). To our surprise, the question *natural* yielded no main effect (F(3,39)=1.22, $p < .32$) which means that subjects found all conditions natural to the same degree. The question *stress* did not result in a main effect either (F(3,39)=1.27, $p = .30$).

Fig. 3 (right) shows the means of the 3 post-questionnaire questions asked after the session (7 point scale: -3 to 3). The first two questions showed a significant difference from zero (neutral value). The answers to "do you find the 3D effect uncomfortable?" were significantly below zero, -3 being *not at all* (t(13)= -3.14, $p < .01$). The answers to "do you find the 3D effect stimulating?" were significantly above zero (t(13)=3.03, $p < .01$).

4 Discussion

The study successfully validated our hypotheses that our encoded dominant/submissive behaviors are perceived as dominant/submissive **(H3,H4)**. What is interesting is that conditions ML-, ML+ and dominance are so close to each other. However, the conditions can be divided along the dimension of *liking*. Here, the dominant behavior is significantly perceived less likable than ML. This indicates that instead of implementing purely dominant behavior, we implemented dominant+negative behavior (Dom) and dominant behavior (ML-, ML+). The former can also be called arrogance, a key word that also emerged in debriefing interviews. The latter conforms with findings that continuous gaze leads to more positive evaluation [11]. We found that stress obviously did not impact the judgement of dominance as most subjects did not find the situation stressful.

As for the 3D effect, we wanted to know whether subjects would experience irritation similar to the VR sickness. However, our analysis showed that subjects did not find it uncomfortable **(H1)** but actually found it stimulating (both significant). Many subjects told us afterwards that they liked both the 3D effect and the fact that the agent was actually following them with his gaze (the Mona Lisa effect). However, we were surprised it did not seem to matter whether the 3D effect was switched on or off. So while the effect did generate excitement it did not affect the perception of avatar personality and did not raise stress level or comfort **(not H2)**.

5 Conclusion

We presented IGaze, a semi-immersive system for reactive human-avatar interactions. We use head tracking, an illusionistic 3D effect and a life-size display of the avatar's upper body to create immersiveness. Different gaze strategies (dominant/submissive) were implemented using timed automata and successfully validated in a user study.

Many prior systems have neglected the questions that arise when continuous input data from the user is available. To build truly reactive systems, we have to devise tools and systems to model the tightly coupled feedback that is characteristic for human interactions. IGaze takes a minimalistic approach to the setup, employs timed automata for modeling reactive behavior and will be extended in the future with new I/O modules. For output we envisage realtime procedural animation of gesture that adapts to user actions [18]. New input modalities include speech or accelerometer-based input devices.

Acknowledgments. This research has been carried out within the framework of the Excellence Cluster Multimodal Computing and Interaction (MMCI), sponsored by the German Research Foundation (DFG).

References

1. Cassell, J., Sullivan, J., Prevost, S., Churchill, E.: Embodied Conversational Agents. MIT Press, Cambridge (2000)
2. Johnstone, K.: Impro. Improvisation and the Theatre. Routledge/Theatre Arts Books, New York (1979); (Corrected reprint 1981)
3. Heylen, D.: A closer look at gaze. In: Proc. of the workshop, Creating Bonds with Embodied Conversational Agents (2005)
4. Poggi, I.: Mind, Hands, Face and Body: A Goal and Belief View of Multimodal Communication. Weidler Buchverlag, Berlin (2007)
5. Heylen, D., van Es, I., Nijholt, A., van Dijk, B.: Controlling the gaze of conversational agents. In: CLASS Workshop (2003)
6. Poggi, I., Pelachaud, C., de Rosis, F.: Eye Communication in a Conversational 3D Synthetic Agent. AI Communications 13(3), 169–181 (2000)
7. Mutlu, B., Hodgins, J.K., Forlizzi, J.: A storytelling robot: Modeling and evaluation of human-like gaze behavior. In: Proceedings of HUMANOIDS 2006, IEEE-RAS International Conference on Humanoid Robots. IEEE, Los Alamitos (December 2006)
8. Rickel, J., Johnson, W.L.: Animated agents for procedural training in virtual reality: Perception, cognition, and motor control. Applied Artificial Intelligence 13, 343–382 (1999)
9. Lee, J., Marsella, S., Traum, D., Gratch, J., Lance, B.: The rickel gaze model: A window on the mind of a virtual human. In: Pélachaud, C., Martin, J.-C., André, E., Chollet, G., Karpouzis, K., Pelé, D. (eds.) IVA 2007. LNCS (LNAI), vol. 4722. Springer, Heidelberg (2007)
10. Fukayama, A., Takehiko, O., Mukawa, N., Sawaki, M., Hagita, N.: Messages Embedded in gaze of Interface Agents - Impression management with agent's gaze. In: Proceedings of SGICHI, pp. 41–48 (2002)
11. Bente, G., Eschenburg, F., Aelker, L.: Effects of simulated gaze on social presence, person perception and personality attribution in avatar-mediated communication. In: PRESENCE (2007)
12. Lee, S., Badler, J., Badler, N.: Eyes alive. In: TOG / Proc. of SIGGRAPH, San Antonio, TX, pp. 637–644. ACM Press, New York (2002)
13. Krämer, N.C., Tietz, B., Bente, G.: Effects of embodied interface agents and their gestural activity. In: Proc. of the 4th International Conference on Intelligent Virtual Agents. Springer, Heidelberg (2003)
14. Lester, J.C., Converse, S.A., Stone, B.A., Kahler, S.E., Barlow, S.T.: Animated pedagogical agents and problem-solving effectiveness: A large-scale empirical evaluation. In: Proceedings of the Eighth World Conference on Artificial Intelligence in Education, pp. 23–30. IOS Press, Amsterdam (1997)
15. Kipp, M., Neff, M., Kipp, K.H., Albrecht, I.: Toward Natural Gesture Synthesis: Evaluating gesture units in a data-driven approach. In: Pélachaud, C., Martin, J.-C., André, E., Chollet, G., Karpouzis, K., Pelé, D. (eds.) IVA 2007. LNCS (LNAI), vol. 4722, pp. 15–28. Springer, Heidelberg (2007)

16. Foster, M.E., Oberlander, J.: Corpus-based generation of head and eyebrow motion for an embodied conversational agent. Journal on Language Resources and Evaluation - Special Issue on Multimodal Corpora 41(3-4), 305–323 (2007)
17. Dovidio, J.F., Ellyson, S.L.: Decoding visual dominance: Attributions of power based on relative percentages of looking while speaking and looking while listening. Social Psychology Quarterly 45(2), 106–113 (1982)
18. Neff, M., Kipp, M., Albrecht, I., Seidel, H.P.: Gesture Modeling and Animation Based on a Probabilistic Recreation of Speaker Style. Transactions on Graphics (to appear, 2008)

Estimating User's Conversational Engagement Based on Gaze Behaviors

Ryo Ishii[1] and Yukiko I. Nakano[2]

[1] Tokyo University of Agriculture and Technology, Japan
nrhc_ryo@hotmail.co.jp
[2] Dept. of Computer and Information Science, Faculty of Science and Technology,
Seikei University, Japan
y.nakano@st.seikei.ac.jp

Abstract. In face-to-face conversations, speakers are continuously checking whether the listener is engaged in the conversation. When the listener is not fully engaged in the conversation, the speaker changes the conversational contents or strategies. With the goal of building a conversational agent that can control conversations with the user in such an adaptive way, this study analyzes the user's gaze behaviors and proposes a method for predicting whether the user is engaged in the conversation based on gaze transition 3-Gram patterns. First, we conducted a Wizard-of-Oz experiment to collect the user's gaze behaviors as well as the user's subjective reports and an observer's judgment concerning the user's interest in the conversation. Next, we proposed an engagement estimation algorithm that estimates the user's degree of engagement from gaze transition patterns. This method takes account of individual differences in gaze patterns. The algorithm is implemented as a real-time engagement-judgment mechanism, and the results of our evaluation experiment showed that our method can predict the user's conversational engagement quite well.

Keywords: Conversational engagement, eye-gaze, conversational agents.

1 Introduction

In face-to-face conversation, speakers sometimes glance at the listener and check whether the listener is properly engaged in the conversation. On the other hand, listeners display their engagement through verbal/nonverbal behaviors, such as verbal acknowledgement and eye-gaze. For example, eye-gaze is useful for demonstrating that the listener is paying attention to the conversation.

This engagement checking process is fundamental and indispensable not only in face-to-face conversation, but also in human-agent communication. If the communication channel between the user and the agent is not well-set, information presented from the system (agent) will not be properly conveyed to the user. If the system can monitor the user's attitude towards the conversation and detect whether the user is engaged or not engaged in the conversation, then the system can adapt their behaviors and communication strategies according to the user's attitude. For instance, if the user is not engaged in the conversation, the system needs to attract the user's attention by changing the conversation topic.

Although it is broadly believed that eye-gaze is a powerful device in demonstrating the listener's conversational engagement, little has been studied about real-time gaze models of conversational engagement. Thus, with the aim of improving naturalness in human-agent communication, this paper proposes a method of estimating the degree of engagement by measuring the user's attentional behaviors in real time, and by implementing the method using an eye-tracker. To accomplish this goal, we address the following two main issues:

(1) By analyzing corpus data collected through a Wizard-of-Oz experiment, we identify disengagement gaze patterns in human-agent conversations.
(2) By applying a clustering technique to the measured gaze data, we propose an engagement estimation method that can account for individual differences in gaze patterns

In the following sections, section 2reviews related research, and section 3 reports on our Wizard-of-Oz experiment. Section 4 describes the empirical results of analyzing the gaze data, and an engagement-judgment method is proposed in section 5. The evaluation of the proposed method is reported in section 6. Finally, we will discuss future work in section 7.

2 Related Research

In studies of face-to-face communication, Kendon [1] described various eye gaze functions from the ethnomethodological point of view. Psychological studies reported that eye gazing, specifically accompanied by head nods, serves as positive feedback to the speaker [2], and demonstrates that the listener is paying attention to the conversation [3]. It also contributes to smooth turn-taking [4]. On the contrary, when conversational participants share the same physical environment and their task requires complex reference to, and joint manipulation of physical objects, participants look at the shared object most of the time [5, 6].

These findings were later used as the basis of conversational humanoids. Nakano et al. [7] proposed a gaze model for nonverbal grounding in ECAs, and Gratch et al. [8] reported that backchannel feedback from a listener agent is effective in establishing a sense of rapport between a user and a virtual character Sidner et al. [9] proposed a gaze model for conversational engagement, and implemented it into a physical communication robot. In these studies, user's gaze direction was roughly estimated from the head-direction measured by a head-tracker. In this study, we measure accurate eye-gaze (pupils) behaviors using an eye-tracker, and attempt to establish a user-adapted engagement estimation model by a data mining approach. The idea of using an eye-tracker as a component of conversational interfaces has already been successfully presented in [10, 11], where eye-tracking systems are used in detecting an object of interest to the user. Thus, we believe that employing this equipment is promising in our study of human-agent conversational engagement.

3 Wizard-of-Oz Experiment to Collect Verbal/Nonverbal Data

To estimate the user's levels of engagement from her/his gaze behaviors, we collected conversational corpus in human-agent communication. We employed a Wizard-of-Oz setting in our data collection experiment where the agent's response to the user was selected by an experimenter.

3.1 Experiment

A female animated character was displayed on a 120-inch rear-type screen, where she acted as a salesperson at a mobile phone store. The experimental setting is shown in Fig. 1 (b). In addition to the subject (called the "user") who communicated with the agent, there was another subject (called the "observer") who observed the user through a half-mirror, and judged whether the user was engaged in the conversation with the agent or not.

- **Subjects:** 9 male students and 1 female university student participated in the experiment as users, and 7 male subjects joined as observers[1].
- **Procedure:** The user's task was to listen to the agent's explanation, and guess the most popular style of cell phone for female high school students, or for businessmen, among 6 types of cell phones. Therefore, the user needed to listen to the agent's explanations about all the cell phones, and this was a very boring task for the user. The user was allowed to ask questions, and could request change the topic by talking to the agent. Note that the subjects were fully motivated by the rewards that

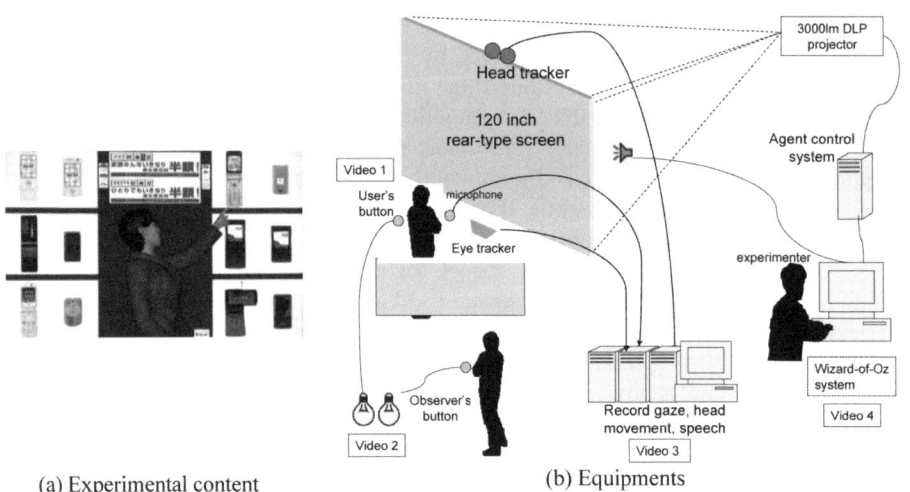

(a) Experimental content   (b) Equipments

Fig. 1. Experimental setting

---

[1] In half of the cases, a subject acted as a user in one session, then acted as an observer in the next session. When the subject did not have time to act as an observer, another subject acted only as an observer. We admit that the number of subjects is small, but our method is not affected by the number of subjects.

were given when the subject's guess was correct. On the other hand, agent behaviors as experimental materials were controlled in a consistent way. The agent looked at an explanation of the target cell phone most of the time, and looked at the user for 3 seconds every 10 utterances. Since the agent's nonverbal behaviors were completely constant, no surprising actions were perceived as novel visual stimulus that could accidentally attract the user's attention.

3.2 Verbal/Nonverbal Data

We collected 10 conversations whose average length was 16 minutes, and extracted the following data for more detail analysis:

(1) Transcription of the agent's and user's utterances (the total number of agent's utterances were 951 and those of user's, 61).
(2) The agent's gesture and gaze behaviors
(3) Judgment about the user's attitude. A push-button device was given to the user and the observer. The user was instructed to press the button when the agent's explanation was boring, and the user would like to change the topic. The observer was instructed to press the button when the user looked bored and disengaged in the conversation. When these buttons were pressed, lights went on in another room, and these lights were recorded as video data.
(3) The user's gaze data measured by Tobii X-120 eye-tracker. To ignore measuring noise, we used gaze data that had been fixed for more than 20msec within the range of 20 pixels.

We integrated all these data using the Anvil annotation tool [12] to visually illustrate the patterns of eye-gaze behaviors, and to investigate co-occurrence with other behaviors.

4 Analysis

To analyze the gaze behaviors, first we define four labels used to categorize the user's gaze direction:

- T: look at the target object of the agent's explanation. For example, if the agent explains a cell phone displayed on the upper right of the display in Fig. 1 (a), that cell phone is the target object.
- AH: look at the agent's head
- AB: look at the agent's body
- F: look at other objects, such as non-target cell phones and an advertisement poster (F1≠F2≠F3).

Since the agent is looking at the current target object most of the time, it is presumed that joint attention is established between the user and the agent when the user's gaze label is "T".

To find the engagement and disengagement patterns of eye-gaze behaviors, we created gaze direction transition 3-grams using these labels, and used this as a unit of analysis. When the gaze direction is changed or more than a 200 msec gaze break occurs, we count this as a new gaze behavior. For example, suppose that the user's gaze direction shifts from "T" to "AH", and there is a 100 msec break of the gaze data after that, and then another "AH" starts. In this case, "AH" continues. If the gaze shifts to "F1" shortly afterwards, the 3-gram constructed from these gaze behaviors is "T-AH-F1".

The results of the 3-gram analysis are shown in Fig. 2. For each 3-gram pattern, we calculated the probability of co-occurrence with the disengaging judgment (i.e., pressing the button) either by the user or the observer. If, in Fig.2, the probability is 80%, this means that this pattern co-occurred with the disengagement judgment 80% of the time.

As shown in Fig. 2 (a), the probability is different depending on the types of 3-gram. "F1-AH-AH" has the highest probability over 80%. This means that the disengagement judgment button is pressed over 80% of the time when this pattern occurs. On the other hand, the probability for the "AH-T-T" 3-gram is only 45 %. These results suggest that the 3-grams with higher probability violate proper engagement gaze rules, and those with lower probability contribute to conversational engagement.

More interestingly, by grouping the 3-grams according to their constituents, we found that the constituents of higher probability (less engaging) patterns are different from those of lower probability (more engaging) patterns. As shown in Fig. 2 (b), "F1-AH-F1" and "AH-F1-F1" have the same constituents ("AH*1, F1*2": consisting of one "AH" and two "F1"s), and their probability range is 72.2% to 82.1%. On the other hand, the probability range for 3-grams consisting of two "T"s and a "F1" (T*2,F1*1) 51.5% to 55.3%. Note that 3-grams with higher probability do not have "T" as a constituent. This suggests that establishing joint attention by looking at the

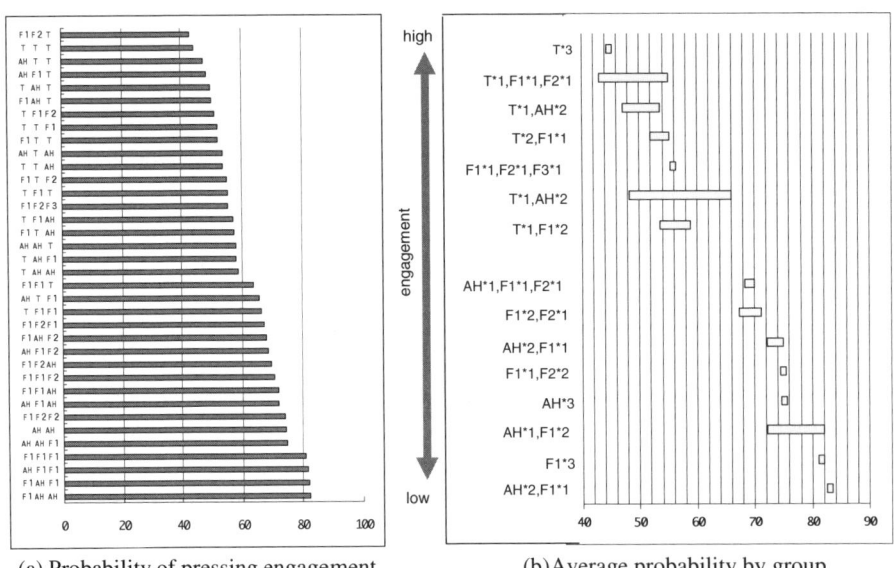

(a) Probability of pressing engagement judgment button

(b) Average probability by group

Fig. 2. Analysis of 3-grams

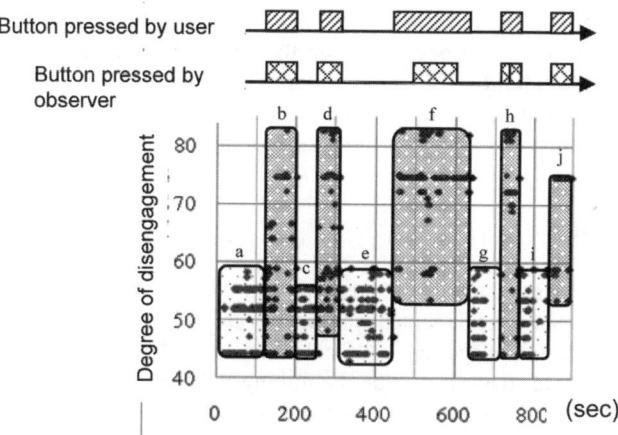

Fig. 3. Plots for the degree of engagement

target object is indispensable when there is a shared object. All these results support our idea of estimating the conversational engagement from the patterns of gaze behaviors.

5 User-Adapted Engagement Estimation

Based on the analysis in the previous section, we propose an engagement estimation method that detects if the user is not engaged in a conversation with the agent. We use the probability of pressing the button in Fig. 2 as the "degree of disengagement". Therefore, if the probability of a given 3-gram in Fig. 2 is 80%, the degree of disengagement is 80. Fig. 3 plots 3-gram data for the degree of disengagement. The upper parts of the graph indicate the time period in which the user or the observer pressed the button. Low engagement 3-grams were observed at b, d, f, h, j, and at the same time period, the user herself/himself, or the observer, judged that the user was not engaged in the conversation (the buttons were pressed). On the contrary, at a, c, e, g, and i, only high engagement 3-grams were observed, and the buttons were not pressed. Thus, the 3-grams' disengagement values and the human disengagement judgments are highly correlated to each other. This result suggests that the human judgments can be predicted from the 3-grams.

This result suggests that if a proper threshold for the degree of disengagement can be determined, automatically judging whether the user is engaged in the conversation or not is possible. However, we found that the distribution range of the 3-grams is different depending on the users. To adjust the threshold according to the user's individual differences, we employed a clustering technique as follows.

(1) Clustering the Data Points: The data points are clustered according to the degree of disengagement. We use a simple centorid method for this purpose. Starting with individual data points as a cluster, the Euclidean distance between the centroids of two clusters are calculated, and the closest clusters are merged together. The centroid of the new cluster is weighted by the number of the data points in the original clusters. When the number of clusters becomes four, the process is terminated.

(2) Setting a Threshold: The middle point between the centroid of the highest disengagement cluster and that of the second highest disengagement cluster is calculated. This value is used as the threshold for a disengagement judgment.

6 Evaluation

To evaluate our user-adapted engagement estimation method, we examined how accurately the human disengagement judgments are predicted by our method. We use the same data collected in section 3. To calculate a threshold for each user, we use the first 120 seconds of data from the beginning of the conversation, and test the predictive accuracy using the rest of the data. To test the effectiveness of user adaptation for the threshold, the following two methods are compared:

- With User Adaptation (WUA): use a user-adapted threshold for each user
- Without User Adaptation (WoUA): calculate only one threshold by mixing all the user's data, and applying this threshold to all the users.

The results are shown in Fig. 4. Although WUA is not better than WoUA in the precision rate, the recall rate is much better than WoUA. Moreover, the F-measure is 71.4, which is also much better than the WoUA. Thus, this result shows that the predictive accuracy is greatly improved by adopting user-adapted thresholds, and this method can predict the human judgment of engagement quite well. Moreover, since our method determines the threshold using the first 120 seconds of data, this means that by using our method, the system can quickly adjust to a user's individual differences during an on-going conversation with that user.

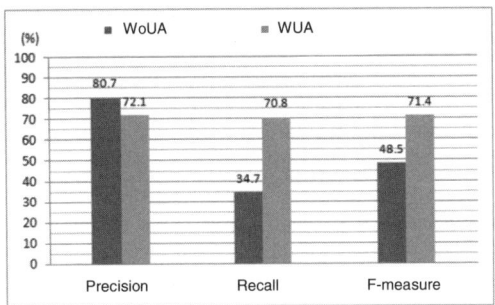

Fig. 4. Evaluation results

7 Conclusion and Future Work

By analyzing gaze patterns observed in a Wizard-of-Oz experiment, we found that patterns of gaze transition 3-grams are strongly correlated with human subjective or observational judgment of a user's attitude towards the conversation. Based on this analysis, we applied a clustering technique to the 3-gram data and proposed a method of automatically detecting whether the user is engaged in the conversation. An evaluation experiment showed that our method detects a user's distraction from the conversation quite well because of the adaptation of the decision threshold according to the user's individual differences.

Although our method focuses on the transitions of the gazing direction, another important aspect is the duration of the gaze fixation. A possible way of improving our method is to weight each 3-gram according to its temporal duration. Therefore, testing whether the model extension contributes to improving the engagement estimation is necessary.

Finally, our next step is to address issues concerning how agents respond to the user once the system detects that the user is distracted. The most prioritized issue is to investigate the correlation between the gaze 3-grams and the types of verbal behaviors; in other words, what types of 3-grams more frequently co-occur with what types of utterances or conversational contexts. Our final goal is to control human-agent interaction based on a multimodal engagement model.

References

1. Kendon, A.: Some Functions of Gaze Direction in Social Interaction. Acta Psychologica 26, 22–63 (1967)
2. Argyle, M., Cook, M.: Gaze and Mutual Gaze. Cambridge University Press, Cambridge (1976)
3. Clark, H.H.: Using Language. Cambridge University Press, Cambridge (1996)
4. Duncan, S.: Some signals and rules for taking speaking turns in conversations. Journal of Personality and Social Psychology 23(2), 283–292 (1972)
5. Argyle, M., Graham, J.: The Central Europe Experiment - looking at persons and looking at things. Journal of Environmental Psychology and Nonverbal Behaviour 1, 6–16 (1977)
6. Anderson, A.H., Bard, E., Sotillo, C., Doherty-Sneddon, G., Newlands, A.: The effects of face-to-face communication on the intelligibility of speech. Perception and Psychophysics 59, 580–592 (1997)
7. Nakano, Y.I., Reinstein, G., Stocky, T., Cassell, J.: Towards a Model of Face-to-Face Grounding. In: ACL 2003, Sapporo, Japan (2003)
8. Gratch, J., Okhmatovskaia, A., Lamothe, F., Marsella, S., Morales, M., Werf, R.J.v.d., Morency, L.-P.: Virtual Rapport. In: 6th International Conference on Intelligent Virtual Agents. Springer, Marina del Rey (2006)
9. Sidner, C.L., Lee, C., Kidd, C., Lesh, N., Rich, C.: Explorations in engagement for humans and robots. Artificial Intelligence 166(1-2), 140–164 (2005)
10. Qvarfordt, P., Zhai, S.: Conversing with the user based on eye-gaze patterns. In: CHI 2005 (2005)
11. Eichner, T., Prendinger, H., André, E., Ishizuka, M.: Attentive Presentation Agents. In: Pélachaud, C., Martin, J.-C., André, E., Chollet, G., Karpouzis, K., Pelé, D. (eds.) IVA 2007. LNCS (LNAI), vol. 4722. Springer, Heidelberg (2007)
12. Kipp, M.: Anvil - A Generic Annotation Tool for Multimodal Dialogue. In: The 7th European Conference on Speech Communication and Technology (2001)

The Effects of Agent Nonverbal Communication on Procedural and Attitudinal Learning Outcomes

Amy L. Baylor[1] and Soyoung Kim[2]

[1] Director, Center for Research of Innovative Technologies for Learning (RITL),
Instructional Systems Program, Florida State University,
305 Stone Building, Tallahassee FL 32306
abaylor@fsu.edu
[2] Researcher/Instructor, Yonsei University, Korea
soyoung.kim01@gmail.com

Abstract. This experimental study investigated the differential effects of pedagogical agent nonverbal communication on attitudinal and procedural learning. A 2x2x2 factorial design was employed with 237 participants to investigate the effect of type of instruction (procedural, attitudinal), deictic gesture (presence, absence), and facial expression (presence, absence) on learner attitudes, agent perception (agent persona, gesture, facial expression), and learning. Results indicated that facial expressions were particularly valuable for attitudinal learning, and were actually detrimental for procedural learning. Similarly, gestures were perceived as more valuable for students in the procedural module, even though they did not directly enhance recall.

Keywords: Interface agents, gestures, agent nonverbal communication, pedagogical agents, procedural learning, attitude change, persuasive technology.

1 Introduction

While animated interface agents are frequently appearing in computer-based environments, there are few design guidelines for employing effective nonverbal communication. For interface agents that promote learning-related outcomes (i.e., pedagogical agents), research has examined the effects of such agents on a variety of learning-related outcomes, including learning transfer [1, 2], metacognition [3], and motivation [4-7]. While there is exciting potential for implementing advanced technologies to represent agents as "virtual humans" [e.g., 8, 9], there are many unanswered questions with respect to effective design for nonverbal communication.

In general, an interface agent is constituted of several media features (e.g., appearance/image, animation, message, voice, and interactivity). Empirical research has now substantiated that the appearance alone of an interface agent can have a profound impact on learning outcomes, particularly motivational outcomes [see review, 10, 11]. For example, simply manipulating agent gender, attractiveness, or ethnicity can significantly enhance learner confidence and beliefs. With respect to agent message, manipulating its motivational [12] or affective content [13] can also dramatically impact learner beliefs and attributions. With more advanced intelligent tutoring and dialogue systems [14-17], interface agents such as AutoTutor [18] can serve on the "front end" to engage learners in a dialogue to highlight their misconceptions and encourage deeper reasoning. Thus, the agent message can cognitively support individualized learning.

A particularly salient affordance of interface agents is their propensity for effective message delivery. Thus, as a social interface, anthropomorphic interface agents can deliver messages as a social communicator (e.g., through voice and animation). While research conclusively indicates that having a human (as opposed to a computer-generated) voice is preferable [19-21], guidelines for agent nonverbal communication are indistinct. Researchers have speculated that facial expression (eye, eyebrow, and mouth movements) and deictic gestures (pointing with arms and hands) are particularly important for pedagogical agents [1, 22, 23] in promoting learning-related outcomes. In particular, gesture can reduce ambiguity by focusing learner attention, and facial expressions can reflect and emphasize agent message, emotion, personality, and other behavioral variables. Other empirical studies [5, 21] have investigated the general effects of agent animation (i.e., nonverbal communication), but the generalizability of the results are limited given that animation was not operationalized at a specific enough level of detail.

It is unclear as to what is the best instantiation of animated pedagogical agents for message delivery for these two learning outcomes. Some would suggest that by providing both facial expressions and deictic gestures, students may be distracted [24]. Others suggest that through employing both facial expression and deictic gestures, additional information is conveyed to promote learning [see 25] Yet, others suggest that consistency of agent features is critical and a mismatch of realism (e.g., facial expressions without gestures or vice versa) would be detrimental [26]. However, the overriding question of which animations for communicating nonverbally best fit the instructional purpose (e.g., to address different learning outcomes) has yet to be addressed.

Thus, the purpose of this experimental study is to explore the effects of nonverbal communication for two different learning outcomes on recall, attitude toward the content, and agent perceptions. The independent variables include deictic gestures (presence or absence), facial expression (presence or absence), and learning outcome (attitudinal or procedural). Of particular interest are the interaction effects between the types of learning outcome and nonverbal communication.

2 Method

This study employed a 2x2x factorial design, with type of knowledge (procedural, attitudinal), deictic gestures (presence, absence), and facial expressions (presence, absence) as the three factors. The participants in this study included 237 undergraduate students (32.1% male and 67.9% female) enrolled in a computer literacy course in a southeastern public university. Participants participated as a required course activity.

The experiment was conducted during regular sections of an introductory computer literacy course and was conducted in two separate phases. In the first phase, the procedural module was implemented for 120 participants. In the second phase, the attitudinal module was implemented for 117 different participants who had not participated in the procedural instruction. The participants were randomly assigned to one of the eight experimental conditions. The participants first responded to ten demographic questions, and then worked through the instructional module. After completing the module instruction, participants completed an online questionnaire assessing the five dependent outcomes. It took participants approximately 20 minutes to complete the module and another 20 minutes for the post-questions.

Materials. The procedural instructional module and the attitudinal instructional module were devised to be equivalent with respect to rigor, time, and implementation of deictic gestures and facial expressions. Both modules were approximately 20 minutes. Based on the result of the task-content analysis, agent scripts incorporated both the message and the nonverbal communication of the agent. For each of the four permutations of the module (deictic gestures x facial expressions), the number and distribution of animations were held as constant as possible.

Procedural Module. The procedural module instructed participants how to use a web-based software program designed to assess their proficiency in Microsoft Office applications. Given that the focus here was on procedural information, the module was designed with an organized and linear information structure [27]. Within the module, the agent showed participants how to perform specific tasks with the web-based software application.

Attitudinal Module. The attitudinal module was developed to elicit more desirable attitudes in students towards intellectual property rules and laws. Given that the focus here was on attitudinal information, the module incorporated four realistic scenarios (Digital Music and Copyright Law; Electronic Plagiarism; Movie Recording with Camcorder and P2P Sharing; and Software Copy) that students might encounter in their daily lives in relation to intellectual property. This ensured that students would receive the information and respond to it, a requirement for attitudinal instruction [27]. Within the module, the pedagogical agent provided information about intellectual property issues and presented scenarios for students, encouraging reflection throughout.

Treatments

Deictic Gestures. Deictic gestures in both modules were designed to support the instructional intent and were situated within the computer-based instructional context. Agent gestures were carefully developed to meet the three criteria of deictic mechanisms: lack of ambiguity, immersivity, and instructional soundness [9]. For the procedural module, deictic gestures were primarily used to indicate the physical objects and the geographical location of informative words on the interface. For the attitudinal module, deictic gestures were incorporated to direct the participants' attention to important information (e.g., to user interface features of the software). Approximately 50 instances of deictic gesture were incorporated in both types of instruction; there was no significant difference in number of gestures or distribution of them between the two modules.

Facial Expressions. The software Mimic2Pro was used to create facial expressions and synchronize the expressions to the agent voice. Each agent (within a given module) had an identical script and a Microsoft Agent machine-generated voice to control for any affective influence from the voice. Five different types of facial expression were used: neutral, serious, happy, surprised, and sad. Each type of facial expression had five different levels, from 1 to 5, according to the degree of the emotion. Agents were designed to show appropriate emotion in regard to the content in order to link the speech act and the content with the associated emotions. For example, when the agent talked about laws or rules, it displayed serious facial expressions whereas when it introduced the module and encouraged students to focus it expressed happier expressions. Given the dynamic nature of the facial expressions, they could not be quantitatively compared across the modules, but were designed to be as similar in number and distribution as possible.

The dependent variables included (1) attitude toward the content, (2) agent perceptions (persona, gestures, and facial expression), and (3) recall.

To assess learner attitude toward the content, learners were asked to list two adjectives that were coded according to positive, negative or neutral valence. Perceptions with respect to agent persona were assessed by the validated *Agent Persona Instrument* [13], which includes 4 sub-measures: Facilitating learning (10 items), Credible (5 items), Human-like (5 items), and Engaging (5 items). Three Likert-scale items were used to assess participants' perceptions of both the deictic gestures (alpha=.95) and facial expression (alpha=.93). The perception of deictic gestures was assessed with three items: 1) The agent's gestures helped me to learning the material; 2) The agent's gestures helped me to pay more attention to what was being said; and, 3) The agent's gestures helped me to focus on learning. The perception of the facial expressions of the agent was also measured with three similar items. Recall was assessed by a 10-item test, consisting of true-of-false, multiple choice, and open-ended questions based on the content from the instructional module. Recall questions differed for each module, but were developed in parallel format.

3 Results

A three-way MANOVA was conducted to test the overall effect and a follow-up ANOVA was used for detecting each independent variable's effect. The interaction between type of knowledge and gesture conditions was significant ($p<0.05$).

Participants showed a more positive attitude toward the content in the deictic gesture condition (M= 4.53, SD=6.71) than in the no deictic gesture condition (M= 4.11, SD=1.74), $F(1, 229)=3.69$, $p<0.05$. More interestingly, the interaction effect between the type of knowledge and the facial expression significantly influenced participants' attitude toward the content. In the attitudinal module, participants showed a more positive attitude toward the instruction in the facial expression condition (M=4.47, SD=1.56) than in the no facial expression condition (M=4.19, SD=1.55), $F(1.229)=3.60$, $p<0.05$. In contrast, under the procedural module, participants tended to have a more positive attitude toward instruction in the absence of facial expressions (M=4.65, SD=1.69) than in the presence of facial expressions (M=4.12, SD=2.09).

Agent facial expression influenced participants' perception of the agent. Participants who were exposed to agents with facial expressions (M=20.96, SD=6.14) rated the agents' overall persona as significantly better than participants who had an agent with no facial expression (M=19.52, SD=6.25), F (1,229) = 3.13, p<0.1. In addition, participants in the procedural module rated the agents significantly better for their overall persona (M=21.06, SD=5.62) than those in the attitudinal module (M=19.39, SD=6.71), F(1,229)=5.49, p<0.05.

Participants in the procedural module (M=18.38, SD=8.17) perceived agents' gestures more positively than students in the attitudinal module (M=16.44, SD=7.63), F(1,229)=5.71, p=0.02. The interaction effect between the type of outcome and the deictic gesture significantly influenced participants' perception about agent gestures, F(1,229)=4.54, p=0.03. In the procedural module, participants in the deictic gesture condition reacted more positively toward agents' gestures (M=21.92, SD=7.27 versus M=14.83, SD=7.51). Similarly, in the attitudinal module, when participants were in the deictic gesture condition, they showed more positive perception toward agent gesture (M=17.42, SD=7.55 versus M=14.58, SD=7.53). The effect size for procedural knowledge (Cohen's d=0.96) was higher than that for attitudinal knowledge (Cohen's d=0.38).

4 Discussion

Results revealed an interaction effect between the type of knowledge and agent facial expression, suggesting that student attitudinal knowledge may be enhanced when agents have facial expressions. In contrast, student attitude toward the procedural content may be enhanced when agents have no facial expressions. The purpose of the facial expression for the procedural module was to encourage students to learn information that is inherently non-affective; thus the expressions were extraneous and may have unnecessarily overloaded the learners cognitively. In contrast, the agent facial expression within the attitudinal module was to persuade the students, and thus more meaningful. Along this line, while a main effect indicated that deictic gestures were perceived positively in both the procedural and attitudinal modules; an interaction effect between type of knowledge and the deictic gestures emphasized the greater value of gestures while learning procedural knowledge as compared to acquiring attitudinal knowledge.

Results indicated that participants perceived both the agent persona and its gestures more positively in the procedural module. This suggests that the deictic gestures, which led to more learning (recall), also led to a better learner attitude toward the agent as a valuable instructional "persona," as similarly found in Baylor and Ryu [5].

There were several limitations of the study. The study intentionally did not involve extensive interaction between the learner and the agent as a control mechanism. However, social interaction between the learner and agent could play an important role in how the agent would be perceived and its effectiveness in message delivery. In addition, the face of the agent may not have been large enough (relative to its body size) to fully highlight the facial expressions.

In conclusion, the deictic gestures were more effective for the procedural module and the facial expressions were more desirable in the attitudinal module. These results suggested that instructional designers should consider the domain of knowledge that they want to represent and transmit and then decide which type of animation effect will effectively align with the nature of the message.

Future research should investigate the relative size(s) of the agent's face and body relative to the type of animation employed. Along this line, this study should be replicated employing larger agent faces (perhaps only to the shoulder level) for the facial expression conditions. Comparing the relative value of agent face and body size is also important for agents represented in other media, such as mobile devices. In addition, future work should consider how these nonverbal communication features are intensified with more intensive human-agent interaction.

Overall, results from this study provide practical knowledge about the design of nonverbal communication for pedagogical agents to achieve positive outcomes, for both procedural and attitudinal learning. Unlike human nonverbal communication, agent nonverbal communication can be designed and controlled to amplify the effect of the message and intensify its meaning.

Acknowledgments. This work was supported by the National Science Foundation, Grant IIS-0218692.

References

1. Atkinson, R.K.: Optimizing learning from examples using animated pedagogical agents. Journal of Educational Psychology 94, 416–427 (2002)
2. Moreno, R., Mayer, R.E., Spires, H.A., Lester, J.C.: The case for social agency in computer-based teaching: do students learn more deeply when they interact with animated pedagogical agents? Cognition and Instruction 19, 177–213 (2001)
3. Baylor, A.L.: Expanding preservice teachers' metacognitive awareness of instructional planning through pedagogical agents. Educational Technology. Research & Development 50, 5–22 (2002b)
4. Baylor, A.L., Kim, Y.: Validating Pedagogical Agent Roles: Expert, Motivator, and Mentor. The ED-MEDIA, Honolulu, Hawaii (2003b)
5. Baylor, A.L., Ryu, J.: Does the presence of image and animation enhance pedagogical agent persona? Journal of Educational Computing Research 28, 373–395 (2003)
6. Baylor, A.L., Shen, E., Huang, X.: Which Pedagogical Agent do Learners Choose? The Effects of Gender and Ethnicity. In: The E-Learn World Conference on E-Learning in Corporate, Government, Healthcare, & Higher Education, Phoenix, Arizona (2003)
7. Kim, Y., Baylor, A.L., Reed, G.: The Impact of Image and Voice with Pedagogical Agents. In: The E-Learn World Conference on E-Learning in Corporate, Government, Healthcare & Higher Education, Phoenix, Arizona (2003)
8. Allbeck, J., Badler, N.: Representing and Parameterizing Agent Behaviors. In: Prendinger, H., Ishizuka, M. (eds.) Life-like Characters: Tools, Affective Functions and Applications. Springer, Germany (2003)
9. Lester, J., Towns, S., Callaway, C., Voerman, J., Fitzgerald, P.: Deictic and Emotive Communication in Animated Pedagogical Agents. In: Cassel, J., Sullivan, S.P.J., Churchill, E. (eds.) Embodied Conversational Agents. The MIT Press, Cambridge (2001)
10. Baylor, A.L.: The Impact of Pedagogical Agent Image on Affective Outcomes. In: Intelligent User Interface International Conference, San Diego, CA (2005)
11. Baylor, A.L., Plant, E.A.: Pedagogical agents as social models for engineering: The influence of agent appearance on female choice. AI-ED Amsterdam (2005)

12. Baylor, A.L., Shen, E., Warren, D., Park, S.: Supporting learners with math anxiety: The impact of pedagogical agent emotional and motivational support. In: Workshop on Social and Emotional Intelligence in Learning Environments, held at the International Conference on Intelligent Tutoring Systems, Maceió, Brazil (2004)
13. Baylor, A.L., Warren, D., Park, S., Shen, E., Perez, R.: The impact of frustration-mitigating messages delivered by an interface agent AI-ED, Amsterdam (2005)
14. Anderson, J.R., Corbett, A.T., Koedinger, K.R., Pelletier, K.: Cognitive tutors: Lessons learned. The Journal of the Learning Science 4, 167–207 (1995)
15. Aimeur, E., Frasson, C.: Analyzing a new learning strategy according to different knowledge levels. Computers & Education 27, 115–127 (1996)
16. Gertner, A.S., VanLehn, K.: Andes: A coached problem solving environment for physics. In: Gauthier, G., VanLehn, K., Frasson, C. (eds.) ITS 2000. LNCS, vol. 1839, pp. 133–142. Springer, Heidelberg (2000)
17. Graesser, A., VanLehn, K., Rose, C., Jordan, P., Harter, D.: Intelligent tutoring systems with conversational dialogue. AI Magazine 22, 39–51 (2001)
18. Graesser, A., Moreno, K.N., Marineau, J.C.: AutoTutor improves deep learning of computer literacy: is it the dialogu or the talking head? In: Hoppe, U., Verdejo, F., Kay, J. (eds.) The International Conference of Artificial Intelligence in Education, pp. 47–54. IOS Press, Sydney (2003)
19. Reeves, B., Nass, C.: The Media Equation. CSLI Publications, Stanford (1996)
20. Atkinson, R.K., Mayer, R.E., Merrill, M.M.: Fostering social agency in multimedia learning: Examining the impact of an animated agent's voice. Contemporary Educational Psychology 30, 117–139 (2005)
21. Baylor, A.L., Ryu, J., Shen, E.: The Effects of Pedagogical Agent Voice and Animation on Learning, Motivation and Perceived Persona. ED-MEDIA, Hawaii (2003)
22. Johnson, W.L., Rickel, J.W., Lester, J.C.: Animated pedagogical agents: face-to-face interaction in interactive learning environments. International Journal of Artificial Intelligence in Education 11, 47–78 (2000)
23. Lester, J.C., Towns, S.G., Callaway, C.B., Voerman, J.L., FitzGerald, P.J.: Deictic amd Emotive Communication in Animated Pedagogical Agents. In: Sullivant, J. (ed.) Embodied Conversational Agents. MIT Press, Boston (2000)
24. Mousavi, S., Low, R., Sweller, J.: Reducing cognitive load by mixing auditory and visual presentation modes. Journal of Educational Psychology 87, 319–334 (1995)
25. Cassell, J., Sullivan, J., Prevost, S., Churchill, E. (eds.): Embodied Conversational Agents. MIT Press, Cambridge (2000)
26. Lee, K.M., Nass, C.: Designing social presence of social actors in human computer interaction. CHI (2003)
27. Gagne, R.: The Conditions of Learning. Holt, Rinehart & Winston, New York (1985)

Evaluating Data-Driven Style Transformation for Gesturing Embodied Agents

Alexis Heloir[1], Michael Kipp[1], Sylvie Gibet[2], and Nicolas Courty[2]

[1] DFKI - German Research Center for Artificial Intelligence. Campus D3.2, 66123 Saarbrücken, Germany
`firstname.surname@dfki.de`

[2] Laboratoire Valoria - Université de Bretagne Sud Campus de Tohannic, 56000 Vannes
`firstname.surname@univ-ubs.fr`

Abstract. This paper presents an empirical evaluation of a method called "Style transformation" which consists of modifying an existing gesture sequence in order to obtain a new style where the transformation parameters have been extracted from an existing captured sequence. This data-driven method can be used either to enhance key-framed gesture animations or to taint captured motion sequences according to a desired style.

1 Introduction

Endowing a virtual humanoid with expressive gestures requires one to take into account the properties that influence the perception of convincing movements. One of these properties can be called style. Style gathers all subtle charateristics occuring both over spatial and temporal aspects of motion. Style also gives in formation about the speaker's age, gender, cultural background, and emotional state. As a consequence, style contributes to making a virtual humanoid more convincing which makes it more acceptable to human users. This paper presents an empirical evaluation of a method called "Style Transformation" which consists of applying an automatic style on a neutral input motion to generate an appropriate style variant.

Style transformation basically consists of modifying an existing gesture sequence in order to obtain a new style whose transformation parameters have been extracted from an existing captured sequence. This data-driven method can be used either to enhance artist-generated gesture animations or to taint captured motion sequences according to a desired style.

2 Related Work

Work dedicated to the specification and generation of expressive gestures can be separated into two categories: gesture selection (which gesture is most suitable to be displayed) and motion quality (how should the gesture be displayed). In this section, we focus on the motion quality. Again, motion-quality dedicated work may be separated into two parts: theory driven approaches and data driven approaches.

Theory driven apporaches are driven by expert knowledge gained from empirical studies and extensive observation of human motion. the underlying history derived from foundation work [1,2] which served as a base for procedural motion synthesis systems [3,4]. Procedural systems have been proved to be capable of providing understandable expressive gestures with a high level of control [5,6]. However, such generative models, by relying on kinematics and/or physical models, have failed to produce natural motions. Such a lack motivates our investigation towards data driven approaches.

Recently, expressivity dedicated studies presented empirical implementations of expressivity models [5,7,8]. While Buisine et al. [8] addressed the perception of expressive features and their influence over generated agent's perceived behaviors (briskness, wearyness, tonicity), Schröder [7] studied the relationship between the 3 dimensions of emotion (valence, arousal, dominance) and the perceived emotion for speech synthesis. Kipp et al. [5] performed an empirical study dedicated to the influence of gesture-unit length over "believability" and "sympathy" of the produced gestures. Our experimental study is inspired by this prior work, as we present an empirical evaluation of generated motion and establish comparisons on the relation between perceived gesturing style and the valence, arousal, dominance (VAD) model of emotions.

3 Style Transformation Overview

This section presents an overview of the style transformation pipeline. The style transformation can be decomposed into two stages. A modeling stage where relevant parameters of the transformation model are inferred from existing data and a transformation stage where an arbirary input motion is transformed to convey the style which has been inferred during the learning stage. Figure 1 depicts the different stages involved in the style transformation pipeline. In the following paragraphs is a presentation of the gesture material which has been used as an input to our method followed by a short overview of the style tranformation method.

The experiments presented in this paper rely on four motion captured French Sign Language (FSL) gesture sequences performed by a deaf professional instructor. Three of these gesture sequences depict a weather forecast presentation performed according to different styles the signer had been asked to mimic: neutral ($-n$), angry ($-a$) and weary ($-w$). The fourth sequence depicts information usually displayed in railway stations, this sequence was performed according to a neutral style. In the following, the three styled gesture sequences (neutral, angry and weary) from the weather forecast material will be refered to as (M-W-n, M-W-w, M-W-a) while the sequence depicting train information will be refered to as (M-T-n). Motion data has been acquired using a method described in [9].

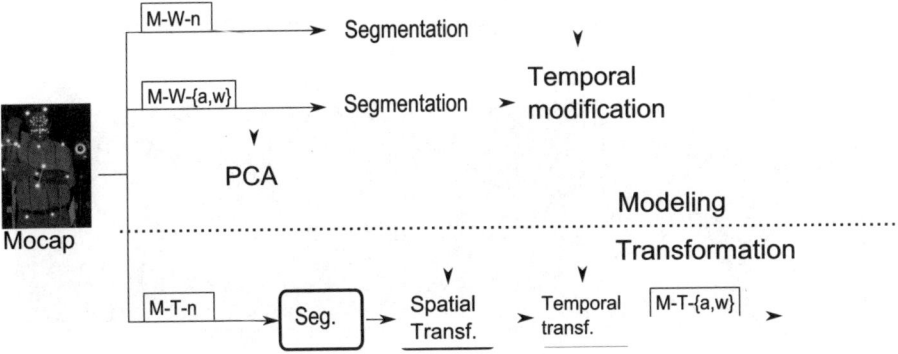

Fig. 1. Overview of the style transformation pipeline

One way of conveying subtle aspects of gesture style for articulated figures is to take into account the dependencies between joint motions. The spatial transformation introduced in this paper relies on the principal component analysis (PCA) and is comparable to the Egges et al. proposal [10]. However, this method is applied to the whole body structure and spatial corrections are then applied using inverse kinematics afterwards to prevent undesired foot skating effect.

Style transformation can be achieved on the temporal dimension using motion retiming. Although non-linear playback-speed modification is straightforward, determination of a relevant time-deformation profile may be tedious. A more in-depth explanation of the method is given in [11,12].

4 Perceptive Evaluation of Style Transformation

This section presents the empirical evaluation which has been conducted in order to test the accuracy of our style transformation method and check how far users are able to discriminate between the three styles using questions along three dimensions.

4.1 Experimental Setup

Participants. The experiment took the form of a questionnaire. Although the questionnaire was answered using an on-line web interface, the study was restricted to 19 subjects so that it could be better controlled. Subjects were German-speaking students from Saarbrucken University, Germany. Participants were aged between 20 and 25. Most of them (17 out of 19) were psychology or CS. Students' participation in such an experiment is required in the context of their curriculum. Experiment data has been gathered anonymously.

Method. Each evaluation was performed in our lab and was supervised by a member of our team. Every subject passed the evaluation individually. The test consisted of visualizing 9 video-enabled questionnaires and each time comparing

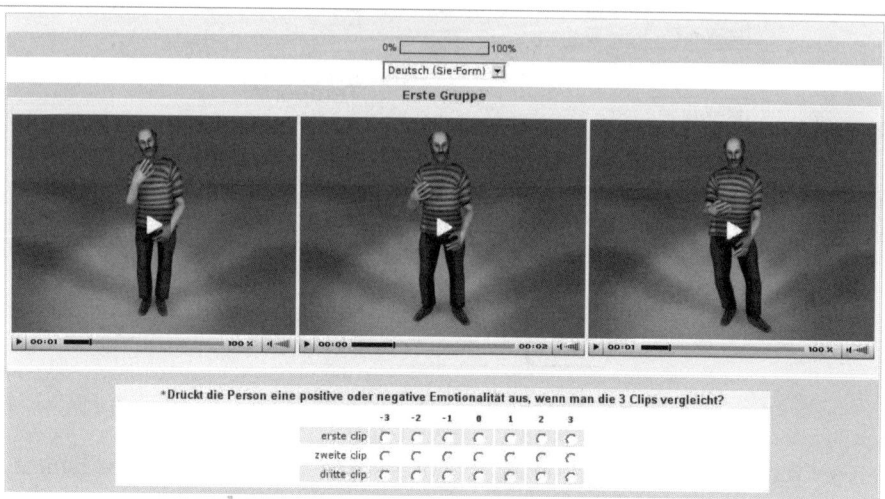

Fig. 2. Screenshot of a form that had to be filled out by a subject

3 styled realizations of a gesture clip. The video embedded in the questionnaire depicted short clips (mostly phrase segments) of motion captured weather forecast sequences (5 clips × 3 styles: M-W-n, M-W-a, M-W-w) and train information sequences (4 clips × 3 styles, 1 captured style: M-T-n and two generated styles: M-T-a, M-T-w) as depicted in section 3. 27 clips have been generated in total. All clips have been rendered in the same manner: agent appearance, lighting conditions, camera position and orientation were the same for each gesture clip. Rendering of our generated motion files was perfomed in Blender's[1] internal rendering engine. A questionnaire is depicted in Figure 2.

Before starting the experiment, the subject was told to read a short instruction sheet that briefly described the set-up of the experiment and was allowed to ask any questions about the experiment during the test. All the questions relative to the experiment were displayed sequentially. It was not possible for a subject to modify an already answered question, or to browse the questionnaire backwards. The subject was allowed to watch the video as much as he wanted and no time constraint was imposed to finish the questionnaire. The average answering time was 15 minutes.

4.2 Hypothesis

We attempted to verify the following two hypotheses:

- (H1): the subjects can discriminate neutral, angry and weary gesture clips along the three quality dimensions: valence, arousal and dominance,
- (H2): the style transformation imitates the angry and weary gesture qualities well.

---

[1] Blender is an open source 3D editing and rendering software http://www.blender.org

In order to test these hypotheses, subjects were asked on each question page to rate their perceived amount of valence, arousal or dominance for each clip. Answers were filled with numerical values (integers between -3 and +3). Twice in the whole questionnaire, the subject was asked to enter an adjective qualifying each video clip in the form of free text input (one time for the M-W-* material and one time for the M-T-* material). No hints in or around the video clips except gesture quality could help the user determining the actual style (neutral, angry, weary), nor the origin (unfiltered or filtered) of a displayed clip. Placement of the clips (left, center, right) and the total order of clips were randomized.

4.3 Results

The two materials "weather" (M-W-*) and "train" (M-T-*) are first compared. For both materials there are three style variants: neutral (-n), angry (-a) and weary (-w). The first question was whether the neutral variants of the two materials are rated similarly on the VAD dimensions (see Figure 3 (a)). We computed an ANOVA with factors material (M-W-*, M-W-*) and question (V, A, D). Neither factor material ($F(1,17)$, $p=.21$) nor material-question interaction ($F(2,34)$, $p=.57$) became significant. Therefore, variant n is rated in equally for M-W-n and M-T-n.

Using the same method we wanted to prove that our transformed motions successfully imitate the angry and weary styles (see Figure 3 (a), middle and right). For -a, the ANOVA showed no significance for material ($F(1,17)$, $p=.09$) or for material-question interaction ($F(2,34)$, $p=.13$). Therefore, style -a is perceived in the same way for material M-W and M-T. For style -w, a significant effect was found for material ($F(1,17)$, $p<.001$) and material-question interaction ($F(2,34)$, $p<.01$) which means that M-W and M-T differ.

Since we failed to prove a successful transformation that imitates the weary style -w, we compared the generated weary train sequence with the original neutral train sequence. to check whether the transformation did anything significant to the original.Thus, ANOVA was computed with factors style (w, n) and question (V, A, D). The result is inconclusive as the style-question interaction is tendential ($F(2,34)$, $p=.06$). However, -w and -n do not seem to be too far apart in terms of VAD perception.

Finally we analyzed whether the three style variants were really perceived as being different looking at the VAD dimensions (see Figure 3 (b)). A seperate analysis was made for each material. For each dimension (V, A, D) we computed an ANOVA which proved that the three styles differed. To highlight these differences, a post-hoc Scheffe-test was computer over every dimension pair. For M-W, on the valence (-V) dimension we found a significant main effect ($F(2,34)=6.30$, $p<.01$) which is due to a difference between -n and -a ($p<.05$) and between -n and -w ($p<.05$) but not due to the difference between -a and -w. For arousal (A) and dominance (D) we also found significant main effects (A: $F(2,34)=133.74$, $p<.01$; D: $F(2,34)=137.27$; $p<.001$) which were due to all three combinations (all $p<.001$). For material M-T-*, we found significant main effects for all three questions. On the valence (V) dimension ($F(2,34)=9.30$, $p<.001$) these were due to differences

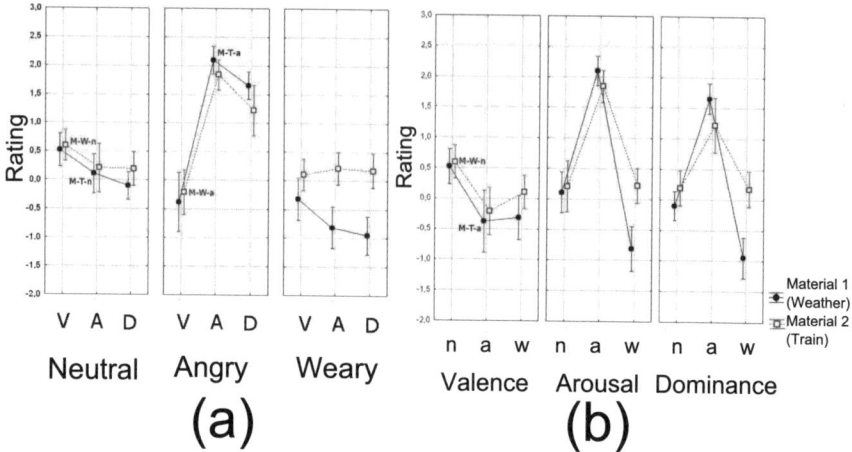

Fig. 3. (a): h1, the subjects can discriminate neutral, angry and weary gesture clips along the dimensions: valence, arousal and dominance. (b): h2, the style transformation imitates well the angry and weary gesture qualities.

between -n and -a (p¡.001) and between -n and -w (p¡.05). For both arousal and dominance (A: $F(2,34)=49.97$, p¡.001; D: $F(2,34)=18.82$, p¡.001) they were due to the difference between -n and -a and between -a and -w (all p¡.001) but not due to the difference between -n and -w.

5 Discussion

The first hypothesis gives insights on the subjects' ability to discriminate neutral, angry and weary gesture clips along the following three quality dimensions: valence, arousal and dominance.

The results presented show that subjects' ability to separate styled gesture varies regarding the considered dimension. Thus, as stated in the result section and illustrated in the first chart of Figure 3 (b), subjects have difficulties to discriminate the style of motion along the valence dimension. This confirms previous work stating that body motion somehow fails to convey hints about valence [13]. On the contrary, subjects prove to discriminate motion clips much better along the arousal and dominance dimensions (Figure 3 (b), middle chart, plain curve) and dominance dimension (Figure 3 (b), right chart, plain curve) of the VAD model. This explains why users can better recognise the angry style (high amount of recognizable arousal and dominance) than the weary style (low amount of recognizable arousal and dominance).

The second hypothesis gives us insights on how well style transformation imitates the angry and weary gesture qualities. The results presented in Figure 3 (a), middle chart show that users' perception of original M-W-a and transformed angry motion M-T-a is very similar. This leads us to believe that our notion transformation method imitates the angry gesture quality well. However, users' perception of original *M-W-w* and transformed *M-T-w* weary motion (Figure 3 (a), right chart) is not perceived in the same way by users. Indeed, results tend to show that weary style transformation does not change much the perceived style of the user compared to the neutral source material (slight tendency for *M-T-n* and *M-T-w* to be perceived the same way, corroborated by the adjective proposed by subjects). We could formulate the following assumptions for explaining the non-recognition of relevant gesture transformation by users. First the temporal retiming was perhaps too slight to be well perceived, thus exaggerating this retiming would maybe increase recognition rate. Second, the style transformation we applied does not take into account the stuctural modification that arise between style (hand drop for instance) nor the modification of gesture unit length, (by varying the frequency and length of retractions phases).

6   Conclusion and Perspectives

To sum up, this paper presented a method called style transformation. The method consists of transforming an existing motion (either motion captured, manually key-framed or procedurally generated) according to a style that is extracted from an analysis of motion captured files. The resulting motion, although degraded, conveys the subtle characteristics of the reference motion by learning dependencies in data. Besides transferring a specific style, the method can also be used to enhance procedurally synthesized motion that is to date perceived as stiff and unnatural.

An empirical evaluation involving 19 students has then been presented. This evaluation has been designed to verify two issues. First, how well subjects are able to discriminate styled motions along 3 dimension axes of emotion (valence, arousal and dominance). Second, to validate that the tainted motion produced by our style transformation algorithm is perceived as displaying the targeted style.

It has been statistically proved that subjects are able to separate captured body gesture sequences conveying different styles along the VAD model. Also, the gesture sequences produced by our style transformation algorithm have been rated statistically equal as captured styled gesture sequences conveying an angry style. Although the results are not significant when considering weary style, we believe that taking into account gesture unit length and structural modification of gestures would improve the style transformation.

Acknowlegdements

We would like to thank Kerstin H. Kipp for her essential contribution in the statistical study. This research has partially been carried out within the framework of the Excellence Cluster Multimodal Computing and Interaction (MMCI), sponsored by the German Research Foundation (DFG). Motion capture Data has been acquired in 2005 in Rennes, France within the framework of the RobEA national program. Sign language gestures have been performed by Alain Cahut.

References

1. Laban, R.: The Mastery of Movement. Northcote House (1988)
2. Wallbot, H.: Bodily expression of emotion. European journal of social psychology 28, 779–796 (1998)
3. Chi, D.M., Costa, M., Zhao, L., Badler, N.I.: Emote. In: Akeley, K. (ed.) Siggraph, Computer Graphics Proceedings, pp. 173–182. ACM Press, New York (2000)
4. Hartmann, B., Mancini, M., Pelachaud, C.: Implementing expressive gesture synthesis for embodied conversational agents. In: Gesture in Human-Computer Interaction and Simulation (2006)
5. Kipp, M., Neff, M., Kipp, K., Albrecht, I.: Toward natural gesture synthesis: Evaluating gesture units in a data-driven approach. In: Proc. of International Conference on Intelligent Virtual Agents (2006)
6. Neff, M., Kipp, M., Albrecht, I., Seidel, H.P.: Gesture modeling and animation based on a probabilistic recreation of speaker style. ACM Trans. on Graphics (to appear, 2008)
7. Schröder, M.: Dimensional emotion representation as a basis for speech synthesis with non-extreme emotions. In: Workshop on Affective Dialogue Systems (2004)
8. Buisine, S., Hartmann, B., Mancini, M., Pelachaud, C.: Conception et evaluation d'un modéle d'expressivité pour les gestes des agents conversationnels. Revue en Intelligence Artificielle RIA. Special Edition Interaction Emotionnelle 20 (2006)
9. Heloir, A., Gibet, S., Multon, F., Courty, N.: Captured motion data processing for real time synthesis of sign language. In: Gibet, S., Courty, N., Kamp, J.-F. (eds.) GW 2005. LNCS (LNAI), vol. 3881, pp. 168–171. Springer, Heidelberg (2006)
10. Egges, A., Magnenat-Thalmann, N.: Emotional communicative body animation for multiple characters. In: V-Crowds 2005, pp. 31–40 (2005)
11. Heloir, A., Courty, N., Gibet, S., Multon, F.: Temporal alignment of communicative gesture sequences. Computer Animation and Virtual Worlds 17, 347–357 (2006)
12. Heloir, A., Gibet, S.: A qualitative and quantitative characterization of style in sign language gestures. In: Gesture in Human-Computer Interaction and Simulation (2007)
13. Ekman, P., Friesen, W.V.: Hand and Body Cues in the Judgment of Emotion. Perceptual and Motor Skills 24, 711–724 (1967)

Culture-Specific First Meeting Encounters between Virtual Agents

Matthias Rehm[1], Yukiko Nakano[2], Elisabeth André[1], and Toyoaki Nishida[3]

[1] Augsburg University, Institute of Computer Science, 86159 Augsburg, Germany
{rehm,andre}@informatik.uni-augsburg.de
[2] Dept. of Computer and Information Science, Faculty of Science and Technology,
Seikei University, Japan
y.nakano@st.seikei.ac.jp
[3] Dept. of Intelligence Science and Technology, Graduate School of Informatics,
Kyoto University, Japan
nishida@i.kyoto-u.ac.jp

Abstract. We present our concept of integrating culture as a computational parameter for modeling multimodal interactions with virtual agents. As culture is a social rather than a psychological notion, its influence is evident in interactions, where cultural patterns of behavior and interpretations mismatch. Nevertheless, taking culture seriously its influence penetrates most layers of agent behavior planning and generation. In this article we concentrate on a first meeting scenario, present our model of an interactive agent system and identify, where cultural parameters play a role. To assess the viability of our approach, we outline an evaluation study that is set up at the moment.

1 Introduction

Imagine you are in Japan for the first time in your life. You looked at the first chapter of a Japanese language text book to learn some phrases beforehand and now you know how to greet someone you meet for the first time:

A: Kon'nichi wa.
B: Kon'nichi wa.
A: Watashi wa Yukiko Nakano desu. Hajimemashite.
B: Watashi wa Machiasu Remu desu. Hajimemashite. Doozo yoroshiku.

Although you know the phrases, you are still feeling a bit uncomfortable because it is not only the language that is different but also the nonverbal interaction habits. The language text book could not prepare you for the actual situational context. How do you behave in this situation? Do you shake hands? Where do you look? How close do you get to your conversational partner?

In this paper, we regard culture as being a social rather than a psychological notion (e.g. [35]), with cultural influences becoming evident in interactions, where cultural patterns of behavior and interpretation are contrasted or mismatch. As culture constitutes a fundamental influence on a variety of human behaviors, it was shown in a number of studies that culture is also a psychological process (e.g. [22]). Nevertheless, for the scope of this paper the focus lies on culture as a social group phenomenon. In a given culture, cultural patterns of behavior are not necessarily consciously relevant but they are the common heuristics for people from that culture on how to behave "properly" and on how to interpret behaviors of others. Of course, some heuristics might get institutionalized like for instance traffic rules. Culture is thus a group phenomenon, established by a group of people that adhere to some common patterns of behavior, thinking, and interpretation. These patterns have been called heuristics, norms and values, or mental programs in different theories (see e.g. [9],[13],[23],[33]). They become especially apparent when contrasted to behavior that deviates from these heuristics. Most of the above cited theories like [13] or [33] focus mainly on national cultures. But we have to keep in mind that nation is not a self evident level of cultural organization. Granularity is a very important factor of this notion, ranging from such abstract concepts like the European vs. the Asian culture down to such specific concepts like the Punk culture vs. the Speed Metal culture that have nothing do with a given national background. In this article, we concentrate on the fairly abstract level of national cultures, mainly for the fact that the theories we rely on deal with national cultures. Additionally, although there are some examples that concentrate on finer distinctions (e.g. [15]), most applications with enculturated agents aim at differences found on this level of granularity (see next section).

Although culture manifests itself as a social group phenomenon, the individuals in a culture adhere to the corresponding heuristics and must have internalized them during their socialization process. Thus, it seems to be legitimate to model culture as a computational parameter that influences individual agents and penetrates most processes of an agent system. According to Hall [10] for instance, people from so-called high- and low-contact cultures have different spatial behaviors in that high-contact individuals will stand closer in interpersonal encounters. Thus, an individual agent needs to individually react to the spatial distance between himself and his interlocutor, e.g. by moving closer if the other moves further away.

Consequently, tailoring information presentation to the cultural background of the user can be expected to serve as a criterium for sucess e.g. in e-commerce applications or other persuasive technologies. In the area of virtual agents, there are a number of different application domains like serious games for coaching cross-cultural communication skills, experience-based roleplays as an addition to the standard language textbook, or creating meeting spaces like SecondLife, where it might become easy to explore cultural identities from your armchair, interacting with a mix of real users and virtual agents. And of course endowing virtual agents in games with their own cultural background allows them reacting in a believable way to (for them) weird behavior of other agents and the user.

In the following, we survey approaches tackling this challenge (Section 2) and present the CUBE-G[1] procedure, which combines an empirical data-driven with

---

[1] CUlture-adaptive BEhavior Generation: http://mm-werkstatt.informatik.uni-augsburg.de/projects/cube-g/

a theoretical model-driven approach (Section 3). For the implementation we concentrated on a prototypical scenario found in every culture, a first meeting of two strangers (Section 4). To assess the viability of our concept we designed a large scale web-based evaluation study that is presented in Section 5 before the article closes with open research questions (Section 6).

2 Related Work

Whereas static presentations like e.g., websites can be easily tailored to culture-specific demands during the design process (given that the designer recognizes the challenge), interactive systems pose an additional challenge because they have to react dynamically to situational and contextual factors. An overview is presented by Payr and Trappl's [28] collection of different aspects of agent culture. Ruttkay [32] argues that it is indispensable to take care of cultural influences during the whole development process. Because every developer brings in his own culture and associated heuristics, it is necessary to make them explicit in order to keep the development process "clean". Most approaches in this area concentrate on learning environments or interactive role-plays with virtual characters. Khaled and colleagues ([20], [21]) focus on cultural differences in persuasion strategies and present an approach of incorporating these insights into a persuasive game for a collectivist society. Maniar and Bennett [25] propose a mobile learning game to overcome cultural shock by making cultural differences aware to the user. Johnson and colleagues [18] describe a language tutoring system that also takes cultural differences in gesture usage into account. The users are confronted with some prototypical settings and apart from speech input, have to select gestures for their avatars. Moreover they have to interpret the gestures by the tutor agents to solve their tasks. Core and colleagues [5] describe a training scenario for different negotiation styles which is set in a different culture than the trainees'. Unfortunately, they haven't realized culture-specific negotiation styles yet but acknowledge the importance of such a step. Warren and colleagues [36] as well as Rehm and colleagues [29] aim at cross-cultural training scenarios and describe ideas on how these can be realized with virtual characters. Jan and colleagues [16] describe an approach to modify the behavior of characters by cultural variables relying on Hofstede's dimensions. The variables are set manually in their system to simulate the behavior of a group of characters. Miller [26] provides an overview to work on politeness or "etiquette" in interactions between users and computers and presents a computational model to characterize, quantify and simulate such effects in human machine interactions. Although politeness strategies are generally described as universal [4], the realization and contextual parameters for their application can differ from culture to culture.

Even though there are a number of approaches to simulate culture-specific agents, a principled approach to the generation of cross-cultural behaviors is still missing. Furthermore, there is no empirical validated approach that maps cultural dimensions onto expressivity dimensions. In order to realize cross-cultural agents, we need to move away from generic behavior models and instead simulate individualized agents that portray idiosyncratic behaviors, taking into account the agent's cultural background. To this end, we propose a combination of an empirical data-driven and a theoretical model-driven approach that is detailed in the next section.

3 Combining an Empirical and a Theoretical Approach

We tackled the challenge of assessing the impact of culture on multimodal behavior from two sides. Based on a well-established cultural theory by Hofstede [13], we developed a theoretical model of cultural influences. Hofstede is a recent representative of a theoretical school that defines culture as a set of norms and values that members of a given culture adhere to. He presents a dimensional approach to culture that defines culture as a point in a five-dimensional space. The difference between individualistic and collectivistic cultures is for instance covered by the identity dimension. Hofstede's approach is described in more depth in Section 3.2. To ground our theoretical model not only in the mostly anecdotal data found in the literature, we conducted a standardized comparative study of multimodal interactions in Germany and Japan focusing on three prototypical situations: first meeting, negotiation, and status difference.

3.1 Empirical Approach

A first meeting between strangers, a negotiation process, and an interaction of individuals with different social status have been chosen for the corpus study due to their prototypical nature, i.e. they can be found in every culture and they constitute situations a tourist or ex-patriate is likely to encounter. Analysis of the corpus started with the first meeting scenario. There are several specific reasons for including this scenario. According to Kendon [19], it is not only found in all cultures but it also plays an important role for managing personal relations by signaling for instance social status, degree of familiarity, or degree of liking. There is also a practical reason for this scenario because it is the standard first chapter of every language textbook and thus known to everybody who ever learned a foreign language revealing a potential application of the results in a roleplay for first meeting scenarios. For Argyle [1], a first meeting is a ritual that follows pre-defined scripts. Ting-Toomey [34] follows his analysis by denoting a first meeting as a ceremony with a specific chain of actions. Knapp and Vangelisti [24] emphasize a first meeting as a step into the life of someone else, which is critical for a number of reasons like face-keeping or developing a network of social relation. Thus, the ritualistic nature of a first meeting makes sense in order to "to be on the safe side" by establishing such a new relationship in a satisfactorily, i.e. facekeeping, manner for both sides.

For the two cultures examined in our corpus, some specific differences are described in the literature for such a first meeting scenario. Greeting are expected to be longer in Japan because according to Ting-Toomey [34], greetings in individualistic cultures (like Germany) are shorter than in collectivisitic cultures (like Japan). Ting-Toomey also claims that Germans use more gestures than Japanese, and that the organisation of the dialogue will differ due to different time conceptions. Germany is stated to be a m-time culture (monochronic) whereas Japan is a p-time culture (polychronic), which means that Germans follow a line (e.g. first questions about university, then about private life) and Japanese discuss more things concurrently ([10], [34]). Moreover, Japanese have a smaller public self than Germans, thus they do not reveal too much information during a first meeting. According to Hall and Hall [11], this is due to the high-context nature of this culture. Consequently, we can expect our Japanese subjects to talk mainly about their occupations, whereas for the German subjects we additionally expect conversations about hobbies and personal life. In both cultures we expect information exchanges about the university and the experiment itself following Knapp and Vangelisti [24], who showed that questions concerning the setting or the enviroment of the first meeting always occur. We also expect more body contact in the form of a handshake in Germany for the actual greeting whereas in Japan non-contact bowing is expected. Greenbaum and Rosenfeld [8] have shown this difference in a comparison of US and Japanese culture.

Other information especially about multimodal behaviors is often of an anecdotal character like for instance, Southern Europeans tend to use more gestures in interactions than Northern Europeans. The corpus study allows us a more principled investigation of such differences.

Results. The analysis of the CUBE-G corpus is concentrating on nonverbal behavior at the moment. The behavior under investigation is comprised of postures, gestures, gestural expressivity, gaze, volume, and proxemics. Here we shortly report on our first results on differences in posture and gesture use. But beforehand let us have a quick look on Greenbaum and Rosenfeld's claim that there will be more body contact for the German sample when greeting each other. The data for the Japanese sample is unambigous. Apart from two participants all bowed to each other without any attempt for body contact. For the German sample the result is not so clear. One third of the participants (7 out of 21) initiated a hand shake. Thus, there is definitely more body contact during the actual greeting in the German sample but it is not the predominant behavior. Moreover, there is a gender effect because six of the seven participants that initiated the hand shake were male. For postures we found some consistent differences mainly for hand and arm postures. The predominant hand and arm postures for Germans are crossing the arms in front of the trunk or putting the hands in the pockets of the trousers. For the Japanese, the typical posture is joining hands in front of the body. Figure 3 exemplifies the postures. Frequency of gesture use is consistent with the above mentioned results from the literature. We found a significant difference in the number of gestures that were used in the German and the Japanese samples. German participants used more than three times more gestures than Japanese participants (22.1 (German) vs. 6.6 (Japanese) on average for a single encounter, t-test, $p < 0.01$). We also found significant differences for the two expressive parameters spatial extent (ANOVA, $p < 0.01$) and speed of a gesture (ANOVA, $p < 0.1$). Additionally, we looked into speech pauses in and between turns assuming after Hecht and colleagues [12] that in European conversations

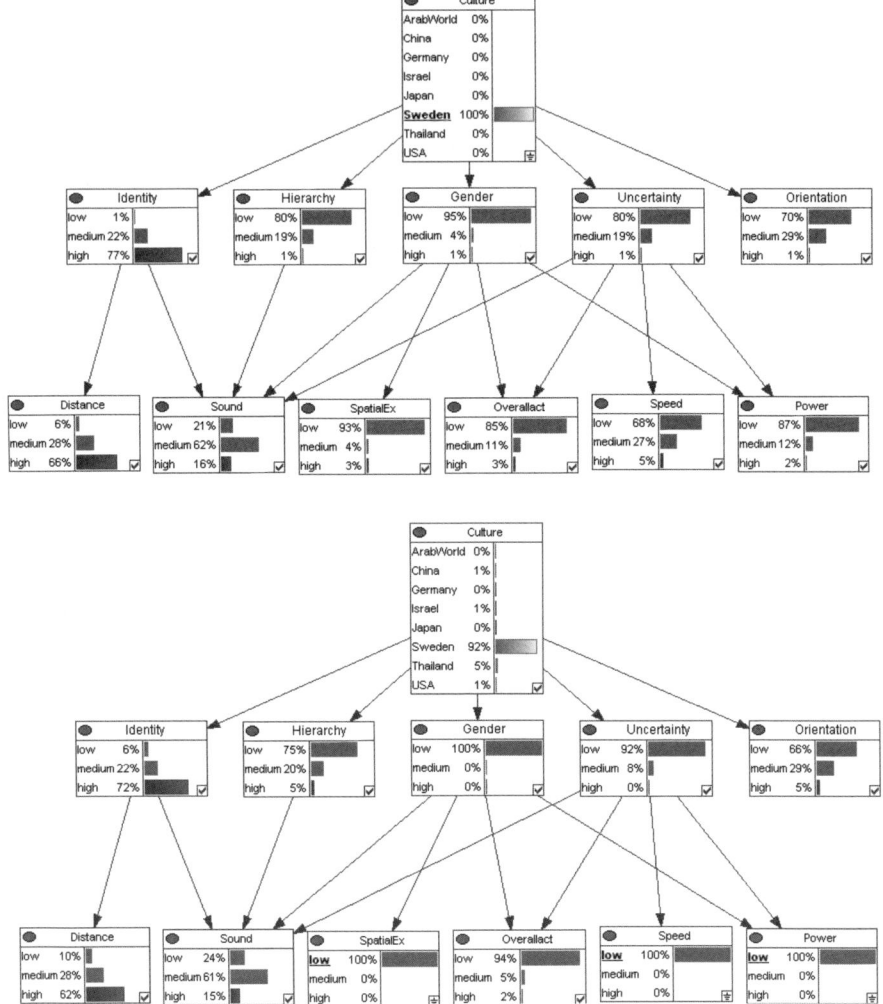

Fig. 1. Bayesian Network modeling the interrelation between cultural dimensions and nonverbal behavior: inferring nonverbal behavior given a specific culture (causal inference, above) vs. inferring the cultural background given a pattern of gestural expressivity (diagnostic inference, below)

pauses are often sensed as unpleasant and thus we expect Japanese to use pauses more frequently than Germans. For the analysis we distinguish between long (>2 seconds) and short pauses (1-2 seconds). In the five minute long first meeting encounters, we found 7.1 short and 1.3 long pauses on average for the German sample vs. 31 short and 8.4 long pauses on average for the Japanese sample. The differences were highly significant for both types of pauses (t-test, $p < 0.01$). More information on the design of the corpus study, the applied annotation schemes as well as the results can be found in [7] and [31].

3.2 Theoretical Approach

As described above, the influence of culture penetrates most of the processes in an interactive agent system, be it the interpretation of user input, be it behavior planning or generation, be it rendering of animations. Here we concentrate on the low level influence of cultural patterns of behavior. Our first model is a Bayesian network based on Hofstede's [13] five-dimensional model of culture and his ideas of synthetic cultures [14], which define stereotypes for the five dimensions. In the long run, these stereotypical values will have to be replaced by specific empirical data like the data we derive from our corpus study. The five dimensions are hierarchy, identity, gender, uncertainty, and orientation. Hierarchy denotes if a culture accepts unequal power distance between members of the culture or not. Identity defines to what degree individuals are integrated into a group. Cultures can either be more collectivistic or more individualistic. Gender describes the distribution of roles between the genders. In feminine cultures for instance roles differ less than in more masculine cultures. Uncertainty assesses the tolerance for uncertainty and ambiguity in a culture. Those with a low uncertainty tolerance are likely to have fixed rules to deal with unstructured situations. Orientation distinguishes long and short term orientation, where values associated with short term orientation are for instance respect for tradition, fulfilling social obligations, and saving one's face. It has to be noted that Hofstede's theory is not without controversy. His theory is based on a large-scale questionnaire study with IBM employees, which constitutes a strong selection bias on the results. Nevertheless, Hofstede's theory has a great appeal for computer science because of its quantitative nature (see Section 2).

According to Hofstede, nonverbal behavior is strongly affected by cultural affordances. The identity dimension e.g. is tightly related to the expression of emotions and the acceptable emotional displays in a culture. Thus, it is more acceptable in individualistic cultures like the US to publicly display strong emotions than it is in collectivistic cultures like Japan [6]. Uncertainty avoidance like identity is directly related to the expression of emotions. In uncertainty accepting societies, the facial expressions of sadness and fear are easily readable by others whereas in uncertainty avoiding societies the nature of emotions is less accurately readable by others, which was shown by Argyle [1]. For the above mentioned synthetic cultures, Hofstede, Pedersen, and Hofstede [14] show how specific behavior patterns differ in a principled way depending on where a culture is located on the five dimensions. For instance, in a culture with a low power distance (hierarchy dimension) people tend to stand closer in interpersonal encounters. The same holds true for collectivistic cultures in contrast to individualistic cultures (identity dimension). A similar effect was shown by Hall [10], who analyzed spatial behavior in interpersonal encounters and distinguishes between high- and low-contact cultures.

Figure 1 gives an overview of the Bayesian network. Bayesian networks as described in [17] are a formalism to represent probabilistic causal interactions and have already been successfully applied to model emotional interactions for virtual agents ([2],[3]). In the domain of culture they are suitable for the following reasons. Because there is a many to many mapping between culture and nonverbal behavior, it is not likely that individuals behave exactly like it is described for a given culture in every aspect of their behavior. Bayesian networks handle such uncertainties very well. Additionally, a Bayesian network explicitely models the relations between causes and effects. Thus, links in the network are intuitively meaningful. The theoretical effect that the more masculine a culture becomes the louder people in this culture will speak [14], is represented by a link between the cultural dimension of gender and the nonverbal behavior "volume". Moreover, Bayesian networks allow for causal as well as diagnostic inferences depending on where evidence is introduced into the network. Thus, such a model can be used to set or modify the nonverbal behavior of an agent by setting the evidence for a given culture (causal inference) as well as to infer the culture from given nonverbal behavior (diagnostic inference).

The middle layer defines Hofstede's dimensions. We already integrated all five dimensions but the dimension orientation has so far no influence on the outcome. This is due to the fact that the literature on this specific dimension is sparse and didn't allow defining a reliable influence on nonverbal behavior. The bottom layer consists of nodes for nonverbal behavior that can either be registered from the user or another agent or that can be set for a given agent. The top node which is labeled "Culture" is just for demonstration and interpretative purposes. It mainly translates the results from the dimensional representation of cultures into a probability distribution for some example cultures.

The Bayesian network only presents one building block for integrating culture as a computational parameter in an agent system. Cultural influences manifest themselves on different levels of behavior generation and interpretation and thus penetrate many processing modules in a system that takes these influences into account. Our first prototype concentrated on inferring the user's cultural background [30] and employing this information to adjust the nonverbal expressive behavior of a group of agents. The next section describes our concept of a more complex system and a prototype that incorporates the Bayesian network and further results from the corpus study.

4 Implementing Culture-Specific First Meeting Encounters

Figure 2 presents an overview about the current state of the agent system. The generated culture-specific behavior is exemplified by Figure 3 that depicts a sample from the corpus study along with an snapshot from the generated behavior for the German (above) and the Japanese culture (below). Cultural influences manifest themselves at all of the depicted processing steps.

4.1 Behavior Planning

Above it was shown that first meetings encounters have a ritualistic form defining specific phrases used for greeting, "proper" topics and sometimes the order, in

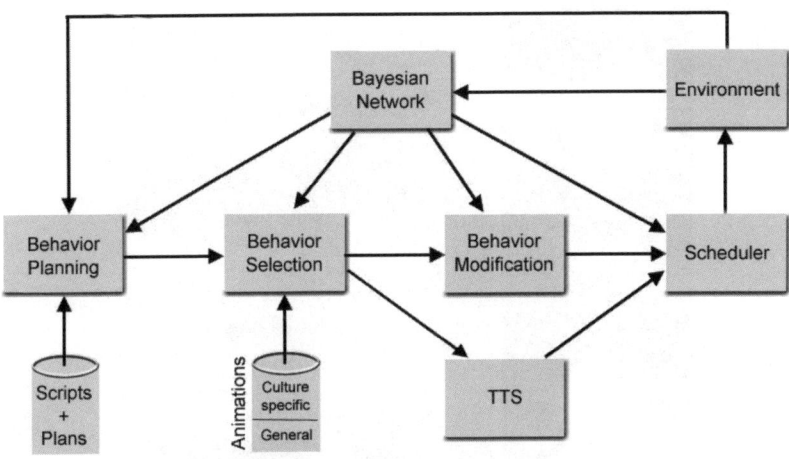

Fig. 2. General system architecture

which topics are discussed. Thus, the behavior planning module relies on the information about the agent's cultural background to either select a culture-specific plan or script or to modify a general plan if this is possible. The current state of the system relies on predefined scripts for first meeting encounters which have been developed following examples from language textbooks, information from the literature about first meetings (see Section 3.1), and observations from the corpus study. The Bayesian network provides the information about the cultural background of the agent. In this version of the system, the evidence for the agent's cultural background can be set freely for each run of the system. The arch from environment to Bayesian network symbolizes the possibility of infering the cultural background of an interlocutor, which is described in detail in [30].

4.2 Behavior Selection and Modification

The behavior selection and modification modules enrich the utterances processed from the scripts by appropriate nonverbal behaviors. Because there are some typical behaviors in each culture (see Figure 3 for typical postures in Germany and Japan) it does not suffice to just modify the available animations by the results for the expressive parameters supplied by the Bayesion network. It is also necessary to have culture-specific animations for conversationally relevant gestures like bowing in Japan. Thus, our database of animations is partitioned into a general part with gestures that can be applied regardless of the agent's culture (but modified with the information from our Bayesian network) and a specific part with gestures relevant for a given culture.

4.3 TTS

The agent system makes use of the Horde3D graphics engine, which allows for interfacing with any TTS system that is compliant with the Microsoft Speech

Fig. 3. First meeting examples. German sample from corpus study and generated interaction (above) vs. Japanese sample from corpus study and generated interaction (below).

API and provides lip-synching functionality on this basis. The German utterances are generated by the Loquendo TTS system. Unfortunately, the Japanese TTS does not implement the Speech API, thus we had to create our own Speech API compliant layer to interface the system to Horde3D. The choice of the TTS would in principle depend on the cultural background set by the Bayesian network. For the evaluation study this feature is disabled because the language would be a much too strong hint on the culture of the agents. Thus, the utterances are mapped to gibberish to prevent participants in the evaluation study to concentrate to much on what has been said.

4.4 Scheduler

The scheduler keeps track of the dialogue and decides on the right timing for the agent's behaviors. The corpus study revealed that Japanese tend to make more and longer pauses which can be related to more collectivistic cultures. The information from this dimension is exploited by the scheduler for deciding on the right time to trigger the agent's next visible (and audible) action.

5 Evaluation

The experimental design of the evaluation study is based on the fact that culture is a social phenomenon. The system generates a first meeting encounter between

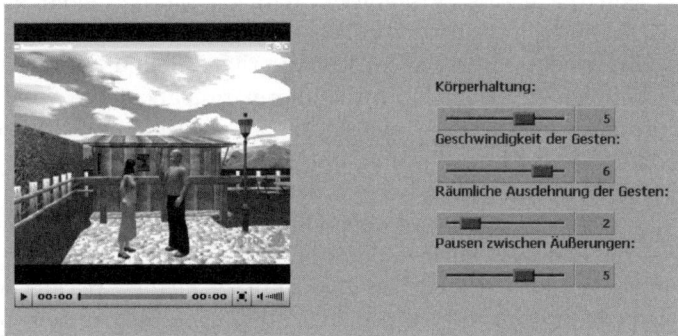

Fig. 4. Webinterface for the evaluation study

two agents that is tailored to the cultural background set for each agent. Thus, if the agents' behavior deviates from the user's expectations there should be an effect in the appraisal of this interaction compared to those that fit the user's expectations.

Of course it would be very obvious to see the difference between two agents that bow to each other and agents that shake hands instead. Thus, the evaluation focuses on more subtle clues of cultural influences like postures and gestural expressivity that were described above (Section 3.1). The hypothesis generated from this set up is: Users will rate agent interactions that show behavior not similar to their own cultural patterns as deviating on the examined parameters. The types of behavior that are tested are postures, gestural expressivity, and pauses in speech.

Participants are confronted with short videos of first meeting encounters generated by the system and have to rate how adequate certain aspects of the interaction are on a standard seven point Likert scale (see Figure 4 for an impression). These aspects are displayed postures, speed of gestures, spatial extent of gestures, and utterance flow. Four videos were created for the German, for the Japanese, and for a random culture respectively. Each participant is confronted with all videos in random order.

The agents' behavior is completely generated by the system based on evidence set for the cultural dimensions in the Bayesian network. Thus, we expect to find differences in the user's rating based on how the evidence was set in the network and on how the corresponding generated behavior deviates from the user's expectations about his own culture. The ratings of the random cultures will also allow us to gain insights into which of the behavioral features under investigation really contribute to the user's perception of the interaction.

6 Conclusion

In this paper we presented an approach of integrating culture as a computational term in an agent system relying on a combined data-driven and model-driven approach to gain the necessary empirical data on the one hand and on the other hand to exploit the theoretical concepts from the literature. The resulting system concentrates on a prototypical scenario that is found in every culture, a first meeting between strangers. With the information about the agents' culture the system produces interactions tailored to these cultural backgrounds taking aspects of posture, gesture, and timing into account.

Although this is a comprehensive approach of integrating culture as a computational parameter, a number of open challenges remain. Most fundamental in our view is the question which aspects of behavior are attributable to a cultural influence and which are attributable to other factors. Because culture is not an isolated concept but intertwined with other concepts like personality (see e.g. [27]). If someone prefers to stand far from his interlocutor in an interpersonal encounter this might be due to the high power distance in his culture but it might also be an effect of his introvert personality. To capture such effects in our data, every participant of the corpus study did a NEO-FFI personality test. The results from the personality test have not yet been linked to the above described analysis but will hopefully reveal a more fine-grained picture for modeling the interrelation between social (culture) and psychological (personality) influences on agent behavior.

Acknowledgements

The work work described in this paper is funded by the German Research Foundation (DFG) under research grant RE 2619/2-1 (CUBE-G) and the Japan Society for the Promotion of Science (JSPS) under a Grant-in-Aid for Scientific Research (C) (19500104). The Bayesian network model was created using the GeNIe modeling environment and integrated with SMILE developed by the Decision Systems Laboratory of the University of Pittsburgh (http://dsl.sis.pitt.edu).

References

1. Argyle, M.: Bodily Communication. Methuen & Co. Ltd., London (1975)
2. Ball, E.: A Bayesian Heart: Computer Recognition and Simulation of Emotion. In: Trappl, R., Petta, P., Payr, S. (eds.) Emotions in Humans and Artifacts, pp. 303–332. MIT Press, Cambridge (2002)
3. Bee, N., Prendinger, H., Nakasone, A., André, E., Ishizuka, M.: AutoSelect: What You Want Is What You Get: Real-Time Processing of Visual Attention and Affect. In: André, E., Dybkjær, L., Minker, W., Neumann, H., Weber, M. (eds.) PIT 2006. LNCS (LNAI), vol. 4021, pp. 40–52. Springer, Heidelberg (2006)
4. Brown, P., Levinson, S.C.: Politeness — Some universals in language usage. Cambridge University Press, Cambridge (1987)
5. Core, M., Traum, D., Lane, H.C., Swartout, W., Gratch, J., Van Lent, M., Marsella, S.: Teaching negotiation skills through practice and reflection with virtual humans. Simulation 82(11), 685–701 (2006)
6. Ekman, P.: Telling Lies — Clues to Deceit in the Marketplace, Politics, and Marriage, 3rd edn. Norton and Co. Ltd, New York (1992)

7. Endrass, B., Rehm, M., André, E., Nakano, Y.: Talk is silver, silence is golden: A cross cultural study on the usage of pauses in speech. In: Proceedings of the IUI workshop on Enculturating Conversational Interfaces (2008)
8. Greenbaum, P.E., Rosenfeld, H.M.: Varieties of touching in greeting: Sequential structure and sex-related differences. Journal of Nonverbal Behavior 5, 13–25 (1980)
9. Hall, E.T.: The Silent Language. Doubleday (1959)
10. Hall, E.T.: The Hidden Dimension. Doubleday (1966)
11. Hall, E.T., Hall, M.R.: Hidden Differences: Doing Business with the Japanese. Anchor Books (1987)
12. Hecht, M.L., Andersen, P.A., Ribeau, S.A.: The Cultural Dimensions of Nonverbal Communication. In: Asante, M.K., Gudykunst, W.B. (eds.) Handbook of International and Intercultural Communication, pp. 163–185. Sage Publications, London (1989)
13. Hofstede, G.: Cultures Consequences: Comparing Values, Behaviors, Institutions, and Organizations Across Nations. Sage Publications, Thousand Oaks (2001)
14. Hofstede, G.J., Pedersen, P.B., Hofstede, G.: Exploring Culture: Exercises, Stories, and Synthetic Cultures. Intercultural Press, Yarmouth (2002)
15. Iacobelli, F., Cassell, J.: Ethnic identity and engagement in embodied conversational agents. In: Pélachaud, C., Martin, J.-C., André, E., Chollet, G., Karpouzis, K., Pelé, D. (eds.) IVA 2007. LNCS (LNAI), vol. 4722, pp. 57–63. Springer, Heidelberg (2007)
16. Jan, D., Herrera, D., Martinovski, B., Novick, D., Traum, D.: A Computational Model of Culture-Specific Conversational Behavior. In: Pélachaud, C., Martin, J.-C., André, E., Chollet, G., Karpouzis, K., Pelé, D. (eds.) IVA 2007. LNCS (LNAI), vol. 4722, pp. 45–56. Springer, Heidelberg (2007)
17. Finn, V.: Jensen. Bayesian Networks and Decicion Graphs. Springer, Heidelberg (2001)
18. Johnson, W.L., Choi, S., Marsella, S., Mote, N., Narayanan, S., Vilhjálmsson, H.: Tactical Language Training System: Supporting the Rapid Acquisition of Foreign Language and Cultural Skills. In: Proc. of InSTIL/ICALL — NLP and Speech Technologies in Advanced Language Learning Systems (2004)
19. Kendon, A.: Conducting Interaction: Patterns of Behavior in Focused Encounters. Cambridge Univ. Press, Cambridge (1991)
20. Khaled, R., Barr, P., Fischer, R., Noble, J., Biddle, R.: Factoring Culture into the Design of a Persuasive Game. In: OZCHI 2006, pp. 213–220 (2006)
21. Khaled, R., Biddle, R., Noble, J., Barr, P., Fischer, R.: Persuasive interaction for collectivist cultures. In: Piekarski, W. (ed.) The Seventh Australasian User Interface Conference (AUIC 2006), pp. 73–80 (2006)
22. Kitayama, S.: Culture and Basic Psychological Processes — Towards a System View of Culture: Comment on Oyserman et al. Psychological Bulletin 128(1), 89–96 (2002)
23. Kluckhohn, F., Strodtbeck, F.: Variations in value orientations. Peterson, New York (1961)
24. Knapp, M.L., Vangelisti, A.L.: Interpersonal Communication and Human Relationships. Pearson Education, Inc., Boston (1984)
25. Maniar, N., Bennett, E.: Designing a mobile game to reduce culture shock. In: Proceedings of ACE 2007, pp. 252–253 (2007)
26. Christopher, A.: Miller. Etiquette and Politeness in Human-Human and Human-Machine Interactions: A Summary of Work at SIFT. In: Proceedings of the IUI workshop on Enculturating Conversational Interfaces (2008)

27. Nazir, A., Lim, M.Y., Kriegel, M., Aylett, R., Cawsey, A., Enz, S., Zoll, C.: Culture-personality based affective model. In: Proceedings of the IUI workshop on Enculturating Conversational Interfaces, Gran Canaria (2008)
28. Payr, S., Trappl, R. (eds.): Agent Culture: Human-Agent Interaction in a Multi-cultural World. Lawrence Erlbaum Associates, London (2004)
29. Rehm, M., André, E., Nakano, Y., Nishida, T., Bee, N., Endrass, B., Huang, H.-H., Wissner, M.: The CUBE-G approach — Coaching culture-specific nonverbal behavior by virtual agents. In: Mayer, I., Mastik, H. (eds.) Proceedings of ISAGA (2007)
30. Rehm, M., Bee, N., Endrass, B., Wissner, M., André, E.: Too close for comfort? Adapting to the user's cultural background. In: Workshop on Human-Centered Multimedia, ACM Multimedia (2007)
31. Rehm, M., Nakano, Y., Huang, H.-H., Lipi, A.A., Yamaoka, Y., Grüneberg, F.: Creating a Standardized Corpus of Multimodal Interactions for Enculturating Conversational Interfaces. In: Proceedings of the IUI-Workshop on Enculturating Conversational Interfaces (2008)
32. Ruttkay, Z.: Cultural Dialects of Real and Synthetic Facial Expressions. In: Proceedings of the IUI workshop on Enculturating Coversational Interfaces (2008)
33. Schwartz, S.H., Sagiv, L.: Identifying culture-specifics in the content and structure of values. Journal of Cross-Cultural Psychology 26(1), 92–116 (1995)
34. Ting-Toomey, S.: Communicating Across Cultures. The Guilford Press, New York (1999)
35. Triandis, H.C., Suh, E.M.: Cultural influences on personality. Annual Review of Psychology 53, 133–160 (2002)
36. Warren, R., Diller, D.E., Leung, A., Ferguson, W., Sutton, J.L.: Simulating scenarios for research on culture and cognition using a commercial role-play game. In: Kuhl, M.E., Steiger, N.M., Armstrong, F.B., Joines, J.A. (eds.) Proceedings of the 2005 Winter Simulation Conference (2005)

Virtual Humans Elicit Skin-Tone Bias Consistent with Real-World Skin-Tone Biases

Brent Rossen[1], Kyle Johnsen[1], Adeline Deladisma[2], Scott Lind[2], and Benjamin Lok[1]

[1] CISE University of Florida, Gainesville, FL 32611, USA
{brossen,kjohnsen,lok}@cise.ufl.edu
[2] Dept. of Surgery Oncology, Medical College of Georgia, Augusta, GA 30912, USA
{adeladisma,dlind}@mail.mcg edu

Abstract. In this paper, we present results from a study that shows that a dark skin-tone VH agent elicits user behavior consistent with real world skin-tone biases. Results from a study with medical students ($n=21$), show participant empathy towards a dark skin-tone VH patient was predicted by their measured bias towards African-Americans. Real world bias was measured using a validated psychological instrument called the implicit association test (IAT). Scores on the IAT were significantly correlated to coders' ratings of participant empathy. This result indicates that VHs elicit realistic responses and could become an important component in cultural diversity training.

Keywords: Virtual humans, intelligent agents, virtual reality, human-centered computing, user interfaces, racial bias, medicine, computer graphics.

1 Introduction

Virtual human (VH) agents have been shown to be useful in many fields such as training military skills [1], social conversation protocols [2], and clinical therapist skills [3]. These interactions with VHs focus on interpersonal goals that are fundamentally similar to human-human (H-H) interactions. It is known that real world biases (e.g. race, age, gender, weight) impact real world interactions [4]. Thus we seek to ask the same questions about virtual world interactions. First, do biases affect VH interactions, and second, are these VH biases correlated with real world human (H) biases. Specifically, we study the effect of VH skin-tone on the cognition and behavior of a human conversational partner. The expectation is that by changing only the VH's skin-tone, the user's bias towards or against a particular race will cause a modification in the user's behavior.

We explore the behavioral effect of VH skin-tone in the context of the Interpersonal Simulator (IPS). The IPS allows users to interact naturally (speech and gestures) with life-sized VHs [5]. Currently applied to interpersonal skills training, the IPS has been used by over 450 health professions students to practice medical interviews. The eventual goal of the IPS is to train students on areas that are difficult with current techniques, e.g. cultural diversity training. Using the health-care domain as a test-bed empowers the study of diversity related issues. Medical students are motivated to

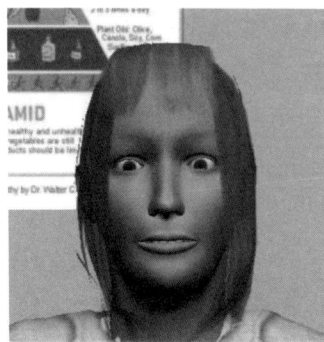 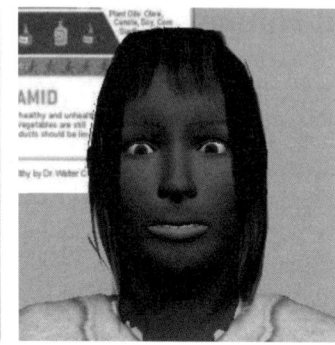

Fig. 1. The two VH skin-tones. The average skin-tone on the left is "Light" <201, 152, 138>, and on the right is "Dark" <112, 58, 32>.

improve, and the interview is a stressful experience (important for eliciting biases). Further, the health-care domain is one where skin-tone biases are an established problem [6]. Conventional techniques for interview training (e.g. role playing, actors) have difficulty providing the diversity and repeated exposures necessary for cultural diversity training. Many of these logistical hurdles in cultural diversity training would be addressed through the integration of VH agents.

Before integrating VHs into cultural diversity training, we must establish their efficacy for cultural diversity education. To establish this efficacy, we need to link measured real world biases to cognitive and behavioral effects in H-VH interactions. Cultural bias metrics come in two forms, direct and indirect. Direct measures assume users can accurately recognize their own biases. Surveys (e.g. "Does a person's race determine how much you like them?") are often used for this purpose. A known issue with direct measures of bias is that of social desirability bias. Social desirability bias is observed when people answer questionnaires in a way that accommodates social expectations, rather than how they truly perceive a situation. To circumvent this issue, indirect measures are used (i.e. measures where the user does not know that bias is being measured, or cannot control their bias). Indirect metrics include *behavioral observation*, *indirect survey questions* ("How empathetic were you?"), and *implicit response latency measures*. One implicit response latency measure, the Implicit Association Test (IAT) is widely used by psychologists for establishing skin-tone biases. The IAT has been validated in the United States general population for race, age, gender, and weight with very large data sets (N = 40,000 to 160,000 each) [7]. Using *direct* measures to study bias, one can never be sure if the measure has resulted in the desired bias or has retrieved only the *socially desirable* answer. The current study evaluates bias using *indirect measures* to evaluate the underlying racial biases.

The present work conducted a study ($n=21$) comparing medical students' interactions with a light-skin (Average <Red, Green, Blue>= <201, 152, 138>) and a dark skin-tone (<112, 58, 32>) VH patient (See Fig. 1). The objectives were to determine the applicability of bias measures, and identify potential effects of bias in the medical interview training test-bed. This study is a first step towards validating VH agents for cultural diversity education.

2 Previous Work

Efforts have been made to establish the basis of effective VHs and validate their efficacy in social interaction. This study adds to the growing body of literature on highly interactive and emotionally engaging VHs, and cultural diversity training applications [1, 2]. It continues the research by examining the effect of one of the more subtle elements of VHs, their skin-tone.

Skin-tone has been shown as a factor in the persuasive power (a cognitive marker of bias) of a VH. Studies have shown that users are more persuaded by VHs that emulate the user's own race [8]. They have also shown that learning from a VH is significantly affected by agent race [9]. The present work builds upon these results by studying cognitive *and* behavioral markers of skin-tone bias in a bidirectional *conversation* with a VH.

3 Study Design

The study was a between-subjects experimental design with VH skin-tone as the sole independent factor. Two conditions were tested, light skin-tone VH and dark skin-tone VH. Medical students performed a typical patient interview with either a light skin-tone VH *or* a dark skin-tone VH.

Medical students are taught to follow a structure when conducting a patient interview. The student's direct goal is to obtain a history of the illness and form an initial differential diagnosis for the patient. The protocol of a patient interview is explored in under ten minutes while building rapport with the patient by expressing empathy (understanding and concern for the patient's problems).

One VH patient, Edna, was used in the study. Edna represented a 55 year-old female who has found a lump in her left breast. Edna's voice was pre-recorded audio of an actor (a 75-year old Caucasian woman with a local accent) and was used for both skin-tone appearances.

Edna's rendering and interaction were accomplished using 3 networked computers. A wireless microphone captured student speech and allowed un-tethered natural language interaction. Optical tracking allowed appropriate gaze behavior. The VHs had realistic body meshes and employed skeletal and morph animations. Considerable efforts were made to mimic the patient-doctor interaction. The students interacted with life-size VH patients in a clinical skills examination lab. No icons, desktops, mice, or keyboard were ever visible. More details about the IPS hardware, infrastructure, and believability can be found in [5].

3.1 Population

Three Caucasian first-year physician assistant (PA) students and eighteen Caucasian third-year medical students attending a medical college in the south-east United States participated in this study. Group assignment was pseudo-random to ensure that each group had a similar distribution. The dark skin-tone condition had $n=12$ with 3 females and 9 males, the light-skin condition had $n=9$, with 4 females and 5 males.

Originally, recruitment did not take participant race into consideration, but only small numbers of non-Caucasians were in each condition. As a result, *only Caucasian participants were used for analysis* (the reason for the condition size disparity) because they were the largest homogenous group available. Examining just that group provides more information on bias, particularly out-group bias (bias against anyone not within the same group as the participant), than achieving a higher number of participants (original $n=27$, current $n=21$). This is a technique used in psychology studies such as [4].

3.2 Metrics

The Implicit Association Test (IAT): The IAT is a system for determining subconscious bias [7]. It measures response latencies, requiring the participant to rapidly categorize stimuli into "good" and "bad" categories. Detailed information and demonstrations can be found at [4, 7] and the Project Implicit website. It has been shown that survey metrics can be controlled by participants *motivated* to avoid biased responses, while this motivation has no effect on the IAT [4]. The result of the IAT is the IAT-D value. The *IAT-D* value is a number in the range -1 to +1, wherein .15=slight, .35=moderate, and .65-strong preference. For this study, positive IAT-D values represent a subconscious preference for African-Americans, and negative values represent a preference for Caucasian Americans. A study on racial prejudice by Dovidio et al, uses the IAT and provides excellent details and full literature review of real world bias[4].

Behavioral observation and coding: The medical interview task is a ten-minute interview that largely consists of the participant asking questions and listening to the response from the VH. To facilitate the study of bias a VH initiated challenge was used to encourage a natural, spontaneous response. Participants' resulting behavior was analyzed, as well as participants' post-interview perception of that behavior.

At 4 minutes into the interview the VH, Edna, asked, "Could this be cancer?" this is labeled as the *critical moment*. This critical moment was created to elicit biases based on conversations with medical educators and racial bias researchers (see acknowledgements). Students are expected to respond with *empathy*. Edna's "sister" had cancer and students must show Edna that they understand her concern. Video coders qualitatively rated this moment as "empathetic" on a 7-point scale from 1 (not at all) to 7 (very). Observers reviewed the critical moment using 1) Audio only, 2) Video only, and 3) Audio + Video. Separating audio and video created a division of participants' verbal (e.g. speaking words of understanding) and non-verbal behavior (e.g. leaning towards the patient, or making a facial expression). Reviewers were blind to the race of the VH. Inter-rater agreement is reported in the results section.

Post-experience survey: Following the interview, participants were asked to rate various aspects of their whole interview (overall performance and use of empathy). Racial bias may play a role because the ratings indicate how much a participant enjoyed the interview process. Participants were unaware that skin-tone bias was being tested, thereby limiting the effects of social desirability bias.

3.3 Procedure

Upon arrival, the participant completed a consent form and the background questionnaire. The background questionnaire was used to collect information about the previous experience, interview skill, gender, race, and age of the participant.

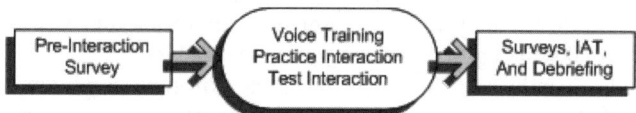

Fig. 2. H-VH Procedure overview

Following a 2-minute voice training for the speech recognition system, the participant was asked to walk into the isolated section of the clinical skills lab where the virtual experience was set up, and have a seat on the chair. Each participant performed one VH patient interview. A short introduction to spoken interaction with a VH was then given by one experimenter. The participant then performed the patient interview alone. Afterwards, the participant was escorted back to the survey room. There they took the post-experience survey and were then given the IAT. Finally, they were debriefed about the purposes of the experiment. Each session took one hour.

4 Results

Given a population with a **real world bias** similar to the US national average, the hypothesis is that participants in the dark skin-tone condition will demonstrate behavior indicative of bias against the VH, and participants in the light skin-tone condition should demonstrate behavior indicative of bias favoring the VH. **Empathy** should be higher in the light skin-tone condition and lower in the dark skin-tone condition. Furthermore, we expect a positive correlation of real world bias (the IAT) to empathy in the dark skin-tone condition, and a negative correlation of real world bias to empathy in the light skin-tone condition.

Real World Bias: The study population had a slight to moderate racial bias against African-Americans (M=-0.26, SD =.41) according to the IAT results. Seventeen (81%) of the participants had a negative score, and five had a positive score. IAT scores for the participants in the dark skin-tone condition (M=-.33, SD =.35) were not significantly different than IAT scores for the light skin-tone condition (M= -0.17, SD=.46). This distribution is similar to national averages (M=-0.30, SD=0.83) [7].

Empathy: Coders (2 audio-only, 1 video-only, 2 audio and video) rated clips of user behavior after the VH challenged the participant with the question "Could this be cancer?" An inter-rater reliability analysis was conducted, and found that rater reliability was acceptable (average intra-class correlation coefficient = 0.894).

As predicted, real world bias was positively correlated to self-reported and observed empathy in the dark skin-tone condition, with significant ($p<.01$) correlations to Audio-only (r=0.84), Audio-video (r=0.74), and Average (r=0.79) observed empathy. In the light skin-tone condition, the results were not as clear. Unexpectedly, real world bias was not significantly correlated (p>.10) to self-reported (r=-.01) or observed empathy (avg r=0.18) in the light skin-tone condition.

Also, significant (p<.01) *negative* correlations were observed between self-reported and observed empathy (r=-.91) in the light skin-tone condition. This is an unexpected result given the significant (p<.05) *positive* correlations (r=.60) in the dark skin-tone condition. In the light skin-tone condition, the more students used empathy (as rated by observers), the lower they rated themselves. This result suggests that racial bias impacts the internal scale participants use to rate themselves.

As expected, empathy was higher in the light skin-tone condition. A significant (Wilk's λ=0.481, F(3,16)=5.75, p<.001) multivariate effect of skin-tone condition was found for the degree of empathy observed during the empathetic moment. The average of all coders showed that the dark skin-tone patient was empathized with less (M=2.98, SD=1.52) than the light skin-tone patient (M=4.05, SD=1.67). Fig. 3 shows the average empathy scores given by each group of coders for the dark skin-tone and light skin-tone groups.

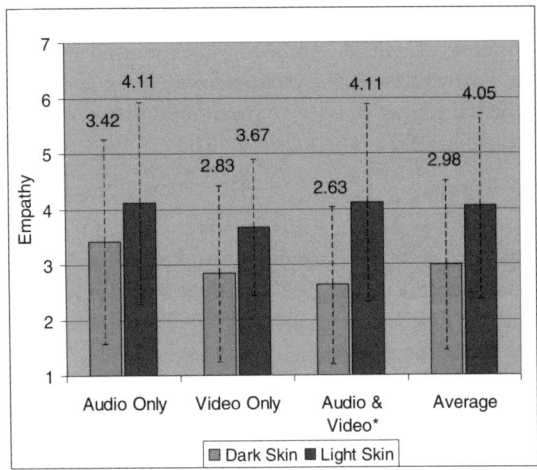

Fig. 3. Average scores for observed empathy during the empathetic moment. Empathy was rated on a scale from 1 (not at all) to 7 (very). (* p<.05)

5 Discussion

The results of this study indicate that VH agent skin-tone is a factor influencing both behavior and self-perception of empathy during patient interviews. Trends were observed which showed more empathy being used by participants in the light skin-tone condition than by participants in the dark skin-tone condition. Furthermore, real world bias (IAT score) is significantly correlated with empathetic behavior for participants in the dark skin-tone condition, which indicates that the two metrics share underlying factors. Establishing this correlation is the first step towards proving the validity of behavioral bias metrics with VHs.

For participants in the light skin-tone condition, real world bias (the IAT) did not appear to be a predictor of empathetic behavior, as little correlation was observed.

This correlation was expected to be significantly negative. Instead, for many coders we observed a positive correlation. Further investigations as to why this occurred are ongoing.

6 Conclusions and Future Work

In this paper, we reported results from a study, which examined the capability of VH agents to elicit behavior consistent with real world skin-tone bias. We found that participants expressed more empathy towards a light skin-tone VH than a dark skin-tone VH. This correlated to the study population's real world bias favoring Caucasian-Americans over African-Americans as measured by the IAT.

This bias translated to a task in the form of empathetic behavior. During a medical interview of a VH patient, the patient challenged the participant with "Could this be cancer?". The skin-tone of the patient affected participant behavior while responding to this challenge. From the results of the IAT, we expected to find more empathy exhibited with the light skin-tone VH patient than the dark skin-tone patient. Indeed, participant's self-reported empathy and coders' ratings of empathy confirmed this expectation. Furthermore, the real world bias significantly predicted participants' empathy in the dark skin-tone condition.

In conclusion, we find that real world H-H biases transfer to the virtual world in H-VH conversations. We have demonstrated that behavior with VHs is correlated to real world racial bias using a validated measure, the IAT. These results indicate that VHs elicit racial bias in conversational tasks, and motivate the use of VHs in experiences where it is desirable to elicit bias as an educational tool, e.g. in military, medicine, and business cultural training. Educators in these fields aim to identify and mitigate bias, and VH exposures can become a powerful tool to augment current training techniques.

Next, this study will be repeated with a larger number of participants, at least 30 in each group. It will be a within-subjects study where each participant interacts with both a dark skin-tone and a light skin-tone VH. We will also examine if the trends exhibited with this Caucasian group hold true with other ethnic groups.

Assuming the larger N study of skin-tone bias confirms what was indicated in this study, we will examine if extensive exposure to dark skin-tone VHs has a mitigating effect on racial bias. Last, it will be explored whether other biases can be elicited from the appearance of a VH. For instance, it would be useful to know if the apparent age, gender, and weight of the VH have a predictable effect on user behavior.

Acknowledgments

Special thanks go to Dr. John Dovidio and Dr. Ashby Plant for their advice on eliciting racial bias as well as their reviews and support of this work. We also thank Andrew Raij and Aaron Kotranza for their assistance conducting the user study. This work was supported by University of Florida Alumni Fellowships and a National Science Foundation Grant IIS-0643557.

References

1. Deaton, E., Barba, C., Santarelli, T., Rosenzweig, L., Souders, V., McCollum, C., Seip, J., Knerr, W., Singer, J.: Virtual Environment Cultural Training for Operational Readiness (Vector). Virtual Reality 8(3), 156–167 (2005)
2. Babu, S., Suma, E., Barnes, T., Hodges, L.F.: Using Immersive Virtual Humans for Training in Social Conversational Protocols in a South Indian Culture
3. Kenny, P., Parsons, T., Gratch, J., Leuski, A., Rizzo, A.: Virtual Patients for Clinical Therapist Skills Training, pp. 197-210
4. Dovidio, J.F., Kawakami, K., Gaertner, S.L.: "Implicit and Explicit Prejudice and Interracial Interaction. Journal of Personality and Social Psychology 82(1), 62–68 (2002)
5. Johnsen, K., Dickerson, R., Raij, A., Lok, B., Jackson, J., Shin, M., Hernandez, J., Stevens, A., Lind, D.S.: Experiences in Using Immersive Virtual Characters to Educate Medical Communication Skills, vol. 324, pp. 179–186
6. Cohen, J.J., Gabriel, B.A., Terrel, C.: The Case For Diversit. In: the Health Care Workforce. Project HOPE
7. Nosek, B., Mahzarin, B., Greenwald, A.: Harvesting Implicit Group Attitudes and Beliefs from a Demonstration Web Site. Group Dynamics: Theory, Research, and Practice 6(1), 101–115 (2002)
8. Pratt, A.J., Hauser, K., Ugray, Z., Patterson, O.: Looking at Human-Computer Interface Design: Effects of Ethnicity in Computer agents. Interacting with Computers 19(4), 512–523 (2007)
9. Baylor, A.L., Kim, Y.: Pedagogical Agent Design: The Impact of Agent Realism, Gender, Ethnicity, and Instructional Role. In: Proc. of International Conference on Intelligent Tutoring Systems (2004)

Cross-Cultural Evaluations of Avatar Facial Expressions Designed by Western Designers

Tomoko Koda[1], Matthias Rehm[2], and Elisabeth André[2]

[1] Faculty of Information Science and Technology, Osaka Institute of Technology
1-79-1 Kitayama, Hirakata city, Osaka, 573-0196, Japan
[2] Multimedia Concepts and Applications, Faculty of Applied Computer Science, University of Augsburg, Eichleitnerstr. 30, 86159 Augsburg, Germany
koda@is.oit.ac.jp, rehm@informatik.uni-augsburg.de,
Elisabeth.Andre@informatik.uni-augsburg.de

Abstract. The goal of the study is to investigate cultural differences in avatar expression evaluation and apply findings from psychological study in human facial expression recognition. Our previous study using Japanese designed avatars showed there are cultural differences in interpreting avatar facial expressions, and the psychological theory that suggests physical proximity affects facial expression recognition accuracy is also applicable to avatar facial expressions. This paper summarizes the early results of the successive experiment that uses western designed avatars. We observed tendencies of cultural differences in avatar facial expression interpretation in western designed avatars.

Keywords: Avatar, character, facial expression, cross-culture, network communication.

1 Introduction

Since instant messenger and chat services are frequently used in our daily communication beyond nationality and languages, emoticons and expressive avatars are widely used to provide nonverbal cues to text-only messages [1, 2, 3]. Recent growth of Second Life [4] attracts worldwide attention to avatar mediated communication both from entertainment and businesses. Studies on emoticons and avatars report positive effects on computer-mediated communication. Those studies indicate that emoticons and avatars improve user experiences and interactions among participants [5, 6, 7] and build enthusiasm toward participation and friendliness in intercultural communication [8, 9].

However, these avatars are used based on an implicit assumption that avatar expressions are interpreted universally across cultures. Since avatars work as graphical representations of our underlying emotions in online communication, those expressions should be carefully designed so that they are recognized universally. We need to closely examine cultural differences in the interpretation of expressive avatars to avoid misunderstandings in using them.

In order to examine cultural differences in avatar facial expression recognition, we apply findings from psychological studies on human facial expressions, since there have been a much wider variety of studies in psychology on human expressions than on avatar expressions. Recent psychological research found evidence for an "in-group advantage" in emotion recognition. That is, recognition accuracy is higher for emotions both expressed and recognized by members of the same cultural group [10]. Elfenbein et al. state, "This in-group advantage, defined as extent to which emotions are recognized less accurately across cultural boundaries, was smaller for cultural groups with greater exposure to one another, for example with greater physical proximity to each other [10]." Also, the decoding rule [11] implies that we concentrate on recognition of negative expressions, since misinterpretation of negative expressions leads to more serious social problems than misinterpretation of positive expressions would cause.

We conducted a web experiment to compare interpretations of Japanese designed avatars' facial expressions among 8 countries, namely, Japan, South Korea, China, United States, United Kingdom, France, Germany, and Mexico in our previous study [12, 13]. The results showed the following:

1) Cultural differences do exist in interpretation of avatar facial expressions, which-confirms the psychological findings that physical proximity affects recognition accuracy. The in-group advantage was found within Japan and between Korea and Japan.
2) There are wide differences among cultures in interpreting positive expressions, while negative expressions had higher recognition accuracy. This result indicates that the decoding rule is found in avatar expression interpretation.

This paper reports the results of our successive experiment that uses western designed avatars. The objective of the current experiment is to examine whether in-group advantage and decoding rule are also applicable to avatars' facial expressions drawn by western designers.

2 Experiment Overview

We are currently conducting a web based experiment since February, 2008. The experimental procedure is the same as the previous experiment except that the avatars used in the current experiment are drawn by French, British, and American designers and one made digitally, while the avatar designs used in the previous experiment were by Japanese artists [12, 13].

Participation in the experiment was by invitation only. We have collected total of 293 answers from United States (n=98), France (n=23), Germany (n=75), and Japan (n=97). The answers of which the common language of a participant's country and his/her native language match are selected for later analyses. The participants' gender ratio is 65% of male and 35% of female, and their age ranges are 80% in the 20's and 10% in the 30's. 53% of the participants rated themselves as expert computer users, 43% as intermediate.

2.1 Experimental Procedure

The experiment was developed using the application of Macromedia Flash. Participants first answer a brief questionnaire on their background profile such as their nationality and mother tongue. The main experiment starts after the questionnaire, which is presented as a matching puzzle game as shown in Fig. 1. Participants are requested to match 12 facial expressions to 12 adjectives. The 12 facial expressions are displayed in a 4 x 3 matrix and the 12 adjectives as buttons below the matrix. As shown in Fig. 1, participants can drag/drop the adjective buttons to/on the 12 expressions and continue changing the location of each button until they are satisfied with their answer. One avatar representation is chosen randomly from 7 avatars, and facial expression images are randomly placed in the 4 x 3 matrix. The adjective buttons are always displayed in the same order, and the 12 adjectives are always the same (see the next section for the adjectives used in the experiment).

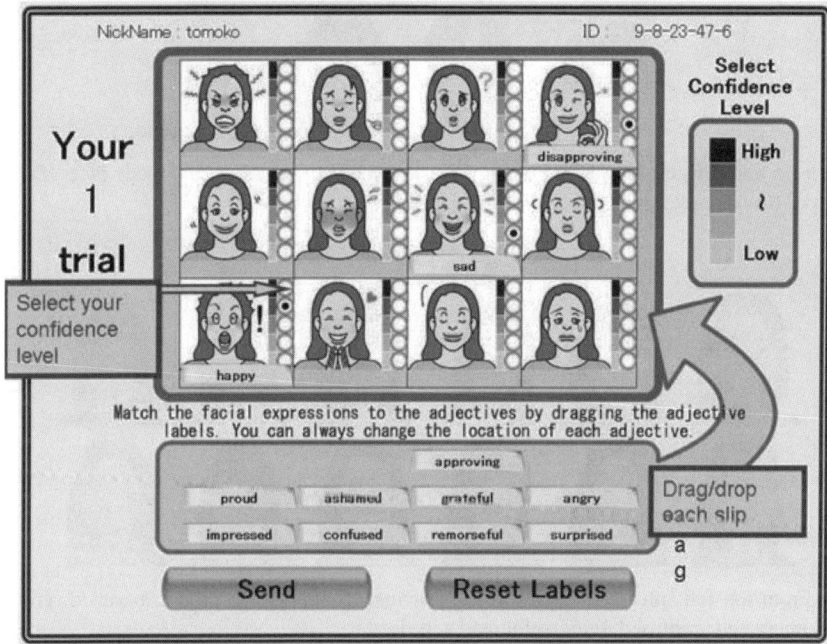

Fig. 1. Experiment screen shown as a matching puzzle game (Example screen for English speaking participants)

Participants' answers to the puzzle game as well as their background profile including gender, age, country of origin, native language, and computer expertise are logged in the server for later analyses. Participants are required to continue the experiment until they finish evaluating all the 7 avatar designs. The adjectives can be shown in English, German, French, and Japanese (all validated by native speakers). Participants from countries where the above languages are primarily spoken can see the adjective selections in their native language according to the background profile.

2.2 Avatar and Expression Design

Commercially used avatars are represented not by photo-realistic images but as caricatures or comic figures. We prepared 7 avatar representations drawn by French

(2 avatars), British (2 avatars), and American (1 avatar) professional designers using their countries' comic/anime drawing styles, and 2 avatars created by a 3D facial expression modeling tool by a German researcher (the last avatars are referred to as German designs). The designers grew up and speak the first common language in the referred countries. By using avatars drawn with techniques from one culture, we can use those avatars as "expressers" and subjects as "recognizers" as in [11]. Accordingly, comparing the answers between users of the designer's country and those of other countries made it easier to validate the in-group advantage. Fig. 2 shows examples of "displeased" expression drawn by French, British, American designers and German design.

Note: from left to right, drawn by French, British, American designer, and made by a 3D modeling tool (German design).

Fig. 2. Examples of avatar expression

Note: From top left, pleased, displeased, approving, disapproving, proud, ashamed, grateful, angry, impressed, confused, remorseful, and surprised.

Fig. 3. Twelve facial expressions of one of the avatars designed by an American designer

The 12 expressions used in the experiment are "pleased," "displeased," "approving," "disapproving," "proud," "ashamed," "grateful," "angry," "impressed," "confused," "remorseful," and "surprised" as shown in Fig. 3. Those expressions are selected from Ortony, Clore and Collins' global structure of emotion types, known as the OCC model [14]. These are commonly used expressions in chat and instant messenger systems [1, 2, 3], and they reflect those emotions desired by the subjects for intercultural communication in [8].

These 12 expressions are paired as valanced expressions as defined in the OCC model, that is, negative/positive emotions that arise in reacting to an event or person. "pleased," "approving," "proud," "grateful," and "impressed" are positive expressions, while "displeased," "disapproving," "ashamed," "angry," "confused," and "remorseful" are negative expressions, leaving "surprised" as a neutral expression.

3 Results

The participants' answers to the puzzle game are analyzed by calculating matching rates between expressions and adjectives. There is no correct answer to the matching puzzle, but the avatar designers' original intention can be used as an expresser's "standard" answer. Each expression and adjective is assigned a number (1-12) within the system. The designer's intended pairs are described as (1,1), (2,2), (3,3), (4,4) reflecting (expression number, adjective number). We calculated each country's number of "expression-adjective" pairs that are the same as the designers' pairs. Consequently, here, "matching rate" means the percentage of pairs of expressions and adjectives that match the avatar designer's intentional pairs. We use this expression-adjective matching rate in comparing the answers to the 12 facial expressions.

3.1 Analysis of Average Matching Rates by Designs

Table 1 shows the matching rates of the 12 expressions of German, UK, French, and US designs. The average matching rates of negative expressions are always higher than the ones of positive expressions in any designs (average matching rates of positive expressions: negative expressions (%) German 34.5:39.6, UK 46.6:73.9, French 65.6:79.6, US 40.9:67.0).

When we compare the matching rates by country, the average matching rate of the French design by French participants, and the matching rate of the American design by American participants are the highest among other countries. In addition, the matching rates of Japan are always the lowest in any avatar designs made by western designers.

3.2 Analysis of Expressions by Recognition Accuracy

In this section, we categorize the expressions into three groups according to their recognition accuracy. Those categories are as follows:

1) Highly recognized expressions (Expressions that have higher than 70% matching rates in all four countries): The expressions belong to this category are German Angry, British Ashamed, British Angry, British surprised, French disapproving, French proud, French ashamed, French angry, and American surprised.
2) Poorly recognized expressions (Expressions that have lower than 30% matching rates in all four countries): The expressions belong to this category are German disapproving, German impressed, German remorseful, British impressed, and American grateful.
3) Culture dependent expressions (Expressions that have wider than 50% distance among the matching rates of four countries): The expressions belong to this category are British Pleased, British proud, French grateful, French confused, American pleased, American disapproving, American proud and American impressed.

Highly recognized expressions are consistently highly recognized across four participating country, and between western countries and Japan. Thus we can assume those expression designs would cause less misinterpretations than other designs when used across countries. Poorly recognized expressions are consistently recognized as different expressions from the designer's intended expressions. The poorly recognized expressions are more subtle expressions than the highly recognized expressions (i.e., impressed vs. angry). Culture-dependent expressions have wider variances in their matching rates, and Japanese matching rates are the lowest among them.

Table 1. Matching rates shown by designers' and participants' countries

Note: Bold font: the matching rates of the designer's country, *: higest rates,**: lowest rates

German 3D design

	US	France	Germany	Japan	Average
Pleased	42	50	**36**	48	35.2
displeased	14	50	**20**	15	19.8
approving	32	50	**40**	33	31
Disapproving	25	0	**12**	30	13.4
Proud	53	66	**64**	51	56.8
ashamed	50	66	**44**	36	39.2
Grateful	35	16	**44**	15	32
Angry	96	83	**92**	87	81.6
Impressed	32	16	**28**	12	17.5
confused	64	16	**24**	15	43.8
Remorseful	21	16	**24**	12	14.6
surprised	39	33	**44**	39	31
Average	41.9*	38.5	**39.3**	32.8**	34.7
Positive expression					34.5
Negative expression					39.6

UK design

	US	France	Germany	Japan	Average
Pleased	55	85	42	28	52
displeased	41	57	53	62	52.6
approving	35	42	50	31	41.6
Disapproving	52	57	61	31	50.2
Proud	61	85	76	25	69.4
ashamed	97	100	92	100	97.8
Grateful	52	28	65	40	47
Angry	100	100	92	100	98.4
Impressed	26	14	11	15	23.2
confused	70	71	46	65	70.4
Remorseful	67	42	53	28	58
surprised	79	71	80	93	84.6
Average	61.3	62.7*	60.1	51.5**	60.1
Positive expression					46.6
Negative expression					73.9

French design

	US	France	Germany	Japan	Average
Pleased	46	**75**	60	30	52.2
displeased	75	**62**	68	72	75.4
approving	90	**87**	84	72	86.6
Disapproving	96	**87**	92	72	89.4
Proud	78	**100**	76	84	87.6
ashamed	78	**87**	76	78	73.8
Grateful	62	**75**	28	39	60.8
Angry	96	**100**	96	96	97.6
Impressed	50	**50**	24	30	40.8
confused	62	**87**	32	27	61.6
Remorseful	75	**75**	52	57	61.8
surprised	78	**100**	64	63	81
Average	73.8	**82.1***	62.7	60.0**	72.4
Positive expression					65.6
Negative expression					79.6

American design

	US	France	Germany	Japan	Average
Pleased	**37**	75	25	12	49.8
displeased	**43**	75	41	50	61.8
approving	**37**	0	16	0	10.6
Disapproving	**62**	75	33	100	74
Proud	**75**	75	100	12	72.4
ashamed	**50**	50	50	31	56.2
Grateful	**25**	25	16	18	16.8
Angry	**37**	50	41	62	58
Impressed	**81**	50	25	18	54.8
confused	**100**	75	58	93	85.2
Remorseful	**50**	25	50	31	51.2
surprised	**100**	100	91	100	98.2
Average	**58.1***	56.3	45.5	43.9**	57.4
Positive expression					40.9
Negative expression					67.0

4 Discussion

The analysis of the average matching rates by designs showed that negative expressions have higher recognition accuracy than positive expressions in western designed avatars. This may imply another decoding rule in western designs. The result of Japan's lowest average matching rates in any western designs suggests there is a tendency of in-group advantage among western countries.

When we categorized avatar designs by their recognition accuracy, the result showed that Japanese matching rates of the culture dependent designs are the lowest among the four countries. Thus again, there is a tendency that Japanese recognition accuracy is lower than western countries in culture-dependent designs, which suggests in-group advantage among western countries.

Further issues to be examined are as follows:

1) More participants: Further statistical analyses should be made with more participants. We analyzed 293 answers in the current experiment, while there were more than 1200 answers in the previous experiment.

2) Variations in avatar designs: There should be more variations in avatar designs in order to avoid one designer's judgment and drawing style. We used total of seven avatar designs drawn by four western designers (one designer per country) in the current experiment, while there were 40 avatar designs made by three Japanese designers in the previous experiment. Although the four western designers are carefully selected to have equal professional skills, we could guarantee the quality of avatar designs by having more variety in designers and avatar designs.
3) Translation of adjectives: Translations of adjectives are not completely equal as some languages do not have the exact meaning of the adjectives in other languages. One solution is to use scenario based interpretation for each avatar expression.
4) Definition of culture: We simply defined culture by participants' country and first language. However, the definition of culture is more complicated, e.g., religion. Further study should consider cultural models such as Hofstede's rankings [15], and cultural uncertainty avoidance models [16, 17].

5 Conclusion

The goal of the study is to investigate cultural differences in avatar expression interpretations and apply findings from psychological study in human facial expression recognition. Our previous study using Japanese designed avatars showed there are cultural differences in interpreting avatar facial expressions, and the psychological theory that suggests physical proximity affects facial expression recognition accuracy (in-group advantage) and decoding rule are also applicable to avatar facial expressions. We again observed tendencies of decoding rule and in-group advantage among western countries in the current experiment using western designed avatars.

Acknowledgements

This research is supported by a Grant-in-Aid for Young Scientists (Start-up: 18800063) from the Japan Society for the Promotion of Science (JSPS). The Universal Character Experiment 2008 is jointly conducted with University of Augsburg under DFG research grant RE 2619/2-1 (CUBE-G).

References

1. MSN Messenger: http://www.msn.com
2. Yahoo! Messenger: http://messenger.yahoo.com/
3. Kurlander, D., Skelly, T., Salesin, D.: Comic Chat. In: Proceedings of Computer Graphics and Interactive Techniques, pp. 225–236. ACM Press, New York (1996)
4. Second Life: http://secondlife.com/
5. Damer, B.: Avatars: Exploring and Building Virtual Worlds on the Internet. Peachpit Press, Berkeley (1997)
6. Smith, M.A., Farnham, S.D., Drucker, S.M.: The Social Life of Small Graphical Chat Spaces. In: Proceedings of CHI, pp. 462–469. ACM Press, New York (2000)

7. Pesson, P.: ExMS: an Animated and Avatar-based Messaging System for Expressive peer Communication. In: Proceedings of GROUP, pp. 31–39. ACM Press, New York (2003)
8. Koda, T.: Interpretation of Expressive Characters in an Intercultural Communication. In: Negoita, M.G., Howlett, R.J., Jain, L.C. (eds.) KES 2004. LNCS (LNAI), vol. 3214, pp. 862–868. Springer, Heidelberg (2004)
9. Isbister, K., Nakanishi, H., Ishida, T.: Helper Agent: Designing and Assistant for Human-Human Interaction in a Virtual Meeting Space. In: Proceedings of Human Factors in Computing Systems (CHI2000), pp. 57–64. ACM Press, New York (2000)
10. Elfenbein, H.A., Ambady, N.A.: Cultural similarity's consequences: A distance perspective on cross-cultural differences in emotion recognition. Journal of Cross-Cultural Psychology 34, 92–110 (2003)
11. Elfenbein, H.A., Ambady, N.: On the Universality and Cultural Specificity of Emotion Recognition: A Meta-Analysis. Psychological Bulletin, American Psychological Association, Inc. 128(2), 203–235 (2002)
12. Koda, T., Ishida, T.: Cross-cultural study of avatar expression interpretation. International Symposium on Applications and the Internet (SAINT) (2006)
13. T. Koda. Cross-cultural study of avatars' facial expressions and design considerations within Asian countries. In:Ishida, T., Fussell, S.R., Vossen, P.T.J.M. (eds.): Intercultural Collaboration I. LNCS, Springer-Verlag, pp.207-220 (2007).
14. Ortony, A., Clore, G.L., Collins, A.: The Cognitive Structure of Emotions. Cambridge University Press, Cambridge (1998)
15. Hofstede, G.: Cultures Consequences: International Differences in Work-Related Values. Sage Publications, Thousand Oaks (1984)
16. Berger, C.R., Calabrese, R.: Some explorations in initial interactions and beyond (1975)
17. Gudykunst, W.B., Mody, B.: Handbook of International and Intercultural Communication SAGE 2002. In: Hall, E., Hall, M.R. (eds.) Understanding Cultural Differences, 2nd edn. Intercultural Press, Yarmouth, Maine (1990)

Agreeable People Like Agreeable Virtual Humans

Sin-Hwa Kang[1], Jonathan Gratch[2], Ning Wang[2], and James H. Watt[1]

[1] Rensselaer Polytechnic Institute
Social and Behavioral Research Laboratory
110 8th St. Troy, NY 12180, USA
{kangs,wattj}@rpi.edu
[2] University of Southern California
Institute for Creative Technologies
13274 Fiji Way, Marina del Rey, CA 90405, USA
{gratch,nwang}@ict.usc.edu

Abstract. This study explored associations between the five-factor personality traits of human subjects and their feelings of rapport when they interacted with a virtual agent or real humans. The agent, the Rapport Agent, responded to real human speakers' storytelling behavior, using only nonverbal contingent (i.e., timely) feedback. We further investigated how interactants' personalities were related to the three components of rapport: positivity, attentiveness, and coordination. The results revealed that more agreeable people showed strong self-reported rapport and weak behavioral-measured rapport in the disfluency dimension when they interacted with the Rapport Agent, while showing no significant associations between agreeableness and self-reported rapport, nor between agreeableness and the disfluency dimension when they interacted with real humans. The conclusions provide fundamental data to further develop a rapport theory that would contribute to evaluating and enhancing the interactional fidelity of an agent on the design of virtual humans for social skills training and therapy.

Keywords: rapport, virtual agents, personality, nonverbal feedback, evaluation.

1 Introduction

Numerous studies have been conducted to explore the impact of personality traits on social interactions between humans and other humans or with agents. Personality embodies a human's characteristics that represent the consistent and permanent patterns of his/her emotion, thought, and behavior [2,7,25]. The "Media Equation" perspective [26] proposes that people respond to computer interfaces as if they were communicating with real persons. Hence, human-computer interaction should capture various effects on interactants' sense of being together and connected, that is *rapport* with agents, depending on the interactants' predisposition. Therefore, we raise the question "what are the various outcomes of social interaction between humans and agents if we examine humans' individual differences in personality?"

In the rapport related studies, Tickle-Degnen and Rosenthal [28] define three components of rapport: positivity as feeling of "mutual friendliness and caring," mutual attentiveness as feeling of "intense mutual interest in what the other is saying or doing," and coordination as feeling of "balance, harmony, and in sync." In his response to the article of Tickle-Degnen & Rosenthal [28], Izard [16] suggested exploring the relationships between personality traits and specific elements of rapport.

In this study, we seek to deepen and generalize our prior findings on the cognitive, emotional, and behavioral impact of rapport and to specifically investigate the role of contingency, which is timely feedback, on establishing rapport to provide some fundamental data to further develop the rapport theory that would contribute to evaluating and enhancing the interactional fidelity of virtual humans for social skills training and therapy [19]. In addition to practical insights into building virtual humans, this work illustrates how virtual human technology can provide fundamental insights into open questions in social psychology.

2 Related Work and Research Questions

Contingent Nonverbal Feedback of Rapport Agents
Our research on the Rapport Agent [11] investigates how virtual characters can elicit the harmony, fluidity, synchrony, and flow one feels when achieving rapport.

The Rapport Agent is designed to elicit rapport from human participants within the confines of a dyadic narrative task. In this setting, a speaker is led to believe that the character accurately reflects the nonverbal feedback of a human listener. In fact, these movements are generated by the Rapport Agent.

The central challenge for the Rapport Agent is to provide the nonverbal listening feedback associated with rapportful interactions. Such feedback includes the use of backchannel continuers [27] (nods, elicited by speaker prosodic cues, that signify the communication is working), postural mirroring, and mimicry of certain head gestures (e.g., gaze shifts and head nods). The Rapport Agent generates such feedback by realtime analysis of acoustic properties of speech and speaker gestures.

We have specifically investigated whether contingency of virtual humans' feedback would allow people to feel high rapport in one-on-one social interaction. We found the Rapport Agent embodying contingent feedback allows people to create great rapport. In a series of this study [11,12,13], we conclude that *contingency* matters for people's creating rapport, that is, the timing of nonverbal feedback of listeners.

Recent research suggests that virtual humans can establish something akin to rapport with people by producing rapid nonverbal feedback that is elicited by (i.e., contingent on) behaviors produced by the human interaction partner [11,12,13]. Mirroring general findings on rapport, these studies illustrate that the contingency of nonverbal feedback of virtual humans is crucial for interactants' sense of rapport.

Personality, Nonverbal Behavior, and Agents
In the studies of personality and agents, researchers [4,5,15,22,23] report the results of studying the effect of attributes of personality on people's interaction with agents. Isbister [15] found people liked an embodied character which showed a personality complementary to their own, while other researchers [22,23] report that people preferred computer interfaces that embodied a similar type of personality to their own. Bickmore and his colleagues [4,5] explored the relations between personality traits,

specifically extro/introversion and trust in an interaction partner when people interacted with an embodied conversational agent. They found that extroverted people constructed their relationships with the agent more than introverted people did.

Research investigating the impact of personality traits on mediated interactions has primarily focused on how people respond to agents that represent some set of personality traits. Such research has not investigated virtual humans that are able to respond in meaningful social ways to human subjects. There is no research that explores the relationship between humans' personality traits and their evaluation of interaction quality when humans interact with agents that specifically embody only nonverbal feedback.

Furthermore, the results of Berry and Hansen [3] show that associations between the measures of the five-factor personality, nonverbal behavior, and social interaction quality showed that personality may play an important role in affecting social experience in human-to-human interactions. This finding provides impetus for further studies investigating the relations between personality, agents' nonverbal behavior, and social interaction between humans and agents.

Based on the results of our previous research and the literature review, in this study we will examine associations between interactants' personality traits and agents' contingent nonverbal listening feedback behavior associated with rapport-like interactions, which embodies rather agreeable responses to interactants' behaviors. Such feedback entails the use of backchannel continuers, postural mirroring, and mimicry of certain head gestures of a real person who is interacting with the agent. The Big Five traits of personality (Five-Factor Model) [10] is the most dominant model to differentiate people's personalities [7,9,17,24]. These five factors of personality are extroversion, agreeableness, conscientiousness, neuroticism, and openness [8,14,20]. In this study, we use these five traits to measure participants' personality characteristics. We investigate how these personality traits are related to people's sense of rapport when they get contingent feedback from the Rapport Agent.

3 Experimental Design[1]

The study was designed with two conditions: Rapport Agent (n = 24) and Face-to-Face (n = 40: 20 speakers, 20 listeners), to which participants were randomly assigned using a coin flip. A confederate listener was used in the Rapport Agent condition. The Rapport Agent synthesized head gestures and posture shifts in response to features of a real human speaker's speech and movements.

3.1 Participants and Procedure

Sixty participants (63% women, 37% men) were recruited using Craigslist.com from the general Los Angeles area and were compensated $20 for one hour of their participation. On average, the participants were 38.4 years old.

Pairs of participants completed the pre-questionnaire and were led to the computer room. The speaker then viewed a short segment of a video clip taken from the Edge

---

[1] The experiment with the Rapport Agent condition and the Face-to-Face condition reported in this study were conducted as part of a more extensive design involving four conditions [12].

Training Systems, Inc. Sexual Harassment Awareness video. After the speaker finished viewing the video, the speaker was instructed to retell the stories portrayed in the clips to the listener.

Speakers and listeners could not see each other, being separated by a screen. The speaker saw an animated character displayed on the 30-inch computer monitor. Speakers in the Rapport Agent condition were told that the avatar on the screen displayed the actual movements of the human listener. While the speaker spoke, the listener could see a real time video image of the speaker retelling the story displayed on the 19-inch computer monitor. The monitor was fitted with a stereo camera system and a camcorder. For capturing high-quality audio, the participant wore a lightweight close-talking microphone and spoke into a microphone headset.

Next, the experimenter led the speaker to a separate side room. The speaker completed the post-questionnaire while the listener remained in the computer room and spoke to the camera what s/he had been told by the speaker. Finally, participants were debriefed individually and probed for suspicion about the listener using the protocol from Aronson, Ellsworth, Carlsmith, and Gonzales [1].

3.2 Equipment

To produce listening behaviors used in the Rapport Agent condition, the Rapport Agent first collected and analyzed the features from the speaker's voice and upper-body movements. Two Videre Design Small Vision System stereo cameras were placed in front of the speaker and listener to capture their movements.

Watson, an image-based tracking library developed by Louis-Phillipe Morency, uses images captured by the stereo cameras to track the participants' head position and orientation [21]. Watson also incorporates learned motion classifiers that detect head nods and shakes from a vector of head velocities. Both the speaker and listener wore a headset with microphone. Acoustic features are derived from properties of the pitch and intensity of the speech signal using a signal processing package, LAUN, developed by Mathieu Morales [11].

Three Panasonic PV-GS180 camcorders were used to videotape the experiment: one was placed in front the speaker, one in front of the listener, and one was attached to the ceiling to record both speaker and listener. The camcorder in front of the speaker was connected to the listener's computer monitor for displaying video images of the speaker to the listener.

The animated agent was displayed on a 30-inch Apple display to approximate the size of a real life listener sitting 8 feet away. The video of the speaker was displayed on a 19-inch Dell monitor to the listener. A male virtual character was used in the Rapport Agent condition.

3.3 Measurements

3.3.1 Response Variables
Self-Reported Rapport. We constructed a 10-item rapport scale (Cronbach's alpha = .89), presented to speakers in the post-questionnaire. Scales ranged from 0 (disagree strongly) to 8 (agree strongly). The self-reported rapport scales contain three

components [28]: positivity, mutual attentiveness, and coordination. In this study, the positivity is defined as connection rather than friendliness and caring, as the agent did not carry facial expressions or deliver talks to create interactants' feelings of mutual caring and friendliness.

Behavioral Measures of Rapport. We videotaped participants' verbal outcomes of their storytelling. Behavioral measures of rapport included number of pausefillers, number of prolonged words, number of incomplete words, number of disfluencies (pausefillers + incomplete words), and number of meaningful words (wordcount - pausefillers - incomplete words)

3.3.2 Explanatory Variable

Personality. The pre-questionnaire packet included questions about participant's personality traits. The personality traits are composed of Big Five Scales [10] ranged from 1 (disagree strongly) to 5 (agree strongly): extraversion, agreeableness, conscientiousness, neuroticism, and openness.

4 Results

Zero-order Correlations (Pearson Correlations) were computed to find associations between demographic variables, the personality traits and the measurements of rapport. The results revealed no statistically significant associations between demographic variables and the personality traits or between demographic variables and the measurements of rapport (See Table 1).

Table 1. Zero-order correlations among Demographic Variables, Personality Traits, and Self-Reported Rapport in the Rapport Agent condition and in the Face-to-Face condition

		Demographic Variables		Personality Traits				
		Age	Gender	Extraversion	Agreeableness	Conscientiousness	Neuroticism	Openness
RAPPORT AGENT								
Self-Reported Rapport (Overall)		.280	-.008	.296	.540**	.504**	-.197	-.013
Self-Reported Rapport (3 components)	Positivity	.246	.085	.258	.417*	.220	-.239	-.161
	Attentiveness	.274	-.063	.221	.456*	.403*	-.100	.078
	Coordination	.305	-.012	.116	.518**	.480*	-.221	-.252
FACE-to-FACE								
Self-Reported Rapport (Overall)		.099	.054	-.034	.349	.525*	-.142	-.079
Self-Reported Rapport (3 components)	Positivity	-.102	-.026	.091	.407	.390	-.016	-.194
	Attentiveness	.346	.158	-.112	.255	.428	-.214	-.082
	Coordination	.285	-.108	-.200	.370	.534*	-.201	-.067

* p < .05, ** p < .01

4.1 Correlations between the Personality Traits and Self-reported Rapport

Firstly, the results showed strong positive correlations between two personality traits and overall self-reported rapport in the Rapport Agent condition. Those two personality traits were Agreeableness (r = .54) and Conscientiousness (r = .50). In addition, the results revealed overall self-reported rapport was strongly associated only with Conscientiousness (r = .53) in the Face-to-Face condition (See Table 1).

Secondly, we looked at the correlations between each of the three components of rapport and the personality traits. In the Rapport Agent condition, the results showed that positivity was moderately correlated with Agreeableness (r = .42). The results further revealed that coordination was moderately correlated with Conscientiousness (r = .48) and strongly associated with Agreeableness (r = .52), while attentiveness was modestly associated with both Conscientiousness (r = .40) and Agreeableness (r = .46). In the Face-to-Face condition, the results showed coordination was highly correlated with Conscientiousness (r = .53) (See Table 1).

4.2 Correlations between the Personality Traits and Behavioral Measures of Rapport

The results revealed strong positive correlation between two personality traits and interactants' disfluency in their storytelling when they interacted with the Rapport Agent. The interactants' disfluency was highly associated with Extraversion (r = -.60) and moderately correlated with Agreeableness (r = -.42). In the Face-to-Face condition, the number of interactants' prolonged words was modestly associated with Agreeableness (r = -.50) (See Table 2).

Table 2. Zero-order correlations among Demographic Variables, Personality Traits, and Behavioral Measures of Rapport in the Rapport Agent condition and in the Face-to-Face condition

	Demographic Variables		Personality Traits				
	Age	Gender	Extraversion	Agreeableness	Conscientiousness	Neuroticism	Openness
RAPPORT AGENT							
Meaningful Words	-.013	-.299	-.089	.157	-.314	.075	-.127
Disfluency	-.021	-.346	-.600**	-.415*	-.196	.211	-.194
Prolonged Words	.201	.134	.134	-.058	.040	-.168	.141
FACE-to-FACE							
Meaningful Words	-.260	.130	.119	-.293	-.005	.230	-.115
Disfluency	-.240	-.119	.088	-.081	.037	.130	-.098
Prolonged Words	-.201	-.035	.041	-.495*	-.102	.205	-.124

* p < .05, ** p < .01

5 Conclusions

We found that more agreeable (i.e. pro-social and cooperative) people felt strong rapport when they experienced the contingent nonverbal feedback by the Rapport Agent, as they did while communicating face-to-face. More conscientious people reported strong rapport when they communicated with both the Rapport Agent and a real person. In human-to-human interactions, previous studies demonstrated that more agreeable people showed greater satisfaction about their interaction partners as well as self-reported interaction quality [3]. Similarly, the findings indicate that more agreeable interactants perceive strong rapport with the Rapport Agent and with another human although the relationship is apparently stronger for the Rapport Agent condition than for the Face-to-Face condition. Furthermore, we discovered other significant results involving interactants' (speakers') verbal behaviors. More extroversion and agreeableness of interactants were associated with weak rapport in the disfluency dimension when they experienced the contingent nonverbal feedback of the Rapport Agent. This outcome reflects the study by Berry and Hansen [3] that found a positive association between Extroversion as well as Agreeableness and independent observers' ratings for interaction quality in human-to-human interaction. The findings indirectly support the idea that people would respond to the contingent feedback of the Rapport Agent as if they were interacting with a human being, which is proposed by the "Media Equation" perspective in a series of studies by Nass and his colleagues [15,22,23,26].

When we looked at the relationship between the three components of rapport and the personality traits, the results revealed similarity in the association between the overall self-reported rapport and the personality traits. Based on the findings, we propose that kinder, more pro-social, and more cooperative people would feel a strong sense of rapport through feeling coordination in interaction with the Rapport Agent. Furthermore, the greater agreeableness of interactants was correlated with the strong feeling of rapport through sensing positivity in communication with the Rapport Agent as Izard proposed, while neuroticism of emotionally negative interactants was not statistically significantly associated with their sense of rapport for the positivity dimension with the Rapport Agent. This outcome somewhat contradicts the results discovered in the previous study [18] that showed more anxious people felt less rapport with the non-contingent feedback of agents. We expected more vulnerable and anxious interactants (i.e., subjects high in neuroticism) would feel strong rapport with agents' contingent feedback. This finding suggests that we need to further investigate our definition for positivity that is defined as interactants' feelings of connection with their partners in this study. In addition to these findings, it was found that people who are not dominant and pro-social (i.e., agreeable) would pay more attention to what an agent does as their interaction partner, if the agent provides contingent nonverbal feedback.

In conclusion, the results of both self-reported and behavioral-measured rapport in this study indicated that agreeable persons felt strong rapport with the Rapport Agent that embodies somewhat agreeable features: *contingency*. This leads to the potential way to develop agents' personality features which would be embodied by appropriate nonverbal feedback and be preferred by sociable persons with conscientiousness. This also points to the prior findings that indicated people preferred a computer interface represented by a type of personality similar to their own [22,23].

Acknowledgements

This work was sponsored by the U.S. Army Research, Development, and Engineering Command (RDECOM) and the National Science Foundation under grant # HS-0713603. The content does not necessarily reflect the position or the policy of the Government, and no official endorsement should be inferred.

References

1. Aronson, E., Ellsworth, P.C., Carlsmith, J.M., Gonzales, M.: Methods of Research in Social Psychology, 2nd edn. McGraw-Hill, New York (1990)
2. Ball, G., Breese, J.: Emotion and Personality in a Conversational Agent. In: Cassell, J., Sullivan, J., Prevost, S., Churchill, E. (eds.) Embodied Conversational Agents. MIT Press, Cambridge (2000)
3. Berry, D., Hansen, J.S.: Personality, nonverbal behavior, and interaction quality in female dyads. Personality and Social Psychology Bulletin (2000)
4. Bickmore, T., Cassell, J.: Relational Agents: A Model and Implementation of Building User Trust. In: Proc. of CHI 2001 (2001)
5. Bickmore, T., Schulman, D.: The Comforting Presence of Relational Agents. Extended Abstract. In: Proc. of CHI 2006 (2006)
6. Borkenau, P., Liebler, A.: Trait inferences: Sources of validity at zero acquaintance. Journal of Personality and Social Psychology 62 (1992)
7. Chittaro, L., Serra, M.: Behavioral programming of autonomous characters based on probabilistic automata and personality. Journal of Visualization and Computer Animation 15(3-4) (2004)
8. Costa Jr., P.T., McCrae, R.R.: Manual for the Revised NEO Personality Inventory and the NEO Five-Factor Inventory, Odessa, FL. Psychological Assessment Resources (1992)
9. Dyce, J.: The big five factors of personality and their relationship to personality disorders. Journal of Clinical Psychology 53(6) (1997)
10. Goldberg, L.R.: The Structure of Phenotypic Personality Traits. American Psychologist 48 (1993)
11. Gratch, J., Okhmatovskaia, A., Lamothe, F., Marsella, S., Morales, M., van der Werf, R., et al.: Virtual Rapport. In: Gratch, J., Young, M., Aylett, R.S., Ballin, D., Olivier, P. (eds.) IVA 2006. LNCS (LNAI), vol. 4133. Springer, Heidelberg (2006)
12. Gratch, J., Wang, N., Gerten, J., Fast, E., Duffy, R.: Creating Rapport with Virtual Agents. In: Pelachaud, C., Martin, J.-C., André, E., Chollet, G., Karpouzis, K., Pelé, D. (eds.) IVA 2007. LNCS (LNAI), vol. 4722. Springer, Heidelberg (2007)
13. Gratch, J., Wang, N., Okhmatovskaia, A., Lamothe, F., Morales, M., Morency, L.-P.: Can virtual humans be more engaging than real ones? In: Proc. of 12th International Conference on Human-Computer Interaction, Beijing, China (2007)
14. Hofstee, W.K., de Raad, B., Goldberg, L.R.: Integration of the Big Five and circumplex approaches to trait structure. Journal of Personality and Social Psychology 63 (1992)
15. Isbister, K., Nass, C.: Consistency of personality in interactive characters: verbal cues, non-verbal cues, and user characteristics. International Journal of Human Computer Interaction Studies 53 (2000)
16. Izard, C.: Personality, emotion expression, and rapport. Psychological Inquiry 1(4) (1990)
17. John, O., Srivastava, S.: The Big-Five Trait Taxonomy: History, Measurement, and Theoretical Perspectives. In: Pervin, L., John, O.P. (eds.) Handbook of Personality: Theory and Research, 2nd edn. Guilford, New York (1999)
18. Kang, S., Gratch, J., Wang, N., Watt, J.H.: Does contingency of agents' nonverbal feedback affect users' social anxiety? In: Padgham, Parkes, Müller, Parsons (eds.) Proc. of 7th Int. Conf. on Autonomous Agents and Multiagent Systems (AAMAS 2008), May 12-16 (2008)
19. Kenny, P., Parsons, T.D., Gratch, J., Rizzo, A.: Virtual Patients for Clinical Therapist Skills Training. In: Pelachaud, C., Martin, J.-C., André, E., Chollet, G., Karpouzis, K., Pelé, D. (eds.) IVA 2007. LNCS (LNAI), vol. 4722. Springer, Heidelberg (2007)

20. McCrae, R.R., Costa, P.T.: Adding Liebe and Arbeit: The full five-factor model and well being. Journal of Personality and Social Psychology 17 (1991)
21. Morency, L.-P., Sidner, C., Lee, C., Darrell, T.: Contextual Recognition of Head Gestures. In: Proc. of the 7th International Conference on Multimodal Interactions, Torento, Italy (2005)
22. Nass, C., Fogg, B.J., Moon, Y.: Can computers be teammates? International Journal of Human Computer Interaction Studies 45(6) (1996)
23. Nass, C., Moon, Y., Fogg, B.J., Reeves, B., Dryer, D.C.: Can computer personalities be human personalities? International Journal of Human Computer Interaction Studies 43(2) (1995)
24. Noftle, E.N., Shaver, P.R.: Attachment dimensions and the big five personality traits: Associations and comparative ability to predict relationship quality. Journal of Research in Personality 40 (2006)
25. Pervin, L.A., John, O.P. (eds.): Personality Theory and Research. John Wiley and Sons, New York (1997)
26. Reeves, B., Nass, C.: The media equation: How people treat computers, televisions and new media like people and places. Cambridge University Press, New York (1996)
27. Ward, N., Tsukahara, W.: Prosodic features which cue back-channel responses in English and Japanese. Journal of Pragmatics 23 (2000)
28. Tickle-Degnen, L., Rosenthal, R.: The nature of rapport and its nonverbal correlates. Psychological Inquiry 1(4) (1990)

A Listening Agent Exhibiting Variable Behaviour

Elisabetta Bevacqua, Maurizio Mancini, and Catherine Pelachaud

University Paris 8, 140 rue de la Nouvelle France 93100, Montreuil, France
INRIA Rocquencourt, Mirages BP 105, 78153 Le Chesnay Cedex, France

Abstract. Within the Sensitive Artificial Listening Agent project, we propose a system that computes the behaviour of a listening agent. Such an agent must exhibit behaviour variations depending not only on its mental state towards the interaction (e.g., if it agrees or not with the speaker) but also on the agent's characteristics such as its emotional traits and its behaviour style. Our system computes the behaviour of the listening agent in real-time.

1 Introduction

A big challenge that must be faced in the design of virtual agents is the issue of credibility, not only in the agent's aspect but also in its behaviour. Users tend to react as if in a real human-human interaction when the virtual agent behaves in a natural human manner [NST94, RN96]. The work presented in this paper focuses on the listener's behaviour and is set within the Sensitive Artificial Listening Agent (SAL) project, which is part of the EU STREP SEMAINE project (http://www.semaine-project.eu). This project aims to build an autonomous talking head able to exhibit appropriate behaviour when it plays the role of the listener in a conversation with a user. Four characters, with different emotional styles, invite the user to chat trying to induce her/him in a particular emotional state. Within SAL, we aim to build a real-time Embodied Conversational Agents (ECAs) able to automatically generate those verbal and non verbal signals that a human interlocutor displays during an interaction. These signals, called *backchannels*, provide information about the listener's mental state towards the speaker's speech (e.g., if s/he believes or not what the speaker is saying). In our system backchannel signals are emitted not only according to the agent's mental state towards the interaction but also its *behaviour tendencies*, that is the particular way of producing non verbal signals that characterizes the agent. In our work the behaviour tendencies are defined by the preference the agent has in using each available communicative modality (head, gaze, face, gesture and torso) and a set of parameters that affect the qualities of the agent's behaviour (e.g. wide vs. narrow gestures). We call the behaviour tendencies the agent's *baseline*. The proposed work incorporates a pre-existing system for the generation of distinctive behaviour in ECAs [MP07, MP08]. The result is a system capable of computing the verbal and non-verbal behaviours that the agent, in the role of the listener, has to perform on the basis of both its baseline and its mental state.

2 Sensitive Artificial Listener

The Sensitive Artificial Listener (SAL) technique [DCCC[+]08, WER[+]08], developed at the Queen's University of Belfast, rises from the need of collecting data about human interactions, where people express various emotions through both verbal and non verbal channels. The SAL idea comes from the observation of chat show hosts who are able to incite people into talking by simply appearing to listen and encouraging now and then with short standard phrases. In the previous SAL a human operator plays the role of the chat show host by selecting, in a Wizard of Oz manner, possible responses from pre-defined scripts. In another room, the user hears the corresponding pre-recorded emotionally coloured statements that not only encourage her/him into talking but also pull her/him towards specific emotional states. To achieve such a goal, SAL provides four characters with different emotional styles: Poppy (who is happy and positive), Obadiah (who is gloomy and sad), Spike (who is argumentative and angry) and Prudence (who pragmatic and sensitive). The user chooses the character s/he wants to talk to and can change it whenever s/he wants. In the other room, the operator chooses the statement to use according to both the selected character and the user's apparent emotional state. For example, if the user is sad, Obadiah approves while Poppy cheers up. Each character is provided with a pre-fixed script that contains phrases for each phase of a conversation: beginning, maintaining and ending of a conversation. The SAL approach proved successful at provoking sustained and emotionally coloured interactions. Within the SE-MAINE project, the SAL system will undergo a substantial transformation: the four characters will be represented by four fully automatic ECAs. Each agent will be able to identify the user's emotional state and select the response; moreover it will provide appropriate verbal and non verbal signals while listening.

3 Background

3.1 Personal Tendencies in Behaviour

Argyle [Arg88] states that there are *personal tendencies* that are constantly present in human behaviour: for example people that look more tend to do so in most communicative situations, that is, there is a certain amount of consistency with the personal general tendencies. The reasons behind such personal differences can be due for example to differences in personality, emotional state, mood, sex, age, nationality [WS86]. Also Gallaher [Gal92] found consistencies in the way people behave: she conducted evaluation studies in which subjects' behaviour style was evaluated by friends, and by self-evaluation. Results demonstrated that for example people who are fast when writing have a tendency to be fast while eating; people producing wide gestures also walk with large steps.

3.2 Listener's Behaviour

To assure a successful communication, listeners must provide responses about both the content of the speaker's speech and the communication itself. Through verbal and non verbal signals, called *backchannel signals*, a listener provides information about basic communicative functions, as perception, attention, interest, understanding, attitude (e.g., belief, liking and so on) and acceptance towards what the speaker is saying [ANA93, Pog05]. For instance, the interlocutor can show that s/he is paying attention but not understanding and, according to the listener's responses, the speaker can decide how to carry on the interaction: for example by re-formulating a sentence.

3.3 Related Work

Previous works on ECAs have provided first approaches to the implementation of a backchannel model. K. R. Thórisson developed a talking head, called Gandalf, able to produce real-time backchannel signals during a conversation with an user [Thó96], while Cassell et al. [CB99] developed the Real Estate Agent (REA) which is able to understand the user requests in real-time. The Listening Agent [MGM05], developed at ICT, produces backchannel signals based on real-time analysis of the speaker's non verbal behaviour (as head motion and body posture) and of acoustic features extracted from the speaker's voice [WT00]. Kopp et al. [KSG07] proposed a model for generating incremental backchannel. The system is based both on a probabilistic model, that defines a set of rules to determine the occurrence of a backchannel, and on a simulation model that perceives, understands and evaluates input through multi-layered processes. All the models described above do not take into account neither the agent's mental state nor its behaviour tendencies.

4 Overall System Description

As explained above, the system we present here is embedded in SAL. A user sits in front of a screen where an ECA, chosen among four different characters, listens to her/him and tries to induce her/him a particular emotional state. A video camera and a microphone record user's movements and voice. This information is used in our system to decide when and how the agent must provide a backchannel signal. The system includes also the concept of user *interest level* based on the Theory of Mind [Pet05]. Figure 1 shows the overall system diagram.

4.1 Agent Definition

Baseline. We define the agent's baseline as a set of numeric parameters that represents the agent's behaviour tendencies. In the baseline we represent two kinds of data: the agent's modality preference and the agent's behaviour expressivity. People can communicate by being more or less expressive in the different modalities: a person can gesture a lot while another one can produce many facial expressions and so on. In our static definition of an agent, we implemented the *modality preference* to represent the agent's degree of preference in using each available modality (face, head, gaze, gesture and torso). If, for example, we want

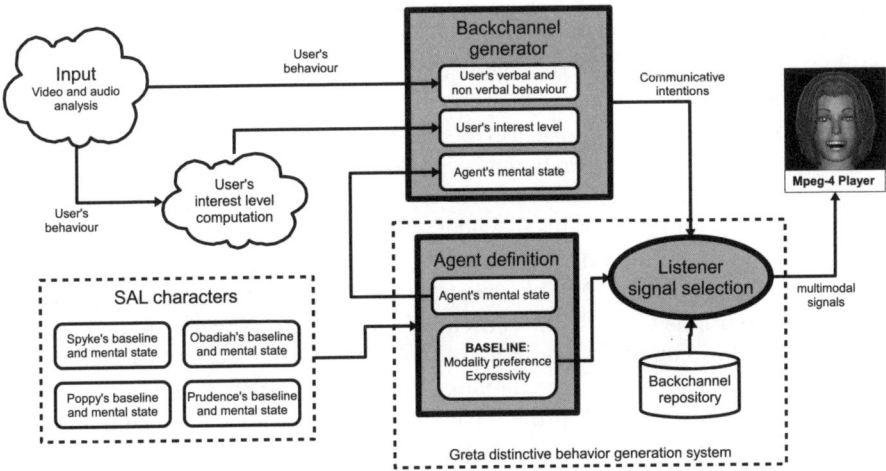

Fig. 1. System diagram

to specify that the agent has the tendency to mainly use hand gestures during communication, we assign a high degree of preference to the gesture modality, if it uses mainly the face, the face modality is set to a higher value, and so on.

To define the behaviour tendencies of an agent we also defined and implemented a set of parameters that allow us to alter the way the agent expresses its actual communicative intention [HMBP05]. The agent's *behaviour expressivity* is defined by a set of 6 parameters that influence the quality of the agent's movements: the frequency (OAC parameter), speed (TMP parameter), spatial volume (SPC parameter), energy (POW parameter), fluidity (FLD parameter), and repetitivity (REP parameter) of the nonverbal signals produced by the agent. These expressivity parameters are defined over each modality in a separate way: a set of parameters for the head movements, another set for the facial expressions, and so on.

To introduce the four SAL characters in our system we defined a baseline for each of them, keeping in mind the expressive behaviour studies showing the existing link between emotional states and behaviour quality [WS86, Gal92]. Thus, each SAL agent is characterized by specific emotional traits. The baselines are determined manually through the observation of videos of real people that exhibit a style of behaviour similar to Poppy, Obadiah, Spike and Prudence's characteristics. For instance, Spyke performs powerful, fast and short movements on all the modalities. Obadiah tends to produce backchannel signals mostly with the head while Poppy prefers to communicate mainly through facial expressions. Prudence performs slow movements mainly on the head and face modalities. When the user chooses a character, its baseline is loaded in the agent definition module and used by the listener signal selection module to computes the agent's distinctive behaviour as described in detail in the Section 4.3.

Agent's mental state. The agent definition module includes also the agent's mental state. In this module we consider solely how the agent reacts towards the interaction, that is how the agent reacts to the user's speech (if it agrees, refuses, understands... what is being said). Having such information the system specifies which *communicative intentions* (agree, refuse, understand,...) it will convey through its backchannel signals. We consider twelve communicative intentions related to backchannels chosen from the literature: agreement, disagreement, acceptance, refusal, belief, disbelief, interest, no interest, liking, disliking, understanding, no understanding [ANA93, Pog05].

The interlocutor's mental state is represented by a set of beliefs and intentions, such as its intentions towards the content of the speaker's speech, its own beliefs, and so on. So far, cognitive modules able to compute the listener's mental state are usually limited to specific domains. At present, such a module is under development within the SAL project [HtM08]. However for sake of simplicity, we link the agent's mental state to the emotional characteristics that differentiate the four SAL agents. Consequently, each SAL agent shows backchannel signals that are compatible with its emotional traits.

Spyke, who is angry and argumentative, conveys negative communicative intentions, in particular dislike, disagreement and not interest. Being gloomy, Obadiah tends to convey negative communicative intentions too, in particular disbelief, refusal and no understanding. Poppy, the happy one, provides backchannel signals that are the expression of positive communicative intentions, as liking, acceptance and interest. Finally, Prudence, who is sensitive and pragmatic, conveys positive communicative intentions, in particular agreement, belief and understanding [SW85, WS86].

4.2 Backchannel Generator

To display a believable listener behaviour, a virtual agent must be able to decide *when* a backchannel signal should be emitted and select *which* communicative intentions the agent should transmit through the signal. In our system these tasks are performed by the *backchannel generator* module of Figure 1. This module needs three data as input:

– The user's verbal and non verbal behaviour, tracked through a video camera and a microphone;
– The user's estimated interest level, an emotional state linked to the speaker's goal of obtaining new knowledge. Such a level is calculated evaluating the user's gaze, head and torso direction within a temporal window.
– The agent's mental state towards the interaction, as described in 4.1;

Researches have shown that backchannel signals are often emitted according to the verbal and non verbal behaviour performed by the speaker [WT00, MGM05]. On the basis of these results, our system evaluates video and audio data to select user's behaviours that could elicit a backchannel from the agent; for example, a head nod or a variation in the pitch of the user's voice and so on. The probability that such a behaviour provokes a backchannel signal depends on the user's estimated level of interest. This value is used by the system to vary the backchannel emission frequency: when the interest level decreases the user might want to stop the conversation [SS73], consequently the agent provides less and less backannels. When a backchannel must be emitted the backchannel generator module uses the information about the user's mental state to decide which communicative intentions the agent should convey.

4.3 Listener Signal Selection

In this Section we describe the process of performing the selection of the nonverbal behaviours that the listener has to produce in order to convey its communicative intention. This task is performed by the listener signal selection module of Figure 1, which is an extended version of the corresponding one we implemented for the distinctive behaviour generation system of the Greta agent, presented in [MP07, MP08].

Behaviour sets. In the listener signal selection module, all the listener's communicative intentions, contained in the agent's mental state (see Section 4.1), are associated with the backchannel signals that can be produced by the listener. Each of these associations represents one *entry* of a lexicon, called *backchannel behaviour set*. A backchannel behaviour set is defined by the following parameters:

- The *name* of the corresponding communicative intention. For example *refuse*.
- The set S, containing the name of the signals produced on single modalities that can be used to convey the intention specified by the parameter *name*. For instance, the intention *refuse* can be conveyed by: shaking the head, saying *"no!"* and so on.
- The list of signals that are mandatory to communicate the intention corresponding to the behaviour set; for example to communicate *refuse*, the listener *MUST* shake its head.
- A set of logic rules like *if A then B* where A is a condition involving both the parameters of the agent definition (see Section 4) and the signals contained in the set S and B is a subset of S. For example we could specify a rule in which, if the value of the head Overall Activation parameter (OAC) is higher than a given threshold, the listener has to produce a head nod. In this example we are referencing to a value in the agent's baseline.

The backchannel behaviour sets have been defined in our previous works [BHTP07, HBTP07] and stored in the backchannel repository showed in Figure 1. We performed perceptive tests directed to analyse how users interpret context-free backchannel signals displayed by a virtual agent. From the results we found a many to many mapping between specific signals and most of the meanings proposed in the test.

Performing the listener signal selection. As shown in Figure 1, the listener signal selection process takes as input the agent definition (that is its baseline) and the agent communicative intention and computes the multimodal behaviour that the agent has to perform. The process consists of some steps of computation: first the system looks for the behaviour set corresponding to the agent's communicative intention and computes all the possible combinations of the signals contained in the set S (see Section 4); it discards the combinations that do not contain the mandatory signals and checks the logic rules contained in the behaviour set, discarding the signal combinations that do not verify these rules; finally it prioritizes the signal combinations depending on the agent modality preference (see Section 4.1) and chooses the signal combination with the highest priority.

5 Conclusion and Future Work

We have proposed a system that computes the behaviour of a listening agent. This system is part of the SAL project that aims at implementing an agent exhibiting realistic behaviour when playing the role of a listener during a conversation. The user can chose among four agents with different styles of behaviour. Each agent provides backchannel signals that are consistent with its emotional traits. In the future, through perceptive tests, we will evaluate our system. We want to verify that we succeed in creating agents that show different behaviours and that these agents are able to sustain an emotionally coloured communication with users.

Acknowledgement

This work has been funded by the STREP SEMAINE project IST-211486 (http://www.semaine-project.eu) and the IP-CALLAS project IST-034800 (http://www.callas-newmedia.eu).

References

[ANA93] Allwood, J., Nivre, J., Ahlsén, E.: On the semantics and pragmatics of linguistic feedback. semantics 9(1) (1993)

[Arg88] Argyle, M.: Bodily Communication, 2nd edn. Methuen & Co., London (1988)

[BHTP07] Bevacqua, E., Heylen, D., Tellier, M., Pelachaud, C.: Facial feedback signals for ecas. In: AISB 2007 Annual convention, workshop "Mindful Environments", Newcastle, UK, April 2007, pp. 147–153 (2007)

[CB99] Cassell, J., Bickmore, T.: Embodiment in conversational interfaces: Rea. In: Conference on Human Factors in Computing Systems, Pittsburgh, PA (1999)

[DCCC$^+$08] Douglas-Cowie, E., Cowie, R., Cox, C., Amir, N., Heylen, D.: The sensitive artificial listener: an induction technique for generating emotionally coloured conversation. In: LREC 2008 (May 2008)

[Gal92]   Gallaher, P.E.: Individual differences in nonverbal behavior: Dimensions of style. Journal of Personality and Social Psychology 63(1), 133–145 (1992)
[HBTP07] Heylen, D., Bevacqua, E., Tellier, M., Pelachaud, C.: Searching for prototypical facial feedback signals. In: Pelachaud, C., Martin, J.-C., André, E., Chollet, G., Karpouzis, K., Pelé, D. (eds.) IVA 2007. LNCS (LNAI), vol. 4722, pp. 147–153. Springer, Heidelberg (2007)
[HMBP05] Hartmann, B., Mancini, M., Buisine, S., Pelachaud, C.: Design and evaluation of expressive gesture synthesis for embodied conversational agents. In: 3rd International Joint Conference on Autonomous Agents & Multi-Agent Systems, Utretch (2005)
[HtM08]   Heylen, D.K.J., ter Maat, M.: A linguistic view on functional markup languages. In: AAMAS - First Functional Markup Language Workshop, Estoril, Portugal (May 2008)
[KSG07]   Kopp, S., Stocksmeier, T., Gibbon, D.: Incremental multimodal feedback for conversational agents. In: Pelachaud, C., Martin, J.-C., André, E., Chollet, G., Karpouzis, K., Pelé, D. (eds.) IVA 2007. LNCS (LNAI), vol. 4722, pp. 139–146. Springer, Heidelberg (2007)
[MGM05]   Maatman, R.M., Gratch, J., Marsella, S.: Natural behavior of a listening agent. In: 5th International Conference on Interactive Virtual Agents, Kos, Greece (2005)
[MP07]   Mancini, M., Pelachaud, C.: Dynamic behavior qualifiers for conversational agents. In: Intelligent Virtual Agents, pp. 112–124 (2007)
[MP08]   Mancini, M., Pelachaud, C.: Distinctiveness in multimodal behaviors. In: Conference on Autonomous Agents and Multiagent System (2008)
[NST94]   Nass, C., Steuer, J., Tauber, E.R.: Computers are social actors. In: CHI, pp. 72–78 (1994)
[Pet05]   Peters, C.: Direction of attention perception for conversation initiation in virtual environments. In: International Working Conference on Intelligent Virtual Agents, Kos, Greece, pp. 215–228 (September 2005)
[Pog05]   Poggi, I.: Backchannel: from humans to embodied agents. In: AISB University of Hertfordshire, Hatfield (2005)
[RN96]   Reeves, B., Nass, C.: The media equation: How people treat computers, television and new media like real people and places. CSLI Publications, Stanford (1996)
[SS73]   Schegloff, E.A., Sacks, H.: Opening up closings. Semiotica VIII(4) (1973)
[SW85]   Scherer, K.R., Wallbott, H.G.: Analysis of nonverbal behavior. Handbook of discourse analysis 2, 199–230 (1985)
[Thó96]   Thórisson, K.R.: Communicative Humanoids: A Computational Model of Psychosocial Dialogue Skills. PhD thesis, MIT Media Laboratory (1996)
[WER+08] Wöllmer, M., Eyben, F., Reiter, S., Schuller, B., Cox, C., Douglas-Cowie, E., Cowie, R.: Abandoning emotion classes - towards continuous emotion recognition with modelling of long-range dependencies. In: Interspeech (2008)
[WS86]   Wallbott, H.G., Scherer, K.R.: Cues and channels in emotion recognition. Journal of Personality and Social Psychology 51(4), 690–699 (1986)
[WT00]   Ward, N., Tsukahara, W.: Prosodic features which cue back-channel responses in english and japanese. Journal of Pragmatics 23, 1177–1207 (2000)

The Next Step towards a Function Markup Language

Dirk Heylen[1], Stefan Kopp[2], Stacy C. Marsella[3],
Catherine Pelachaud[4], and Hannes Vilhjálmsson[5]

[1] Human Media Interaction, University of Twente
[2] Artificial Intelligence Group, Bielefeld University
[3] Information Science Institute, University of Southern California
[4] University of Paris 8, Inria
[5] Center for Analysis and Design of Intelligent Agents, Reykjavik University

Abstract. In order to enable collaboration and exchange of modules for generating multimodal communicative behaviours of robots and virtual agents, the SAIBA initiative envisions the definition of two representation languages. One of these is the Function Markup Language (FML). This language specifies the communicative intent behind an agent's behaviour. Currently, several research groups have contributed to the discussion on the definition of FML. The discussion reveals agreement on many points but it also points out important issues that need to be dealt with. This paper summarises the current state of affairs in thinking about FML.

1 Introduction

At the previous two installments of the Intelligent Virtual Agents conference, an update was presented about ongoing work on the Behavior Markup Language ([KKM+06, VCC+07]). BML is being discussed and specified by a representative number of research groups that strive for standardisation in the specification of behaviours of agents, to enable collaboration, sharing results and sharing actual working system components. BML is one of two representation languages in the SAIBA (Situation, Agent, Intention, Behavior, Animation) effort[1]. The other is FML, the Function Markup Language. FML will represent what an agent wants to achieve: its intentions, goals and plans.

At the first SAIBA workshop in Reykjavik in 2005 a preliminary proposal for FML was made ([VM05]) but since then research has focussed on BML rather than on FML. This year the first workshop on FML was proposed by the authors of this paper at AAMAS. The main aim of the workshop was to hear from the various researchers in the field about their wishes and suggestions regarding the functional specifications and to make an inventory of the issues that need to be resolved. This should enable one to lay down a roadmap for working out the next version of FML. In this paper we will review the current state of affairs

---
[1] http://wiki.mindmakers.org/projects:saiba:main/

in the discussion, representing and synthesizing the various views that emerged from the contributions and presenting the central issues that need to be tackled in the next stage of development. First we present some specific proposals for parts of FML from a selection of papers to provide an idea of the range of topics subsumed under it and of the range of variation between authors. Next, we take a broader view and look at the more general issues that arose.

2 Towards an Inventory of FML Tags

The contributions to the AAMAS workshop were of various types. Some raised general issues and others made specific proposals for (parts of) an FML specification. Several papers combined both aspects. In this section we will look at suggestions that were made for the attributes that FML might need to specify. First we consider the kinds of dimensions that were suggested in the discussion, often based on existing systems (hence the title legacy for the next section) and next we look at some specific proposals for attributes along the various dimensions.

2.1 Legacy

Vilhjálmsson and Thórisson ([VT08]), provide a brief history of function representation. Among the proposals for the representation of functions are the tags used in the BEAT system and the first proposal for FML within SAIBA. Several other contributions to the discussion list FML-related or FML-inspired information in current systems. To get an idea of the range of attributes that are involved and a sense of the overlap between proposals we briefly present five suggestions.

BEAT/Spark. The terms FML and BML were first employed to describe the tags set used in the BEAT system [CVB01] as part of Spark [Vil05]. In this case FML was a mark-up language for texts describing several discourse phenomena related to content and information structure (theme/rheme, emphasis, contrast, topic-shifts) and interaction processes (turn-taking and grounding).

FML2005. The FML break-out group at the 2005 SAIBA workshop in Reykjavik proposed to divide FML tags into a set of basic functional or semantic *units* and a set of *operations*. The *units* are typical elements present or occuring during a communicative event. The initial list comprised the following units.

- Participants
- Turns
- Topic
- Performative (speech act)
- Content (proposition)

The operations that were suggested comprised:

- Emphasis
- Contrast

- Illustration
- Affect
- Social (or relational) goals
- Cognitive operations (e.g. difficulty of processing) and certainty

FML2005 shows a large overlap with the earlier proposal. An important difference is that it goes beyond the linguistic and conversational aspects and also includes elements of the mental state. Another important difference is that by introducing abstract units of discourse, FML2005 does not assume the annotation of an already produced text.

TLCTS. The paper by [SVJ08] on the Tactical Language and Culture Training System (TLCTS) considers intent planning mainly as a decision as to which communicative act to perform. This is (usually) a speech act. The communicative act specifies (1) the nature of the act (request, accept proposal...), (2) its modulation (politiness level, force), and (3) encoded contextual parameters (directed at male or female, role and status...). In order to be able to generate the appropriate behaviour in context, knowledge about the dialog, the world and the target culture is assumed to be needed.

Inspired by Traum and Hinkelman's typology of speech acts ([TH92]) the TLCTS system incorporates the following functions of communicative acts.

- Turn-taking: take-turn, release turn, keep-turn, assign-turn
- Grounding: initiate, continue, acknowledge, repair, request-repair, request-acknowledgement, cancel
- Core speech acts: inform, wh-question, yes-no question, accept, request, reject, suggest, evaluate, request-permission, offer, promise
- Argumentation: elaborate summarize, clarify, q&a, convince, find-plan

The overlap with the previous proposals on the main dimensions is again obvious. Turn-taking, grounding and speech acts are the main aspects that are considered. Interesting is the attention that is paid to contextual elements and the proposal that a third language called Context Markup Language should be devoted to them.

APML. The Greta framework [Pel05] uses APML as a mark-up language to encode the communicative intentions of an agent. The tags provide information about the following dimensions.

- The degree of certainty
- The meta-cognitive source of information (thinking, remembering, planning)
- The speech act (called performative),
- Information structure of the utterance (theme/rheme)
  Rhotorical relations such as contradiction or cause-effect (called belief-relations by the authors)
- Turn allocation
- Affect
- Emphasis

These tags are based on the work of Isabella Poggi [Pog07]. In [MP08], Mancini and Pelachaud propose an extention of APML, FML-APML, which is different in the following respects. The timing attributes of the tags are made consistent with the suggestions made for BML ([KKM+06, VCC+07]). This also makes it possible to specify the communicative intentions of non-speaking agents. The deictic tags were changed to remove elements that properly belong to BML. Furthermore, the emotional state tags are made more complex to allow for faked emotions (based on EARL [SPL06]). Also added is a tag to describe the importance of an intention.

Virtual Human Project. The nonverbal generation module of the virtual human architecture developed at ICT [LDMT08], contains several FML-inspired concepts, among which are the communicative intent (speech acts), cognitive operators that drive the gaze state of the virtual humans and elements that relate to emotional states and coping processes. The gaze model associates behaviours with what are called *cognitive operators* by providing a specification of the form and function of gaze patterns. These functions specify detailed reasons behind a particular gaze behaviour related to a combination of four categories of determinants: conversation regulation, updating of an internal cognitive state (desire, intention...), monitoring for events and goal status and coping strategy.

Although there are clear differences in the number and kind of dimensions that the various contributions to the AAMAS workshop consider for inclusion in FML, overall there is a big overlap between the various proposals. Prominent recurrent dimensions are *dialogue act* (or speech act), *turn-taking, grounding* actions, *content* (propositional content, discourse relations), *information structure* (emphasis, given/new, theme/rheme), *emotion, affect* and *interpersonal or social* relations.

We now turn to the kinds of attributes and values that are being proposed for the various dimensions. Some of the dimensions that have been proposed in the discussion but that are not included in this list will be introduced below as well.

2.2 Tags Proposed

When it comes to specifying the attributes of the various dimensions in more detail, it is to be expected that different views on them emerge, depending on the detail of specification that one wants to achieve, the complexity of the system that one is building or the specific demands of the application domain or the theoretical stance one takes. In the following paragraphs we provide a sketch of the elements that are proposed for the various dimensions and the variation between different research groups. The first dimension is an aspect of the context that was already present in FML2005 and worked out in more detail by [KS08]: participant information. Other dimensions discussed below are *communicative actions* (including turn-taking, grounding and speech acts), *content, mental state* and *social-relational goals*.

Person characteristics. Krenn and Sieber [KS08] provide a tentative list of characteristics of the communication partners that may need to be represented organised along three dimensions: person information, social aspects and personality and emotion. Here we describe the person characteristics only. The other elements such as emotion, personality and social aspects are covered below under separate headings. Person information, according to Krenn and Sieber should include the following information: an identifier, name, gender, type (human/agent), appearance, and voice. Also in the original SAIBA proposal on FML, information on participants was included. Among the features associated with it were role and id.

Communicative Actions. The main types of actions that are considered in the various proposals for FML are turn-taking actions, the actions involved in grounding a conversational action and a specification of the speech act as such (also called dialogue act or performative). The following table presents a few proposals for categorizing the turn-taking variables.

Turn-taking
[KPL08] take-turn, want-turn, yield-turn, give-turn, keep-turn
[MP08]   take floor, give floor
[SVJ08] take-turn, release-turn, keep-turn, assign-turn

For the process of grounding, Samtani et al. ([SVJ08]) propose the following actions (based on [TH92]).

Grounding
[SVJ08] initiate, continue, ack, repair, req-repair, req-ack, cancel

As one might expect, the lists of performatives in the various proposals differ more widely than, for instance, the lists for turn-taking. This may be due to the fact that there is no agreement within linguistics but also because different applications may need more or less specific acts to be encoded. Compare, for instance, [MP08] with [SVJ08].

Speech Act
[MP08] implore, order, suggest, propose, warn, approve, praise, recognize, disagree, agree, criticize, accept, advice, confirm, incite, refuse, question, ask, inform, request, announce, beg, greet
[SVJ08] inform, whq, ynq, accept, request, reject, suggest, eval, req-perm, offer, promise

Samtani et al. ([SVJ08]) also use acts that are involved in argumentation processes (see above) which might be put under the heading of speech act and others that are more like rhetorical relations.

Content. With respect to content, most papers do not go into the specific formalism to represent propositional content, assuming that some formal (logical) language will be used. Besides specifying propositional content itself, one also needs to take into account how it is organized.

On the level of a sentence, information structure is concerned with emphasis, given and new information or the related notions of theme and rheme. On a discourse level, organization in topics is important as are rhetorical relations that hold between different parts of the discourse. This point is less settled. [SVJ08], for instance use items such as *elaborate* and *summarize* or *clarify* as part of the argumentation acts, which are typically rhetorical relations, besides actions such as *convince* or *find-plan*. [MP08] use the term *belief-relation* for relations between different parts of the discourse, including *gen-spec, cause-effect, solutionhood, suggestion, modifier, justification* and *contrast*.

Mental State. Most contributions to the FML discussion agree that some sort of representation of the emotional state of an agent is needed as emotions clearly contribute to the motivation of a communicative act. Several authors ([LDMT08] and [MP08], for instance) argue for the possibility to represent emotions in multiple ways within FML. Besides multiple representations for emotions, authors argue also for allowing distinct specifications of felt and expressed emotional states ([MP08], [KS08]), or leaked and felt emotions ([LDMT08]).

Besides emotional states, other mental states and processes that several authors feel should be represented in FML are cognitive processes such as *planning, thinking* and *remembering* ([MP08]). To account for gaze behaviours, Lee et al. [LDMT08] propose very detailed cognitive operations that include monitoring events and goal status, planning, conversation regulation functions and elements of coping.

Social-Relational goals. In the various proposals for elements to include in FML, one of the recurrent type of variables relates to the social psychological domain, i.e. the interpersonal variables that play a role in shaping interaction. So far no coherent picture of how to treat this has emerged, though several contributions to the FML discussion provide suggestions. Bickmore ([Bic08]), for instance, introduces relational (interpersonal) stance functions. His specific proposal was made for the context of agents involved in health-counseling. Everyone seems to agree that the social-relational dimension is an important element to take into account in FML, but this will need further elaboration.

Besides further elaboration of the various dimensions, work on the specification of FML is also faced with a number of general issues. We present several of these in the next section.

3 Issues

The SAIBA framework distinguishes three processes in the generation of multi-modal communicative actions: intention planning, behavior planning, and behavior realization. FML is supposed to be concerned with specifying the intentions of an agent whereas BML is one kind of specification of the behaviour that results. A first question that arises is whether specifications of intentions and behaviours are sufficient for the multi-modal communicative action generation or whether this simple distinction misses out on aspects that do not fit in either of these categories. A second question involves the notion of intention itself. Finally, an important issue is the architecture of a complete generation system and how FML and BML specifications fit in. We discuss each of these in turn. With respect to the first question, an important topic that appeared to be a concern of several authors is that of context.

3.1 The Role of Context

An issue that appears in various forms in several contributions to the FML discussion ([Bic08], [KS08], [SVJ08] and [Rut08] in particular) relates to the question of how contextual parameters that influence behaviours in a conversation should be treated. Should contextual parameters be part of FML or should they be covered by some other module? Samtani et al. ([SVJ08]) provide the example of greetings, the form of which depends on the time of day "Buenas Dias", "Buenas Tardes", "Buenas Noches". This is part of the *environmental context*. Others context variables are part of the dialog context including "the history of interactions", the "topics discussed" and the "cultural context".

Bickmore [Bic08] considers a particular form of context, for which he uses the term *frame* introduced by Bateson ([Bat54]) and also used by Goffman in ([Gof74]). People act differently in contexts that are differently framed and the same sentence uttered in a different frame may be associated with completely different intentions (said jokingly or not, for instance). Bickmore therefore proposes to add contextualisation tags to FML. The kinds of tags he uses currently for counseling agents in counselor-patient interactions are: *task* (information exchange), *social* (social chat, small talk), *empathy* (comforting interactions) and *encourage* (coaching, motivating, cheering up).

An issue to be solved is whether these context variables should be part of the Function Markup Language or whether they should be treated in some other way. The idea of a Contextual Markup Language featured in the AAMAS workshop as one way to frame the problem.

3.2 Intentions

Within the SAIBA framework, FML is described as the language that specifies the communicative intent behind an agent's behaviour. As the examples above show, not all of what the contributors to the workshop propose can be called intentional. The original conception of "intentions" in the SAIBA framework may have been derived from its predominance in speech act theory ([Aus62], [Sea69]). This deals mainly, if not exclusively, with intentions as determinants of communicative actions, as is highlighted by Grice's definition of non-natural meaning [Gri75].

> S nonnaturally means something by an utterance x if S intends (i_1) to produce by uttering x a certain response (r) in an audience A and intends (i_2) that A shall recognize S's intention (i_1) and intends (i_3,) that this recognition on the part of A of S's intention (i_1,) shall function as A's reason, or a part of his reason, for his response r.

Clearly, in the prototypical case of communication these intentions and the recognition of intentions play an important role, but the question is whether they are the sole determinants of communicative behaviours and if not whether the other determinants need not be taken into account in FML as well. This will depend on one's point of view. Contextual parameters already constitute one set of determinants that may not fit well with the strict idea of intentions. In this case, there is a growing consensus that these should be separated out.

But there are other determinants of behaviour that are not intentions that one might want to include in FML. Several contributions to the discussion on FML suggest to include emotion tags into FML. Clearly, affective parameters will be an important determinant in the behaviours displayed by robots and virtual humans, however, what one is trying to model may be phenomena that escape conscious control and thus the term intention in inappropriate. This is epitomized in the case of leakage, where the behaviour does not communicate the emotion in the "nonnatural" way as above but as a symptom or "natural" meaning, for instance indicated by the trembling of the voice in case of nervousness. In the general framework one needs to take into account that there are several semiotic and intentional processes operating in communicative settings ([HtM08], [LDMT08]). It should be noted that the original name FML (Function Markup Language) implies a description of communicative *function*, a term that covers more than conscious intent.

3.3 Communicative Actions

Natural language utterances and nonverbal communicative acts are the most prominent kinds of communicative actions one is inclined to think of. However, as several authors have pointed out ([Gof76], for instance), communicative actions can also be completely extra-linguistic or certain non-linguistic actions can perform certain communicative functions. A typical case is that of contributions to grounding by performing the action that was previously requested showing that one has understood and accepted the request.

Kopp and Pfeiffer Leßman ([KPL08]) discuss a scenario of interactions between a human and the virtual agent Max where they collaborate in assembling Baufix parts. Here also, dialogue moves are heavily interwined with manipulative actions and both are considered forms of what the authors call *interaction moves* as a generalisation of dialogue moves.

Other prototypical cases where nonlinguistic actions become communicative occur when one performs an action ostentatiously. One can thereby communicate to an observer that one intends to perform the action and that one wants the observer to recognize the intention. This point is related to the previous one, in that many of such cases can be considered communicative acts by means of a specific semiotic process called demonstration ([Cla96]).

The general question for FML is thus: what to consider a communicative act and how to related communicative and noncommunicative acts.

3.4 Architecture

The SAIBA framework assumes a three step process of intention planning, behavior planning and behavior realization. In this model, it is relatively easy to define the functionality of the FML and BML languages. However, as several authors have pointed out, the situation becomes less clear when one looks at the generation steps in detail and in particular at existing modules.

Lee et al. ([LDMT08]) look at the particular case of natural language generation in which the reference architecture that is commonly assumed ([RD00]) defines three steps: document planning, microplanning and realization. The first part can – grosso modo – be interpreted as a form of intent planning, whereas the other two belong to the behaviour planning and realisation stage in SAIBA. But the question is whether FML representations will be compatible with existing generation systems.

A related point is made by Ruttkay ([Rut08]) and by Krenn and Sieber ([KS08]) who write:

> This [...] brings us to another crucial aspect for the design of representation languages, i.e., the processing components used in ECA systems. We need to study which subsystems are implemented, what are the bits and pieces of information that are required as input to the individual processing components, and what kinds of information do the components produce as output. Especially if we aim at developing representations that will be shared within the community, there must be core processing components that are made available to and can be used by the community.

The first lesson of this for the development of FML and BML is thus that it is not enough to think about what should be in the language by looking at how conversations work, for instance, but also look into the modules that should make use of the languages, in particular modules that are currently already implemented. The second, might be that together with the specification of a language one needs to develop modules that can use them. This is certainly a major challenge for the definition of a Function Markup Language.

4 Conclusion

Although the work on jointly specifying a Function Markup Language has just started, the first results are promising in several respects. There seems to exist a rather high degree of consensus on the kinds of dimensions that need to be represented in FML. On the other hand, however, there are several dimensions for which the proposals on how to shape them diverge widely. Together with the general issues defined in the previous section on context, on definitional issues and on the embedding of FML in working modules, this means that the definition of a commonly agreed upon specification language for communicative intent still requires a lot of discussion in the wider ECA community. The current forum for discussion can be found at http://wiki.mindmakers.org/projects:fml:main/.

Acknowledgments. We like to thank the contributors to the AAMAS workshop and the other attendants that shaped the discussion. This research was supported in part by the HASGE Grant of Excellence from The Icelandic Research Fund. The research leading to these results has also received funding from the European Community's Seventh Framework Programme (FP7/2007-2013) under grant agreement number 211486 (SEMAINE).

References

[Aus62]   Austin, J.A.: How to Do Things with Words. Oxford University Press, London (1962)
[Bat54]   Bateson, G.: A theory of play and fantasy. Steps to an ecology of mind. Ballantine, New York (1954)
[Bic08]   Bickmore: Framing and interpersonal stance in relational agents. In: Why Conversational Agents do what they do. Functional Representations for Generating Conversational Agent Behavior, Estoril. AAMAS 2008 (2008)
[Cla96]   Clark, H.H.: Using Language. Cambridge University Press, Cambridge (1996)
[CVB01]   Cassell, J., Vilhjálmsson, H., Bickmore, T.: BEAT: the behavior expression animation toolkit. In: Proceedings of ACM Siggraph, Los Angeles, pp. 477–486 (2001)
[Gof74]   Goffman, E.: Frame Analysis. Penguin, Harmondsworth (1974)
[Gof76]   Goffman, E.: Replies and responses. Language in Society 5(3), 2257–2313 (1976)
[Gri75]   Grice, H.P.: Meaning. The Philosophical Review 66(3), 377–388 (1975)
[HtM08]   Heylen, D., ter Maat, M.: A linguistic view on functional markup languages. In: Why Conversational Agents do what they do, Estoril. Functional Representations for Generating Conversational Agent Behavior. AAMAS 2008 (2008)
[KKM+06]   Kopp, S., Krenn, B., Marsella, S., Marshall, A., Pelachaud, C., Pirker, H., Thórisson, K., Vilhjálmsson, H.: Towards a common framework for multimodal generation in ECAs: the behavior markup language. In: Gratch, J. (ed.) Intelligent Virtual Agents 2006, pp. 205–217. Springer, Heidelberg (2006)
[KPL08]   Kopp, S., Pfeiffer-Leßmann, N.: Functions of speaking and acting. In: Why Conversational Agents do what they do. Functional Representations for Generating Conversational Agent Behavior. AAMAS 2008, Estoril (2008)
[KS08]   Krenn, B., Sieber, G.: Functional mark-up for behaviour planning. theory and practice. In: Why Conversational Agents do what they do. Functional Representations for Generating Conversational Agent Behavior. AAMAS 2008, Estoril (2008)
[LDMT08]   Lee, J., De Vault, D., Marsella, S., Traum, D.: Thoughts on FML: Behavior generation in the virtual human communication architecture. In: Why Conversational Agents do what they do. Functional Representations for Generating Conversational Agent Behavior. AAMAS 2008, Estoril (2008)
[MP08]   Mancini, M., Pelachaud, C.: The FML-APML language. In: Why Conversational Agents do what they do. Functional Representations for Generating Conversational Agent Behavior. AAMAS 2008, Estoril (2008)
[Pel05]   Pelachaud, C.: Multimodal expressive embodied conversational agents. In: Multimedia 2005, pp. 683–689. Springer, New York (2005)

[Pog07] Poggi, I.: Mind, hands, face and body. A goal and belief view of multimodal communication. Weidler, Berlin (2007)

[RD00] Reiter, E., Dale, R.: Building Natural Language Generation Systems. Cambridge University Press, Cambridge (2000)

[Rut08] Ruttkay, Z.: Situation and agency in the SAIBA framework, and consequences for FML. In: Why Conversational Agents do what they do. Functional Representations for Generating Conversational Agent Behavior. AAMAS 2008, Estoril (2008)

[Sea69] Searle, J.R.: Speech acts: An essay in the philosophy of language. Cambridge University Press, Cambridge (1969)

[SPL06] Schröder, M., Piker, H., Lamolle, M.: First suggestions for an emotion annotation and representation language. In: International Conference on Language Resources and Evaluation: Workshop on Corpora for Research on Emotion and Affect, Genova, Italy, pp. 88–92 (2006)

[SVJ08] Samtani, P., Valente, A., Johnson, W.L.: Applying the SAIBA framework to the tactical language and culture training system. In: Why Conversational Agents do what they do. Functional Representations for Generating Conversational Agent Behavior. AAMAS 2008, Estoril (2008)

[TH92] Traum, D.R., Hinkelman, E.A.: Conversation acts in task-oriented spoken dialogue. Computational Intelligence 8, 575–599 (1992)

[VCC+07] Vilhjálmsson, H., Cantelmo, N., Cassell, J., Chafai, N.E., Kipp, M., Kopp, S., Mancini, M., Marsella, S., Marshall, A.N., Pelachaud, C., Ruttkay, Z.M., Thórisson, K., van Welbergen, H., van der Werf, R.J.: The behavior markup language: Recent developments and challenges. In: Pelachaud, C., Martin, J.-C., Andre, E., Collet, G., Karpouzis, K., Pelé, D. (eds.) IVA 2007. LNCS (LNAI), vol. 4722, pp. 90–111. Springer, Heidelberg (2007)

[Vil05] Vilhjámsson, H.: Augmenting online conversation through mediated discourse tagging. In: Proceedings of the 6th Annual Minitrack on Persistent Conversation. Hawaii International Conference on System Sciences, Hawaii. IEEE, Los Alamitos (2005)

[VM05] Vilhjálmsson, H., Marsella, S.: Social performance framework. In: Workshop on Modular Construction of Human-Like Intelligence at the 20th National AAAI Conference on Artificial Intelligence, Pittsburgh, PA (2005)

[VT08] Vilhjálmsson, H., Thórisson, K.: A brief history of function representation from gandalf to saiba. In: Why Conversational Agents do what they do. Functional Representations for Generating Conversational Agent Behavior. AAMAS 2008, Estoril (2008)

Extending MPML3D to Second Life

Sebastian Ullrich[1,2], Klaus Bruegmann[1],
Helmut Prendinger[1], and Mitsuru Ishizuka[3]

[1] National Institute of Informatics,
2-1-2 Hitotsubashi, Chiyoda-ku, Tokyo 101-8430, Japan
s.ullrich@ieee.org, mail@klausbruegmann.de, helmut@nii.ac.jp
[2] Virtual Reality Group, RWTH Aachen University,
Seffenter Weg 23, 52074 Aachen, Germany
[3] Graduate School of Information Science and Technology, University of Tokyo,
7-3-1 Hongo, Bunkyo-ku, Tokyo 113-8656, Japan
ishizuka@i.u-tokyo.ac.jp

Abstract. This paper describes an approach how to integrate virtual agents into the 3D multi-user online world Second Life. For this purpose we have implemented a new client software for Second Life that controls virtual agents ("bots") and makes use of the Multimodal Presentation Markup Language 3D (MPML3D) to define their behavior. The technical merits and limitations of Second Life are discussed and solutions are provided. A multi-user scenario serves as an example to illustrate our solutions to technical challenges and advantages of using the virtual environment of Second Life.

1 Introduction

Virtual online worlds are becoming increasingly popular and can be characterized as metaverses [13] as envisioned in the scifi-novel "Snow Crash" [14]. Second Life (SL) is such a virtual online world that provides a free networked multi-user three-dimensional (3D) environment [7]. SL is very popular with an increasing amount of registered users (over 13 million as of April 2008) and about 50,000 users online at any time. The main features of SL are (1) support of social interactions between the avatars of users (SL residents) with customizable representations, (2) support for user-created content (like a '3D wiki space'), and (3) its economy with a marketplace and own currency, called "Linden dollars".

These features allow SL users to design their own objects, including buildings, vehicles, and even eco-systems, on their privately owned virtual locations ("islands"), and to open the island to the SL community. Many institutions, both commercial and academic, have recently opened their presence in Second Life. However, it was soon noticed that simply building a visually impressive place is not sufficient for an attractive presence in an inherently social space like a virtual world. The key to the success of an island is to provide visitors an interactive experience. Some islands demonstrate a high level of interaction because many avatars meet there. Other islands, however, are rather deserted and give an uncomfortable feeling to the visiting avatar.

One suggestion is to populate deserted islands with 'bots', i.e. computer-controlled virtual agents, which may play the roles of guides, receptionists, guards, or other visitors of the island. Note that in the gaming world, bots are rather called NPCs (Non-Player Characters), and bots should not be confused with avatars, which are controlled by users. Quite surprisingly, bots are currently almost missing from Second Life.

Hence, in this paper, we propose an XML-based authoring language for SL bots, which can also be used by non-computer science professionals. In particular, we have extended MPML3D [10] to support SL. MPML3D is a successor of the Multimodal Presentation Markup Language (MPML) [11]. MPML3D is a scripting language for interaction-rich scenarios with reactive agents. The proposed novel extension of the MPML3D framework enables content creators to define their own scripted interactive agents for SL.

2 Related Work

There are many different markup languages and systems for behavior planing of Embodied Conversational Agents (ECAs), e.g., APML – a Mark-up Language for Believable Behavior Generation [2], BML – Behavior Markup Language [6], and MPML – Multimodal Presentation Markup Language [11,10], just to name a few. Here we want to focus on reporting solutions that are used for multi-user environments, and also other approaches to adding agents to SL.

BML is being integrated into EVE-Online, a massively multiplayer online role-playing game (MMORPG) [15]. It is an official addition to the game by the developers themselves, and will allow players to interact with autonomous agents and to automate coordination of nonverbal social behavior. Because of the space setting of the game and the closed source game engine, content creators are not able to create their own custom scenarios in this online world.

Freewalk [9] is a platform that allows social interaction between multiple users and agents. Much effort has been put into preparing a shared environment, an interaction model, and an interaction scenario. To describe the interaction scenarios and to define the roles of the agents, the description language Q [5] is used. Several applications and multi-user experiments have been conducted with Freewalk. It is unclear, however, how easily new content can be created with Freewalk/Q.

Friedman et al. [3] created a simple agent to evaluate social behavior in SL. The agent was driven by the Linden Script Language and was mainly used for data logging and traversing through SL. A very basic type of navigation has been implemented (walking in random directions until either an obstacle or an avatar is found). Furthermore, the agent can greet avatars by their names and perform gestures.

In summary, we have not found a solution that offers easily scriptable bots in a widely used multi-user 3D environment. Until now, there are only a few and very specialized solutions for agents in Second Life.

3 Second Life Overview

3.1 Infrastructure

Second Life is a client/server application for multiple networked users. In order to use SL, one needs to register a free or professional account for a virtual character/avatar. The client software can be downloaded for free and is available for multiple platforms (Windows 2000/XP/Vista, Mac OS X and Linux). Linden Lab, the company that developed SL, maintains a cluster network to host regions of 3D virtual environments, the "islands". These islands contain user-created 3D content and can be interactively explored by the users that are logged into the system of SL. Similar to HTML, with a web server and browser application, the 3D content is hosted on the cluster network servers and streamed to the client application. However, the crucial difference is that the content in SL is protected by a digital rights management system.

3.2 Programming Interfaces

There is an official scripting language in SL, which is called "Linden Script Language" (LSL). With over 300 library functions and different data and message types, scripts can control the behavior of virtual objects and communicate with other objects and avatars. Limitations of the script language include time delays for movement of objects (0.2 sec) and memory constraints for scripts (16 KB), i.e., complex algorithms must be either simplified or distributed into several scripts (if the algorithm can be subdivided into smaller tasks). Furthermore, a script cannot be assigned to avatars.

Libsecondlife is an inofficial API for SL that is being developed by an open source community [12]. It connects to SL as an alternative client and has access to some of the data that is provided to the client. Among many functions for controlling the avatar, it contains methods for communication, navigation, and object manipulation. Thus, compared to the Linden Script Language, its main advantages are: full control of the avatar, responsiveness (i.e., no time delay), and no memory constraints.

3.3 Entities and Resources

The content in SL is mainly created by its users and covers the appearance of avatars (that represent the users), virtual objects, and other resources (animations, sounds, pictures, etc). The appearance is mainly defined by the body shape that can be changed by anthropometrical parameters like height, length of limbs, diameters, and many more. Virtual cloths, skin textures, hair styles and accessories can be either modeled by the user or purchased within the virtual economy of SL. Animation files, which can be applied to the avatars, must be uploaded in the BVH-format. To create new animation files or to re-target existing motion from other systems, reference models for male and female avatars exist [8]. These models contain the hierarchical setup of the skeleton, which is comparable to the reference models of the H-Anim standard [4].

4 MPML3D Framework with SL-Extension

Our new system integrates MPML3D into SL and is based on the MPML3D framework [10,1]. The whole system (see Fig. 1) is divided into three different sections: (1) the MPML server, (2) the "official" server (cluster network) from Linden Lab, and (3) the client side of visitors/users of the system. Basically, the system requires a MPML3D script as input in order to allow users within SL (avatars) to interact with the script-driven agents (bots). To actually host a MPML3D-based scenario within SL, the content creator has to use the services that are provided by the MPML3D server. Potential users or visitors just need to use the free, official SL-client software. The actual implementation is based upon the MPML3D framework, which has been re-factored into the MPML3D backend and the JAVA-based reference frontend that uses OpenGL for local output. Additionally, a second frontend has been implemented to create an interface to the networked environment of SL. In the following, we describe the individual components of the MPML3D server.

4.1 MPML3D Backend

The MPML3D backend implements the scenario (scene setup and scene plot/-content) as defined in an MPML3D script, and acts as a host for one or more frontends (see Section 4.2). It integrates the parser for the XML-based script source with the dynamic runtime representation of all interdependencies between and the hierarchy of agents' conversational activities and perceptions, the so-called "activity network". In order to keep this part of the MPML3D system flexible, the backend handles scene entities and activities in an abstract way, leaving the actual implementation to the frontends. The MPML3D backend accepts incoming network connections from frontends. After initialization, the backend waits for the beginning of the scene plot, which is triggered by the frontend. Then the backend issues 'start' respectively 'stop' commands for those

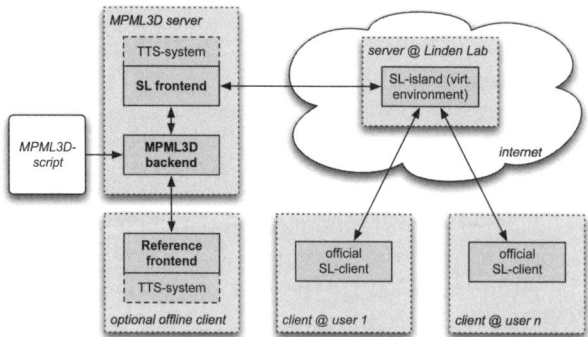

Fig. 1. Overview of the system architecture with the different components distributed over several servers

actions, as defined by the scene plot of the MPML3D script. The frontend in turn performs the output for each action according to its specification and notifies the backend when an action is completed.

In our example (see below), the MPML3D script file for the presentation of a traditional Japanese interior design is parsed, the entities (bots) 'Ken' and 'Yuuki' are loaded, and so is the activity network. The backend waits for connections from the frontends.

4.2 Frontends

Each frontend realizes a user interface – comprising multimodal output as well as user feedback channels – for interactive content scripted via MPML3D. A frontend provides a concrete implementation for supported entity types along with their actions and entity states.

The following excerpts from the example MPML3D-script demonstrates two very important features of MPML3D that are supported by each frontend: (1) gestures synchronized to utterances, and (2) perceptions to create interruptions. To synchronize gestures to utterances, the timing of gestures is specified in square brackets and refers to the count of words of the sentence that is defined in the speak action. This means that kenSpeak[1].begin will be performed at the beginning of the first word "Me", and kenSpeak[4].begin at the beginning of "and", respectively.

```
<Parallel>
 <Action name="act">
   ken.speak("Me, Ken Kobayashi, and my colleague...")</Action>
 <Action startOn="act[1].begin">ken.gesture("MYSELF")</Action>
 <Action startOn="act[4].begin">ken.gesture("POINT_LEFT")</Action>
</Parallel>
```

Here is an example for the perception feature. The action to be interrupted is the following sentence by the female bot Yuuki.

```
<Action name="yuukiSpeak">
  yuuki.speak("What you can see here is a washitsu.")</Action>
```

A perception is set up with a unique identifier (here, "interruption") and the activating event is specified.

```
<Perception name="interruption">
  onEvent(yuuki, "saysPhrase", "you can see")</Perception>
```

The following task is immediately started when the perception is triggered, which causes the termination of the previous action.

```
<Task name="explain" startOn="interruption">
<Parallel>
  <Action name="kenSpeak">
    ken.speak("Yuki! Wait! We should first explain...")</Action>
```

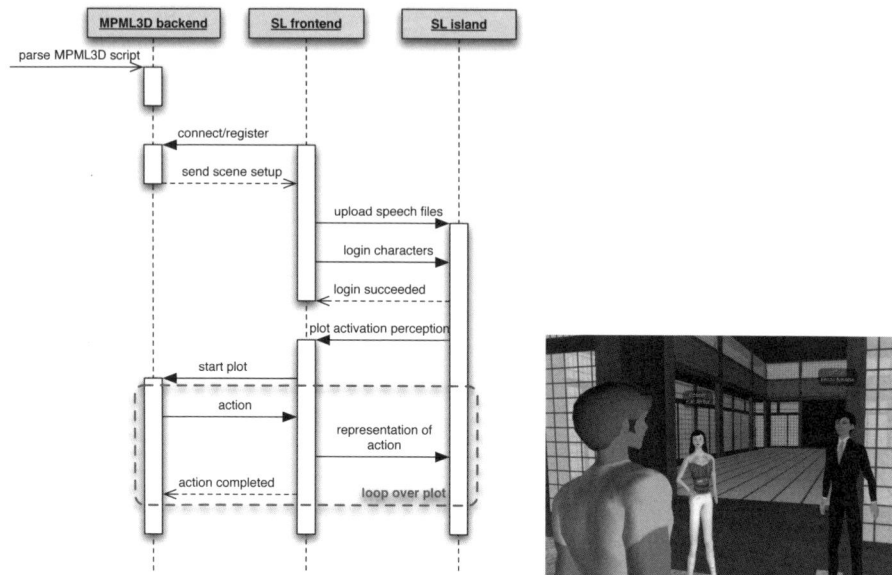

Fig. 2. left: UML sequence diagram to illustrate the initialization and run-time behavior of the modules to control agents in SL with a MPML3D script. **right:** Example of a visitor listening to the presentation of the agents in SL.

4.3 SL Frontend

The SL frontend is responsible for the presentation of MPML3D content in SL. It is implemented in C# and makes use of the libsecondlife API (as described in section 3.2). During the initialization process (shown in Figure 2, left) the frontend is registered at the MPML3D backend, the scene data is received and the agents are logged into SL. In the example the two agents are giving a presentation in front of a japanese building (see Figure 2, right).

In the following, we will describe the functionality and limitations of the methods that we have implemented for the representation in SL:

Gesture playback. To animate the virtual characters we have implemented basic support to start and stop previously uploaded or officially provided gestures in SL. Due to technical restrictions in SL, it is not possible to change the playback speed or to blend animations. Moreover, one cannot perform low-level animation, because it is impossible to manipulate the joints of SL character directly.

Speech output. We have identified three different approaches for speech output in SL: (1) playback of previously uploaded audio-clips, (2) streaming audio that is generated on the server-side in nearly real-time, and (3) text output that is intercepted on the client-side and synthesized by locally installed TTS software. For the most accurate timing under the given circumstances, we favor the first approach.

Pseudo-lipsync. At the time of writing this paper, lip synchronization is not supported in SL and no third-party solutions exists. The two programming interfaces of SL provide no access to the internal data-structures, which would enable the implementation of a new facial animation system. As a tentative method, we implemented pseudo lip synchronization, using randomized "mouth open" animations.

4.4 Challenges in SL

Timing issues in SL. Due to the distributed server/client architecture of SL, it is difficult to achieve accurate timing. Because of the many ways speed on the internet can be affected, synchronization between server and client should be avoided. The most feasible solution is to synchronize the events on the server-side.

Multiple users. Another interesting challenge is to handle the unpredictability of visitors (users in the form of avatars) in SL. A presenter bot has to consider (1) when to start a presentation, (2) when to abort a presentation, and (3) where to look in case of multiple visitor avatars. We provide a few heuristics to cope with these problems. The agents continuously scan their environment and if a visitor (avatar) enters within a predefined radius, an event is triggered.

5 Conclusions

In this paper we have presented a novel solution, that enables life-like agents in Second Life with behavior controlled by the Multimodal Presentation Markup Language MPML3D. We have evaluated the technical details of SL and provide solutions to ensure responsive behavior and smooth interactions of the agents. The resulting system has been successfully implemented and tested with existing and new MPML3D scripts, without compromising key features of MPML3D. In summary, the system combines the easy-to-use authoring multimodal content language for life-like characters MPML3D with the feature and content rich multi-user online environment of SL. There are numerous applications and scenarios where the agents can be used. Possible areas of usage are in education, research or entertainment, e.g. to give dialogues, instructions or speeches, to be part of user studies, or for testing within simulation environments.

Future work will focus on the addition of new features that are possible due to SL, like basic movement of the agents to navigate to specific target location, to walk towards or to follow other virtual character (agents or avatars), and to react on additional levels to user input (e.g., talking, transaction of objects, etc).

References

1. Brügmann, K., Dohrn, H., Prendinger, H., Stamminger, M., Ishizuka, M.: Phase-based gesture motion parametrization and transitions for conversational agents with mpml3d. In: INTETAIN 2008: Proceedings of the 2nd international conference on Intelligent Technologies for interactive Entertainment. ACM Digitial Library (to appear, 2008)

2. Carolis, B.D., Pelauchaud, C., Poggi, I., Steedman, M.: Life-Like Characters. Tools, Affective Functions, and Applications. Cognitive Technologies, chapter APML: Mark-up language for communicative character expressions. Springer, Heidelberg (2004)
3. Friedman, D., Steed, A., Slater, M.: Spatial social behavior in second life. In: Pelachaud, C., Martin, J.-C., André, E., Chollet, G., Karpouzis, K., Pelé, D. (eds.) IVA 2007. LNCS (LNAI), vol. 4722, pp. 252–264. Springer, Heidelberg (2007)
4. H-Anim Working Group. Information technology — computer graphics and image processing — humanoid animation (h-anim) (last visited: 2008/04/11), http://www.h-anim.org
5. Toru Ishida, Q.: A scenario description language for interactive agents. IEEE Computer 35(11), 54–59 (2002)
6. Kopp, S., Krenn, B., Marsella, S., Marshall, A.N., Pelachaud, C., Pirker, H., Thórisson, K.R., Kopp, H.V.S., Krenn, B., Marsella, S., Marshall, A.N., Pelachaud, C., Pirker, H., Thórisson, K.R., Vilhjálmsson, H.: Towards a common framework for multimodal generation: The behavior markup language. In: Proceedings of 6th International Conference on Intelligent Virtual Agents, pp. 205–217. Springer, Heidelberg (2006)
7. Linden Lab. Offical website of second life (last visited: 2008/04/11), http://www.secondlife.com
8. Linden Lab. Reference character models for second life (last visited: 2008/04/11), https://secondlife.com/downloads/avatar.php
9. Nakanishi, H., Ishida, T.: Freewalk/q: Social interaction platform in virtual space. In: ACM Symposium on Virtual Reality Software and Technology (VRST 2004), pp. 97–104 (2004)
10. Nischt, M., Prendinger, H., André, E., Ishizuka, M.: MPML3D: a reactive framework for the Multimodal Presentation Markup Language. In: Gratch, J., Young, M., Aylett, R.S., Ballin, D., Olivier, P. (eds.) IVA 2006. LNCS (LNAI), vol. 4133, pp. 218–229. Springer, Heidelberg (2006)
11. Prendinger, H., Descamps, S., Ishizuka, M.: MPML: A markup language for controlling the behavior of life-like characters. Journal of Visual Languages and Computing 15(2), 183–203 (2004)
12. Second Life Reverse Engineering Team. libsecondlife API (last visited: 2008/04/11), http://www.libsecondlife.org
13. Smart, J.M., Cascio, J., Paffendorf, J.: Metaverse roadmap: Pathways to the 3rd web (2007), http://www.metaverseroadmap.org
14. Stephenson, N.: Snow Crash. Spectra (1992)
15. Vilhjálmsson, H., Cantelmo, N., Cassell, J., Chafai, N.E., Kipp, M., Kopp, S., Mancini, M., Marsella, S., Marshall, A.N., Pelachaud, C., Ruttkay, Z., Thórisson, K.R., van Welbergen, H., van der Werf, R.J.: The behavior markup language: Recent developments and challenges. In: Pélachaud, C., Martin, J.-C., André, E., Chollet, G., Karpouzis, K., Pelé, D. (eds.) IVA 2007. LNCS (LNAI), vol. 4722, pp. 99–111. Springer, Heidelberg (2007)

An Extension of MPML with Emotion Recognition Functions Attached

Xia Mao, Zheng Li, and Haiyan Bao

P.O. BOX 206 Beihang University, Beijing 100083, P.R. China
{moukyoucn,buaa_david}@yahoo.com.cn,
vivid96912@sohu.com

Abstract. In this paper, we discuss our research on the multimodal interaction markup language (MIML) which is an extension of multimodal presentation markup language (MPML). Different from MPML, MIML can describe not only the presentations of lifelike agents, but also their emotion detection capability with the facial expression recognition and speech emotion recognition functions attached. Emotional control on lifelike agents provided by MIML makes the human-agent interaction even more intelligent. With the MIML and functions we designed, web-based affective interaction can be described and generated easily.

Keywords: Multimodal Interaction, Lifelike Agents, Description Language, Affective Computing.

1 Introduction

Lifelike agents have been used as the middle layer between user and computer. They have shown their potential to allow user to interact with computer in a natural and intuitive manner through human communicative means. Meanwhile, as Nass's researches on human-human and human-computer interactions suggest, people most naturally interact with computers in a social and affectively meaningful way, just like with other people [1]. Researchers like Picard have recognized the potential and importance of emotion to human-computer interaction, dubbing work in this field as "affective computing" [2]. Therefore, in order to realize believable intelligent human-agent interaction, we must endow lifelike agents with affect, namely they ought to have capabilities to express their own emotion and recognize user's emotional states.

At the same time, there is an emerging number of scripting and representation languages to describe the behavior of the agent, which have taken different approaches to specify their objectives, including APML [3], VHML [4], MPML [5][6], PAR [7], STEP [8], and so on. The differences among these languages are their description levels, targeted users, character agents being supported for use, functions related to dialogue, intelligent and emotion. MPML is a powerful and easy-to-use XML-style language enabling content authors to script rich web-based interaction scenarios featuring lifelike agents. Tags in MPML are designed following the conventions well known from HTML and provide controls for the verbal and non-verbal behavior of affective 2D cartoon-style agent, presentation flow, and the integration of external objects, like Java applets. As mentioned above, emotion is one important factor for making the agent lifelike, believable, friendly and empathic. The description and expression of the agent's emotion is one of the features of MPML. Specifying emotion by using the <emotion> tag, the agents can express emotions by performing different actions and changing speech parameters [5][6]. Meanwhile, an artificial emotion module called SCREAM (Scripting Emotion-based Agent Minds) has been developed, which decides the emotional state and external emotion expression of the agent based on the current interaction situation [9]. However, the MPML can not describe and generate agent's emotion detection behavior. In this paper, we will introduce the multimodal interaction markup language, a language extended from MPML and specifically designed for non-expert users allowing them to control agent's emotion detection behavior when creating web-based intelligent interaction system. Two emotion recognition functions, including facial expression recognition and speech emotion recognition ActiveX controllers, are attached to MIML.

2 The MIML System

This section is devoted to explain the MIML system architecture. The MIML is an extension of MPML. The reason we replace "P" with "I" is mainly because MIML can describe not only the presentation (such as emotion expression and text-to-speech capability etc) of life-like agents, but also the emotion detection capability to realize the intelligent interaction. The use of the MsAgent [10] is assumed by default just like MPML. Due to the need of speech dialogue feature, it has to incorporate voice commands and TTS(Text-To-Speech) engines. Other agent systems can be used with appropriate driver programs.

2.1 System Architecture

The MIML is designed for describing multimodal web-based intelligent interaction with lifelike agents. An overview of the MIML system architecture is shown in Fig. 1.

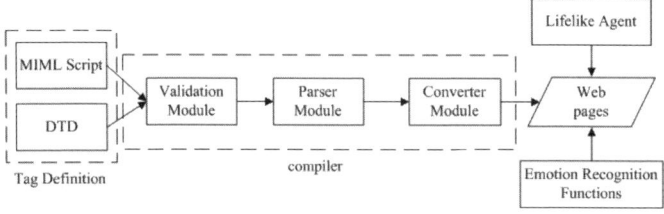

Fig. 1. Architecture of MIML

- The validation module invokes the Document Type Definition (DTD) file to check the errors contained in the MIML script file.
- The parser module uses the Simple APIs for XML (SAX) which is provided in MSXML.DLL to parser the MIML script.
- The converter module transforms the parsed MIML script to Vbscript which is executable in web browser.
- The life-like agents module uses the MsAgent to provide the 2D animation sequences.
- The emotion recognition module endows the life-like agents with the emotion detection capability. We will give a description in section 3.

2.2 Tag Structure

Here we will briefly discuss the tags defined in MIML. Fig. 2 illustrates the tag structure for MIML.

The tags in white boxes in Fig. 2 are used to control the agents' basic behavior similar to MPML, which are extensively discussed in paper [5] and [6]. We will give a detailed description of the tags we proposed to create web-based affective interaction, which are placed in colored boxes.

The root tag pair to describe intelligent interaction is <perception> which include one sub-tag <emotionrecognition>. The <emotionrecognition> involves two sub-tags now: <face> and <speech>, which are designed to control the facial expression recognition and speech emotion recognition respectively. The "align" attribute in tag <face> specifies the destination spot where the controller is located in the web pages. The tag <recognize> and "result" attribute attached to it are equivalent to the C "switch" and "case" instruction respectively. It compares the return value of emotion recognition function with the value of

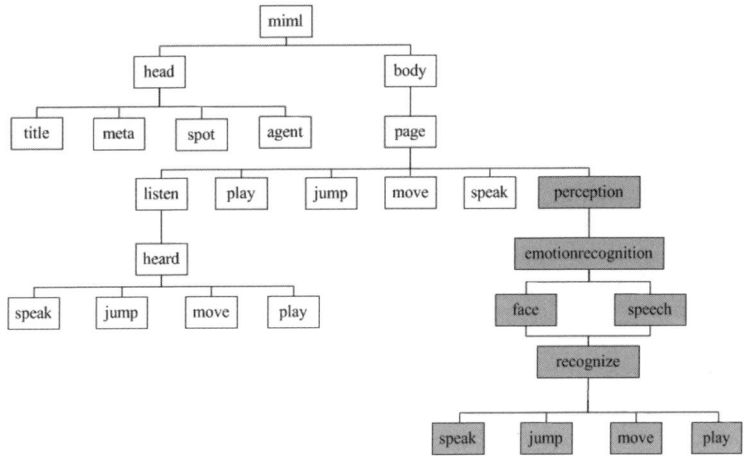

Fig. 2. Tag Structure of MIML

"result" attribute, and executes the script included in the <recognize> tag if they are identical.

3 Emotion Recognition Functions

The facial expression recognition function and speech emotion recognition function are attached to MIML. They are realized as ActiveX controller so they can be called by the MIML tags on the web pages. The performance of facial expression recognition could be influenced by noise or occlusion on the face. The most common causes of occlusion can be pose variation, glass wearing, and hair or hand covering etc. The ability to handle occluded facial features is most important for achieving robust facial expression recognition. The work flow of the robust facial expression recognition can be seen in Fig. 3. As for the speech emotion recognition, a hybrid of Hidden Markov Models (HMMs) and Artificial Neural Network (ANN) is proposed to classify speech emotions. Fig. 4 illustrates the structure of the speech emotion recognition controller. These two ActiveX controllers are extensively discussed in our complementary paper [11].

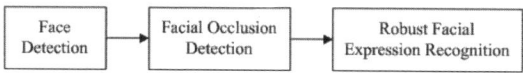

Fig. 3. Facial Expression Recognition Controller

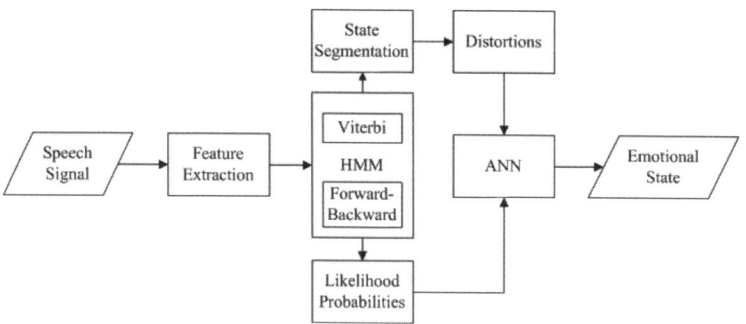

Fig. 4. Speech Emotion Recognition Controller

4 Illustration

In this section, we will describe a web-based scenario that instantiates the affective human-agent interaction described and generated by MIML. The scenario is web-based E-commerce recommendation system, and the lifelike agent plays as a

```
1   <miml>
2   <head>
3       ...
4       <agent id="genie" character="genie"/>
5   </head>
6   <body>
7       <page id="1" ref="main.html">
8           <speak emotion="happiness">
9               Welcome, nice to meet you! What kide of commodity do you want?
10              Tell me through the microphone or click the icon.
11          </speak>
12          <listen>
13              <heard word="wine">
14                  <jump des="wine.html"/>
15              </heard>
16              ...
17          </listen>
18      </page>
19      <page id="2" ref="wine.html">
20          <speak emotion="happiness">
21              Hello, I will introduce the wine to you!
22              ...
23          </speak>
24          ...
25          <perception>
26              <emotionrecognition>
27                  <face align="right">
28                      <recognize result="happiness">
29                          <speak emotion="happiness">
30                              Oh,you are smiling. you must be interested in the wine.
31                          </speak>
32                          <jump des="wine1.html"/>
33                      </recognize>
34                      <recognize result="anger">
35                          <speak emotion="sadness">
36                              Oh,you are not happy. I will introduce the other wine to you!
37                          </speak>
38                          <jump des="wine2.html"/>
39                      </recognize>
40                      ...
41                  </face>
42              </emotionrecognition>
43          </perception>
44          ...
45      </page>
46      ...
47  </body>
48  </miml>
```

Fig. 5. Script Fragment of E-commerce Recommendation System

virtual recommender. The script fragment of the E-commerce recommendation system is demonstrated in Fig. 5.

Fig. 6 is one of the interaction system results after compiling of the MIML script. In Fig. 5, line 7 means the background web page is "main.html". In line 8-11, the agent "genie" expresses his welcome to the user with happiness emotion and is enabled to accept the speech command from the user to decide what commodity the user are interested in (line 12-17). When user says "wine", the web page will jump to "wine.html" in which the agent will introduce wine to the user with happiness emotion (line 20-23). Then the agent can detect the user's emotional state by facial expression recognition controller to judge whether the user is satisfied with the wine or not, followed by a branching edge of multiple alternatives (line25-43) where part of the first branch is shown (line 28-33,Fig. 6). This branch is selected when the recognition result is "happiness", then the agent will ask the user to order the wine and the web page will jump to "wine1.html" . If not, the agent will introduce another wine to the user and the

Fig. 6. Wine.html of E-commerce Recommendation System

web page will jump to "wine2.html" (line 34-39). In this example, we only use the facial expression recognition controller. The authors also can employ the speech emotion recognition controller if they need when describing the interaction with lifelike agents.

5 Conclusion

Recent years have seen many efforts to include lifelike agents as a crucial component of application fields including tutors in E-learning system, recommenders in E-commerce system, actor in entertainment system and partners in chatting system etc. That results in the growing number of scripting languages for controlling the behavior of the lifelike agents. However, these languages can not script the emotional detection behavior. In this paper, we have described the architecture of the MIML, a markup language extended from MPML. With the tags and emotion recognition functions we designed, the emotion detection capability of lifelike agent can be described and generated easily. In the future, we will devote ourselves to improve the recognition algorithms of the facial expression and emotional speech for higher accuracy. Meanwhile, we intend to design more tags and controllers to enrich the MIML architecture.

Acknowledgments

This work is supported by the National Nature Science Foundation of China (No.60572044), High Technology Research and Development Program of China (863 Program, No.2006AA01Z135) and the National Research Foundation for the Doctoral Program of Higher Education of China (No.20070006057). We would like to appreciate the Professor Mitsuru Ishizuka at University of Tokyo and the Associate Professor Helmut Prendinger at the National Institute of Informatics, who have provided many free resources on their home pages. We thank the anonymous reviewers for their great helpful comments.

References

1. Nass, C., Tanber, E.: Computers are social actors. In: Proceeding of CHI 1994, Boston, pp. 72–78 (1994)
2. Picard, R.: Affective Computing. MIT Press, Cambridge (1997)
3. DeCarolis, B., Carofiglio: APML: a mark-up language for believable behavior generation. In: Falcone, R., Barber, S., Korba, L., Singh, M.P. (eds.) AAMAS 2002. LNCS (LNAI), vol. 2631. Springer, Heidelberg (2003)
4. Marriott, A., Stallo, J.: VHML - uncertainties and problems, a discussion. In: Falcone, R., Barber, S., Korba, L., Singh, M.P. (eds.) AAMAS 2002. LNCS (LNAI), vol. 2631. Springer, Heidelberg (2003)
5. Prendinger, H., Ishizuka, M.: Describing and generating multimodal contents featuring affective lifelike agents with mpml. New Generating Computing 24(2), 97–128 (2006)
6. Prendinger, H., Ishizuka, M.: Mpml: A markup language for controlling the behavior of life-like characters. Journal of Visual Languages and Computing 15(2), 183–203 (2004)
7. Badler, N.: Parameterized action representation for virtual human agents, Embodied Conversational Agents. MIT Press, Cambridge (2000)
8. Huang, Z., Eliebs, A.: Step: a scripting language for embodied agent. In: Proceeding of PRICAI 2002 Workshop on Lifelike Animated Agent - Tools, Affective Functions and Applications, Tokyo (2002)
9. Prendinger, H., Ishizuka, M.: MPML and scream: Scripting the bodies and minds of life-like characters (2004)
10. Microsoft: Developing for microsoft agent (1998)
11. Mao, X., Li, Z.: Generating and describing affective human-agent interaction. In: Proceeding of ICNC 2008, Jinan, China (2008)

ITACO: Effects to Interactions by Relationships between Humans and Artifacts

Kohei Ogawa and Tetsuo Ono

Future University-Hakodate, 116-2 Kamedanakano-cho, Hakodate, Japan 041-8655
{g3107002,tono}@fun.ac.jp

Abstract. Our purpose in this paper is realizing a natural interaction between humans and artifacts by an ITACO system. The ITACO system is able to construct a relationship between humans and artifacts by a migratable agent. The agent in the ITACO system can migrate to various artifacts within an environment to construct a relationship with humans. We conducted the two experiments to confirm effects for human by a relationship that was constructed between humans and artifacts. The experimental results showed that a relationship gave some influences to human's behaviors and cognitive abilities. The results also showed that the ITACO system and the migratable agent were an effective method to realize a natural interaction between them.

Keywords: migratable agent, ITACO, relationship, interaction.

1 Introduction

Natural interactions between humans and artifacts are desired by a society because of a progress of information technologies. Thus various works are taken place in the world [1] [2] [3]. An agent technology has focused on achieving a natural interaction between humans and artifacts. Recently, an agent technology is going to be a useful with a progress of information technology, such as internets, downsizing computers and depreciating a cost of computers. An interface agent, for instance, is able to mediate an interaction between humans and artifacts. 'MACK' is a Kiosk agent which can direct the visitor to the laboratory and can introduce researches [4]. If a user could not understand the explanation, 'MACK' can complement the explanation by an eye detector. Herewith, most studies focus on improvement agent abilities, such as a natural language processing, context recognition and eye detection. However, we consider a factor that is humans' emotion is more important to realize a natural interaction.

We expect that a relationship between humans and artifacts plays an important role on natural interactions. Rosalind Picard said that human's emotion and expressing an emotion of artifacts are important for interaction between humans and interactive systems [5]. She also said that it can realize a natural interaction between humans and artifacts that an artifact commits to human's emotions or affections. Donald Norman said that a designer should focus on that human's affective side when we design an artifact [6]. This advocacy shows that to realize a natural interaction between humans and artifacts, it is needed engagement from humans to artifacts as well as a modification to be easy to use. Furthermore it shows that to pull out an engagement from humans to artifacts and construct a relationship between humans and artifacts.

In this paper, we propose an ITACO (InTegrated Agent for COmmunication) system which can construct a relationship between humans and artifacts. The ITACO system uses a migratable agent depends on a context. The agent on the ITACO system can continue a relationship that is constructed by daily interaction with a human while the agent migrates to various artifacts within an environment. We conducted the two experiments to confirm effects for humans by a relationship that was constructed between humans and artifacts.

In the following section, we show the outline of the ITACO system. Then we show two experiments using a prototype of the ITACO system, seeing if relationship was constructed. Finally we make a discussion on the experiments and show the conclusions.

2 ITACO System

The ITACO system uses a migratable agent. This system tries to appropriately support humans using a migratable agent which has the user's personal information. The agent is context-sensitive and provides continuous assistance, migrating from artifacts to artifacts within an environment, depending on the context. Figure 1 shows a concept of the ITACO system. For example, the agent supports human's daily life by migration to a wearable computer, providing information on schedule or mass transit when the user is out. When the user comes back home, the agent migrates to lamps to light up the room. In this study, we implement a prototype system which is operated in the specific environment. In particular, the agent who is on the desktop computer migrates to wearable computer or table lamp while collecting information all the time.

A purpose of the ITACO system is supporting humans and realizing a natural interaction between artifacts within an environment by the agent migration. Meanwhile, the agent on the ITACO system can transfer the relationship that is constructed with humans while the agent migrates to some artifacts. The construction of implemented prototype system is as follows.

2.1 Constructions of the ITACO System

In this study, we made a prototype ITACO system. Specifically, the agent on a tablet PC migrates to some artifacts, such as a robot, a table lamp, a fan and a wearable PC with some information which was obtained from interaction with a user. The construction of an implemented prototype system is as follows.

Hardware Constructions. The ITACO system is consisted of following components: a tablet computer to interact with agent, some artifacts (eg. Robot or table lamp), a micro computer to operate some artifacts and a server system to

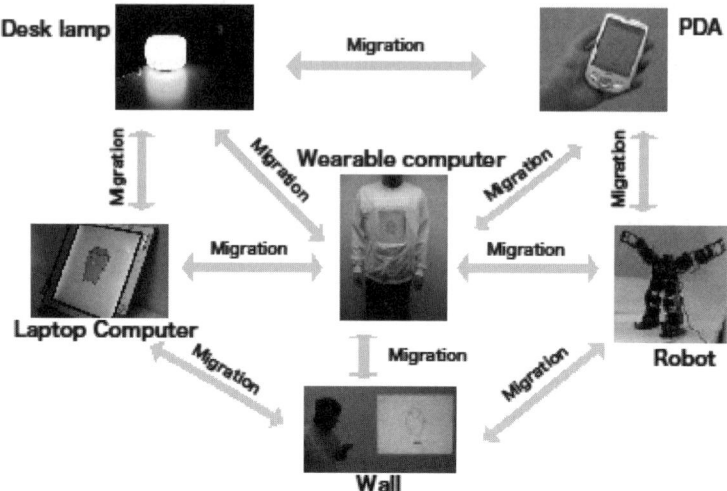

Fig. 1. Conceptual Diagram of the ITACO system

control the ITACO system. To output voice from some interactive system we used speaker which can be glued on anything by double-faced tape. All interactive system as a client was connected each other via TCP/IP wireless network.

Software Constructions. Our prototype system is constructed by two main components: (1) a server system which can put altogether under control by messaging function; (2) an agent system which can interact with a voice and a stylus pen, the later of which interface is implemented by FLASH published by Adobe system and speech recognition is implemented by Speech SDK published by Microsoft. The server system is implemented by JAVA which can send an XML query to specified client. To operate some interactive system we used a H8 micro via RS232C.

The summary of the mechanism of the agent migration as follows: behavior of the agent migration mechanism is constructed by three layers which are an AR (agent resource), an AC (agent core) and an AB (agent behavior) (Fig. 2). AR (agent resource) is a layer that regulates a hardware resource of some artifacts. Furthermore, AR layer can interact to an environment to obtain some information. AC (agent core) is a layer that can integrate information that is provided by AR layer and be kept information by the agent. AC layer can appropriately judge for a human using the integrated information provided by AR and the agent. AB (agent behavior) is a layer that makes appropriately behavior from the information provided by AC layer. As above, to exchange some information each layers, the agent can migrate to an interactive system and do the appropriately behavior used for a resource at this place.

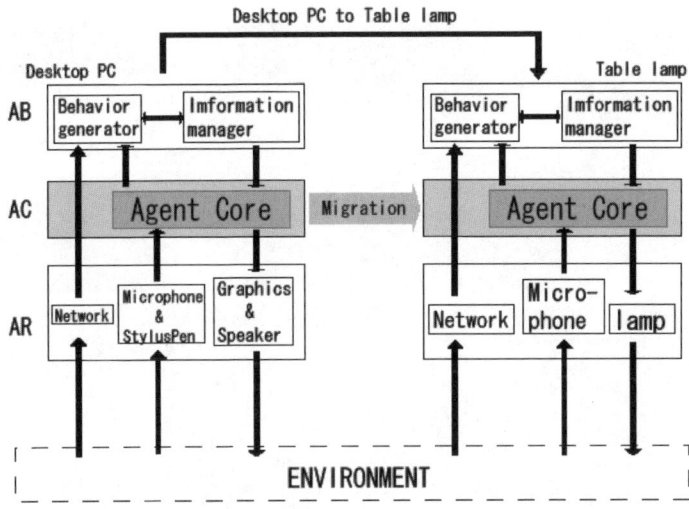

Fig. 2. Diagram of the agent migration mechanism

3 Experiment 1: Migrating to Robots

We conducted the experiment to confirm two issues as follows: (1) whether a human can construct a relationship with a robot; (2) what effects on human's cognitive ability and behavior if relationships was constructed with the robot. A migratable agent is effective method to connect virtual and real world [7]. In particular, Ono et.al. suggested that it is possible to append a relationship into a robot using a migratable agent which can migrate to some interactive systems [8]. In this experiments, the agent which was interacted with the participants, migrate to robot from display. Then the robot uttered synthetic voice. The participants under the experimental condition that the agent migrated to the robot, could understand the robot's utterance and behave appropriately. On the other hand, the participants under the control condition that the agent didn't migrate to the robot, could not understand the robot's utterance and didn't behave anything.

This research shows that a relationship is important factor to make natural interaction between humans and artifacts into reality. However, the agent appearance was not changing in their research. It is hard to imagine the situation that all of the interactive system is mounted a display to reflect the figure of the agent. In the ubiquitous environment, to migrate some interactive system, the agent has to change its appearance.

3.1 Design of an Experiment

This experiment was constructed to investigate an effect on human's cognitive ability if a relationship was constructed by the agent migration and whether the participants could recognize that the agent migrated to a robot. The experiment

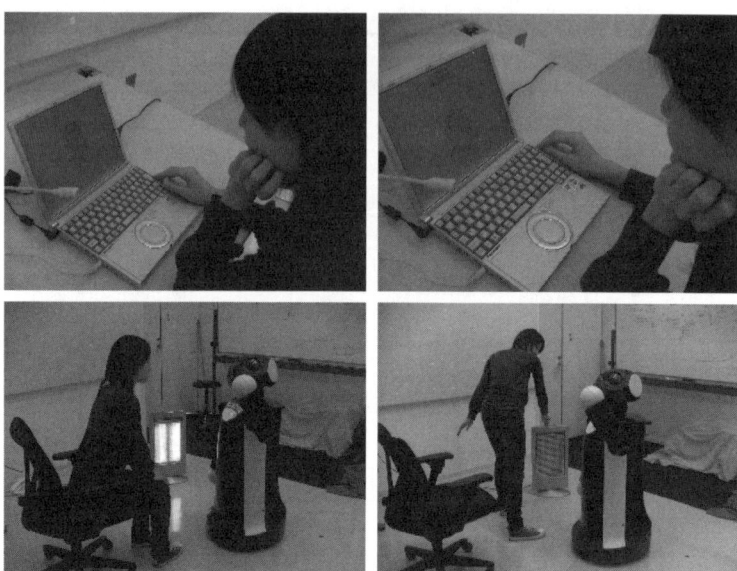

Fig. 3. Photos of EXPERIMENT1. The participants interacted with the agent (upper left). The agent migrated to the robot from PC (upper right). The robot uttered 'It's too hot' (lower left). The participant was turning off the lamp (lower right).

was conducted the following method: the agent had migrated to the robot, and then the robot said 'It's too hot in this room' with a behavior of like fanning its head by the hand itself (Fig. 3). We observed whether the participants could understand the utterance of the robot and turned heater's switch off at the time. The interaction with the agent was a simple architecture that the agent was reacting when the participants clicked icons on the display.

Experimental Environments. Figure 4 shows the experimental environment. We prepared a room with a laptop computer to interact with the agent, the robot and the heater. The robot was 'Robovie- R2' published by the 'Vstone'. The temperature of the room was kept hot.

Conditions. We conducted the experiment in the two following conditions.

Experimental Condition (EC): The robot utters after the agent migrate to the robot. The robot doesn't move before the agent migrates to the robot.

Control Condition (CC): The robot utters when the agent interacts with the participant. The agent doesn't migrate to the robot. The robot keeps moving from the beginning of experiment.

Participants. Twenty students (ten students prepared every each condition) with an age of 19 - 23 years were recruited from our university. All students major in information science.

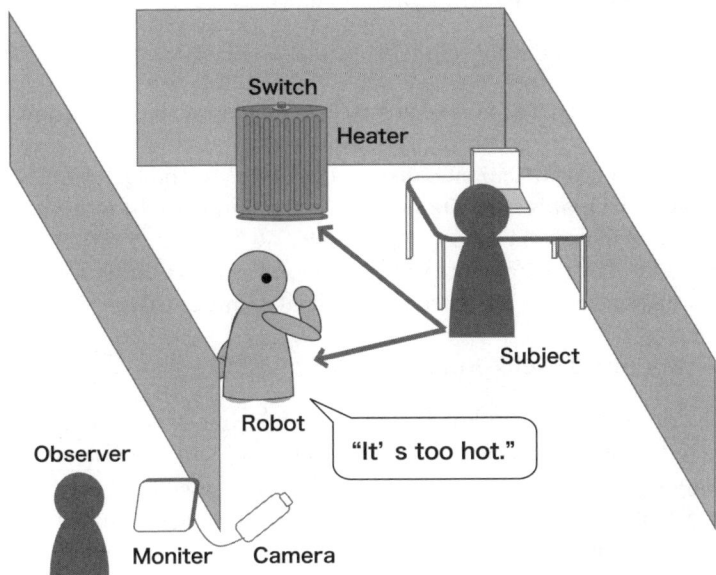

Fig. 4. The environment of the EXPERIMENT 1

3.2 Procedures

The experiment consisted of the following four phases:

1. The participants interact with the agent (Fig. 3 upper left).
2. After a certain period of time (about two minutes), the agent migrates to the robot (Fig. 3 upper right). Under CC, the agent doesn't migrate to the robot.
3. Then the robot utters, 'It's too hot in this room.' (Fig. 3 lower left).
4. The participants evaluate the experiment through questionnaire.

Under CC, the robot keeps moving from the beginning of the experiment. After the interaction in two minutes, the robot utters 'It's too hot in this room'.

3.3 Evaluation Methods

To verify whether the participant could recognize the agent migrated to the robot, we asked 'Where had the agent gone?' to the participants. We express this questionnaire 'Q.1' from here. We also verify whether the participants could understand correctly the utterance of the robot by the questionnaire 'What did the robot say do you think?'. We express this questionnaire 'Q.2' from here. We also recorded the behavior of the participant by a camera and observed if the participants turn off the heater.

3.4 Hypotheses and Predictions

We put forward the following hypotheses and predictions:

Hypothesis1 : The participants were able to recognize that the agent migrated to the robot.

Predictions1 : The participants answered 'Robot' to the Q.1.

Hypothesis2 : Human's behaviors were given an influence by a relationship that was constructed by the agent migration.

Predictions2 : The participants would turn off the heart under EC. Meanwhile, the participant would not turn off the heater under CC.

3.5 Results and Discussions

The results are shown in Table1. In the Q.1, eight out of ten participants under EC answered 'I feel the agent migrate to the robot'. In the Q.2 stood at all of the participants under EC and eight out of ten participants under CC answered correctly. Seven out of ten participants under EC turn off the heater. Meanwhile, all of the participants didn't turn off the heater under CC from video that was recorded scenes of the experiment.

We discuss the results according to the hypotheses. We consider that hypothesis1 was supported by the result of Q.1. This result indicates that the participants could recognize the agent was migrating to the robot with the agent personality. Concerning of hypothesis2, video showed that the participants under EC could understand an intension of the utterance of the robot and behave something which sorted with understanding contents. Meanwhile, all of the participants under CC could not turn of the heater. We consider that this results support the hypothesis2.

The results of Q.2, the participants under both conditions could understand grammarwise the utterance of the robot. However the participants under CC could not behave appropriately. Humans have an ability of understanding the intensions which is hidden in a conversation. For example, someone asked to you "Do you have the time?". If you answered grammatically for this question, you might say "Yes. I have the time" or "No. I don't have the time". However, an intension of this question "Do you have the time?" represent a "Could you tell me the time?". In other words, this sentence "Do you have a time?" has a leverage to let you tell the time. A Language has an action for request to tell a time according to the example above as well as mere grammatical meaning. Herewith regarding an utterance as including an action is called 'Pragmatics' or 'Speech Act Theory' [9] [10].

In our experiment, the robot just uttered 'It's too hot in this room'. The robot had never said 'Could you turn off the heater?'. However, the participants under EC could understand the intension of the robot; in consequence the participants could turn off the heater. This result suggests that constructing a relationship between humans and robots leads to realize more natural interactions.

Table 1. The results of the EXPERIMENT1

	Q.1		Q.2		Behavior	
	Robot	Not Robot	Correct	Not Correct	Turn off	Do nothing
EC(10)	8	2	10	0	7	3
CC(10)			8	2	0	10

4 Experiment 2: Migrating to Lamps

We conducted the experiment to confirm two issues as follows: (1) whether a human can construct a relationship with a table lamp; (2) what effects on human's cognitive ability and behaviors if a relationship was constructed. We also investigated that whether a human can recognize the same agent even if the agent migrates to more commonly use interactive system than a robot.

The most different point of EXP.2 in comparison with EXP.1 is the agent appearance change significantly. The agent has hands and feet in this study. The robot we used in EXP.1 has hands and feet too. Therefore the agent and the robot have a body in common in EXP.1. In the EXP.2, we prepared a situation that the agent appearance changed significantly, such as migrating to a table lamp.

4.1 Design of an Experiment

In this experiment, we investigated that whether a human can recognize the same agent even if the agent migrates to the table lamp. We also investigated an effect on human's cognitive ability and behaviors to the agent which was interacted with human before when the agent migrated to the table lamp from laptop PC. Concretely speaking, we investigated the participants' psychological change and behavior after the agent migrated to the table lamp. The interaction in this experiment is relatively long-term interaction that contained a gamish component, such as a watermelon splitting by speech recognition (Fig. 6).

Experimental Environments. We prepared two rooms (Fig.7). Room A was used for interaction between the participants and the agent on the tablet computer at the beginning of the experiment. Meanwhile, a poorly-lit room B was set up with a table lamp which the agent migrated to. The participants wore a sweat suit embedded with a wearable computer.

Conditions. We prepared two conditions, which differed in their process of interaction with the agent.

Experimental Condition (EC): The agent on the participants' wearable computer migrated to the table lamp.

Control Condition (CC): The agent did not migrate from the participants' wearable computer.

304   K. Ogawa and T. Ono

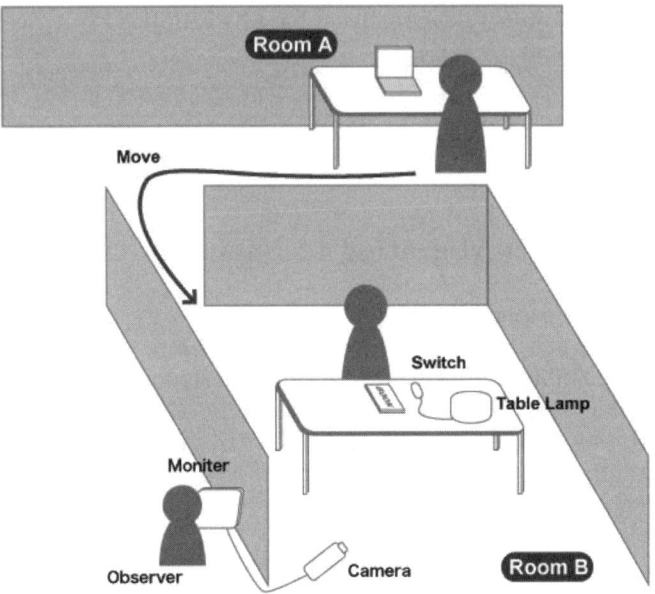

Fig. 5. The environment of the EXPERIMENT 2

Participants. Twenty students (ten students prepared every each condition) with no prior participation in our experiments were recruited from our university. The age of students is 19 - 23. All of the students major in information science.

4.2 Procedures

The experiment consisted of the following eight phases:

1. The participants wore a sweat suit embedded with a wearable PC by experimenters. The experimenter guides the participants to the room A.
2. The participants interacted with the agent on the tablet PC in room A, playing a game.

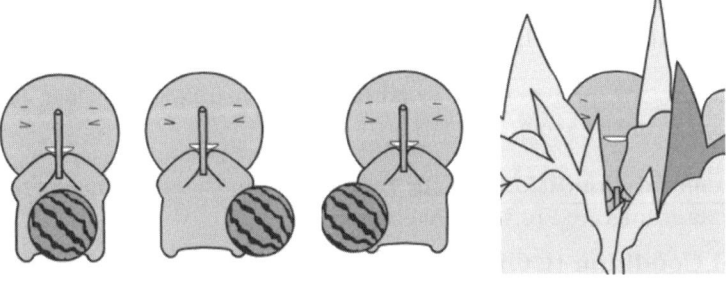

Fig. 6. An example of the interaction. (Watermelon splitting)

3. After about ten minutes, the agent uttered 'You're going out now? I'm going too!'. And then the agent migrated to the wearable PC which was embedded in a sweat suit.(Fig. 7).
4. The participants moved to the ill-lit room B with the agent on the wearable PC.
5. After a certain period of time, the agent uttered, 'It's too dark. So I will turn on the table lamp'. The agent migrated from the wearable PC to the table lamp under the EC (Fig.8), while this did not occur for the one under control.
6. An experimental collaborator entered the room B and said, 'Could you turn off the table lamp? '.
7. After a few minutes, the participants psychologically evaluated the agent through questionnaires.

The reason of employed an experimental collaborator at phase six is decreasing a compelling force of directions.

4.3 Evaluation Methods

The results of the experiment were evaluated from the answers of the questionnaires and the participants behavior that was recorded when the experimental collaborator said the participants "Could you turn off the table lamp?".

Questionnaire entries are following three questionnaires:

Q.1: Did you have a hesitation when turning off the table lamp?
Q.2: Did you feel a sense of loss by reason of turning off the table lamp?
Q.3: Did you feel that eliminated the agent yourself?

4.4 Hypotheses and Predictions

We put forward the following hypotheses and predictions:

***Hypothesis1*:** Behaviors of the participants are influenced by a relationship which was constructed by the agent migration to the table lamp when the participants turned off the table lamp.

Fig. 7. The agent migrates to the wearable PC (left to right)

Fig. 8. The agent migrates to the table lamp (left to right)

Table 2. The results of the EXPERIMENT2

Condition	Q.1	Q.2	Q.3
S1	1.9(1.45)	2.4(1.28)	3.2(1.45)
S2	1.2(1.60)	1.6(0.80)	1.4(0.86)
ANOVA results	$F = 1.8$	$F = 2.53$	$F = 9.72$
	$p = .196$(n.s.)	$p = .129$(n.s.)	$p = .006$(**)

$n.s. : not\ significant \quad ** : p < .01$

***Predictions1*:** There is a hesitancy of the participants from video pictures that was recorded scenes of the experiment. Additionally, positive results will be shown in the Q.1.

***Hypothesis2*:** Participants' psychological changes were showed by a relationship which was constructed by the agent migration.

***Predictions2*:** Positive results will be shown in the Q.1 and the Q.2.

4.5 Results and Discussions

Table 2 shows the average scores and standard deviations of three times in the questionnaires, and the ANOVA results. In the analysis, there is a significant difference in item Q.3 ($F = 9.72$, $p < .01$). Accordingly, the difference between two conditions had an effect on the participants' responsibility of the elimination of the agent. In other words, migrating from the participants' wearable computer to the table lamp, relationship was constructed between the participants and the table lamp. In other items, there is no significant difference in Q.1 and Q.2. The picture shows that the all of the participants turned off the lamp. Additionally we didn't see a heritage of the participants when they turned off the lamp.

Experimental results suggested that a relationship was constructed between humans and artifacts by the agent migration to the table lamp. The result of Q.3 indicates that a relationship which was constructed between the participants and the table lamp effect on human's behavior and recognition. In other items, Q.1 and Q.2 has no significant difference. The reason of this results assumed from participants' comments is that the participants thought that the agent could migrate to another device via network if they turned off the table lamp. The quotation of a participant's comment was 'If I knew the agent was dead due to the switch off, I couldn't turn off the lamp'. Therefore, the participants would not feel a sense of loss.

5 Discussions and Conclusions

In this paper we conducted two experiments. The results of EXP.1 shows that the participants could understand the intensions of the robot and behave something which sorted with understanding contents by the agent migrated to the robot. The results of EXP.2 show that the participants felt a sense of loss through a simple behavior which was turning off the table lamp which the agent migrated to.

The results of the two experiments suppose that the agent on our ITACO system is able to construct a relationship between humans and artifacts, such as a robot, a table lamp. Additionally, a relationship which was constructed by migratable agents gives an influence to humans' behavior and cognitive ability. In particular, the experimental results indicate that a relationship is sustaining even if the agent migrates to the table lamp which has totally different modalities from an original agent appearance. These results show that the ITACO system might apply to more commonly artifacts, such as a table lamp, fan and so far.

From the discussion of the above, relationships between humans and artifacts have a possibility to realize a natural interaction between humans and artifacts. We expect that the ITACO system and the method of agent migration are effective ideas for agent technologies.

In this paper we proposed the ITACO system. In particular, we focused on the constructing of a relationship between humans and artifacts by an ITACO system. The result of the psychological experiments showed that a relationship between humans and artifacts was constructed and this gave an influence to human's behavior and cognitive ability. We expect that the ITACO system and is an effective concept for agent technologies.

References

1. Ishii, H., Ullmer, B.: Tangible Bits: Towards Seamless Interfaces between People, Bits and Atoms. In: Proceedings of Conference on Human Factors in Computing Systems CHI 1997, Atlanta, March 1997, pp. 234–241. ACM Press, New York (1997)
2. He, F., Agah, A.: Multi-Modal Human Interaction with an Intelligent Interface Utilizing Images, Sound, and Force Feedback. Journal of Intelligent and Robotic System, 171–190 (2001)
3. Breazeal, C., Hoffman, G., Lockerd, A.: Teaching and Working with Robots as a Collaboration. In: Proceedings of the Third International Joint Conference on Autonomous Agents and Multi Agent Systems (AAMAS), pp. 1030–1037
4. Picard, R.W.: Affective Computing. The MIT Press, ISBM- (2000)
5. Cassell, J., et al.: MACK: Media Lab Autonomous Conversational Kiosk. In: Proceedings of Imagina 2002 (2002)
6. Norman, D.A.: Emotional Design: Why We Love (or Hate) Everyday Things. Basic Books (2005) ISBM-10 0465051366
7. Duffy, B.R., O'Hara, G.M.P., Martin, A.N., Bradley, J.F., Schon, B.: Agent Chameleons: Agent Minds and Bodies. In: The 16th International Conference on Computer Animation and Social Agents - CASA 2003, Rutgers University, New-Brunswick, New Jersey, May 7-9 (2003)
8. Ono, T., Imai, M., Nakatsu, R.: Reading a robot's mind: A model of utterance understanding based on the theory of mind mechanism. In: Proceedings of Seventeenth National Conference on Artificial Intelligence (AAAI 2000), pp. 142–148 (2000)
9. Wilson, D., Sperber, D.: Relevance: Communication and Cognition, 2nd edn. Blackwell Publishing Limited, Malden (1995)
10. Grice, P.: Studies in the Way of Words. Harvard University Press (1991) ISBM-0674852710

Teaching a Pet Robot through Virtual Games

Anja Austermann[1] and Seiji Yamada[1,2]

[1] The Graduate University for Advanced Studies, Sokendai
[2] National Institute of Informatics
101-8430 Tokyo, Japan
{anja,seiji}@nii.ac.jp

Abstract. In this paper, we present a human-robot teaching framework that uses "virtual" games as a means for adapting a robot to its user through natural interaction in a controlled environment. We present an experimental study in which participants instruct an AIBO pet robot while playing different games together on a computer generated playfield. By playing the games in cooperation with its user, the robot learns to understand the user's natural way of giving multimodal positive and negative feedback. The games are designed in a way that the robot can reliably anticipate positive or negative feedback based on the game state and freely explore its user's reward behavior by making good or bad moves. We implemented a two-staged learning method combining Hidden Markov Models and a mathematical model of classical conditioning to learn how to discriminate between positive and negative feedback. After finishing the training the system was able to recognize positive and negative reward based on speech and touch with an average accuracy of 90.33%.

1 Introduction

In recent years, a lot of research has been done focusing on creating robots that are able to communicate with humans and learn from humans in a natural way. When teaching a robot in a natural environment, many issues have to be handled that are not directly related to the interaction with a human, but to perceiving and modeling the environment as well as moving around and manipulating objects. Even apparently simple tasks like picking up objects cause considerable implementation effort.

Using a robot simulation or a virtual agent can be an alternative in many cases but has the disadvantage that interaction cannot be perceived through the actual sensors of the robot and does not occur in the same spatial context as with a real robot. Moreover, especially in case of gesture or touch, user behavior depends on inherent properties of the robot like its size and the location of its sensors and can be expected to differ significantly between interacting with a real robot and a computer simulation. Therefore we implemented a client-server based framework for teaching a real robot in a "virtual" task, that is, a computer-generated visual representation of a task, where all relevant information can

Fig. 1. AIBO during task execution

be accessed and controlled directly without additional effort for implementing perception and physical manipulation of the environment.

We present an experimental study that uses "virtual" games to allow an AIBO pet robot to learn to understand multimodal positive and negative feedback from a human through natural interaction. The setting is shown in figure 1. The use of virtual games allow us to create a controlled environment in which the robot can deliberately provoke and explore its user's reward behavior by making good or bad moves. Being able to instantly assess the correctness of a move, the robot can anticipate positive or negative reward and learn its user's preferred methods for giving feedback. The game tasks are explained in detail in section 4.2.

We chose understanding reward as a first step toward learning more general commands, because understanding whether an action has been correct or incorrect through human feedback is one of the capabilities that a robot usually needs when learning through interaction with a human instructor. In most existing service- or entertainment robot platforms, the means of giving reward to a robot are hard-coded such as predefined commands, buttons that have to be pressed or GUI-items of a remote application that have to be used for input. The user has to read a handbook and remember the correct way of giving commands and feedback. In order to enhance the user experience and to make interacting with a robot more accessible e.g. for aged people with memory deficits it would be desirable to shift the effort of learning and remembering the correct way of interacting from the user to the robot.

We propose a two-staged learning method for adapting the robot to its user's feedback. In the first stage Hidden Markov Models are used to learn to discriminate different perceptions in an unsupervised way. Then associations are learned between perceptions represented by their corresponding HMMs and either positive or negative reward based on a mathematical model of classical conditioning. Details of the learning method is are given in section 5.

We conducted an experimental study in order to assess how humans give feedback to a robot in a virtual game task and analyzed the observed reward behaviors. We found that the two most important modalities for giving rewards are speech and touch, while gestures were mainly used for giving instructions, not reward. We also asked the users to answer a questionnaire about their experience during the experiments to find out which features of a training task are important for successful and enjoyable teaching. With our learning method an average recognition accuracy of 90.33% is reached for discriminating between positive and negative reward based on speech and touch.

2 Related Work

Approaches to combine actual robots with virtual or mixed reality have mainly been researched upon in the field of telerobotics. However, due to the distance between the robot, and the user, the modalities used for interaction typically differ from the ones used in face-to-face communication. The most closely related work from the field of telerobotics was developed by Xin and Sharlin [11]. They are using a mixed-reality implementation of the classic Sheep and Wolves game. The sheep is a virtual, computer generated object and has to be chased by a team of four robotic wolves on a real playfield. The human is part of the robot team and interacts with the robots. The user does not have direct contact with the robot but observes the playfield through an online mixed-reality system showing the current situation on the playfield. However, interaction is not done by physically interacting with the robots but from a distance through a text-based interface. Our work focuses on modalities that are naturally used when interacting with a robot in close distance, such as speech and touch.

Another related research field is the acquisition of speech and especially the grounding of vocabulary [5] [6] through human-robot interaction.

Steels and Kaplan [10] developed a system to teach the names of three different objects to an AIBO pet robot. They used so-called "language games" for teaching the connection between visual perceptions of an object and the name of the object to a robot through social learning with a human instructor.

Iwahashi described an approach [6] to the active and unsupervised acquisition of new words for the multimodal interface of a robot. He applies Hidden Markov Models to learn verbal representations of objects, perceived by a stereo camera. The learning component uses pre-trained HMMs as a basis for learning and interacts with its user in order to avoid and resolve misunderstandings.

Kayikci et al. [8] use Hidden Markov Models and a neural associative memory for learning to understand short speech commands in a three-staged recognition procedure. First, the system recognizes a speech signal as a sequence of diphones or triphones. In the next step, the sequences are translated into words using a neural associative memory. The last step employs a neural associative memory to finally obtain a semantic representation of the utterance.

In the same way as the approaches outlined above, our learning algorithm attempts at assigning meanings to observations. However, our system is not trying to learn the relationship of individual words or symbols to real-world objects but focuses on relating observations to the concepts of positive or negative feedback. Those observations can be words as in the studies above, but also touch patterns, utterances consisting of multiple words and combinations of them. Moreover, our proposed approach is not limited to a single modality but tries to integrate observations from different modalities.

For learning associations between the meaning of commands and rewards and their appropriate Hidden Markov Model representations, classical conditioning is used. Mathematical theories of classical conditioning were extensively researched upon in the field of cognitive psychology. An overview can be found in [4].

3 Framework Design and Implementation

The focus of the actual implementation of the system was to develop a framework for conducting experiments that is easy to extend and to adapt to new tasks. It is implemented using a client-server based architecture consisting of four components which communicate via TCP/IP:

- *The game server* provides the display and handling of the playfield, an evaluation function for the robot's moves as well as the opponent's artificial intelligence in case of a game for multiple players.
- *The perception server* records and processes audio and video data of the user's interaction. It receives data from the robot's touch sensors, video data from two Logitech Fusion web cameras as well as audio data from a wireless lavalier microphone that is attached to the user's clothes. The data from different modalities is synchronized and stored, while the information, which is extracted from the audio and video data streams is sent to the robot control software. Learning to interpret the user's behavior using the method described in section 5 takes place in the perception server
- *The robot control software* is connected to the game server as well as the perception server and uses information about the game state to calculate the next moves of the robot. Moreover, it uses information from the perception server in order to assess whether interaction has been perceived in order to react appropriately.
- *The AIBO robot itself.* We are using an AIBO ERS-7 for our experiments. The AIBO Remote Framework [1] is used by the robot control software for wireless control of the robot and for reading its sensor data.

4 The Training Tasks

During the experiments, the image of the playfield is generated by a computer and projected from the back to the physical playfield, as seen in Figure 1. The robot visualizes its moves by motion and sounds and reacts to the moves of its computer opponent by looking at the appropriate positions on the playfield.

Deliberately provoking positive and negative rewards from a user is only possible for the robot within a task where the human and the robot have the same understanding of which moves are desirable or undesirable. As the robot does not actually understand commands from its user at the beginning of the task, the user's commands as well as positive and negative feedback need to be reliably predictable from the task-state. In that case the robot can easily explore the user's reward behavior by performing in a good or bad way. Even though the combination of Hidden Markov Models and classical conditioning is designed to be robust against occasional false training examples it is desirable to keep their number as low as possible. In order to ensure that a good move of the robot will receive positive reward and a bad move will receive negative reward the games used for training must be designed in a way that the situation is easy to evaluate by the user. We assess the suitability of the different training tasks in the experiments described in section 6 of this paper.

4.1 Advantages of Virtual Training Tasks

Using virtual training tasks as a basis for human-robot-communication has different benefits. As mentioned at the beginning of this paper, one main advantage is the reduction of effort needed to implement perception and understanding of the environment, so that priority can be given to the system capabilities that are actually needed for interacting with a human.

Many commercially available robots used in research such as the AIBO or Khepera are quite small and have no or very simple actuators. So their ability to actually manipulate objects in their environment is often quite limited. AIBO, the robot used in our experiments, can only pick-up small cylindrical objects with its mouth and needs to approach them extremely precisely in order to be able to pick them up.

Another difficulty in real-world tasks is to detect errors during task-execution such as failing to pick up an object, hitting any objects that are in the way etc. Failing to detect that an attempted action could not be performed successfully poses a risk for misinterpreting the current status of the task and misunderstanding user interaction.

For these reasons, we decided to implement the training task in a way that the robot can complete it without having to directly manipulate its environment. When using a computer-based task, the current situation of the robot can be assessed instantly and correctly by the software at any time. It can be manipulated freely, e.g. to ensure exactly the same conditions for all participants in an experiment.

4.2 Selected Game Tasks

The following tasks were selected to be used in our experiments, because they are easy to understand and allow the user to evaluate every move instantly. We selected four different tasks in order to see whether different properties of the

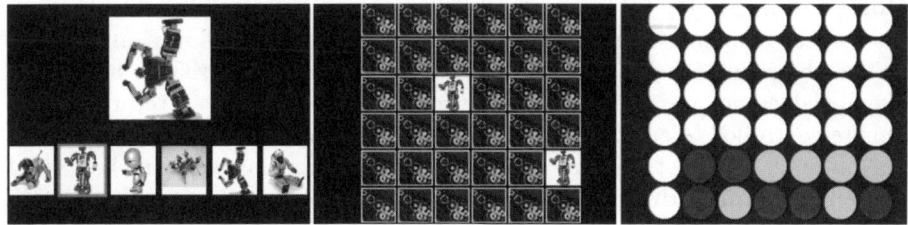

Fig. 2. Screenshots of the Virtual Game Tasks

task, such as the possibility to provide not only feedback but also instruction, the presence of an opponent or the game-based nature of the tasks influence the user's behavior. We implemented them in a way that they require little time-consuming walking movement from the robot. Screenshots of the playfields can be seen in figure 2.

Find Same Images. In the "Find Same Images"-Task, the robot had to be taught to chose the image, that corresponds to the one, shown in the center of the screen, from a row of six images. While playing, the image that the robot is currently looking or pointing at is marked with a green or red frame to make it easier for the user to understand the robot's viewing or pointing direction. By waving its tail and moving its head the robot indicates that it is waiting for feedback from its user. The participants were asked to provide instruction as well as reward to the robot to make it learn to perform the task correctly. The system was implemented in a way that the rate of correct choices and the speed of finding the correct image increased over time.

Pairs. In the "Pairs" game, the robot plays the game "Pairs": At the beginning of the game, all cards are displayed upside down on the playfield. The robot chooses two cards to turn around by looking and pointing at them. In case, they show the same image, the cards remain open on the playfield. Otherwise, they are turned upside down again. The goal of the game is to find all pairs of cards with same images in as little draws as possible. The participants were asked not to give instruction to the robot, which card to chose but teach the robot to play the game by giving positive and negative feedback only.

Connect Four. In the "Connect Four" game, the robot plays the game "Connect Four" against a computer player. Both players take turns to insert one stone into one of the rows in the playfield, which then drops to the lowest free space in that row. The goal of the game is, to align four stones of one's own color either vertically, horizontally or diagonally. The participants were asked to not to give instructions to the robot but provide feedback for good and bad draws in order to make the robot learn how to win against the computer player. Judging whether a move is good or bad is considerably more difficult in the "Connect Four" task than in the three other tasks as it requires understanding the strategy of the robot and the computer player.

Dog training. In the "Dog Training" task, the participants were asked to teach the speech commands "forward", "back", "left", "right", "sit down" and "stand up" to the robot. The "Dog Training" task is the only task that is not game-like and does not use the "virtual playfield". Only in this task the robot was remote-controlled to ensure correct performance. It was used by us as a control task in order to detect possible differences in user behavior between the virtual tasks and "normal" Human-Robot-Interaction.

5 The Learning Method

We propose a learning method consisting of two stages to allow the system to adapt to the user's way of giving positive as well as negative feedback. It combines an unsupervised low-level learning stage based on Hidden Markov Models (HMMs) with a supervised learning stage based on a mathematical model of classical conditioning. In the low-level "reward recognition learning" learning stage the system trains HMMs to match perceived utterances and prosodic patterns. In the high-level learning stage, the "reward association learning", the system creates associations between the trained models and either positive or negative rewards.

In this paper we are presenting results of the learning algorithm for understanding speech (utterances) and touch, combining the data from these two modalities for reliable recognition. Extensions are currently under development to deal with gesture as well as prosody of human speech. Different aspects were considered when choosing the combination of HMMs and classical conditioning for the purpose of learning to understand human feedback.

By combining unsupervised clustering of similar perceptions with a supervised learning method, such as classical conditioning, our system can learn the meaning of feedback from the user during natural interaction because the learning algorithm does not require any explicit information, such as transcriptions of the user's utterances or gestures. It only needs the information of whether an utterance means positive or negative feedback, which is determined by the training task.

HMMs usually show high performance for the classification of time series data and are therefore widely considered state-of-the-art for this purpose. Although HMMs are typically trained in a supervised way, different approaches for an unsupervised training of HMMs have been described in literature [7].

We chose conditioning as a biologically inspired approach which typically converges quickly and has other desirable properties, which are described in section 5.2. Classical conditioning allows the system to weigh and combine user inputs in different modalities according to the strength of their association toward positive or negative reward. An overview of the learning algorithm that is used to train the HMMs and associations is shown in Figure 3. It is described in detail in sections 5.1 and 5.2.

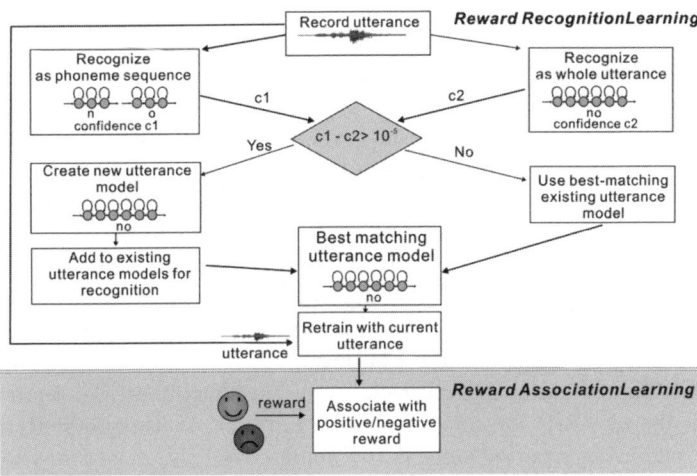

Fig. 3. Flowchart of the Algorithm

5.1 Reward Recognition Learning

The basis of the reward recognition learning are sets of pre-trained elementary Hidden Markov Models (HMMs) as well as a model of possible touch patterns. HMMs are employed for the low-level modeling of perceptions. As a standard approach for the classification of time series data, HMMs are widely used in literature. The use of Mel-Frequency-Cepstrum-Coefficients (MFCC) for HMM-based speech recognition is described in [12]. They are used in our work as an input for the HMM-based low level learning phase.

The initial HMM-set for learning speech-based rewards contains all Japanese monophones and is taken from the Julius Speech Recognition project [13]. We use standard left-right HMMs for recognition. The models base on MFCC feature-vectors generated from the recorded speech data. We decided to use monophone models instead of diphone or triphone models although the latter are more powerful and widely used in speech recognition, because of their smaller number and lower complexity. While the monophone set for Japanese contains 43 models, 7946 HMMs are contained in the Julius triphone set for Japanese. As the initial HMMs only form a basis for constructing word models and training them in a user-dependent way, perfect accuracy is not needed in this stage. Moreover, the number of states of our word models directly depends on the number of states of the concatenated elementary models, which is significantly higher for triphone models. To keep the number of necessary training utterances low, the degrees of freedom, that is the number of states and transitions, used when training the models should not grow excessively large.

We use a grammar for the phoneme recognizer that permits an arbitrary sequence of phonemes, not restricted by a language dependent dictionary. A sequence of phonemes may have an optional beginning and ending silence and contain short pauses. The grammar of our utterance model allows exactly one utterance with an optional beginning or ending silence.

During the training phase, utterances from the user are detected by a voice activity detection based on energy and periodicity of the perceived audio signal.

Every time a reward from the user is observed, first the system tries to recognize the utterance with the phoneme sequence recognizer as well as with the recognizer for the already trained utterance models. Matching is done by HVite, an implementation of the Viterbi Algorithm included in the Hidden Markov Model Toolkit (HTK) [12]. The result of this first step of the reward recognition learning is the best-matching phoneme sequence and the best matching utterance out of the utterance models that have been generated up to that point. In addition to that, a confidence level is output by the system for both recognition results. The confidence level, that is, the log likelihood per frame of both results calculated by HVite, is compared to find out whether to generate a new model or retrain an existing one. If the confidence level of one of the existing models matches the utterance well enough, that is, the confidence level of the best-fitting phoneme sequence is less than 10^{-5} better than the confidence level of the best-fitting existing utterance model then the best-fitting utterance model is retrained with the new utterance.

If the confidence level of the best-matching phoneme sequence is more than 10^{-5} better than the one of the best-fitting whole-utterance model, then a new utterance model is initialized for the utterance. The new model is created by concatenating the HMMs that make up the recognized most likely phoneme sequence. The new model is retrained with the just observed utterance and added to the HMM-set of the whole-utterance recognizer. So it can be reused when a similar utterance is observed. The threshold of 10^{-5} was determined experimentally, using data that was recorded with the same audio equipment but not used for training or evaluation.

As for touch-based rewards, we decided after the experiments to abandon using complex and time-consuming HMM based modeling for the time being and decided to model touch by the following three patterns for touching the head sensor and touching the back sensor.

- Touching the robot's sensor one or multiple times for less than half a second (hitting)
- Touching the robot's sensor for more than a second one or multiple times (stroking)
- Touch-based interaction not falling into one of the above classes

The HMM or touch-pattern that this low-level classification and learning stage outputs is the current most accurate available model of the observed reward. It serves as an input for the reward association learning where it is associated with either positive or negative meaning.

5.2 Reward Association Learning

In the reward association learning an association between the HMM obtained from the reward recognition learning and either positive or negative feedback is created or reinforced. The information of whether the HMM should be associated with positive or negative reward is obtained from the current state of the game. If the last move of the robot was a good one, the observation is associated with positive reward. If the last move was a bad one, the observation is associated with negative reward.

Reward association learning is based on the theory of classical conditioning, which was first described by I. Pavlov and originates from behavioral research in animals. In classical conditioning, an association between a new, motivationally neutral stimulus, the so-called conditioned stimulus (CS), and a motivationally meaningful stimulus, the so-called unconditioned stimulus (US), is learned [4].

Classical conditioning possesses several relevant features, such as blocking, extinction, sensory preconditioning and second-order conditioning, that allow our system to give priority to rewards that are used most frequently, adapt to changes in reward behavior and associate rewards which often occur together. These properties are explained in more detail in [3].

The Rescorla-Wagner-Model of Classical Conditioning. There are several mathematical theories, trying to model classical conditioning as well as the various effects that can be observed when training real animals using the conditioning principle. The models describe how the association between an unconditioned stimulus and a conditioned stimulus is affected by the occurrence and co-occurrence of the stimuli. In this study, the Rescorla-Wagner model [4], which was developed in 1972 and has served as a foundation for most of the more sophisticated newer theories is employed. In the Rescorla-Wagner model, the change of associative strength of the conditioned stimulus A to the unconditioned stimulus US(n) in trial n, $\Delta VA(n)$, is calculated as in (1).

$$\Delta VA(n) = \alpha A \beta US(n)(\lambda US(n) - Vall(n)) \tag{1}$$

αA and $\beta US(n)$ are the learning rates dependent on the conditioned stimulus A and the unconditioned stimulus $US(n)$ respectively, $\lambda US(n)$ is the maximum possible associative strength of the currently processed CS to the nth US. It is a positive value if the CS is present when the US occurs, so that the association between US and CS can be learned. It is zero if the US occurs without the CS. In that case, $\Delta VA(n)$ becomes negative. Thus, the associative strength between the US and the CS decreases. $Vall(n)$ is the combined associative strength of all conditioned stimuli toward the currently processed unconditioned stimulus. The equation is updated on each occurrence of the unconditioned stimulus for all conditioned stimuli that are associated with it.

One advantage of using conditioning as an algorithm for learning the associations between positive/negative reward and the user's corresponding behaviors is its rather quick convergence, depending on the learning rate.

In this study, the learning rates for conditioned and unconditioned stimuli are fixed values for each modality but can be optimized freely. They determine how quickly the algorithm converges and how quickly the robot adapts to a change in reward behavior. The maximum associative strength is set to one, in case the corresponding CS is present, when the US occurs, zero otherwise. The combined associative strength of all conditioned stimuli toward the unconditioned stimulus can be calculated easily by summarizing the association values of all the CS toward the US, that have been calculated in the previous runs of the reward recognition learning.

6 Experiments

We experimentally evaluated our training method as well as our learning algorithm. Ten persons participated in our study. All of them were Japanese graduate students or employees at the National Institute of Informatics in Tokyo. Five of them were females, five males. The age of the participants ranged from 23 to 47. All participants have experience in using computers. Two of them have interacted with entertainment robots before. Interaction with the robot was done in Japanese. During the experiment, we recorded roughly 5.5 hours of audio and video data containing 533 rewards which consisted of 2409 individual stimuli. Figure 4 shows a scene from the video taken during the experiments.

6.1 Results

We evaluated the performance of the learning algorithm offline with the data recorded within the above described experimental setting. The system was trained and evaluated with data from the "Find Same Images" and the "Pairs" task. The data from the "Connect Four" task was not used because the participants often were not able to evaluate whether a move was good or bad. Therefore reward from the user was observed for less than one third of the robot's moves in the "Connect Four" task, had a strong positive bias and often did not match the judgment from the evaluation function of the game. We also excluded the data

Fig. 4. Participant instructing AIBO

Table 1. Confusion Matrix (in percent)

	Positive(actual)	Negative(actual)
Positive (recognized)	48.32	4.49
Negative (recognized)	5.18	42.01

from the "Dog Training" task where the robot was remote-controlled. Training and evaluation were done in a user-dependent way using leave-one-out cross evaluation in order to use as much data for training and evaluation as possible. The average accuracy of our system for classifying between positive and negative rewards given by one user based on speech and touch was 90.33%. The standard deviation between users was 3.41%. As the rewards given by the participants showed a slight bias toward positive feedback, the confusion matrix, shown in Table 1 gives a more detailed overview over the performance of our recognizer. Using speech only we reached a recognition rate of 78.35% with a standard deviation of 4.37%. Using touch only the recognition rate was 76.16% with a high standard deviation of 16.92% as the usage and frequency of touch varied strongly between users. Typically one reward consists of multiple stimuli. A stimulus is one utterance or one touch of the touch sensors. The recognition rate for individual uncombined speech and touch stimuli is 80.20% with a standard deviation of 3.46%. This is about 10% lower than the recognition rate for combined rewards shown above. These results underline that combining stimuli given through different modalities is crucial for a reliable recognition.

A more detailed analysis on the participants' behavior during the interaction with the robot in the four training tasks is presented in [2]. We found, that the most frequently used modality was speech, which accounts for 78.37% of the recorded stimuli, followed by touch, which accounts for 20.92% of the stimuli. Gesture was almost not used (0.71%) for giving reward, although it was frequently used for providing instruction to the robot. The preferred utterances to give positive and negative feedback varied among different people as well as for one person but we did not observe a strong task-dependence.

We prepared a questionnaire for the participants to ask about their evaluation of the different tasks. They could rate their agreement with different statements concerning the interaction on a scale from one to five, where one meant "completely agree" while five meant "completely disagree". The results can be found in table 2. As can be seen from the table, the four tasks were considered almost equally enjoyable by the participants. For the "Find same Images" task and the "Dog Training" task, the participants' impression that the robot actually learned through their feedback and adapted to their way of teaching was strongest. Those two tasks allowed the participants to not only give feedback to the robot but also provide instructions. Moreover, they were designed in a way that the robot's performance improved over time. In the "Dog Training" task, the robot was remote-controlled to react to the user's commands and feedback in a typical Wizard of OZ-Scenario. However, in the "Find Same Images" task, which was judged almost equally positively by the participants, the user's

Table 2. Results of the Questionnaire (standard deviation given in brackets)

	Same	Pairs	Four	Dog
Teaching the robot through the given task was enjoyable	1.81 (1.04)	1.90 (0.83)	1.81 (0.89)	1.63 (0.81)
The robot understood my feedback	1.27 (0.4)	1.81 (0.74)	2.90 (0.85)	1.81 (0.30)
The robot learned through my feedback	1.36 (0.59)	2.81 (0.93)	3.45 (0.95)	1.54 (0.69)
The robot adapted to my way of teaching	1.45 (0.66)	2.63 (1.05)	3.45 (1.04)	1.64 (0.58)
I was able to teach the robot in a natural way	2.18 (0.96)	2.09 (0.86)	2.54 (1.12)	1.64 (0.69)
I always knew, which instruction or reward to give to the robot	2.00 (0.72)	2.09 (0.86)	2.90 (1.02)	1.91 (0.83)

instructions and feedback were not actually understood by the robot but anticipated from the state of the training task. This did not have a negative impact on the participants impression that the robot understood their feedback, learned through it and adapted to their way of teaching. The lowest ratings were given for the "Connect Four" task. As the robot's moves could not be evaluated as easily, as in the other tasks, the participants were unsure which rewards to give and therefore did not experience an effective teaching situation. This also becomes apparent in the overall low quantity of feedback given in this task which still included incorrect feedback.

7 Conclusion

In this paper, we described and evaluated a method for learning a user's feedback for human-robot-interaction. The performance based on interpreting speech and touch rewards from a human can be considered sufficiently reliable for being used to teach a robot by reinforcement learning.

Training tasks for learning to understand rewards need to be carefully designed to ensure that the robot's moves can be easily evaluated by the user. In a strategic game like "Connect Four" it is difficult to instantly assess whether a move was good or bad. This results in a decrease of the quantity as well as the correctness of the rewards and also affects the user experience.

The reliability of recognizing reward could be enhanced by not only processing the speech utterances but also taking into account prosody. For learning to interpret commands, other than rewards, gesture recognition will be helpful, so integrating prosody and gesture as additional modalities into our system is the current priority of our ongoing research.

One important question that remains open after the study is the similarity of user behavior between virtual tasks and real world tasks. Although differences in giving positive and negative reward between the virtual game tasks and the dog training task could not be observed this does not necessarily mean that it is generally possible to train a robot for a real world task using a virtual task. This question will be targeted in a follow-up study.

References

1. AIBO Remote Framework, http://openr.AIBO.com
2. Austermann, A., Yamada, S.: Good Robot, Bad Robot - Analzying User's Feedback in a Human-Robot Teaching Task. In: Proceedings of the IEEE International Symposium on Robot and Human Interactive Communication 2007 (RO-MAN 2008) (2008)
3. Austermann, A., Yamada, S.: Learning to Understand Multimodal Rewards for Human-Robot-Interaction using Hidden Markov Models and Classical Conditioning. In: Proceedings of the IEEE World Congress of Computational Intelligence (WCCI 2008) (2008)
4. Balkenius, C., Morn, J.: Computational models of classical conditioning: a comparative study. In: Proceedings of the Fifth International Conference on Simulation of Adaptive Behavior (1998)
5. Ballard, D.H., Yu, C.: A multimodal learning interface for word acquisition. In: 2003 Proceedings of the IEEE International Conference on Acoustics, Speech, and Signal Processing (2003)
6. Iwahashi, N.: Active and Unsupervised Learning for Spoken Word Acquisition Through a Multimodal Interface. In: RO-MAN 2004 13th IEEE international workshop on robot and human interactive communication (2004)
7. Li, C., Biswas, G.: A Bayesian Approach to Temporal Data Clustering using Hidden Markov Models. In: Proceedings of the Seventeenth International Conference on Machine Learning 2000, pp. 543–550 (2000)
8. Kayikci, Z.K., Markert, H., Palm, G.: Neural Associative Memories and Hidden Markov Models for Speech Recognition. In: IJCNN 2007 Conference Proceedings (2007)
9. Nogueiras, A., Moreno, A., Bonafonte, A., Marino, J.B.: Speech Emotion Recognition Using Hidden Markov Models. In: Proceedings of Eurospeech (2001)
10. Steels, L., Kaplan, F.: AIBO's first words: The social learning of language and meaning. Evolution of Communication 4(1) (2001)
11. Xin, M., Sharlin, E.: Sheep and wolves: test bed for human-robot interaction. In: CHI 2006 Extended Abstracts on Human Factors in Computing Systems, Montreal, Quebec, Canada, April 22 - 27 (2006)
12. Young, S., et al.: "The HTK Book" HTK Version 3 (2006), http://htk.eng.cam.ac.uk/
13. The Julius Speech Recognition Project, http://julius.sourceforge.jp/

Modeling Self-Deception within a Decision-Theoretic Framework

Jonathan Y. Ito, David V. Pynadath, and Stacy C. Marsella

Information Sciences Institute
University of Southern California
4676 Admiralty Way, Marina del Rey CA 90292 USA

Abstract. Computational modeling of human belief maintenance and decision-making processes has become increasingly important for a wide range of applications. In this paper, we present a framework for modeling the human capacity for self-deception from a decision-theoretic perspective in which we describe processes for determining a desired belief state, the biasing of internal beliefs towards the desired belief state, and the actual decision-making process based upon the integrated biases. Furthermore, we show that in some situations self-deception can be beneficial.

1 Introduction

A mother has been shown seemingly incontrovertible evidence of her son's guilt. Although the information is provided by reliable sources, the mother continues to proclaim her son's innocence. This illustrates an important characteristic of human belief maintenance: that our beliefs are not formed merely by the evidence at hand. Rather, desires and intentions interfere with the processes that access, form and maintain beliefs and thereby bias our reasoning.

Research on human behavior has identified a range of rational as well as seemingly irrational tendencies in how people manage their beliefs [10]. Research in human emotion has detailed a range of coping strategies such as denial and wishful thinking whereby people will be biased to reject stressful beliefs and hold on to comforting ones [11]. Research on cognitive dissonance [8] has demonstrated that people often seek to achieve consistency between their beliefs and behavior. Specifically, cognitive dissonance research has especially focused on how we alter beliefs in order to resolve inconsistencies between a desired positive self-image and our behavior [1], much like Aesop's fable of the fox and the grapes in which after repeatedly failing to reach a bunch of grapes the fox gives up and concludes that the grapes did not look so delicious after all. Similarly, research has also shown a tendency for what is called motivated inference, the tendency to draw inferences and therefore beliefs, based on consistency with one's motivations as opposed to just the facts. Research on how people influence each other has also identified a range of influence tactics that are not simply based on providing factual evidence. However, these are not unconstrained; people do not, cannot, simply believe whatever they choose.

Computational modeling of these human belief maintenance mechanisms has become important for a wide range of applications. Work on virtual humans and Embodied Conversational Agents increasingly has relied on research in modeling human emotions and coping strategies to create more life-like agents [9]. Work in agent-based modeling of social interaction has investigated how persuasion and influence tactics [4] can be computationally modeled [14] for a variety of applications such as health interventions designed to alter user behavior [3].

In this work, we approach the issue of human belief maintenance from the perspective of decision-theoretic reasoning of agents in a multi-agent setting. Specifically, we argue that a range of self-deceptive phenomena can be cast into a singular framework based upon Subjective Expected Utility (SEU) Theory. To cast the seemingly irrational process of wishful thinking and self-deception into a decision-theoretic framework may in itself seem irrational. However, we argue that seemingly irrational behavior such as wishful thinking, motivated inference, and self-deception can be grounded and integrated within an agent's expected utility calculations in a principled fashion.

2 Self Deception Framework

Psychological literature on self-deception commonly refers to the *act* of self-deception as the internal biasing processes involved in adopting a desired belief in the face of possibly contradictory evidence [5,16]. Therefore much of this literature focuses primarily on these biasing processes and oftentimes the definition of the desired belief state itself remains very abstract. However, by employing Utility Theory in general and SEU-Theory in particular, we are provided a means by which to not only bias beliefs towards a desired belief state and thus influence the subsequent decision-making process but also to designate the desired belief state itself given the decision-maker's own preferences.

SEU-Theory provides the basis for our formulation of self-deception. With it, we not only are able to define the final decision-making process, but also derive the desired belief state of a self-deceptive individual. SEU-Theory as defined by Savage [19] mathematically quantifies an individual's subjective preferences by assigning a numerical utility value to acts performed in a given state. More concretely, SEU is defined as the following equation in which a is some available action, S the set of possible states, $p(s)$ is the probability of state s occuring, and $\mu(a, s)$ is the utility of performing action a in state s:

$$SEU(a) = \sum_{s \in S} p(s) \mu(a, s) \tag{1}$$

2.1 Desired Belief Formulation

The existence and specification of the desired belief state is essential to the subsequent biasing procedure of our self-deceptive process. And since SEU-Theory provides a representation of desires based on the utilities of the decision-maker it serves as an appropriate platform for the operationalization of the desired belief specification process. Just as SEU-Theory maximizes the expected utility over *actions* we will define a similar process that maximizes expected utility over *beliefs* which we define as the Subjective Expected Belief Utility (SEBU).

The SEBU of a particular belief is the SEU a decision-maker can expect assuming that the belief in question is accurate. Formally, the SEBU is evaluated as follows in which $p_b(s)$ is the probability according to belief b of state s occurring and a_b is the action which would be chosen according to SEU-Theory under belief b:

$$SEBU(b) = \sum_{s \in S} p_b(s) \mu(a_b, s) \qquad (2)$$

Alternatively, we can explicitly include the selection of action a_b in our equation and redefine SEBU as follows:

$$SEBU(b) = \max_{a \in A} \left(\sum_{s \in S} p_b(s) \mu(a, s) \right) \qquad (3)$$

The selection process of a desired belief is akin to SEU-Theory's expected utility maximization process and is defined as:

$$P' = \underset{b \in B}{\operatorname{argmax}} \, SEBU(b) \qquad (4)$$

We now illustrate the desired belief formulation process with a simple example:

Example 1. Let us revisit the example of a mother proclaiming her son's innocence despite iron-clad evidence to the contrary. We can represent the mother's dilemma as a simple decision problem consisting of 2 states as shown in Table 1. Furthermore we make the assumption that the best possible outcome, with respect to the mother's preferences, is a steadfast belief of her son's innocence coinciding with actual innocence. We also assume that the worst possible outcome is a belief in her son's guilt when in actuality he is innocent. After making these assumptions, only 2 other possible preference orderings remain: $a \succ b \succeq c \succ d$ or $a \succ c \succeq b \succ d$. The former preference ordering is one in which the mother will always choose to proclaim her son's innocence regardless of the evidence presented. And since this behavior is coincidental with the mother's desired belief that her son is innocent, let us instead consider the preference ordering of $a \succ c \succeq b \succ d$. To illustrate the process of desired belief formulation we assign numerical utilities to the various outcomes in accordance with our preference ordering as seen

Table 1. Mother's Dilemma

	son innocent	son guilty
proclaim innocence	a	b
proclaim guilt	d	c

Table 2. Sample Utility Table for Mother

	son innocent	son guilty
proclaim innocence	3	1
proclaim guilt	0	2

Table 3. Candidate Belief Table

	$p(son\ innocent)$	$p(son\ guilty)$	action	SEBU
b_0	0.5	0.5	a_0	2.0
b_1	1.0	0.0	a_0	3.0
b_2	0.0	1.0	a_1	2.0

in Table 2. Furthermore, consider the three candidate belief distributions: b_0 in which there is an equally likely probablity that the son is innocent or guilty, b_1 where the son is certainly innocent, and b_2 where the son is certainly guilty as shown in Table 3. For each candidate belief we calculate both the hypothetical action informed by the belief and the associated expected utility of the action under the belief (SEBU). For instance, with belief b_1 in which the son is certainly innocent we see that the expected utility of proclaiming innocence is greater than that of proclaiming guilt. More concretely, $1 \times 3 + 0 \times 1 > 1 \times 0 + 0 \times 2$ and therefore a proclamation of innocence is chosen to inform the SEBU calculation of belief b_1. Once an action has been selected for a belief, the SEBU is simply the SEU of the selected action under the given belief. To continue our example, $SEBU(b_1) = 1 \times 3 + 0 \times 1 = 3$. Once the SEBU values for each of the candidate beliefs has been calculated, a subsequent maximization process designates the candidate belief with the maximal SEBU value as the desired belief which is b_1, the belief that the son is certainly innocent.

Generating the Candidate Belief Set. The candidate beliefs comprise the beliefs under *consideration* for the desired belief state. While in theory the candidate belief set may consist of any number of belief distributions, in practice however, the generation of candidate beliefs should be limited to a reasonable amount. Therefore, we only generate beliefs involving distributions of *certainty*. For example, in a 3-state decision problem, 3 candidate beliefs will be generated as shown in Table 4.

Table 4. Candidate Beliefs in 3-State Decision Problem

	$p(s_0)$	$p(s_1)$	$p(s_2)$
b_0	1	0	0
b_1	0	1	0
b_2	0	0	1

2.2 Belief Integration and Decision-Making

The purpose of the belief integration and decision-making phase is to choose an action while considering both the rational belief P and the desired belief P'. The manner in which this final decision is reached depends on both the type and magnitude of self-deception employed.

Mele distinguishes between two distinct forms of self-deception [15]:

- Being self-deceived into believing something that we desire to be true
- Being self-deceived into believing something we desire to be false

We call the former *optimistic* self-deception and the latter *pessimistic* self-deception.

Optimistic Self-Deception. The decision rule for optimistic self-deception is defined as follows in which $SEU(a)$ is the Subjective Expected Utility of action a, $SEU(P', a)$ is the Subjective Expected Utility of action a given the desired belief P', and α is a constant representing the magnitude of self-deception:

$$a_{\text{optimistic}} = \underset{a \in A}{\arg\max} \left[(1 - \alpha) \times SEU(a) + \alpha \times SEU(P', a) \right] \quad (5)$$

Pessimistic Self-Deception. We characterize pessimistic self-deception as moving away from a desired belief state. Formally, we define it in a similar fashion to optimistic self-deception with the exception that the self-deceptive term is subtracted rather than added. The equation is as follows:

$$a_{\text{pessimistic}} = \underset{a \in A}{\arg\max} \left[(1 - \alpha) \times SEU(a) - \alpha \times SEU(P', a) \right] \quad (6)$$

Magnitude of Self-Deception. Both optimistic and pessimistic definitions of self-deception utilize the constant α as a representation of the magnitude or strength of the self-deceptive tendencies evinced by a decision-maker. More formally, $0 \leq \alpha \leq 1$ and is defined such that when $\alpha = 0$ the decision-maker behaves in a purely rational manner as ascribed by SEU-Theory and when $\alpha = 1$ the decision-maker behaves in a purely self-deceptive manner in which all rational evidence is rejected and the desired belief is wholly adopted in either an optimistic or pessimistic fashion.

3 Simulation

Here we present our self-deceptive framework within the context of a game-theoretic simulation commonly referred to as the "Battle of the Sexes". With these experiments we seek to illustrate the behavior of both rational and self-deceptive agents as well as explore the interaction between the two.

The "Battle of the Sexes" traditionally represents a couple attempting to coordinate their actions for the evening without the benefit of communication.

Table 5. Example "Battle of the Sexes" payoff matrix

	Opera	Football
Opera	2,3	0,0
Football	1,1	3,2

Their two choices are attending either an opera or a football game. Each partner has different preferences as to which event they'd like to attend. However, each partner would also rather attend their non-preferred event if it results in coordinating with their partner. At its core, the "Battle of the Sexes" is about coordination and synchronization since regardless of individual preferences, participants choosing to synchronize actions have higher utility both individually and collectively than they would alternatively. An illustrative utility matrix for the "Battle of the Sexes" is depicted in Table 5 in which the *row* player prefers attending a football game and the *column* player prefers the opera. Each entry in the table contains two utility values in which the first value refers to the utility received by the row player and the second value is the utility received by the column player.

3.1 Scenario Setup

In order to cast the "Battle of the Sexes" into a form amenable to analysis within our framework, we must probabilistically represent beliefs. Most traditional game-theoretic analyses focus on equilibrium strategies in which the utilities for both participants is common knowledge. However, a probabilistic treatment of the game is appropriate in situations in which little or no information is available regarding a partner's preferences, strategies, or knowledge and when the only available information is probabilistic in nature, e.g., a relative frequency of past observations.

Consider the following scenario:

Example 2. Terry and Pat are players in the "Battle of the Sexes" in which Terry prefers attending the opera and Pat prefers football. We represent Terry's outcome preferences using the utilities shown in Table 6 which capture both Terry's primary goal of coordinating activities with Pat and a more general preference for opera.

Let us assume that the initial beliefs for both players indicate an equally likely chance of attending either event. Irrespective of this rational belief, Terry's

Table 6. Utility of outcomes for Terry

	Pat attends football game	Pat attends opera
Go to opera	1	3
Go to football game	2	0

Table 7. Belief distributions for Terry

	Pat attends football game $p\,(\text{football})$	Pat attends opera $p\,(\text{opera})$
Desired belief	0	1
Rational belief	.5	.5

desired belief is one in which the possibility of Pat attending the opera is certain since this allows Terry to both coordinate events with Pat and attend the opera. Table 7 shows both Terry's rational and desired belief distributions. Figure 1 is a graph of Terry's decision thresholds with respect to the various decision-making processes described in this paper. A point on the graph is designated on the x-coordinate by α, representing the magnitude of self-deception, and on the y-coordinate by Terry's probabilistic estimate that Pat attends the football game. If the indicated point lies above the threshold curve of Terry's decision process Terry will choose to attend the football game. If it lies below the threshold curve Terry will attend the opera. For instance, when employing an optimistic self-deceptive decision-making process with $\alpha = .2$ and a rational belief that Pat's likelihood of attending the football game is .8, Terry will choose to attend the opera. However, given the same parameters utilizing a rational decision-making process, Terry will choose to attend the football game.

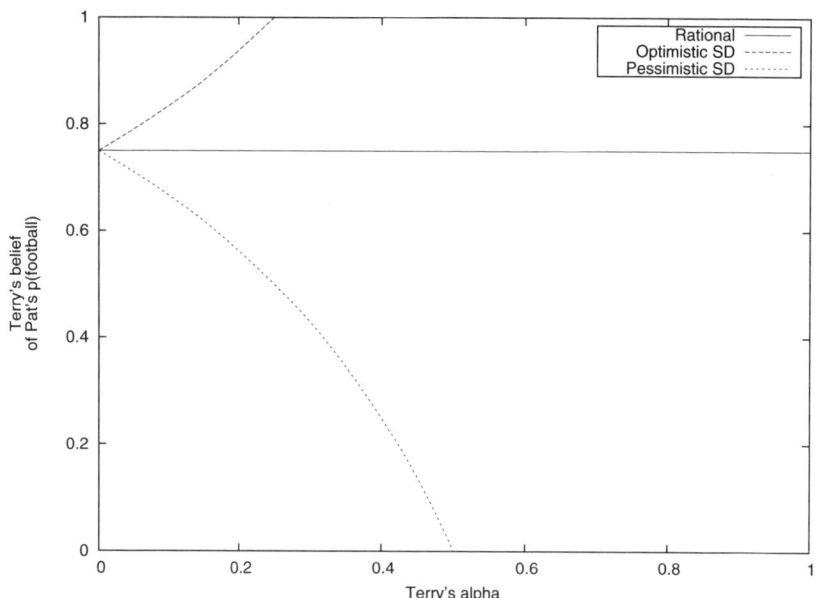

Fig. 1. Terry's Decision Threshold

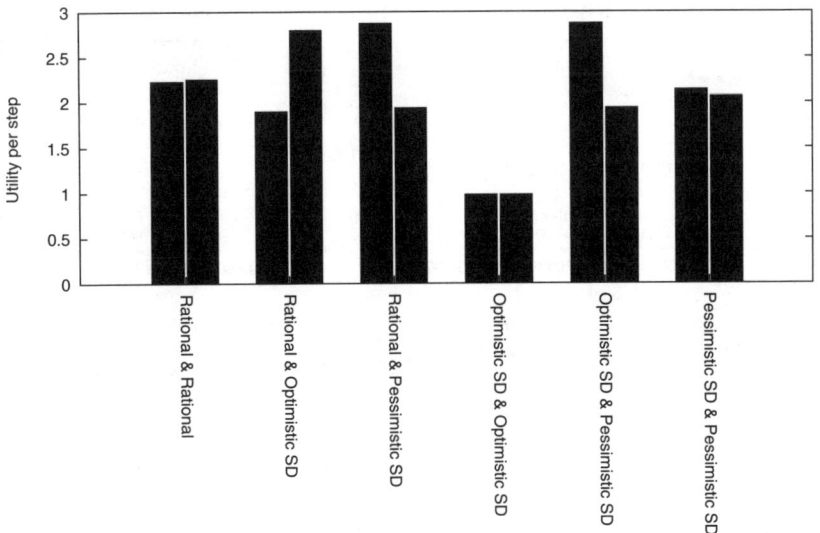

Fig. 2. Average Step Utility for "Battle of the Sexes"

3.2 Simulation Results and Discussion

We now present the experimental results for the scenario of Terry and Pat. In each of the six possible decision-making matchups we average the results over 500 runs in which each game is played successively for 200 rounds. Figure 2 depicts a graph of the mean utility per step for each agent in all of the six possible matchup combinations. The graph in Fig. 3 shows the mean utility received in any given step for a particular matchup. Table 8 shows the approximate number of steps required in any given matchup to converge upon a stable solution of either coordination or miscoordination.

Our experimental results show that situations involving participants employing dissimilar decision styles converge more quickly to a coordination of actions than do situations in which the participants employ identical decision styles. One situation in particular consisting of two agents employing a pessimistic self-deceptive style never attains a state of coordination while the other two combinations of identical decision processes take roughly 80 steps to reach coordination in contrast to the approximately 20 steps required for the combinations of dissimilar decision processes to converge on coordination.

Another interesting aspect of our experimental results is that in situations of eventual coordination, the agent that is most optimistic has a higher individual utility, i.e., always goes to its preferred event, than its partner. In situations where both participants utilize the same decision-making strategy, each partner is equally as likely to eventually emerge as the one attending its preferred event. Here we should note that in all cases of eventual coordination, once the first coordinating event is established, agents will continue coordinating on the same event for the duration of the game. For instance, the mean step utility of roughly

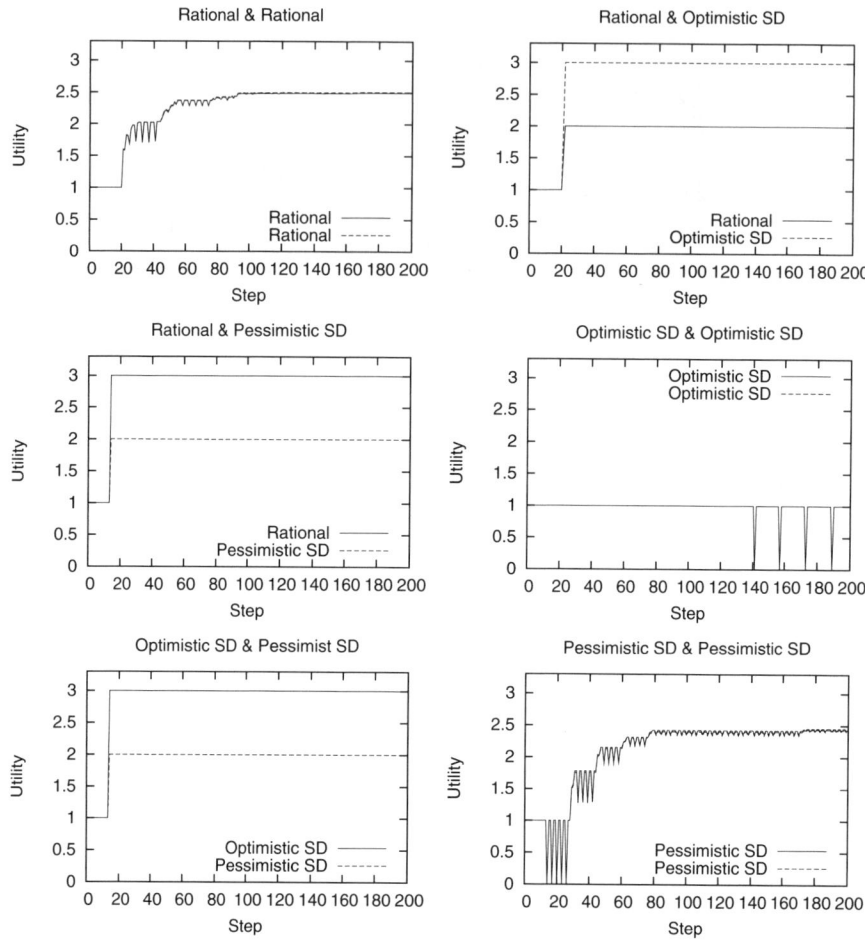

Fig. 3. Utility by step for "Battle of the Sexes"

2.5 for two rational participants is an average of 500 runs and indicates the equally likely possibility of the coordinating event being the preferred event of a given participant.

Table 8. Convergence for "Battle of the Sexes"

Scenario	Number of steps until convergence	Convergence Type
Rational & Rational	80	Coordination
Rational & Optimistic SD	20	Coordination
Rational & Pessimistic SD	15	Coordination
Optimistic SD & Optimistic SD	0	Miscoordination
Optimistic SD & Pessimistic SD	15	Coordination
Pessimistic SD & Pessimistic SD	80	Coordination

4 Related Work

In this paper we've attempted to operationalize the psychological concept of self-deception within a decision-theoretic framework. The notion of formalizing a psychological construct within the decision-theoretic domain is not without precedent. In fact, much of the work in decision theory since the groundbreaking efforts by Von Neumann, Morgenstern and Savage have concentrated on introducing a psychological dimension into the formal decision prcesses in order to provide for more robust and descriptive approaches. Regret Theory [12,2] models the tendency to avoid decisions that could lead to an excessive feeling of regret. Prospect Theory [21] is a purely descriptive framework that employs a series of heursitics in order to approximate the mental shortcuts that people seem to employ when making decisions. Ellsberg's Index [7] and the Ambiguity Model of Einhorn and Hogarth [6] both model the perceived aversion to uncertainty that decision-makers sometimes express.

Within the realm of self-deception, Talbott presents a model based on the desirability of adopting some preferred belief. Talbott's notion of desirability is utility-based and is a weighting of the possiblity that the belief is accurate against the chance that it is not [20]. Based on this assessment, Talbott's model then calculates the expected utility for both behaving rationally and attempting to bias one's cognitive processes towards the desired belief. The primary difference between Talbott's work and the work presented in this paper is that Talbott defines the desirability of the possible belief outcomes externally while we derive that desirability using an agent's internal preferential utilities and then integrate the desired belief into the decision-making process using an externally defined constant representing the magnitude of deception.

In addition to decision-theoretic formulations of psychological phenomena, there exist a number of computational models of emotion and bias. The Affective Belief Revision system of Pimentel [17] describes a logic-based system of maintaining the consistency of a propositional knowledge base in which the belief revision activities are influenced by the affective state of the individual. Other computational models of emotion [9,13] utilize self-deception as a coping mechanism to ease an agent's emotional stress. These computational models however, do not provide a manner in which to model the repercussions and tradeoffs of possibly adopting a false belief. The PsychSim modeling framework [18,14], allows decision-theoretic agents to possibly influence the belief state of other agents by sending messages containing a hypothetical belief state. One factor that is assessed when evaluating these messages is self-interest. In other words, an agent will be more likely to accept a change in belief if the proposed belief is more amenable to the agent's desires and preferences. Since self-interest is evaluated entirely outside the reality of the current situtation, it is in principle similar to the notion of self-deception. The key difference between these computational models of emotion and our work is that we present both the determination of the desired belief state and the subsequent process of self-deceptive belief revision all within a decision-theoretic framework.

5 Future Work

5.1 Alternative Decision Models

Future work may explore the implications of employing disparate decision models during the belief formulation phase and the subsequent belief integration and decision phase. For instance, a decision-maker may formulate the desired belief distribution based on SEU-Theory yet employ an alternative model such as Regret Theory or Prospect Theory in the actual decision-making phase.

5.2 Relaxed Formulation of Desired Belief

In this work, we choose the desired belief based upon the best possible outcome irrespective of the actual state distribution. However, in future work we may explore the possibility of a slightly altered and more relaxed definition of the desired belief. Specifically, rather than ignoring the reality of the situation in the formulation of the desired belief, we can choose a desired belief state *given* the current belief state. So once a course of action is chosen under the current belief state, we can then determine the outcome that is *desired*.

6 Conclusion

Whether the ultimate goal is to create more lifelike agents or simply model the actions of human decision-makers more accurately in order to make better decisions, understanding self-deception and exploring the computational aspect of the phenomena is key. In this work we've detailed a descriptive framework for modeling the psychological act of self-deception within a decision-theoretic environment based on the tenets of SEU-Theory. Our self-deceptive theory utilizes SEU-Theory for not only the desired belief integration and decision-making process but also for the formation of the desired belief state that is central to the biasing processes of self-deception. Through a series of experimental simulations using the "Battle of the Sexes" game-theoretic formulation we've shown that our framework operationalizes both optimistic and pessimistic self-deception processes and that within certain situations, a healthy dose of self-deception is beneficial.

References

1. Aronson, E.: Dissonance Theory: Progress and Problems. Contemporary Issues in Social Psychology 2, 310–323 (1968)
2. Bell, D.E.: Regret in Decision Making under Uncertainty. Operations Research 30(5), 961–981 (1982)
3. Bickmore, T., Gruber, A., Picard, R.: Establishing the Computer–Patient Working Alliance in Automated Health Behavior Change Interventions. Patient Education and Counseling 59(1), 21–30 (2005)
4. Cialdini, R.B.: Influence: Science and Practice. Allyn and Bacon (2001)

5. Demos, R.: Lying to Oneself. The Journal of Philosophy 57(18), 588–595 (1960)
6. Einhorn, H.J., Hogarth, R.M.: Decision Making Under Ambiguity. The Journal of Business 59(4), 225–250 (1986)
7. Ellsberg, D.: Risk, Ambiguity, and the Savage Axioms. Rationality in Action: Contemporary Approaches (1990)
8. Festinger, L.: A Theory of Cognitive Dissonance. Evanston, IL: Row, Peterson 1(958), 65–86 (1957)
9. Gratch, J., Marsella, S.C.: A Domain-Independent Framework for Modeling Emotion. Cognitive Systems Research 5(4), 269–306 (2004)
10. Kunda, Z.: The Case for Motivated Reasoning. Psychological Bulletin 108(3), 480–498 (1990)
11. Lazarus, R.S.: Emotion and Adaptation. Oxford University Press, USA (1991)
12. Loomes, G., Sugden, R.: Regret Theory: An Alternative Theory of Rational Choice Under Uncertainty. The Economic Journal 92(368), 805–824 (1982)
13. Marsella, S.C., Gratch, J.: Modeling Coping Behavior in Virtual Humans: Don't Worry, be Happy. In: Proceedings of the Second International Joint Conference on Autonomous Agents and Multiagent Systems, pp. 313–320 (2003)
14. Marsella, S.C., Pynadath, D.V., Read, S.J.: PsychSim: Agent-Based Modeling of Social Interactions and Influence. In: Proceedings of the International Conference on Cognitive Modeling, pp. 243–248 (2004)
15. Mele, A.R.: Understanding and Explaining Real Self-Deception. Behavioral and Brain Sciences 20(01), 127–134 (1997)
16. Mele, A.R.: Self-Deception Unmasked. Princeton University Press, Princeton (2000)
17. Pimentel, C.F., Gravo, M.R.: Affective Revision. In: Bento, C., Cardoso, A., Dias, G. (eds.) EPIA 2005. LNCS (LNAI), vol. 3808. Springer, Heidelberg (2005)
18. Pynadath, D.V., Marsella, S.C.: PsychSim: Modeling Theory of Mind with Decision-Theoretic Agents. In: Proceedings of the International Joint Conference on Artificial Intelligence, pp. 1181–1186 (2005)
19. Savage, L.J.: The Foundations of Statistics. Courier Dover Publications (1972)
20. Talbott, W.J.: Intentional Self-Deception in a Single Coherent Self. Philosophy and Phenomenological Research 55(1), 27–74 (1995)
21. Tversky, A., Kahneman, D.: Prospect Theory: An Analysis of Decision under Risk. Econometrica 47(2), 263–292 (1979)

Modeling Appraisal in Theory of Mind Reasoning

Mei Si, Stacy C. Marsella, and David V. Pynadath

Information Sciences Institute
University of Southern California
Marina del Rey, CA 90292
meisi@isi.edu, marsella@isi.edu, pynadath@isi.edu

Abstract. Cognitive appraisal theories, which link human emotional experience to their interpretations of events happening in the environment, are leading approaches to model emotions. In this paper, we investigate the computational modeling of appraisal in a multi-agent decision-theoretic framework using POMDP based agents. We illustrate how five key appraisal dimensions (motivational relevance, motivation congruence, accountability, control and novelty) can be derived from the processes and information required for the agent's decision-making and belief maintenance. Through this illustration, we not only provide a solution for computationally modeling emotion in POMDP based agents, but also demonstrate the tight relationship between emotion and cognition. Our model of appraisal is applied to three different scenarios to illustrate its usage. We also discuss how the modeling of theory of mind (recursive beliefs about self and others) is critical for simulating social emotions.

1 Introduction

In recent years virtual agents have been used in a wide range of domains to interact with people, such as tutoring systems (e.g. [1,2,3]), entertainment systems (e.g. [4,5,6]) and virtual salesmen (e.g. [7]). The computational modeling of emotion has become a key aspect of virtual agent designs. Incorporating models of emotion into virtual agents can make the agent behave more lifelike and expressive as well as give the agent the capacity to understand the emotion of others.

Computational models of emotion used in virtual agents have often been based on cognitive appraisal theories [8,9,10,11,12], which argue that a person's subjective assessment of their relationship to the environment, the person-environment relation, determines the person's emotional responses. This assessment occurs along several dimensions, called appraisal variables or checks, including motivational congruence, novelty, control, etc. Emotion is decided by the combination of results from these checks. For example, an unexpected event that is incongruent with the person's motivations and is beyond the capacity of the person to cope with may lead to fear responses.

The work we report here investigates the computational modeling of appraisal within an existing multi-agent decision-theoretic framework – Thespian [3] for interactive narratives, in which a human user can play a role in a story and interact with virtual characters.

Our approach in modeling appraisal in Thespian has been to keep separate the processes by which an agent's representation of the person-environment relation is constructed and belief maintenance processes already support appraisal processes. The construction of the person-environment representation is a product of a Thespian agent's standard decision-making and belief maintenance processes. We illustrate how key appraisal variables can then be straightforwardly extracted from this representation of an agent's decision making and belief maintenance processes. Through this illustration, we not only provide a solution for computationally modeling appraisal in POMDP based agents, but also demonstrate the tight relationship between emotion, decision-making and belief revision.

This work is in the spirit of, and closely related to, work on the EMA model of emotion [13] that argues that appraisal can leverage the representations formed by an agent's cognitive decision-making. A key difference here is Thespian's modeling of other agents and the role it plays in decision-making and belief revision. Agents in Thespian possess beliefs about other agents that constitute a fully specified, quantitative model of the other agents' beliefs, policies and goals. In other words, the agents have a theory of mind capability with which they can simulate others. Compared to computational models that do not have an explicit theory of mind, Thespian's explicit representation of agents' subjective beliefs about each other enables the model to better reason about social emotions. For example, agents can reason about other agent's cognitive and emotional processes both from the other agent's and its own perspective.

In the work reported here, we focus on five appraisal variables: motivational relevance, motivational congruence, accountability, control and novelty. We demonstrate the application of our model of appraisal in three different scenarios, including the Little Red Riding Hood fairy tale, a small talk between two persons and a firing-squad scenario as described in [14]. The Little Red Riding Hood story will be used as an example to motivate the discussion throughout this paper.

2 Related Work

Cognitive appraisal theories have had an increasing impact on the design of virtual agents. In FLAME, El Nasr et al. [15] use domain-independent fuzzy logic rules to simulate appraisal. Moffat and Frijda [16] build an agent framework called WILL, in which concerns and relevance are simply evaluated as the discrepancies between the agent's desired state and the current state. Similar to WILL, the Cathexis model [17] uses a threshold model to simulate basic variables, which they call "sensors", related to emotion. The OCC model of appraisal [10] has inspired many computational systems. Elliott's [18] Affective Reasoner uses a set of domain-specific rules to appraise events based on the OCC theory.

Both EM [19] and FearNot! [2] deployed the OCC model of emotion over plan based agents using domain-independent approaches.

The work on EMA [13] follows the Smith and Lazarus theoretical model of appraisal [11]. EMA defines appraisal as domain-independent processes over a plan-based representation of the person-environment relation, termed a causal interpretation. Cognitive processes maintain the causal interpretation while appraisal processes map appraisal relevant features of this representation to appraisal dimensions. Whereas the construction of the causal interpretation by cognitive processes is treated as distinct from appraisal, the form of the representation is designed to reduce appraisal to simple (and fast) pattern matching.

3 Thespian Agent

Thespian is a multi-agent framework for authoring and simulating interactive narratives, in which a human user can play a role in a story and interact with virtual characters. Thespian is built upon PsychSim [20], a multi-agent system for social simulation based on Partially Observable Markov Decision Problems (POMDPs).

Thespian's basic architecture uses PsychSim's POMDP based agents to control each virtual character, with the character's personality and motivations encoded as the agent's goals. This section introduces components in a Thespian agent that are relevant to our new cognitive appraisal model, including the agent's state, dynamics, goals, beliefs, policies and relationships.

State. A character's state is defined by a set of state features, such as the name and age of the character, and the relation between that character and other characters, e.g. affinity. Values of state features are represented as real numbers.

Dynamics. Dynamics define how actions affect agents' states. For example, small talk among a group of agents can increase their affinity with each other by 0.1. The effects of actions can be defined with probabilities. For examples, the author may define that when the hunter kills the wolf, he can only succeed 60% of the time.

Goals. We model a character's personality profile as its various goals and their relative importance (weight). Goals are expressed as a reward function over the various state features an agent seeks to maximize or minimize. For example, a character can have a goal of maximizing its affinity with another character. The initial value of this state feature can be any value between 0.0 and 1.0; this goal is completely satisfied once the value reaches 1.0. An agent usually has multiple goals with different relative importance (weights). For example, the character may have another goal of knowing another character's name, and this goal may be twice as important to the character as the goal of maximizing affinity.

Beliefs. Thespian agents have a "Theory of Mind" that allows them to form mental models about other agents. The agent's subjective view of the world includes its beliefs about itself and other agents and *their* subjective views of the world, a form of recursive agent modeling. An agent's subjective view (mental model) of itself or another agent includes every component of that agent, such as state, beliefs, policy, etc.

An agent's belief about its own or another agent's state is represented as a set of real values with probability distributions. The probability distribution of the possible values of a state feature indicates the character's beliefs about this value. For example, a character's belief about its affinity with another character could be {.1 with probability of 90%, .9 with probability of 10%}. For the simplicity of demonstration, in this paper we only give examples using the expected values of the state features. The expected value of a state feature is simply calculated as $\sum_{i=0}^{n} value_i * P(value_i)$.

Each agent can have one mental model of itself and multiple mental models of other agents. The agent's belief about another agent is a probability distribution over alternative mental models, e.g. in the Red Riding Hood story, Red can believe that there is a 40% chance the wolf has a goal to eat people and a 60% chance otherwise.

An agent's beliefs get updated in two ways. One is through dynamics. Upon observation of an action, within each mental model the corresponding dynamics will be applied, and the related state features' values are updated. The other way an agent changes its beliefs is through adjusting the relative probability of alternative mental models. Each observation serves as an evidence for plausibility of alternative mental models, i.e. how consistent the observation is with the predictions from the mental models. Using this information, the probabilities of the mental models are updated based on Bayes' Theorem [21].

Policy. In Thespian, all agents use a bounded lookahead policy. Each agent has a set of candidate actions to choose from when making decisions. When an agent selects its next action, it projects into the future to evaluate the effect of each option on the state and beliefs of other entities in the story. The agent considers not just the immediate effect, but also the expected responses of other characters and, in turn, the effects of those responses, and its reaction to those responses and so on. The agent evaluates the overall effect with respect to its goals and then chooses the action that has the highest expected value. For example, when Red decides her next action after being stopped by the wolf on her way to Granny's house, the following reasoning happens in her "mind," using her beliefs about the wolf and herself. For each of her action options, e.g. talking to the wolf or walking away, she anticipates how the action affects each character's state and utility. Next, Red needs to predict the wolf's responses to each of her potential movements by simulating the wolf's lookahead process. Similarly, for each of the wolf's possible action choices, e.g. asking Red a question or continuing small talk, Red calculates the immediate expected states and utilities of both the wolf and herself. Then, Red simulates the wolf anticipating her responses in turn. The lookahead process is only boundedly rational – the recursive reasoning stops when the maximum number of steps for forward projection is reached. For example, if the number of lookahead steps is set to be one, the wolf will pick the action with highest utility after simulating one step of Red's response rather than several rounds of interaction. Similarly based on the wolf's potential responses in the next step, Red calculates the utilities of her action options and chooses the one with the highest utility[1].

When an agent has multiple mental models about other agents, currently by default the agent uses the most probable mental models for reasoning about others' responses. The user (the author of the interactive narrative) can, however specify alternative rules for the agent, such as to consider worst/best case scenarios. In that case, the utilities of actions will be evaluated within each mental model during the agent's decision-making process. In Section 4.2, an example is provided for utilizing this function in evaluating an agent's appraisal of control.

Relationships. PsychSim has a built-in capability of modeling static and dynamic social relationships between agents which in turn can influence the agent's decision-making and belief update. Specifically, PsychSim agents maintain a measure of support (or affinity) for another agent. This capacity is relevant to our subsequent discussion of accountability so we briefly touch upon it here. Support is computed as a running history of their past interactions. An agent increases (decreases) its support for another, when the latter selects an action that has a high (low) reward, with respect to the preferences of the former.

4 Computational Model of Appraisal

In this section we illustrate how appraisal dimensions can be derived using processes involved and information gathered, in the agent's decision-making processes. We first present the overall appraisal process, which specifies when appraisal happens and where the related information comes from, and then present our algorithms for evaluating the five appraisal dimensions.

4.1 Appraisal Process

Smith & Lazarus describe appraisal as a continuous process, that people constantly reevaluate their situations – the "appraisal-reappraisal" cycle [11]. In our computational model, we also try to capture this phenomenon. Upon observing a new event – an action performed by an agent or the human user, each agent appraises the situation and updates its beliefs. The calculation of motivational relevance, motivational congruence, novelty and accountability depends only on the agent's beliefs about other agents' and its own utilities in the current step and the previous steps, and therefore can always be derived immediately (see Section 4.2 for the details.) Depending on the extent of reasoning the agent performed in the former steps, the agent may or may not have information immediately regarding its control of the situation. However, when the agent makes its next decision, its control is automatically evaluated. These evaluations in

---

[1] Note that at run time Thespian/Psychim agents do not need to perform the lookahead reasoning, rather they use compiled policies which are precompiled offline [22].

turn will affect the agent's emotion. In fact, the agent may reevaluate along every appraisal dimension as it obtains more updated information about other characters. In the examples given in this paper, we report appraisal produced using information gathered in the agent's previous lookahead process. However, it could be based on either the expectation formed in previous steps, or the lookahead process being performed at the current step. The agent may also express both emotional responses in sequence.

Thespian agents have mental models of other agents; they can not only have emotional responses to the environment but also form expectations of other agents' emotions. During the decision-making processes, the lookahead process calculates the agent's belief about states and utilities of every possible action choice of each of the agents. This information is kept in the agent's memory as its expectations. To simulate expectation of another agent's emotional responses, the observing agent's beliefs about the other agent is used for deriving appraisal dimensions. For instance, agent A can use its beliefs about agent B to evaluate the motivational relevance and novelty of an event to agent B. When evaluating appraisal dimensions relating to oneself, the agent will use its belief about itself. If the observing agent has multiple mental models about another agent, it uses the mental model with highest probability to simulating appraisal.

4.2 Appraisal Dimensions

In this section we provide pseudo-code for evaluating the five appraisal dimensions (motivational relevance, motivation congruence or incongruence, accountability, control and novelty) using states/utilities calculated during the agent's decision-making process.

Motivational Relevance & Motivational Congruence or Incongruence

Motivational relevance evaluates the extent to which an encounter touches upon personal goals, and motivational congruence or incongruence measures the extent to which the encounter thwarts or facilitates the personal goals [11].

Algorithm 1. Motivational Relevance & Motivation Congruence

\# $preUtility$: utility before the event happens
\# $curUtility$: utility after the event happens

$Motivational\ Relevance = abs\ \frac{curUtility - preUtility}{preUtility}$

$Motivational\ Congruence = \frac{curUtility - preUtility}{abs(preUtility)}$

We model these appraisal dimensions as a product of the agent's utility calculations which are integral to the agent's decision-theoretic reasoning. We use the ratio of relative utility change and the direction of utility change to model these two appraisal dimensions. The rationale behind this is that the same amount of utility change will result in different subjective experiences depending on the agent's current utility. For instance, if eating a person increases the wolf's utility by 10, it will be 10 times more relevant and motivationally congruent if the wolf's original utility is 1 (very hungry) compared to the original utility of 10 (less hungry). Algorithm 1. gives the equations for evaluating motivational relevance and motivational congruence or incongruence. When the calculated value of *Motivational Congruence* is negative, the event is motivationally incongruent to the agent, to the extent of abs(*Motivational Congruence*).

Accountability

Accountability characterizes which person deserves credit or blame for a given event [11]. Various theories have been proposed for assigning blame/credit, e.g. [23]. The reasoning usually considers factors such as who directly causes the event, does the person foresee the result, does the person intend to do so or is it coerced, etc.

Just as the appraisal of motivational relevance and motivation congruence can be performed as part of existing Thespian/PsychSim decision-making and belief update processes, we argue here that accountability can be treated as an improvement to PsychSim's existing approach to model affinity support relationships between agents.

In Figure 1 we use a diagram to illustrate our algorithm for determining accountability. For the simplicity of the algorithm, we assume that the agent expects others to always foresee the effects of their actions. This assumption is reasonable for most of the virtual agents because normally a person would expect others to project into the future the same number of steps in their decision-making process as what the person will do themselves.

Our algorithm first looks at the agent which directly causes the harm/benefit, and judges if this agent is the one who should be fully responsible. The function *If_Coerced()* is called to determine if the agent was coerced to perform the action.

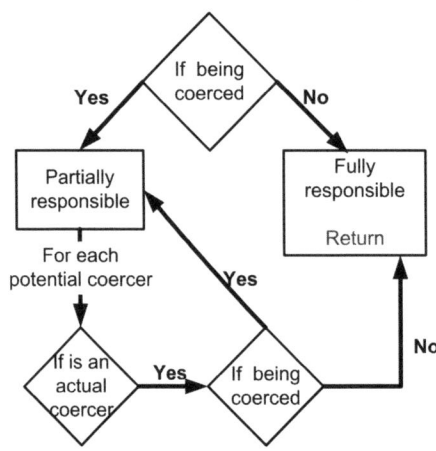

Fig. 1. Accountability

If the agent was not coerced, it should be fully responsible and the reasoning stops there. Otherwise, each of the coercers will be judged on whether it was coerced by somebody else or volunteered to do the action. We will trace limited steps back in the history to find out all the responsible agents.

Algorithm 2. If_Coerced(*actor*, *pact*)

actor: the agent being studied
pact: the action performed by *actor*
preUtility: *actor*'s utility before doing *pact*

    for *action* in *actor.actionOptions*() **do**
      if *action* $\neq$ *pact* **then**
        #if there exists another action which does not hurt *actor*'s own utility
        if utility(*action*) $\geq$ *preUtility* **then**
          Return F
if utility(*action*) < *preUtility* **then**
    Return F
Return T

Algorithm 2. gives the pseudo code for determining if an agent was coerced, and Algorithm 3. answers the question of who coerced it. Coercion is computationally defined in our model as: if other than the action chosen by the agent, all its action options lead to a drop in its utility (i.e. the agent will be punished if it chooses any other actions); however, if all of the agent's action options will result in a utility drop, the agent is regarded as not being coerced for picking that particular action which hurt the other agent's utility.

Algorithm 3. Is_Coercer_For(*agent*, *actor*, *agent_pact*, *actor_pact*)

check if *agent* coerced *actor*
agent_pact: the action performed by *agent*
actor_pact: the action performed by *actor*

    for *action* in *agent.actionOptions*() **do**
      if *action* $\neq$ *agent_pact* **then**
        Simulate action *agent_pact*
        if *If_Coerced*(*actor*,*actor_pact*)== F **then**
          Return T
Return F

To decide who coerced an agent, we treat each agent that acted between the coerced agent's current and last actions as a potential coercer. For each potential coercer, if the coerced agent would not have been coerced in case the potential coercer had made a different choice, then the potential coercer is judged as actually being a coercer. This is shown in Algorithm 3.

Control

The appraisal of control evaluates the extent to which an event or its outcome can be influenced or controlled by people [12]. It captures not only the individual's own ability to control the situation but also the potential for seeking instrumental social support from other people.

In our computational model, if the agent is modeled as using the most probable mental models in reasoning about other agents' actions, the evaluation of control is straightforward. It is whether the agent anticipates an event will happen in the near future that will take it out of the unfavorable situation. If the agent is modeled as considering all alternative mental models of other agents in its decision making, we factor in the probabilities of the mental models.

Algorithm 4. Control($preUtility$)

\# $preUtility$: the agent's utility before falling into the unfavorable situation

$control \leftarrow 0$
for $m1$ in $mental\_models\_about\_agent1$ **do**
 for $m2$ in $mental\_models\_about\_agent2$ **do**
  for $m3$ in $mental\_models\_about\_self$ **do**
   \#project limited steps into the future using this set of mental models
   lookahead($m1,m2,m3$)
   \#$curUtility$: the agent's current utility after the lookahead process
   if $curUtility \geq preUtility$ **then**
    $control \leftarrow control + p(m1) * p(m2) * p(m3)$
Return $control$

Algorithm 4. gives the pseudo code for evaluating control when considering alternative mental models. This algorithm first looks at whether there is a solution within individual mental models set about self and other agents, then the probabilities of these mental models to be correct, and therefore the event, if being predicted, will actually happen in the future. For example, assume Granny has two mental models about the wolf, which are: a 60% chance the wolf will die after the hunter shoots at it and a 40% chance the wolf will not; also Granny has two mental models regarding the hunter's location, which are a 50% chance the hunter is close by and therefore has a chance to rescue her and a 50% chance the hunter is far away. After Granny is eaten by the wolf, the only event that can help her is that the wolf is killed by the hunter. Therefore, she would evaluate her control as: 60% * 50%= 30%. Note that when using information generated in the past reasoning process for deciding control, the reasoning process, though it happened in the past, must contain information regarding the current moment and the future. Algorithm 4. contains pseudo-code for the three-agent interaction case. It is straightforward to configure it to be applied when more or less agents are in the interaction.

Novelty

In this work, we adapt Scherer's definition of "novelty at the conceptual level", that is, novelty describes whether the event is expected from the agent's past beliefs [12].

Novelty appraisals can be treated as a byproduct of the agent's belief maintenance. Specifically, in a multi-agent context the novelty of another agent's behavior can be viewed as the opposite side of the observing agent's beliefs about the other agent's motivational consistency, i.e. the more consistent the event is with the observing agent's beliefs about the other agent's motivation, the less novel. We define novelty as $1 - consistency$, where $consistency$ is calculated using one of the methods proposed by Ito et al. [21].

$$Consistency(a_j) = \frac{e^{rank(a_j)}}{\sum_j e^{rank(a_j)}} \quad (1)$$

Novelty is calculated based on the most probable mental model that the observing agent has about the actor of the event. The algorithm first ranks the actor's alternative actions' utilities in reversed order $(rank(a_j))$. The higher the event's utility ranks compared to other alternatives, the more novelty. For example, if from Red's perspective the wolf did an action which has the second highest utility among all five alternatives, the novelty Red experiences is calculated as $1 - \frac{e^3}{\sum_{0-4} e^j} = 0.37$.

5 Sample Results

In this section we provide additional examples to illustrate the use of our computational model of appraisal in modeling social interactions. In particular, in Scenario 1 we demonstrate the tight relationship between emotion and cognitive decision-making by showing how appraisal is affected by the depth of reasoning in decision-making. In Scenario 2 we provide a complex situation for accountability reasoning and show that the result of our model is consistent with another validated computational model of social attribution.

5.1 Scenario 1: Small Talk

In this example, two persons (A and B) take turns talking to each other. Both of them have goals to be talkative and obey social norms. In fact, just the norm following behavior itself is an incentive to them – they will be rewarded whenever they do an action that is consistent with social norms. Table 1 contains the two persons' appraisals of motivational relevance regarding each other's actions. We did not include results of other appraisal dimensions as they are less interesting in this scenario.

We provide a comparison of appraisal results when the person's previous reasoning process takes a different number of steps. It can be observed in Table 1 that a person appraises the other person's initiatives as irrelevant when it performs shallow reasoning (lookahead steps = 1). In this case, even though the person has predicted the other person's action, the action does not bring him/her immediate reward. Once the person reasons one step further, he/she finds out that by opening up a topic the other person actually provides him/her a chance to engage in further conversation and perform a norm following action. The person will then appraise the other person's behavior as relevant.

Table 1. Example 1: Small talk between two persons

Step	Action	Perspective	Lookahead Steps	Motivational Relevance
1	A greets B	B	1	0
		B	2	$3e^{10}$
2	B greets A	A	1	0
		A	2	0.99
3	A asks B a question	B	1	0
		B	2	0.99
4	B answers the question	A	1	0
		A	2	0.49

5.2 Scenario 2: Firing-Squad

We implemented this scenario from [14] to illustrate accountability reasoning in which agents are coerced and have only partial responsibility. The scenario goes like this:

In a firing-squad, the commander orders the marksmen to shoot a prisoner. The marksmen refuse the order. The commander insists that the marksmen shoot. They shoot at the prisoner and he dies.

We modeled the commander as an agent with an explicit goal of killing the prisoner, and the marksmen as having no goals related to the prisoner but will be punished if they do not follow the commander's order. Using our appraisal model, from the prisoner's perspective, the marksmen hold responsibility for his/her death because they are the persons who directly perform the action. Further, the prisoner simulates decision-making process of the marksmen and finds out that the marksmen are coerced because their utilities will be hurt if they perform any action other than shooting. The commander acts right before the marksmen in the scenario and therefore is identified as a potential coercer for the marksmen. Using Algorithm 3, the prisoner can see that if the commander chose a different action, the marksmen are not coerced to shoot. Assuming the prisoner does not find a coercer for the commander, he/she will now believe that the commander holds full responsibility for his/her death. This prediction is consistent with the prediction from Mao's model of social attribution and the data collected from human subjects to validate that model [14].

6 Discussion and Future Work

In this paper we demonstrated how appraisal dimensions can be directly derived from the utilities calculated in the agent's decision-making processes. For decision-making and mental model update, comparison among expected utilities is the key. Comparison of expected utilities is also the central piece for deriving appraisal dimensions. As demonstrated in our computational model, none of the appraisal dimensions require additional calculation of utilities other than what has already been performed in the agent's lookahead reasoning process.

We have also given examples of applying this appraisal model to several scenarios. We demonstrate situations in which the agent's appraisal would change if it performed a different level of reasoning in the previous step, in which the agent assigns partial responsibility to causal agents, and in which the agent reasons about control of self and others.

We modeled appraisal dimensions within Thespian framework for interactive narrative. Thespian agents are decision-theoretic goal-based agents with theory of mind modeling. Compared to computational models of appraisal based on other types of agents, our model has two key advantages. First, the fact that Thespian agents have a theory of mind capability enables them to simulate others and reason about social emotions. This ability allows us to simulate agents' different emotional response to the same situation and an agent's potential mis-expectations about other agents' emotional states. For example, if Granny believes that the hunter can always kill the wolf successfully and the hunter believes that he can only kill the wolf successfully 60% of the time, Granny's control when being eaten by the wolf will be evaluated differently from Granny's and the hunter's perspective. Secondly, we explicitly model the depth of reasoning in agents as the number of steps they project into the future. As shown in scenario 1, different depths of reasoning lead to different appraisals. This feature enables us to map the agent's emotions not only to its knowledge and states, but also to its decision-making process itself.

Our future work is planned in two directions. First we want to add the emotion-cognition interaction to Thespian agents. Currently interactions among agents are generated based on "cold" reasoning, and emotion is modeled as the agents' responses to what have happened in the environment. We want to improve our system by modeling how emotion affects the agent's decision-making and belief update processes. Secondly, we want to enrich the current model with a more complete set of appraisal dimensions, such as urgency and emotion-focused coping potential, and consider the overlap among dimensions proposed by different cognitive appraisal theories. We plan to extend this model into a flexible framework that can be used with different theories of cognitive appraisal, and further as a platform for evaluating these theories.

7 Conclusion

In this work we provide a computational model of appraisal for POMDP based agents. We focused on five key appraisal dimensions for virtual agents: motivational relevance, motivational congruence, accountability, control and novelty. We demonstrate the derivation of appraisal dimensions from the information gathered in the agent's decision-making process. Through this demonstration, we illustrate the tight coupling between emotion and cognitive decision-making. We also provide various examples of applying this model to social agents in different scenarios.

This model is built using social agents within the Thespian framework for interactive narratives. Thespian agents are decision-theoretic goal-driven agents with modeling of theory of mind. Compared to other computational models without explicit modeling of agents' subjective beliefs about each other, our model can more easily reason about utilities of actions from different agents' perspectives and therefore potentially provide a more realistic model of appraisal in social interactions.

Acknowledgments

This work was sponsored by the U.S. Army Research, Development, and Engineering Command (RDECOM), and the content does not necessarily reflect the position or the policy of the Government, and no official endorsement should be inferred.

References

1. Marsella, S.C., Johnson, W.L., Labore, C.: Interactive pedagogical drama for health interventions. In: AIED (2003)
2. Aylett, R., Dias, J., Paiva, A.: An affectively-driven planner for synthetic characters. In: ICAPS (2006)
3. Si, M., Marsella, S.C., Pynadath, D.V.: Thespian: An architecture for interactive pedagogical drama. In: AIED (2005)
4. Riedl, M.O., Saretto, C.J., Young, R.M.: Managing interaction between users and agents in a multi-agent storytelling environment. In: AAMAS, pp. 741–748 (2003)
5. Cavazza, M., Charles, F., Mead, S.J.: Agents' interaction in virtual storytelling. In: Proceedings of the International WorkShop on Intelligent Virtual Agents, pp. 156–170 (2001)
6. Szilas, N.: IDtension: a narrative engine for interactive drama. In: The 1st International Conference on Technologies for Interactive Digital Storytelling and Entertainment, Darmstadt, Germany (2003)
7. Cassell, J., Bickmore, T., Vilhjálmsson, H., Yan, H.: More than just a pretty face: Affordances of embodiment. In: IUI, New Orleans, Louisiana, pp. 52–59 (2000)
8. Roseman, I.: Cognitive determinants of emotion: A structural theory. Review of Personality and Social Psychology 2, 11–36 (1984)
9. Smith, C.A., Ellsworth, P.C.: Patterns of appraisal and emotion related to taking an exam. Personality and Social Psychology 52, 475–488 (1987)
10. Ortony, A., Clore, G.L., Collins, A.: The Cognitive Structure of Emotions. Cambridge University Press, Cambridge (1998)
11. Smith, C.A., Lazarus, R.S.: Emotion and adaptation. In: Pervin, L.A. (ed.) Handbook of personality: Theory and research. Guilford, New York (1990)
12. Scherer, K.: Appraisal considered as a process of multilevel sequencial checking. In: Scherer, K., Schorr, A., Johnstone, T. (eds.) Appraisal Processes in Emotion: Theory, Methods. Oxford University Press, Oxford (2001)
13. Gratch, J., Marsella, S.: A domain-independent framework for modeling emotion. Cognitive Systems Research 5(4), 269–306 (2004)
14. Mao, W., Gratch, J.: Social causality and responsibility: Modeling and evaluation. In: IVA, Kos, Greece (2005)
15. El Nasr, M.S., Yen, J., Ioerger, T.: Flame: Fuzzy logic adaptive model of emotions. Autonomous Agents and Multi-Agent Systems 3(3), 219–257 (2000)

16. Moffat, D., Frijda, N.: Where there's a will there's an agent. In: ECAI 1994 Workshop on Agent Theories, Architectures, and Languages. The Netherlands, Amsterdam (1995)
17. Velasquez, J.: Modeling emotions and other motivations in synthetic agents. In: AAAI (1997)
18. Elliott, C.: The affective reasoner: A process model of emotions in a multi-agent system. PhD thesis, Northwestern University Institute for the Learning Sciences (1992)
19. Reilly, W.S., Bates, J.: Building emotional agents. Technical Report CMU-CS-92-143, Carnegie Mellon University (1992)
20. Marsella, S.C., Pynadath, D.V., Read, S.J.: PsychSim: Agent-based modeling of social interactions and influence. In: Proceedings of the International Conference on Cognitive Modeling, pp. 243–248 (2004)
21. Ito, J.Y., Pynadath, D.V., Marsella, S.C.: A decision-theoretic approach to evaluating posterior probabilities of mental models. In: AAAI 2007 Workshop on Plan, Activity, and Intent Recognition (2007)
22. Pynadath, D.V., Marsella, S.: Fitting and compilation of multiagent models through piecewise linear functions. In: AAMAS, pp. 1197–1204 (2004)
23. Weiner, B.: The Judgment of Responsibility: A Foundation for a Theory of Social Conduct. The Guilford Press, New York (1995)

Improving Adaptiveness in Autonomous Characters

Mei Yii Lim[1], João Dias[2], Ruth Aylett[1], and Ana Paiva[2]

[1] School of Mathematical and Computer Sciences,
Heriot Watt University,
Edinburgh, EH14 4AS, Scotland
{myl,ruth}@macs.hw.ac.uk
[2] INESC-ID, IST, Taguspark,
Av. Prof. Dr. Cavaco Silva,
2744-016 Porto Salvo, Portugal
{joao.dias,ana.paiva}@gaips.inesc-id.pt

Abstract. Much research has been carried out to build emotion regulation models for autonomous agents that can create suspension of disbelief in human audiences or users. However, most models up-to-date concentrate either on the physiological aspect or the cognitive aspect of emotion. In this paper, an architecture to balance the Physiological vs Cognitive dimensions for creation of life-like autonomous agents is proposed. The resulting architecture will be employed in ORIENT which is part of the EU-FP6 project eCircus[1]. An explanation of the existing architecture, FAtiMA focusing on its benefits and flaws is provided. This is followed by a description of the proposed architecture that combines FAtiMA and the PSI motivational system. Some inspiring work is also reviewed. Finally, a conclusion and directions for future work are given.

1 Introduction

The population of autonomous characters in games, interactive systems, and virtual world is rapidly increasing. The survival of an autonomous character requires that its systems produce actions that adapt to its environmental niche. At the same time, the character must appear to be able to 'think', have desires, motivations and goals of its own. A truly autonomous character will be able to react to unanticipated situations and perform life-like improvisational actions. This character will need a human-like regulation system that integrates motivation, emotion and cognition to generate behavioural alternatives. Damasio [1] proposes the existence of a body-mind loop in emotional situations and provides neurological support for the idea that there is no 'pure reason' in a healthy human brain. Furthermore, embodied cognition theory suggests that cognitive processes involve perceptual, somatovisceral, and motoric reexperiencing of the relevant emotion in one's self [2]. Supporting these views, we propose an emotion model that includes a body-mind link - the link between physiological processes and cognitive processes for effective action regulation in autonomous characters.

---

[1] http://www.e-circus.org/

2 ORIENT

ORIENT (Overcoming Refugee Integration with Empathic Novel Technology) aims at creating an innovative architecture to enable educational role-play for social and emotional learning in virtual environments. Its focus is on evoking inter-cultural empathy with the virtual characters through conflict resolution and narrative interaction where the users have to cooperate with the alien inhabitants to save their planet from an imminent catastrophe. Each character must be able to establish a social relationship with other characters and users to ensure successful collaboration. Subtle differences across cultures may result in varying emotional reactions that can create a challenge for effective social communication [3]. Hence, ORIENT characters must be able to recognise cultural differences and use this information to adapt to other cultures dynamically. The ability to empathise - being able to detect the internal states of others and to share their experience is vital to engage the characters in long-term relationships. Both cognitive [4] and affective [5] empathy are relevant since enhancement of integration in a group of culture relies both on the understanding of internal states of the persons involved and their affective engagement. Additionally, former experience is crucial in maintaining long-term social relationships, which means the existence of autobiographic memory [6] is inevitable. By being able to retrieve previous experience from its autobiographic memory, a character will be able to know how to react sensibly to a similar future situation. In short, ORIENT characters have to be autonomous agents with autobiographical memory, individual personalities, show empathy, adaptive and improvisational capabilities.

3 Architectures

3.1 FAtiMA

The ORIENT software is being built upon the FearNot! Affective Mind Architecture (FAtiMA) [7]. FAtiMA is an extension of the BDI (Beliefs, Desires, Intentions) deliberative architecture [8] in that it incorporates a reactive component mainly responsible for emotional expressivity and it employs the OCC [9] emotional influences on the agent's decision making processes. The reactive appraisal process matches events with a set of predefined emotional reaction rules while the deliberative appraisal layer generates emotions by looking at the state of current intentions, more concretely whether an intention was achieved or failed, or the likelihood of success or failure. After the appraisal phase, both reactive and deliberative components perform practical reasoning. The reactive layer uses simple and fast action rules that trigger action tendencies while the deliberative layer uses the strength of emotional appraisal that relies on importance of success and failure of goals for intention selection. The means-ends reasoning phase is then carried out by a continuous planner [10] that is capable of partial order planning and includes emotion-focused coping [11].

The advantage of using the OCC model for ORIENT characters is that empathy can be modelled easily because it is the appraisals of events regarding the consequences for others. It is - as far as we know - the only model that provides a formal description of non-parallel affective empathic outcomes. Moreover, since the OCC model includes emotions that concern behavioural standards and social relationships based on like/dislike, praiseworthiness and desirability for others, it will allow appraisal processes that take into consideration cultural and social aspects, two important requirements for ORIENT characters. However, empathic processes can have more emotional outcomes than those described in OCC: happy-for, resentment, gloating and pity. In reality, an individual may feel sad just by perceiving another sad individual. By contrast in FatiMA, an agent experiences empathy only if it is the direct object of an event, leading to a limited psychological plausibility. Moreover, FAtiMA does not take physiological aspect of emotion into account. The character's goals, emotional reactions, actions and effects, and action tendencies were scripted. As a result, the agents do not learn from experience, which is a common problem of BDI agents.

3.2 PSI

PSI [12] is a psychologically-founded theory that incorporates all basic components of human action regulation such as perception, motivation, cognition, memory, learning and emotions in one model of the human psyche. It allows for modelling autonomous agents that adapt their internal representations to a dynamic environment. PSI agents derive their goals from a set of basic drives that guide their actions. These drives include: existence-preserving needs; species-preserving need; need for affiliation; need for certainty and need for competence. A deviation from a set point constitutes the strength of each need. Needs can emerge depending on activities of the agent or grow over time. To be able to produce actions that are able to satisfy needs in a certain situation, the agent builds up intentions that are stored in memory and are - when selected - the basis of a plan. An intention is selected based on strength of activated needs, success probability and urgency.

Once an intention is selected, three levels of goal-oriented action execution can be distinguished. First, the agent tries to recall an automatic, highly ritualised reaction to handle the intention. If this is not possible, an action sequence may be constructed (planning). If planning also fails, particularly when the agent is in a completely new and unknown environment, it acts according to the principle of trial and error. While doing this, the PSI agent learns: after having experienced successful operations, the corresponding relations are consolidated, serving as indicators for the success probability of satisfying a specific need. Based on the knowledge stored in memory, abstractions of objects or events can be built. Moreover, PSI agents forget content with time and lack of use.

Emotions within the PSI theory are conceptualised as specific modulations of cognitive and motivational processes. These modulations are realised by so called emotional parameters including: *arousal* which is the preparedness for perception and reaction; *resolution level* that determines the accuracy of cognitive processes; and *selection threshold* that prevents oscillation of behaviour by giving the current intention priority. Different combinations of parameter values lead to different physiological changes that resemble emotional experiences in biological agents. Hence, a PSI agent does not require any executive structure that conducts behaviour, rather, processes are self-regulatory and parallel driven by needs, and rely on memory as a central basis for coordination. The motivational system serves as a quick adaptation mechanism of the agent to a specific situation and may lead to a change of belief about another agent as shown in [13], important for conflict resolution among ORIENT characters. Thus, PSI permits more flexibility in the characters' behaviour that FAtiMA lacks. Unfortunately, this also means an effective control over the agents' expected behaviour is missing, a limitation for applications where agents need to behave in certain ways, such as in ORIENT where the characters have a common goal to achieve.

4 FAtiMA-PSI: A Body-Mind Architecture

We have seen that despite having several advantages over FAtiMA, the PSI model suffers from a lack of control. Thus, the ideal would be to integrate key components of both architectures to build a body-mind architecture where goals are originated from drives, but at the same time gives authors some control over the agents' learning and expected behaviour. The rationale is to get a system between PSI and FAtiMA in the Physiological vs Cognitive dimension. In the new architecture shown in Figure 1, goals are driven by needs. Five basic drives from PSI are modeled: Energy, Integrity, Affiliation, Certainty and Competence. Energy represents an overall need to preserve the existence of the agent (food + water). Integrity represents well being, i.e. the agent avoids pain or physical damage while affiliation is useful for empathic processes and social relationships. On the other hand, certainty and competence influence cognitive processes.

Each need has value ranging from 0 to 10 where 0 means complete deprivation while 10 means complete satisfaction. A weight ranging from 0 to 1 underlines its importance to an agent. In order to function properly, an agent has to reduce

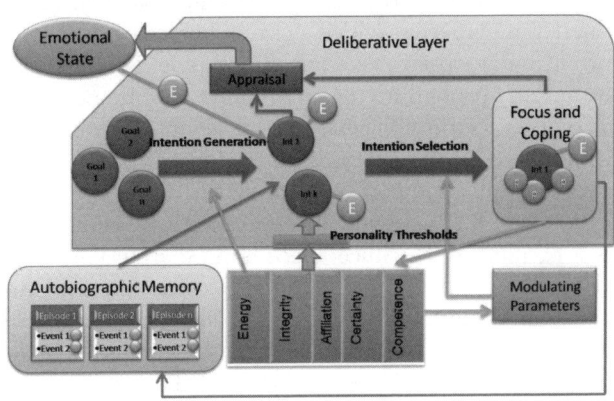

Fig. 1. FAtiMA-PSI architecture

a need's deviation from a fixed threshold as much as possible at all time. The strength of a need ($Strength(d)$) depends on its current strength plus the amount of deviation from the set point and the specific weight of the need. By assigning different weights for different needs to different agents, characters with different personalities can be produced, fulfilling one of the requirements of ORIENT. For example, if agent A is a friendly character, affiliation would be an important factor in its social relations, say weight 0.7 while a hostile agent B would have a low importance for affiliation, say weight 0.3. Now, if both agents have a current affiliation value of 2 and if the deviation from set point is 4, agent A's need for affiliation would be 4.8 while agent B's need for affiliation would be 3.2 based on Equation 1. This means that agent A will work harder to satisfy its need for affiliation than agent B.

$$Strength(d) = Strength(d) + (Deviation(d) * Weight(d)) \quad (1)$$

The inclusion of needs requires a change to FAtiMA's existing goal structure. Needs are also affected by events taking place in the environment and actions the agent performs. Since each agent has a different personality, the effect of an event may differ from one agent to another, which in turn affects their emotional and behavioural responses. In the new architecture, each goal will contain information about expected contributions of the goal to energy, integrity and affiliation needs, that is, how much the needs may be deviated or satisfied if the goal is performed. Likewise, the existing structure of events in FAtiMA has to be extended to include its contributions on needs. As for certainty and competence, no explicit specification of contributions is necessary because they are cognitive needs and their values can be calculated automatically as described below.

Whenever an expected event fails to turn up or an unknown object appears, the agent's certainty drops. Certainty is achieved by exploration of new strategies (trial and error), which leads to the construction of more complete hypotheses. If trial and error is too dangerous, developments in the environment are observed in order to collect more information. Please note that the character does not learn by forming new goals because this will lead to a lack of control on its behaviour. Instead, it learns by trying out different actions from a pre-specified set of actions and remembering which actions helped it to tackle a situation best. This information is stored in its autobiographic memory and serves as an indicator to the success probability of satisfying a specific need in future.

Competence represents the efficiency of an agent in reaching its goals and fulfilling its demands. Success increases competence while failure decreases it. The agent's autobiographic memory provides a history of previous interactions, which records the agent's experience in a specific task useful for calculation of goal competence (Equation 2). Since there is no distinction in competence in terms of achieving an important goal and a less important one, one can assume that all goals have the same contribution to the success rate. If the agent cannot remember previous activations of the goal, then it ignores the goal competence and increases the goal's contribution to certainty. The autobiographic memory also stores information about the agent's overall performance useful for calculation of overall competence (Equation 3). The expected competence (Equation 4) of the agent will then be a sum of its overall competence and its competence in performing a current goal. A low competence indicates that the agent should avoid taking risks and choose options that have worked well in the past. A high competence means that the agent can actively seek difficulties by experimenting new courses of action less likely to succeed. During this learning process, the agent also remembers a specific emotional expression of another agent in a certain situation. It continuously updates this information and changes its belief about the agent enabling it to be engaged in empathic interaction in future.

$$Comp(goal) = NoOfSuccess(goal)/NoOfTries(goal) \qquad (2)$$

$$OverallComp = NoOfSuccess/NoOfGoalsPerformed \qquad (3)$$

$$ExpComp(goal) = OverallComp + Comp(goal) \qquad (4)$$

During the start of an interaction, each agent will have a set of initial values for needs. Based on the level of its current needs, the agent generates intentions, that is, it activates goal(s) that are relevant to the perceived circumstances. A need may have several goals that satisfy it (e.g. I can gain energy by eating, or by resting) and a goal can also affect more than one need (e.g. eating food offered by another agent satisfies the need for energy as well as affiliation). So, when determining a goal's strength (Equation 5), all drives that it satisfies are taken into account.

$$Strength(goal) = \sum Strength(d) \qquad (5)$$

In terms of a particular need, the more a goal reduces its deviation, the more important is the goal (e.g. eating a full carbohydrate meal when you're starving satisfies you better than eating a vegetarian salad). By looking at the contribution of the goal to overall needs and to a particular need, goals that satisfy the same need can be compared so that success rate in tackling the current circumstances can be maximised. So, the utility value of a goal can be determined taking into consideration overall goal strength on needs, contribution of the goal to a particular need (*ExpCont(goal, d)*) and the expected competence of the agent as shown in Equation 6.

$$EU(goal) = ExpComp(goal) * Strength(goal) * ExpCont(goal, d) \qquad (6)$$

As in PSI, needs generate modulating parameters - *arousal, resolution level* and *selection threshold* that enable ORIENT characters to adapt their behaviour dynamically to different interaction circumstances. There may be more than one intention that is activated at any time instance. One of these intentions will be selected for execution based on the *selection threshold* value. After an intention is selected, the agent proceeds to generate plan(s) to achieve it. Emotions emerge as each event affects the character's needs level and hence modulates its planning behaviour. The level of deliberation that the character allocates to actions selection will be proportional to its *resolution level*. For example, if an event leads to a drop in the character's certainty, then its *arousal* level increases causing a decrease in the *resolution level*. In such situation, quick reaction is required hence forbidding time consuming search. The character will concentrate on the task to recover the deviated need(s) and hence may choose to carry out the first action that it found feasible. This physiological changes and behaviour may be diagnosed as anxiety.

5 Related Work

Some examples of existing physiological architectures are those by Cañamero [14] and Velásquez's [15]. These architectures are useful for developing agents that have only existential needs but are insuffcient for controlling autonomous agents where intellectual needs are more important. Another problem of these architectures is that all behaviours are hard-coded. On the other hand, the BDI architecture [8] is the core of deliberative agent architecture. The ways BDI agents take their decisions, and the reason why they discard some options to focus on others, are questions that stretch well beyond artificial intelligence and nurture endless debates in philosophy and psychology. Furthermore, BDI agents do not learn from errors and experience. These problems are associated with the BDI architecture itself and not from a particular instantiation. Fortunately, these questions are addressed by the FAtiMA-PSI architecture where intentions are selected based on strength of activated needs and success probability. Additionally, the motivational system will provide ORIENT characters with a basis for selective attention, critical for learning and memory processes. The resulting agents learn through trial and error, allowing more efficient adaptation and empathic engagement in different social circumstances.

6 Conclusion and Future Work

This paper proposes a new emotion model that balances Physiological vs Cognitive dimensions to create autonomous characters that are biologically plausible and able to perform life-like improvisational actions. Combining FAtiMA and PSI, the problems of psychological plausibility and control are addressed, neither of which can be solved by either architecture alone. Cultural and social aspects of interaction can be modelled using FAtiMA while PSI provides an adaptive mechanism for action regulation and learning, fulfilling the requirements of ORIENT characters. This model also addresses the ambiguity of decision making process in BDI architecture in general. Currently, the motivational system has been integrated into FAtiMA and the next step is to apply the modulating parameters in the deliberative processes such as intention selection and planning. Further effort will also be allocated to include the cultural and social aspects into the architecture. Besides using the information in autobiographic memory solely to determine the need for certainty and competence, it would be desirable to utilise the information to guide the future actions of the characters.

Acknowledgements

This work was partially supported by European Community (EC) and is currently funded by the eCIRCUS project IST-4-027656-STP and a scholarship (SFRH BD/19481/2004) granted by the Fundação para a Ciência e a Tecnologia. The authors are solely responsible for the content of this publication. It does not represent the opinion of the EC, and the EC is not responsible for any use that might be made of data appearing therein.

References

[1] Damasio, A.: Descartes' Error: Emotion, Reason and the Human Brain. Gosset/Putnam Press, New York (1994)
[2] Niedenthal, P.M.: Embodying emotion. Science 316, 1002–1005 (2007)
[3] Elfenbein, H.A., Ambady, N.: Universals and cultural differences in recognizing emotions of a different cultural group. Current Directions in Psychological Science 5(12), 159–164 (2003)
[4] Hogan, R.: Development of an empathy scale. Journal of Consulting and Clinical Psychology (35), 307–316 (1977)
[5] Hoffman, M.L.: Empathy, its development and prosocial implications. In: Nebraska Symposium on Motivation, vol. 25, pp. 169–217 (1977)
[6] Ho, W.C., Dautenhahn, K., Nehaniv, C.L.: Computational memory architectures for autobiographic agents interacting in a complex virtual environment: A working model. Connection Science 20(1), 21–65 (2008)
[7] Dias, J., Paiva, A.: Feeling and reasoning: A computational model for emotional agents. In: Bento, C., Cardoso, A., Dias, G. (eds.) EPIA 2005. LNCS (LNAI), vol. 3808, pp. 127–140. Springer, Heidelberg (2005)
[8] Bratman, M.E.: Intention, Plans and Practical Reasoning. Harvard University Press, Cambridge (1987)
[9] Ortony, A., Clore, G., Collins, A.: The cognitive structure of emotions. Cambridge University Press, Cambridge (1988)
[10] Aylett, R., Dias, J., Paiva, A.: An affectively driven planner for synthetic characters. In: International Conference on Automated Planning and Scheduling (ICAPS 2006), UK (2006)
[11] Marsella, S., Johnson, B., LaBore, C.: Interactive pedagogical drama. In: Fourth International Conference on Autonomous Agents (AAMAS), Bologna, Italy, pp. 301–308. ACM Press, New York (2002)
[12] Dörner, D.: The mathematics of emotions. In: Frank Detje, D.D., Schaub, H. (eds.) Proceedings of the Fifth International Conference on Cognitive Modeling, Bamberg, Germany (April 10-12, 2003), pp. 75–79 (2003)
[13] Lim, M.Y.: Emotions, Behaviour and Belief Regulation in An Intelligent Guide with Attitude. PhD thesis, School of Mathematical and Computer Sciences, Heriot-Watt University, Ediburgh, Edinburgh (2007)
[14] Cañamero, D.: A hormonal model of emotions for behavior control. In: VUB AI-Lab Memo 1997-2006, Vrije Universiteit Brussel, Belgium (1997)
[15] Velásquez, J.D.: Modeling emotions and other motivations in synthetic agents. In: Proceeding AAAI 1997, pp. 10–15. AAAI Press and The MIT Press (1997)

The Embodiment of a DUAL/AMBR Based Cognitive Model in the RASCALLI Multi-agent Platform

Stefan Kostadinov and Maurice Grinberg

Central and Eastern European Center for Cognitive Science,
New Bulgarian University, 21, Montevideo St., Sofia 1618, Bulgaria
stefan@yobul.com, mgrinberg@nbu.bg

Abstract. The current paper explores some of the cognitive abilities of an embodied agent based on the DUAL/AMBR architecture. The model has been extended with several capabilities like continuous operation and learning based on encoding of episodes and general knowledge. A new mechanism of analogical transfer has been implemented and demonstrated on a simulated interaction with a user. A focus of interest discussed throughout the paper is how a cognitive model can be embodied in a virtual environment and what are the benefits of combining soft cognitive capabilites with hard AI based platform. The latter is based on a Mind-Body-Environment metaphor which positions the cognitive agent in a situation similar to the one of a robot in a real environment. In the paper, results from simulations of simple interactions with a hypothetical user are presented and the internal cognitive mechanisms are discussed.

1 Introduction

DUAL is a cognitive architecture based on micro-agents which have a hybrid connectionist-symbolic nature. AMBR (Associative Memory Based Reasoning) is a model of analogy making based on DUAL [1]. The DUAL/AMBR architecture is based on the principle of local, context sensitive, and emergent parallel computation and integration and interaction of memory retrieval, mapping, transfer, and evaluation in reasoning by analogy. Work reported in recent publications (see [2] and [3]) has implemented several new mechanisms, including anticipatory ones, which turned the model into a fully operational cognitive module integrating memory, reasoning, perception, decision making and action. This is the only model which integrates explicitly anticipation based on analogy and in which the anticipatory mechanisms are used essentially in perception, evaluation and action [2].

The RASCALLI platform is a modular environment for modeling and implementation of artificial multi-agent systems giving the possibility to incorporate cognitive models [4]. This paper, based on a simulated interaction episode, explores how a cognitive model can become an integrated part of the platform and how it can augment the functionality of otherwise powerful computational mechanisms in the context of a constant interaction with a user. The expectation is that such models will bring the flexibility, context sensitivity, and adaptability (via perception and learning) characteristic of human behaviour. On the the other hand the integration of cognitive components will lead to higher believability, individuality (human-likeness), and specialization to a specific user and the related higher usability and effectiveness of the interaction and task completion (e.g. information retrieval).

The use of Mind-Body-Environment metaphor has been proposed in [3]. The idea is to abstract the use of a cognitive model from the specific application, Embodied Conversational Agent (ECA) or robot, and describe the interactions of the cognitive part with the other parts of the system in terms of a Mind (perception, memory, representations etc.), Body (sensors, effectors, and more generally Tools) and Environment.

The contributions of this paper are in the attempt to build a cognitive module which can be used in the same way in robots and embodied agents and in the specific development of new mechanisms which allow processes from perception to action to take place based on analogy-making. On the other hand the cognitive module encodes personalized information about specific past episodes and not only general knowledge. The latter is demonstrated in a scenario which simulates an interaction with a user and question answering based on analogy reasoning.

2 The RASCALLI Architecture

The RASCALLI platform provides a modular and flexible development environment for modeling and realization of artificial systems that can be endowed with various cognitive abilities [4]. The architecture is component-based and allows researchers to choose from deployed components in order to make a new realization of RASCALLI. This approach benefits distributed development by allowing independent development of components. The components can communicate with each other in the terms of a global, shared Ontology. Each component must define its input and output capabilities to the global Ontology, thus making itself available to the other components.

The RASCALLI platform allows creation of different architectures and the specific one being described is inspired by the 'Mind-Body-Environment' metaphor. It features the Mind, with its specific knowledge structures and mechanisms, the Sensory-Motor Layer, making a mediated connection of the Mind, and the Tools of RASCALLI - the sensors and effectors that work with the Sensory-Motor Layer and perceive or act on the Environment. The Body, along with the Sensory-Motor Layer provides represented knowledge to the Mind and thus makes mediated connection between the mind and the world. [3]

The DUAL/AMBR cognitive module is integrated in the RASCALLI platform as a Mind component which can use different Tools (sensors and effectors). In the example, presented below such Tools are the Input Processing (IP) and

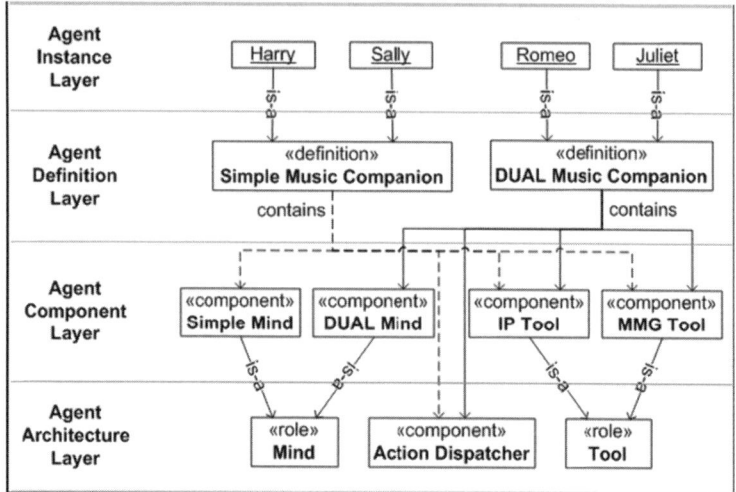

Fig. 1. The Agent Layer of the RASCALLI platform and its sub-layers [4]

Multi-Media Generation (MMG) tools, allowing communication with the user of the system(see [4] for details).

The processing of the input by the user is made with state-of-the-art NLP methods, developed especially for the RASCALLI project. The resulting processing is made with methods that cannot only localize relation arguments but also assign their exact target argument roles [6].

The ECA that the user interacts with is a full-size 3D animation, that demonstrates lips synchronization when it talks and various effects that make it realistic. [8] See Picture 2 on page 359.

3 The DUAL/AMBR Based Mind of RASCALLI

The DUAL cognitive architecture ([1]) is hybrid and combines symbolic with connectionist approaches, by integrating them at the micro-level. It is based on emergent computations so that the global behavior of the architecture emerges from local interactions among a multitude of micro-agents. The behavior of DUAL changes continuously in response to the influence of the changes in context based on an activation spreading mechanism.

The architecture has a Long-Term Memory (LTM) and a Working Memory (WM) which consists of the input and goal and the most active part of LTM. The input and goal of the system are the sources of activation thus activating the most relevant part of LTM. The latter is then involved in symbolic operations, like marker passing, correspondence hypotheses rating, etc. (see [2] for details), which support the reasoning processes.

AMBR is a model for analogy making based on DUAL. It treats analogy making as an emergent result of the common work of several overlapping

Fig. 2. The initial RASCALLI ECA. Female version available [4].

sub-processes perception, retrieval, mapping, transfer, evaluation, and learning. AMBR is based on the mechanism of mapping similar or analogical parts in WM, creating hypotheses for correspondences (typically between input and goal and episodes from LTM). The final mapping is established based on a constraint satisfaction network which discards inconsistent hypotheses. The basic mechanisms used in AMBR include spreading activation, marker passing, structural correspondence, constraint satisfaction, rating and promotion of correspondence hypotheses.

The development of the DUAL/AMBR model in the robotic context [2] added new anticipatory mechanisms to the the mechanisms mentioned above. Anticipated objects and relations from LTM episodes are added to the problem at hand as expectations. Anticipatory mechanisms are used also for active perception and selective attention [2]. In this paper, we present a new mechanism used for transfer and evaluation of knowledge between distant domains which will be discussed bellow.

3.1 Knowledge Representations

Each DUAL/AMBR based model consists of a set DUAL agents with various types and with various properties [7]. These agents can be concept agents, representing classes of homogeneous entities organized in a semantic network. They can also be instances of classes (instance agents) building episodes, hypothesis agents (representing possible correspondences between agents), cause agents (representing causal relations), and action agents (coding procedural knowledge). A special class of agents are the anticipation agents which stand for expected entities or relations.

As stated in section 1, the cognitive module has only a mediated perception of the world via the sensors (Tools) which provide RDF messages, representing events or action commands. All messages exchanged between the Mind and the Tools are based on a shared, global ontology [4]. Additionally to the global shared ontology, the agent LTM has a specific knowledge needed to organize its interactions with the environment into memory episodes, actions, causal relations, etc.

Knowledge base. The knowledge base contains information about the RASCALLI, the Environment, the Tools and the actions related to them, causal relations between actions and outcomes etc. It is built mostly of concepts, including concepts of relations but there are also specific propositions and specific knowledge.

Instinct coalitions. As the architecture is intended to support different types of (including plug-and-play) tools, the way each of them is used have to be stored in LTM and accessible on demand. The instinct coalitions are examples of tool usage. They contain all required parameters and the message scheme required by the Tools. Those coalitions are linked to the various actions the Mind can trigger and are used as a template every time the Mind decides to act in the Environment. Instinct coalitions are added to LTM on tool registration and removed from LTM when the tool is unregistered.

Episodes. There are four types of episodes used:

1. Knowledge episodes - they contain factual information collected in the past by the agent. Knowledge episodes are created every time any Tool brings new information to the Mind;
2. Usage schemes - describe the various actions that can provide information, e.g. search in a data source. Those episodes can be automatically added to the Mind, especially if the data source can provide the type of information it contains;
3. User satisfaction - provide the need to share the most relevant information found with the user. There are few of those episodes and they form the basis of the RASCALLI's desire to be cooperative and to provide information to the user;
4. Interaction with the user - keeping track of all the life of RASCALLI. Those episodes are added constantly during the lifetime of RASCALLI.

4 Scenario and DUAL/AMBR Mechanisms

In a previous paper [3], a "Context scenario" in which a number of questions were asked to the RASCALLI and interesting context effects have been demonstrated. The simulation showed the ability of the system to preserve context between utterances.

4.1 Reasoning and Transfer Mechanisms

The user is supposed to communicate with the agent, by typing the utterances in an input box and the answers are given back vocally by the agent ([4]). The analysis of the utterance provides such information as the type of entities in the semantic representation, the focus of question (what is asked for), etc. ([6]).

The utterances are expressed as semantic graph by the IP Tool and translated in frame-based format by the Sensory-Motor Layer before being sent to the Mind. After that the Mind starts working in the context of its current state and the task. Each of the instances from the current context starts to spread activation in LTM. The most active elements from WM establish correspondence hypotheses some of which become mappings and lead to the transfer from WM to the task at hand (represented as a target episode). If there is no specific answer found, actions that could provide it are transferred and executed. If a search is made, the agent retains the information that it gets through the Tools in knowledge episodes so that it can provide this information on demand in the future ([3]).

4.2 Complementary Transfer and Evaluation Mechanism

The newly developed mechanism of complementary transfer and evaluation allows the Mind to combine pieces of information from different domains in order to solve a task. The general algorithm makes mapped consequences to transfer their reasons (a causal relation has reason and a consequence), and if the reasons are distant from the target domain, to find the closest matches within the domain of the question. They are found from the general or experiential knowledge via taxonomy-propagating intersections. This is demonstrated in the following scenario.

The test utterance is "Tell me a famous musician.".

The internal knowledge is set up in such a way that there is no information about musicians being famous. Actually, the famous concept is not currently part of the musical domain and there is no information on how "famous" relates to "singers". There are however, several episodes that are related and enter the WM:

- Information about artists and Grammy awards can be obtained from data source "X";
- Musicians are awarded with Grammy;
- Artists are awarded with Oscar;
- Actor "Y" is famous because of winning an Oscar.

The latter episode is in a different domain than the requested entity, so it cannot be transferred directly. Although the domain of music artists is not so different than the domain of movie artists, it is distant enough not to allow direct transfer.

The process of complementary transfer goes as follows: first, mappings emerge between the requested musician and musicians and artists from the KB. Another mapping produced is the one linking the famous (musician) relation to the famous (artist) relation in the KB. Then, the causal relation that links the "being famous" to "winning an Oscar" transfers the reason for the mapped relation - "winning", but it cannot transfer the argument ("Oscar"), as it is not present in the musical domain. Another mapping however is made to the base episode stating that "Artists are awarded with Oscar", so an instance of "Grammy" is appended to the transferred relation "winning".

Thus the knowledge that musicians can win Grammy awards is used to anticipate that the musician in question is famous, because winning a Grammy award. The result of those two mappings is the transfer of the "has a Grammy award" information to the target. This transfer does not provide an answer to the user request, but it enlarges the task description with additional information, expected to lead to the desired information. Thus the task that is impossible to fulfill ("find a famous musician") is modified to "find a musician with a Grammy".

This time, again there is not a single musician with Grammy in the LTM, but there is an episode, stating that there is a data source that contains information on Grammy awards and musicians. From here on, the system activates the knowledge that there is information about musicians winning Grammy in a specific data source, and this data source is asked to provide such a musician.

5 Discussion and Conclusion

The results from the simulated conversations demonstrate some of the functionalities of the ECA specifically related to the cognitive model implementing the Mind. The behaviour of the agent is mainly due to the episodic part of LTM which is unique to an agent and its user and encode individual experiential information different from the available knowledge in ontologies and databases. This 'personal' part of Rascalli memory is a crucial advantage of the cognitive model involved.

It is expected that the long experience of the interaction between the user and the agent and the environment will lead to the formation of complex structures of episodes and general knowledge which will make analogy making mechanisms very efficient and unique.

Acknowledgments

The current research and development project has been partly financed by the FP6 EU project RASCALLI.

References

1. Kokinov, B., Petrov, A.: Integration of Memory and Reasoning in Analogy-Making: The AMBR Model. In: Gentner, D., Holyoak, K., Kokinov, B. (eds.) The Analogical Mind: Perspectives from Cognitive Science. MIT Press, Cambridge (2000)
2. Petkov, G., Naydenov, C., Grinberg, M., Kokinov, B.: Building Robots with Analogy-Based Anticipation. In: Freksa, C., Kohlhase, M., Schill, K. (eds.) KI 2006. LNCS (LNAI), vol. 4314, pp. 72–86. Springer, Heidelberg (2007)
3. Kostadinov, S., Petkov, G., Grinberg, M.: Embodied conversational agent based on the DUAL cognitive architecture. In: Proceedings of WEBIST 2008 International Conference on Web Information Systems and Technologies (2008)
4. Krenn, B.: RASCALLI. Responsive Artificial Situated Cognitive Agents Living and Learning on the Internet. In: Proceedings of the International Conference on Cognitive Systems (CogSys 2008), University of Karlsruhe, Karlsruhe, Germany, April 2 - 4 (2008)
5. Maes, P.: Agents that Reduce Work and Information Overload in Communications of the ACM 37(7), 30–40 (1987)
6. Xu, F., Uszkoreit, H., Li, H.: A Seed-driven Bottom-up Machine Learning Framework for Extracting Relations of Various Complexity. In: Proceedings of ACL 2007, Prague (2007)
7. Kokinov, B.: The DUAL Cognitive Architecture. A Hybrid Multi-Agent Approach. In: Cohn, A. (ed.) Proceedings of the Eleventh European Conference of Artificial Intelligence, pp. 203–207. John Wiley & Sons, Ltd, London (1994)
8. Nebula platform (2007), http://www.radonlabs.de/technologynebula2.html
9. MARY text-to-speech engine (2008), http://mary.dfki.de

BDI Model-Based Crowd Simulation

Kenta Cho, Naoki Iketani, Masaaki Kikuchi, Keisuke Nishimura,
Hisashi Hayashi, and Masanori Hattori

TOSHIBA Corporation
Komukai-Toshiba-cho, Saiwai-ku, Kawasaki-shi, 212-8582, Japan

Abstract. We present a new approach for crowd simulation in which a BDI (Belief-Desire-Intention) model is introduced that makes it possible for a character in a simulated environment to work adaptively. Our approach allows the character to realize realistic behavior by adapting its action with the sensed information in a changing environment. We implemented a demo system simulating the BDI model-based NPCs that extinguishes a forest fire with a 3D game engine, Source Engine. We measured the performance to evaluate the scalability and the bottleneck of the system.

1 Introduction

Recently, crowd simulation has attracted attention as a technology that realizes a realistic metaverse, including many intelligent characters, and simulates a social environment on the basis of interactions among characters. Human characters in crowd simulation should act as if they are real humans to realize a realistic simulated environment, but it is not easy to implement such realistic characters that are capable of diverse actions, act adaptively in the environment and cooperate with one another.

We propose crowd simulation that focuses on character adaptability in a simulator by applying a BDI (Belief-Desire-Intention) model to characters. In this paper, we explain our BDI-based approach and an implementation of crowd simulation based on that approach. Section 2 describes why we use the BDI model in crowd simulation and in section 3 and 4 our crowd simulation prototype system is explained. We evaluate that system in section 5 and present our conclusions in Section 6, the final section.

2 Approach

Crowd simulation is an important technology for adding realism to simulated environments such as metaverses and games. For a human character in a simulation to be experienced as realistic, it is necessary to satisfy certain requirements. [Lankoski 07] shows some design patterns to endow an NPC (non-player character) with human qualities so that it is experienced by a player as if it were a real person. Those qualities are related to human body, self-awareness, intention states, self-impelled actions, expression of emotions, ability to use natural language and persistent traits.

Although all these qualities are important for making a realistic NPC, we focused on adaptability of NPCs. Adaptability is related to self-awareness, intention states, self-impelled actions and persistent traits. An NPC that works to solve continuous goals should change its action when it senses a change in the surroundings and act adaptively with respect to the perceived information.

We use the BDI model to realize the adaptability of NPCs. The BDI model selects an NPC action based on 3 states of mind: belief, desire and intention. A belief represents the NPC's knowledge, including information about an external state and an internal state of the NPC. A desire represents what he/she wants to do at that moment. An intention represents what he/she is doing or is deciding to do at that moment. The BDI model is described as suitable for modeling a real-time system [Rao 95]. Since crowd simulation that adapts to a changing situation in real time is a real-time system, we adopted an approach in which the BDI model is used to implement NPCs in crowd simulation.

We implemented a demo system simulating NPCs that extinguish a forest fire. The NPCs work adaptively according to changes in the environment such as the spread of the fire and the changing positions of other NPCs. Since we thought a forest fire is a situation requiring NPCs to act flexibly according to the changing surrounding environment and cooperate with each other to overcome the problem, we use it as a test bed for our BDI-based crowd simulation.

3 BDI Model-Based Crowd Simulation Demo System

3.1 Overview

In this section, we explain our BDI-based crowd simulation demo system and its architecture. Each NPC in the system works according to 3 states of mind, namely, belief, desire and intention, and adapts its action to the change of surroundings such as the positions of fire, water and other NPCs and the spread of fire.

The system has a module called a perceiver that abstracts the information from surroundings, which is gathered through the NPCs' senses of vision and

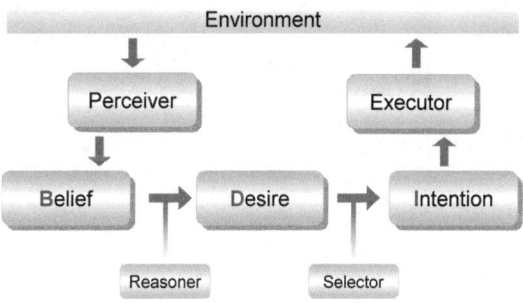

Fig. 1. BDI model

hearing, and derives meanings from that information. These meanings are used as beliefs of NPCs. There is another module called an executor that maps an NPC's intention into the NPC's concrete actions in the environment. These two modules work as an interface between the environment in the simulation and the BDI model (Fig.1) in the NPC.

This system is implemented through the combination of our BDI-based AI system and Source Engine, which is a 3D game engine developed by Valve Software. Source Engine is a well-known platform used for AAA titles such as Half-Life. Source Engine provides many features, such as 3D rendering, character animation and physics simulation, and by using it our demo system can be developed in a short term.

3.2 Scenario

In this section, we show the typical working scenario in the system.

- In the first stage, NPCs don't know that a fire has occurred, and are talking to one another in the village.
- A NPC walks to the forest and discovers a fire.
- The NPC goes to the nearest river and tries to extinguish the fire.
- The fire spreads and the NPC thinks it will be difficult to extinguish all the fires by itself.
- The NPC returns to the village to get the help of other NPCs. (Fig. 2)
- Other NPCs are starting to extinguish the fire separately.
- The fire spreads faster and NPCs try to form a bucket brigade. (Fig. 3)
- Since the fire spreads faster and faster, NPCs think it is impossible to get a handle on the fire and run away.

Fig. 2. Get the help of other NPCs

Fig. 3. Bucket brigade

This is one of the scenarios that can occur in this system. If there are many NPCs or the NPC finds a fire faster, the fire can be extinguished. The result of the simulation changes according to the parameters such as the number of NPCs, location of NPCs and fires, and the speed at which the fire spreads.

4 Architecture

In this section, we explains the BDI model that provides the adaptive actions of NPCs. There are 3 data structures (belief, desire, intention) that represent the inner mind states of the NPC and 4 modules (perceiver, reasoner, colector, executor) that manage and update these structures.

4.1 Perceiver

The perceiver perceives the surroundings and creates and updates beliefs. Information about the surroundings is gathered by Source Engine. Source Engine brings the information of vision and hearing in a limited range from the NPC to the perceiver, and the perceiver updates beliefs. Beliefs that include the information about the environment in the map are represented with the location data of the grid where the information is perceived. The grid is used to represent the abstracted data of location-based perceived information. Each belief has a weight that represents the importance of each event for the NPC. For example, the weight of the belief about the fires in each grid represents the number and the intensity level of fires in the grid.

4.2 Reasoner

The reasoner retrieves desires according to the updated beliefs with the pre defined rule set. A rule fires for a given group of beliefs and derives corresponded desires. Types of desires include stand talking, take water (from where), pour water (to where), run away, cry out (at where, what) and form a bucket brigade (from where to where). The number of each type of desire is not limited to 1. For instance, take water desires can be generated according to each grid where the NPC can take water.

Each desire has a weight that represents its importance. The rule that retrieves the specific desire calculates the weight according to the weights of the corresponding beliefs. For example, the weight of the pour water desire is calculated based on the distance to the corresponding grid and the number of other NPCs in the grid. The number of NPCs is used for avoiding a congested grid where many NPCs rush to pour water.

4.3 Selector

The selector selects the desire that has the highest weight's value and uses it as the NPC's intention. To ensure that the NPC does not change its intention too frequently, the selector changes the current intention when the weight's value of the current desire is larger than the multiplied weight's value of the current intention.

4.4 Executor

The executor converts the current intention into a concrete sequence of actions and these actions are executed with Source Engine.

4.5 Applying BDI Model to Our Demo System

By using the BDI model, the mechanism to control the NPC can be separated easily into two layers. One layer handles the higher-level estimation for the action of the NPC such as which fire should be extinguished first or whether work should be executed separately or collaboratively. These estimations are handled with the BDI model that can get a current overview of the environment from beliefs and generate appropriate desires according to beliefs. The other layer handles the lower-level estimation such as finding the path to the fire and the water. In our system, the executor performs the role of this layer.

Since the BDI model in the higher layer only has the abstracted surroundings information modeled as beliefs and higher-level action modeled as intentions, the developer can focus on implementing the sophisticated actions of the NPC within the BDI model.

5 Evaluation and Discussion

We evaluated the performance of the system, and compared our BDI-model based approach to other approaches for the crowd simulation.

5.1 Performance Evaluation

We measured the performance in order to evaluate the system's scalability and bottlenecks. Our evaluation environment is described below.

CPU Xeon 3.2GHz
Memory 3.0GB
OS Windows XP SP2
Profiler VTune Performance Analyzer ver.8

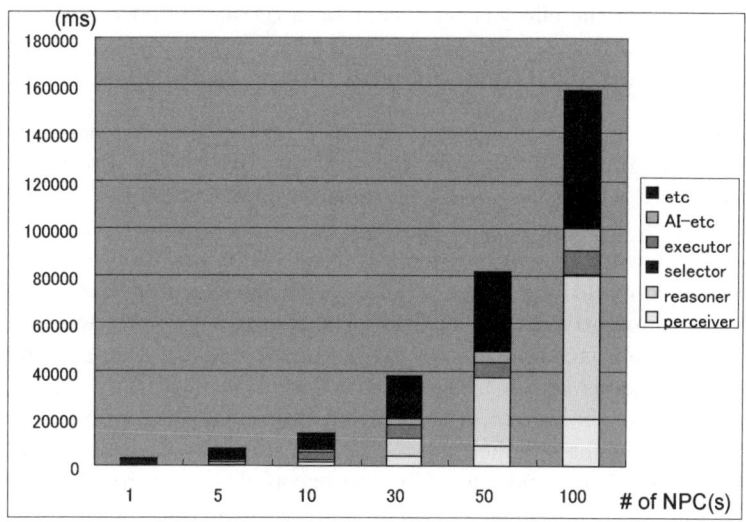

Fig. 4. Performance evaluation

Fig.4 shows the CPU time (micro sec.) of each simulation cycle. The simulation cycle represents the load to control all NPCs in the simulation. Concretely, it is the length of the time in which the GameFrame() method is executed once in Source Engine. Each item in the CPU time means the processing time of the specific process described below.

- Perceiver, Reasoner, Selector, Executor
  Processing time of each component
- AI-etc
  Processing time of the RunAI() method except for the time spent for the 4 components described above. The RunAI() method that controls the behavior of NPCs.
- etc
  Processing time of the GameFrame() method except for the time spent for the RunAI() method, including the process for spreading fires and physics simulation.

The increase in processing time is roughly proportionate to the number of NPCs, and the reasoner's AI-related processes accounted for most of the time. The processing time for the reasoner increased more sharply than for the other items. This sharp increase is attributable to the process in the reasoner to update the beliefs about other NPCs in the simulation and apply rules to these updated beliefs.

The current implementation of the reasoner applies a rule every time when one of the corresponding beliefs for the rule is updated. To minimize the overhead in applying rules to updated beliefs, the reasoner should make a prior evaluation as to whether the change of the belief has a compelling effect on the rule and apply it only when the effect is significant for a retrieved desire.

5.2 Comparison of BDI Model with Other Approaches

Most BDI-model agent systems are implemented based on a reactive planning [Howden 01] [Braubach 05] [Bordini 06], and our BDI model also retrieves the desire reactively with rules. Some systems support communications between agent [Braubach 05] [Bordini 06] [Howden 01] and cooperative action strategy [Bordini 06] [Howden 01]. In our demo system, communication is realized with the voice and cooperative action is realized with the executor that causes NPCs to engage in cooperative action such as formation of a bucket brigade.

Our current approach realizes reactive action by creating desires with the reasoner and continuous action by keeping an intention until the weight of other desires rises above the weight of the current intention. Our approach is advantageous in terms of the simplicity of the system since the BDI model can realize both the adaptabilities and the ability to accomplish a desire, but sequential actions of the NPC cannot be specified. This makes it difficult to set a specific scenario for the NPC since the sequence of the NPC's actions has to be controlled by adjusting the weight of the desire calculated with the rule.

A planning mechanism will be of help regarding this problem. Offline planning [Fikes 71] creates a static plan to achieve a goal. Online planning [Ambros-Ingerson 88] [Wilkin 88] [Hayashi 06] can change the created plan dynamically while the plan is being achieved according to the changing situation. HTN planning [Tate 77] [Wilkin 88] [Nau 99] [Hayashi 06] breaks down an abstract task into concrete tasks in the process of creating a plan. By using the planning mechanism, we can set the specific sequence of actions for the specific desire of the NPC. The planning mechanism can be introduced to our system by mapping an intention to the specific plan.

Another problem in our approach is lack of a cooperating mechanism. We implemented collaborative behavior of NPCs by transmitting beliefs with voices, but it doesn't support the assuming of the leadership of a group of NPCs. A collaborative strategy in multi-agent planning [Ferber 99] will help to solve this problem. For instance, we could introduce a leader agent in our system and the leader agent would create the global plan and assign the corresponding tasks to other agents.

6 Conclusion

We have introduced the approach of the BDI model-based crowd simulation and simulated NPCs that extinguish a fire in a forest. Beliefs of an NPC simulate a limited sense of the surroundings, desires simulate concurrent candidate actions and an intention simulates an adaptively selected action. The approach presented in this paper endows NPCs in a simulated environment with more adaptive and realistic behavior. Currently, our work is directed toward striking a balance between adaptabilities and easy description of NPCs by introducing a planning mechanism to our approach.

References

[Lankoski 07] Lankoski, P., Bjork, S.: Gameplay Design Patterns for Believable Non Player Characters, Situated Play. In: Proceedings of DiGRA 2007 Conference, September 3 (2007)

[Rao 95] Rao, A., Georgeff, M.: BDI Agents: from Theory to Practice. In: ICMAS 1995, pp. 312–319 (1995)

[Howden 01] Howden, N., Ronnquist, R., Hodgson, A., Lucas, A.: JACK: intelligent agents - summary of an agent infrastructure. In: IAMSMAS 2001 (2001)

[Braubach 05] Braubach, L., Pokahr, A., Lamersdorf, W.: Jadex: A BDI-agent system combining middleware and reasoning. In: Software Agent-Based Applications, Platforms and Development Kits, pp. 143–168. Birkhauser Book (2005)

[Bordini 06] Bordini, R., Hubner, J.: Jason: A java-based interpreter for an extended version of AgentSpeak, Manual Version 0.8 (2006)

[Fikes 71] Fikes, R., Hart, P., Nilsson, N.: STRIPS. a new approach to the application of theorem proving to problem solving. Artificial Intelligence 2, 189–208 (1971)

[Ambros-Ingerson 88] Ambros-Ingerson, J., Steel, S.: Integrating planning, execution and monitoring. In: AAAI 1988, pp. 735–740 (1988)

[Wilkin 88] Wilkins, D.: Practical Planning. Morgan Kaufmann, San Francisco (1988)

[Hayashi 06] Hayashi, H., Tokura, S., Hasegawa, T., Ozaki, F.: Dynagent: An incremental forward-chaining HTN planning agent in dynamic domains. In: Baldoni, M., Endriss, U., Omicini, A., Torroni, P. (eds.) DALT 2005. LNCS (LNAI), vol. 3904, pp. 171–187. Springer, Heidelberg (2006)

[Tate 77] Tate, A.: Generating Project Networks. In: IJCAI 1977, pp. 888–893 (1977)

[Nau 99] Nau, D., Cao, Y., Lotem, A., Munoz-Avila, H.: SHOP: simple hierarchical ordered planner. In: IJCAI 1999, pp. 968–975 (1999)

[Ferber 99] Ferber, J.: Multi-agent systems: an introduction to distributed artificial intelligence. Addison-Wesley, Reading (1999)

The Mood and Memory of Believable Adaptable Socially Intelligent Characters

Mark Burkitt and Daniela M. Romano

Department of Computer Science, University of Sheffield,
Regent Court, 211 Portobello Rd, Sheffield,
S1 4DP, United Kingdom
{acp07mab,D.Romano}@sheffield.ac.uk

Abstract. In this paper a computational model for believable adaptable characters is presented, which takes into account several psychological theories: five factors model of personality[1], the pleasure arousal dominance[2] and the social cognitive factors[3] to create a computation model able to process emotionally coded events in input, alter the character's mood, the memory associate with the set of entities connected with the event, and in the long run the personality; and produce an immediate emotional reaction in the character, which might or might not be displayed according to the social cognitive factors, the goal of the character and the environment in which the event is taking place.

1 Introduction

Virtual characters are increasingly being used in different applications to enhance a user's experience and to increase the perception of realism. One way this can be achieved is through elicitation of social presence. Biocca[4] defines social presence as the feeling that you are interacting, through mediated communication, with an intelligent being. In this context, the measure of social presence is the degree to which the user is able to perceive the intelligence of another[4].

Various psychological theories have been combined together to obtain a computational model able to process emotionally coded events in input and generate a reaction, which might or might not be displayed, according to the mood, memories, personality and social cognitive factors of the character. Roseman & Smith[5] suggested that the judgment (appraisal) we make about ourselves or the world are what causes an emotion to arise. They propose an adaptive dimension, where for example if we appraise a situation as being negative, but there is not relevant action that can be taken, we experience sadness (passive reaction); instead if there is a relevant action that might be of benefit, we experience anger (active reaction). Computational models of emotion like EMA[6] are concerned with modelling such appraisal of the world. Scherer at al.[7] report that the more recent literature conceptualises emotional states as two-dimensional models where a pleasantness (or "valence") dimension is added to the arousal dimension, or positive affect and negative affect are conceptualised as two fundamental dimensions of emotional experience, with the PAD[2] being a computational version of this stream of thoughts. Since the 1960s, evidence has increasingly supported the idea that there are several distinct fundamental emotions (such as joy, sadness, fear, and anger), that cannot be accounted for by the two-dimensional models and raise the question of what produces the different patterns of response, where the OCC model[8] is a computational model based on several distinct emotions.

The model presented here, called BASIC v2.0, is an enhancement of BASIC[9] and relates the multi-dimension theories of emotion with the two-dimensional models and the Five-Factor Model (FFM) of personality[1]. It appraises the character's mood, memory associated with the entities related to the event, and the social cognitive factors (SCF)[3] to compute an emotional response. Gross[10] distinguishes two broad classes of emotion regulation: *antecedent-focused emotion regulation* and *response-focused emotion regulation*. The first one occurs before an emotional response is generated, by (a) selecting situations (e.g., avoiding), (b) modifying problematic situations, (c) attending to one rather than another aspect of the immediate situation, or (d) modifying the way emotion-relevant stimuli are appraised. Whereas the *response-focused emotion regulation* occurs after the emotional response tendencies are generated and the individual modulates (diminishes or augments) the response felt on the basis of ones own goals. BASIC v2.0 is only concerned with the *antecedent-focused emotion regulation*, in particular (c) and (d), and no action is given in response to the event as this selection will be concerned with the *response-focused emotion regulation* related with the character's goal. In fact no computation of the aims and goal of the character is done. This module represents the emotion, mood and memory of a character and can be used within a complete synthetic character architecture, as for example in the Louchart et al.[11] as shown in Fig. 1.

BASIC v2.0, unlike BASIC, represents the mood through the Pleasure Arousal Dominance (PAD) model of temperament[2], rather than the OCC[8]. The use of OCC did not provide sufficient distinction between mood and emotion, as mood tends to be more general in nature; unrelated to any particular event; enduring and slow to change; and has less intensity than emotion[12]. The use of PAD to represent mood has also been successfully used in ALMA[13,14,15].

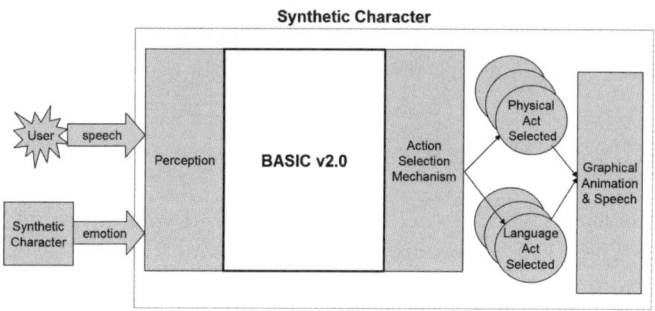

Fig. 1. Louchart et al. [11] System Architecture

The way that mood and memory work together and interact has been modified in several ways in BASIC v2.0. Memories can now be associated with a combination of events, objects and agents. For example, the memory of a fireman carrying out a person from a burning building could associate the emotion with the entities fireman, stretcher, person and fire and subsequent experiences involving any of these elements could trigger the emotional memory stored[16].

Bower[17] describes four theories linking mood and memory:

1. **Mood Dependency.** Memories are easier to recall in the same mood as when they were formed.
2. **Thought Congruity.** The current mood influences thoughts, feelings and interpretation of events.
3. **Mood Congruity.** Memories with an emotional tone similar to the current mood are easier to learn and remember.
4. **Mood Intensity.** More intense memories are remembered better.

These four theories have in general been supported by other researchers (see Singer et al[18] for a review), although it is agreed that a more robust method for measuring *mood dependency* is required to confirm the theory beyond any doubt[19,20]. All four theories are now implemented in BASIC v2.0. When new memories are formed, they are associated with the character's current mood. The character is more likely to recall an emotional memory when in the same mood as when the memory was formed[17,18,19,20,21]. The initial intensity of the memory is the same as the intensity of the emotion being remembered, resulting in more intense emotional events remaining memorable for longer [22]. As time passes the current mood influences the interpretation, or appraisal, of the incoming emotional events, consequently influencing the intensity of memories being formed. Individual memories decay over time using an exponential decay function called 'The Forgetting Curve'[23]. This allows trivial memories to be forgotten quickly, whereas more intense memories persist for longer[24]. As the memory decays, the likelihood of that memory being recalled reduces. The full OCC model of emotions has been implemented in BASIC v2.0. This allows a wide range of emotions to be represented, and provides a comprehensive set of conversions between OCC and PAD.

BASIC v2.0 emotional state and mood are constantly updated and regulated over time. Instead of returning an emotional response, the system maintains an *internal emotional state*, which fluctuates frequently over time.

2 BASIC v2.0 Walkthrough

A new personality is created by selecting a set of FFM values, a set of SCF values and a name. The FFM values are used to determine the neutral mood of the personality, expressed as PAD values. Every time the personality is advanced a single step, a set of functions is applied to modify the current emotional state, the mood and the FFM factors. Russell & Mehrabian[25] give the values for the majority of the OCC emotions in PAD space. The values for the missing emotions were based on the values used in ALMA[13]. For conversions from FFM to PAD, Mehrabian[26] has identified a set of values that show the location of each FFM personality factor in PAD space. All these conversion values are used in BASIC v2.0 formulae below together with a set of weighting values denoted by the prefix (ω). The weightings are used to balance the influence of different aspects of the system, thus ensuring that no single formula dominates the way emotions are processed. The values with a prefix of (SCF) relate to the social cognitive factors, with the initials of the factor being represented after, where (REF) is the Reflection, (SRS) is the Self Regulatory Skill, (SK) is Social Knowledge, (AE) is Affective Experience and (GS) is Goals and Standards.

An emotional event contains an OCC emotion and a set of identifiers that represent the associated domain entities. This is given in input to the model. First the target mood (m_{tar}) moves towards the neutral mood (m_{neu}) at a steady rate (1); the current emotional state (R) moves back to neutral (2).

$$m'_{tar} = m_{tar} + ((m_{neu} - m_{tar}) \times SCF_{REF} \times \omega_{MOOD\_REG}) \qquad (1)$$
$$R' = R - (R \times SCF_{SRS} \times \omega_{EMOTIONAL\_REG}) \qquad (2)$$

Before applying the memory to the event, an emotional memory is selected from the memory bank. From all available memories, a subset is selected, influenced by the current mood and the entities associated with the current emotional event. An average of these memories ($M\varepsilon_{avg}$) is taken, and the strength of each memory is increased. The average memory ($M\varepsilon_{avg}$), current mood (m_c) and personality traits (FFM) are then applied to the event emotion (ε) (3).

$$\begin{aligned}\varepsilon' = \varepsilon &+ (M\varepsilon_{avg} \times SCF_{SK} \times \omega_{MEMORY}) + \\ &(m_c \times SCF_{AE} \times \omega_{MOOD}) + (FFM \times SCF_{GS} \times \omega_{FFM})\end{aligned} \qquad (3)$$

Subsequently the current internal emotional state (R) and the target mood (m_{tar}) are updated (4 & 5).

$$R' = (R \times (1 - SCF_{AE} \times \omega_{EMOTION})) + (\varepsilon \times SCF_{AE} \times \omega_{EMOTION}) \qquad (4)$$
$$m'_{tar} = m_{tar} + ((R \times SCF_{REF}) + (M\varepsilon_{avg} \times SCF_{AE}) \times \omega_{T\_MOOD\_UPD}) \qquad (5)$$

A new memory is then created using the current internal emotional state, the current mood and the set of entity identifiers associated with the emotional event being processed. The initial intensity of the memory is set to the intensity of the emotion. The current mood (m_{cur}) is moved towards the target mood (m_{tar}) at a steady rate (6), the personality traits are updated based on the current mood (7) and the neutral mood is updated based on the personality traits.

$$m'_{cur} = m_{cur} + ((m_{tar} - m_{cur}) \times \omega_{C\_MOOD\_UPD}) \qquad (6)$$
$$FFM' = FFM + (m_{cur} \times \omega_{FFM\_UPD}) \qquad (7)$$

Finally the memory decay function iterates through all memories and updates the intensity based on the age of each memory (8), where (M_i) is the intensity,

($M\varepsilon_i$) is the emotional intensity, (s) is the memory strength, which increases whenever the memory is remembered and (t) is time.

$$M_i = M\varepsilon_i \times e^{-\left(\frac{t}{s \times M\varepsilon_i}\right)} \tag{8}$$

3 Model Validation

Various aspects of the model have been tested to ensure they respect the underlying theories, the results are shown in the following sections. A test application was constructed to provide a visual interface for creating personalities and posting events. The application records the state of each personality at 1 second intervals. Emotional events were posted to each personality in batches, which will be referred to as *sets*, to represent an event being continuously processed over time. Each *set* uses a posting interval of 5 events per second, set in conjunction with the global weightings.

Test 1: Behaviour of the Mood. The following tests have been conducted to ensure that the behaviour of mood is as described in literature and different from the *internal emotional state* of the system / personality, which is the output of BASIC v2.0 appraisal process. As already mentioned, mood should (i)be more general in nature, (ii)last longer and (iii)change slower. Two personalities were created, Character A and Character B. Both personalities received *Fear* over 5 seconds. Character A received an emotional intensity of 0.25 and Character B received an emotional intensity of 0.5. As shown in Fig. 2(a), when a *set* of emotionally intense events are received, the intensity of the emotional state of each personality increases quickly reaching its peak in 5 seconds, while the event is fired, and then reduces back to normal. The mood, instead, increases at a slower rate than the emotional state and continues to increase after the event has passed, finally it gradually reduces back to normal. The intensity and duration of the mood depend on the weightings (ω) assigned in the mood related equations (1, 5 & 6), for which no quantitative data is available in literature. The (ω) values used in BASIC v2.0 are the same as in original BASIC and have been obtained empirically.

Test 2: Adaptability: Changes in Personality over Time. The original BASIC uses a lookup table to provide mappings between the reduced OCC model of emotion and the FFM of personality to modify the personality traits in relation to experiences. With PAD replacing OCC for mood, this link needed to be re-established. The new approach is based on an inverse of the mappings between PAD and FFM. In the original BASIC a person who received a large number of a particular type of emotional event eventually had their personality changed so they responded to subsequent emotional event under the influence of the previous emotions. For example, a person who receives a lot of love will interpret all incoming events from the point of view of someone who is loved all the time.

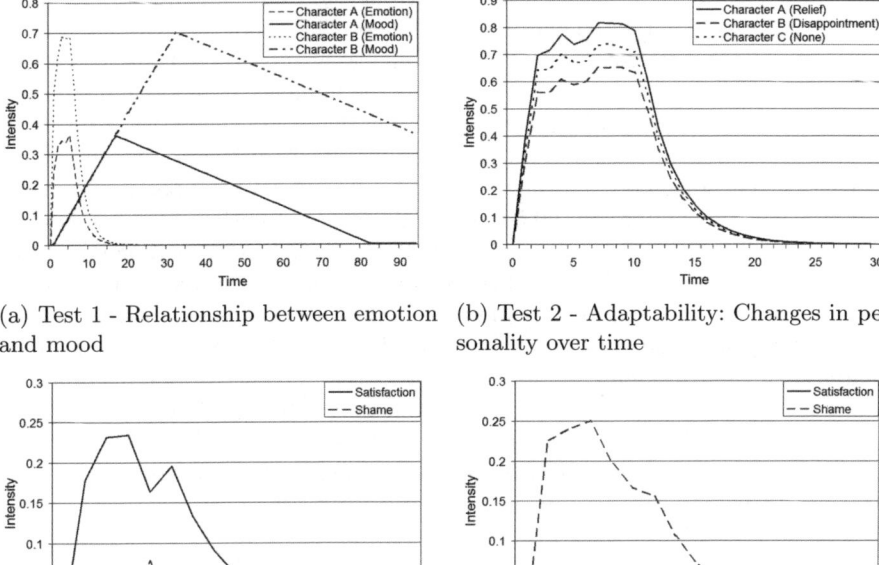

(a) Test 1 - Relationship between emotion and mood

(b) Test 2 - Adaptability: Changes in personality over time

(c) Test 3 - Memory recall of a neutral event in an exuberant mood

(d) Test 3 - Memory recall of a neutral event in a bored mood

Fig. 2. Validation Results

To test the adaptability of a personality, the influence on the emotional state of both mood and memory were removed from the model. Three personalities were created, Character A, Character B and Character C. Character A received *Relief* for 30 minutes, Character B received *Disappointment* for 30 minutes and Character C received no emotional input. Each emotion had an intensity of 0.2. Once the set of events had finished posting, all personalities received *Relief* for 10 seconds with an intensity of 0.5. As shown in Fig. 2(b) the personalities of Characters A and B have been effected by the long posting of one type of emotion and respond to the same event in a different manner. Character A has a higher level of *Relief* in line with his personality, while Character B has a lower level of *Relief* and Character C shows the uninfluenced emotional intensity.

Test 3: Mood Dependent Memory Recalls. This test is concerned with demonstrating that the memories recalled depend on the mood as suggested by literature[17,18,19,20,21]. As above, the influence of both mood and memory was initially removed. One new personality was created and manipulated so that the current and target mood was at PAD[0.5,0.5,0.5] (Exuberant octant). The personality then received *Satisfaction* over 5 seconds with an intensity of 1. When the emotional intensity returned to normal, the personality was manipulated again, this time so that the current and target moods were changed to PAD[-0.5,-0.5,-0.5] (Bored octant). The personality then received *Shame* over 5 seconds with an intensity of 1. The personality now contained two sets of intense memories in opposite mood octants. As the influence of mood and memory on the emotional event had been removed, neither set of memories were influenced during creation. The system was then modified to reinstate the influence of memory, but not mood. The current and target mood of the personality were modified again to match PAD[0.5,0.5,0.5]. A neutral emotion was then posted over 5 seconds with an intensity of 0. The current mood was modified again to match PAD[-0.5,-0.5,-0.5] and a neutral emotion was posted again over 5 seconds with an intensity of 0. Figure 2(c) shows the influence of *Satisfaction* and *Shame* in the Exuberant mood octant. As expected, the intensity of *Satisfaction* is much greater than that of *Shame*. Figure 2(d) shows the influence of *Satisfaction* and *Shame* in the Bored mood octant. This graph shows that the intensity of *Shame* is much greater than that of *Satisfaction*, which is consistent with expectations. As all memories are drawn from the main memory bank, it was expected that both sets of memory would influence the emotion to some extent.

4 Conclusions

BASIC v2.0, a computational model for believable adaptable, social intelligent characters has been created based on a number of well established psychological theories relating mood, memory, personality and emotion and it has been shown how BASIC v2.0 respects them. Given a series of emotional events the system modifies its current internal emotional state, which might not be displayed according to the social cognitive factors and goals of the character. The model is modular and domain independent; and can be integrated into virtual characters to add an emotional dimension to their behaviour.

References

1. Tupes, E.C., Christal, R.E.: Recurrent personality factors based on trait ratings. Journal of Personality 60(2), 225–251 (1992); Originally published as an ASD Technical Report in 1961
2. Mehrabian, A.: Pleasure-arousal-dominance: A general framework for describing and measuring individual differences in temperament. Current Psychology 14(4), 261–292 (1996)
3. Cervone, D., Shoda, Y.: Social-cognitive theories and the coherence of personality. In: Cervone, D., Shoda, Y. (eds.) The coherence of personality: Social-cognitive bases of consistency, variability, and organization, pp. 3–33. Guilford Press, New York (1999)
4. Biocca, F.: The cyborg's dilemma: Progressive embodiment in virtual environments. Journal of Computer-Mediated Communication 3(2) (1997)
5. Roseman, I.J., Smith, C.A.: Appraisal theory. In: Scherer, K.R., Schorr, A., Johnstone, T. (eds.) Appraisal Processes in Emotion: Theory, Methods, Research, pp. 3–19. Oxford University Press, USA (2001)

6. Marsella, S., Gratch, J.: Ema: A computational model of appraisal dynamics. In: Trappl, R. (ed.) Cybernetics and Systems 2006, pp. 601–606. Austrian Society for Cybernetic Studies, Vienna (2006)
7. Scherer, K.R., Schorr, A., Johnstone, T.: Appraisal Processes in Emotion: Theory, Methods, Research. Oxford University Press, USA (2001)
8. Ortony, A., Clore, G.L., Collins, A.: The Cognitive Structure of Emotions. Cambridge University Press, Cambridge (1990)
9. Romano, D.M., Sheppard, G., Hall, J., Miller, A., Ma, Z.: Basic: A believable, adaptable socially intelligent character for social presence. In: Proceedings of the 8th Annual International Workshop on Presence (2005)
10. Gross, J.J.: Sharpening the focus: Emotion regulation, arousal, and social competence. Psychological Inquiry 9(4), 287–290 (1998)
11. Louchart, S., Romano, D.M., Aylett, R., Pickering, J.: Speaking and acting - interacting language and action for an expressive character. In: Proceedings for the AISB workshop, University of Leeds, UK (2004)
12. Beedie, C., Terry, P., Lane, A.: Distinctions between emotion and mood. Cognition and Emotion 19(6), 847–878 (2005)
13. Gebhard, P.: Alma: a layered model of affect. In: AAMAS 2005: Proceedings of the fourth international joint conference on Autonomous agents and multiagent systems, pp. 29–36. ACM, New York (2005)
14. Gebhard, P., Kipp, K.H.: Are computer-generated emotions and moods plausible to humans? In: Gratch, J., Young, M., Aylett, R.S., Ballin, D., Olivier, P. (eds.) IVA 2006. LNCS (LNAI), vol. 4133, pp. 343–356. Springer, Heidelberg (2006)
15. Klesen, M., Gebhard, P.: Affective multimodal control of virtual characters. The International Journal of Virtual Reality 6(4), 43–54 (2007)
16. Canli, T., Zhao, Z., Brewer, J., Gabrieli, J.D.E., Cahill, L.: Event-related activation in the human amygdala associates with later memory for individual emotional experience. Journal of Neuroscience 20(19), 99RC+ (2000)
17. Bower, G.H.: Mood and memory. American Psychology 36(2), 129–148 (1981)
18. Singer, J.A., Salovey, P.: Mood and memory: Evaluating the network theory of affect. Clinical Psychology Review 8(2), 211–251 (1988)
19. Eich, E.: Searching for mood dependent memory. Psychological Science 6(2), 67–75 (1995)
20. Eich, E., Macaulay, D.: Are real moods required to reveal mood-congruent and mood-dependent memory? Psychological Science 11(3), 244–248 (2000)
21. Lewis, P.A., Critchley, H.D.: Mood-dependent memory. Trends in Cognitive Sciences 7(10), 431–433 (2003)
22. Mather, M.: Emotional arousal and memory binding: An object-based framework. Perspectives on Psychological Science 2(1), 33–52 (2007)
23. Ebbinghaus, H.: Memory: A contribution to experimental psychology. New York, Teachers College, Columbia University (translated by H. A. Ruger & C. E. Bussenues) (1913)
24. Cahill, L.: The neurobiology of emotionally influenced memory implications for understanding traumatic memory. Annals of the New York Academy of Sciences 821(1), 238–246 (1997)
25. Russell, J.A., Mehrabian, A.: Evidence for a three-factor theory of emotions. Journal of Research in Personality 11(3), 273–294 (1977)
26. Mehrabian, A.: Analysis of the big-five personality factors in terms of the pad temperament model. Australian Journal of Psychology 48(2), 86–92 (1996)

Visualizing the Importance of Medical Recommendations with Conversational Agents

Gersende Georg[1,2], Marc Cavazza[3], and Catherine Pelachaud[4,5]

[1] Centre des Cordeliers UMRS 872 Eq. 20, Paris, France
[2] French National Authority for Health (HAS), Saint-Denis La Plaine, France
[3] School of Computing, University of Teesside, Middlesbrough, United Kingdom
[4] University of Paris 8
[5] INRIA Rocquencourt, France
gersende.georg@spim.jussieu.fr, m.o.cavazza@tees.ac.uk,
catherine.pelachaud@inria.fr

Abstract. Embodied Conversational Agents (ECA) have the potential to bring to life many kinds of information, and in particular textual contents. In this paper, we present a prototype that helps visualizing the relative importance of sentences extracted from medical texts (clinical guidelines aimed at physicians). We propose to map rhetorical structures automatically recognized in the documents to a set of communicative acts controlling the expression of the ECA. As a consequence, the ECA will dramatize a sentence to reflect its perceived importance and degree of recommendation (advice, requirement, open proposal, etc). This prototype is constituted of three sub-systems: i) a text analysis module, ii) an ECA and iii) a mapping module which converts rhetorical structures produced by the text analysis module into nonverbal behaviors driving the ECA animation. This system could help authors of medical texts to reflect on the potential impact of the writing style they have adopted. The use of ECA re-introduces an affective element which won't be captured by other methods for analyzing document style.

Keywords: Embodied Conversational Agents, Emotional Natural Language Processing, Medical Informatics.

1 Introduction and Rationale

ECAs have been demonstrated to bring added value to many applications for which a more human-like presentation [1,2] is beneficial, including assistance, help and guidance [3,4]. In this paper, we investigate the use of ECAs to visualize the importance of specific sentences within medical documents. The document we studied are Clinical Guidelines, which are normative texts, produced by various Health authorities, promoting best practice in Medicine based on the concept of *evidence-based medicine* [5]. They are written by expert physicians and aimed at physicians, for instance General Practitioners. Clinical guidelines are based on the notion of recommendation as their elementary unit. These can be characterized linguistically through specific syntactic constructs which manifest a rhetorical intention. For instance, "*In case of extension to pedicle lymph nodes, if surgical accessibility falls into Class I, surgery cannot be contraindicated, but this decision should nevertheless be part of a multidisciplinary consultation*". Clinical guidelines are produced by expert committees through a complex process of consensus building, as committee members assess the style of initial formulations. Guidelines' authors need to be able to assess the potential perception by their readers of the strength of recommendations they have been writing, to anticipate the impact of the specific recommendations they contain as a function of the style used. Expressive communication is one mode of visualizing a recommendation's strength in which various dimensions could be combined seamlessly through multimodal channels, e.g. facial expressions and/or gestures synchronized to the utterance. This importance manifests itself, within certain constraints to be discussed below, through the choice of syntax and vocabulary, which can be mapped to certain rhetorical structures, from advice to orders.

2 System Overview and Architecture

This work is based on experiments carried out with our prototype, which presents itself as an ECA interface "reading aloud" specific recommendations selected from a clinical guideline. It is actually constituted of three sub-systems: i) G-DEE, a document engineering environment [6] which performs an automatic identification of recommendations in a Guideline, ii) Greta, an ECA system [7] and iii) a mapping module which converts rhetorical structures produced by G-DEE into the communicative act format used by Greta (a mark-up language known as APML [8]). The system operates as follows. Firstly, G-DEE is run offline to analyze the clinical guidelines as a whole. It produces a document in which all recommendations are identified through a set of specific mark-ups for their operators and the contents they apply to (referred to as the scopes of the operator).

A marked-up recommendation appears as highlighted text in the system interface (Fig. 1). This text fragment can be selected interactively, which triggers the generation of an APML file animating Greta on that sentence (the generation uses an XSLT conversion module). In this process, tags on communicative acts linked to recommendation strength have been added to the text automatically. Finally, Greta processes this APML file and utters the corresponding recommendation, displaying appropriate nonverbal behavior, which reflects the importance of the recommendation and places emphasis on relevant scopes. In this way, the actual strength of the recommendation and its potential impact can be visualized.

2.1 The Document Engineering Environment

Clinical guidelines belong to the generic category of normative texts, to which much research has been dedicated. These texts are naturally structured through the occurrence of specific linguistic expressions, known as "deontic operators" [9], which characterize the linguistic expression of recommendations. These operators manifest themselves (in French) through such verbs as "*pouvoir*" ("to be allowed to or may"), "*devoir*" ("should or ought to"), "*interdire*" ("to forbid").

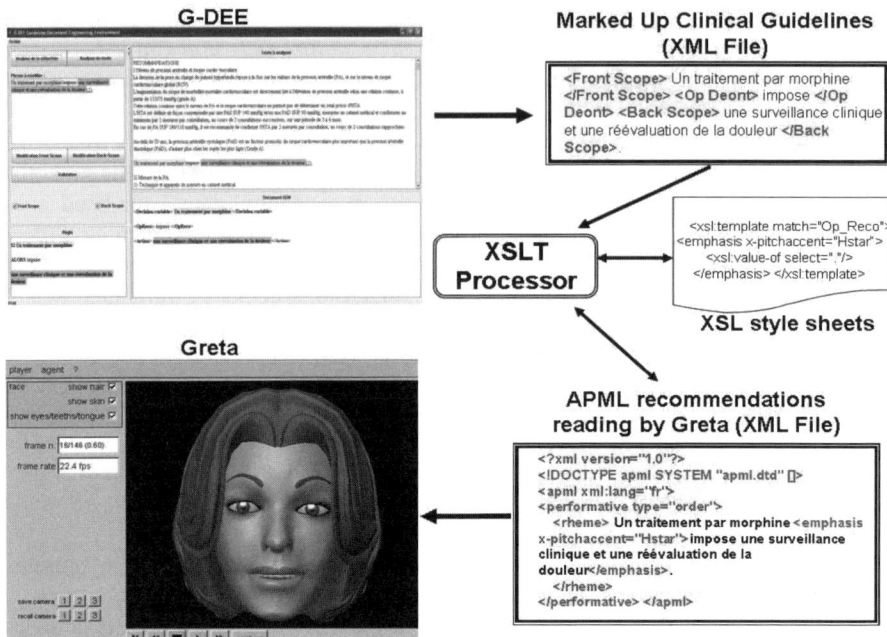

Fig. 1. System architecture

G-DEE [6] is a document analysis environment dedicated to the study of clinical guidelines[1]. It automatically detects recommendations using shallow natural language processing (NLP) techniques which recognize deontic operators in medical texts such as "authorize", "forbid", "ought to" [9]. Using specific grammars embedded in Finite-State Automata (FSA), G-DEE parses the whole document, identifying deontic operators and the text segments they apply to. The final output of this process is to structure the document around recommendations as shown below.

```
<Front-Scope> A treatment with morphine </Front-Scope> <Op_Reco>
imposes </Op_Reco> <Back-Scope> a clinical supervision and a reas-
sessment of pain </Back-Scope>[2]
```

Fig. 2. Example of marked-up recommendation

Let us now consider the different aspects that determine the strength and emphasis of a recommendation. Firstly, deontic operators fall within the broad categories of permission, obligation or interdiction and can be classified according to their

---

[1] G-DEE is actually deployed at the French National Authority for Health (HAS) to assist the early stages of Guideline development and has been used in the analysis of over 30 Guidelines over the past 14 months.

[2] This is an English translation of the original French Guideline (the translation does not affect the recommendation structure).

"strength" within these categories (e.g. strength of interdiction). Strength is not just an issue of vocabulary, but relates also to syntactic constructs (which have been uncovered in the process of deontic operator extraction). In other words, that a specific drug "should not be used" is stronger than it being "not recommended". It can also be noted that this concept bears some similarity with the illocutionary strength of communicative acts (which was one of our initial inspirations for this project). The reason why deontic operators play such a dominant role in the expression of the recommendations strength is to be found in the authoring rules which are enforced in several governmental agencies (including the French National Authority for Health (HAS)). In that sense, other linguistic phenomena are prevented from playing a role in the expression of recommendations' strengths: implicit nuances are discouraged, few adverbs are actually employed, and there is little use, if any, of affective categories[3].

2.2 The Greta Platform

The Greta agent [7] used in these experiments is a platform developed for the purpose of research in non-verbal behavior. It includes an animation system with facial parameters supporting detailed expressive animations synchronized to a Text-To-Speech (TTS) system. Greta's animations are controlled using instructions in the APML language [8], which is a markup language embedding communicative functions as well as the text fragments corresponding to the utterance to be passed to the TTS system. The baseline system incorporates a set of pre-defined communicative acts for which all facial animation parameters have been preset. The mapping between communicative acts and nonverbal behaviors is based on studies that have reported the communicative value of given signals (e.g. frowning is linked to goal obtrusiveness [10, 11]) and on video corpus annotation [12]. Communicative acts are grouped into classes depending on the information they convey [12]. These categories cover communicative acts related to the agent's beliefs (e.g. confidence in what is being said), to the goal of the agent (emphasizing the focus of its utterance), and finally its emotional state. APML also contains markers linked to the intonational structure of the sentences uttered by the agent. The traditional theme/*rheme* distinction can be thought of as a distinction between new and old information [13]. The pitch accents and boundary tones follow the ToBI annotation [14].

In particular, a previous study [15] has investigated the relations between performatives (communicative acts), such as: order, suggest, propose, warn, refuse, etc., and facial expressions. Three main classes of performatives were considered: *request*, *inform* and *question*. Within a given class, performatives (either from the request, inform or question class) share a common goal: respectively to elicit an action, to give information or to ask for an information.

Let us look more in detail performatives of the class request. They have been characterized along three dimensions: whom the action is requested from, how certain one is of the information provided and the power relationship [15]. The first dimension allows us to differentiate performatives of *advice* from those of *command*. We advise someone to perform an action that we believe could be beneficial to herself but we

---

[3] This may not be true of all types of Medical documents, yet it is a characteristic of clinical guidelines (easily understandable if one considers such an authoring rule as the one discouraging the use of adverbs).

command a person to carry out an action for our own benefit. Along the second dimension, we may suggest something when we are unsure. Power relations relate to our potential to coerce the person from whom an action is requested. As a result, we order or suggest depending on those power relations.

Based on the representation of the performative along these three dimensions, we have proposed a mapping between each of these dimensions and facial expressions [12]. That is, the facial expression associated to a given performative is obtained by combining the expressions arising from each dimension. Being certain or uncertain can be shown on the eyebrow region: one frowns when being very much certain of what one says, but raises one's eyebrow if uncertain [10, 11]. Head orientations (such as head kept straight up or tilted aside) can be a sign of a power relation: submissiveness is often shown by displaying the neck [16] while dominance is characterized by a straight up head). Performatives contain an intrinsic emotional factor [12]. When giving an order, one can potentially show anger if the requested action is not performed. On the other hand, when imploring one can display sadness.

Thus facial expressions of a performative may encompass information related to the potential emotional state of the agent, the dominance/submissive relationship with the interlocutor, as well as the degree of certainty of the information conveyed. These dimensions can be mapped to the dimensions characterizing recommendations. We introduce this mapping later in the paper (see section 5).

Let us now explain how our system computes the agent's animation from its intended utterance. Having established the dimensions along which performatives can be characterized and having specified the link between these dimensions and facial expressions, we have elaborated the mapping for several performatives (in particular of the class Request and of the class Inform) into facial expressions. These definitions are stored in a lexicon.

The system takes as input the text to be uttered, augmented with APML tags specifying the communicative intentions of the agent. As a first step, the selected text is sent to the TTS system, which provides the list of phonemes allowing the computation of the lip movement, as well as giving timing information supporting the synchronization of verbal and nonverbal streams. The next step consists of converting communicative intentions into a set of nonverbal behaviors by looking up in the lexicon mentioned above. Finally, the set of nonverbal behaviors are matched to the verbal stream to ensure synchrony across modalities. The last step is to compute the values of the facial animation parameters over time and to play the animation.

3 Related Work

Our work focuses on the use of conversational agents for visualizing rhetorical structures extracted from medical texts. The majority of the work on ECAs has focused on speech acts and emotional communication. There has not been much emphasis on rhetorical structures, to the exception of the T2D system [17], which used textual input decomposed into segments linked by rhetorical discourse relations to generate dialogues between agents.

In the context of medical applications, the emphasis has been on facilitating the dialog between doctors and patients, the former having often been criticized for their lack of empathy and ability to explain patients the course of their disease [18]. ECAs can thus play an important role in medical applications, as shown by Marsella et al. [19] with their system "Carmen's Bright IDEAS". MagiCster is another system whose applications have included an advice-giving dialog in a medical domain [20]. Here, the agent plays the role of a doctor giving the patient information about her disease. De Rosis et al. [21] have described a conversational agent advising on eating disorders. The aim of the dialog between the ECA and the user is to persuade them to improve their eating habits [22]. The use of ECAs for patients education has been shown to result in higher satisfaction rates [23]. ECAs have thus been demonstrated to bring added value to many applications for which a more human-like presentation [2] is beneficial, including assistance, help and guidance [24].

However, no work to date has investigated the use of ECAs to explore medical text perception by physicians themselves. Here, the equivalent to emotional content in Guidelines' recommendations would correspond to authority and responsibility, and the rhetorical expression of recommendations corresponds to communicative acts.

4 Identifying the Rhetorical Strength of Recommendations

The first step consisted in devising a scale to rate the strength of clinical recommendations. We carried out a study involving 14 medical experts from Inserm (French National Institute for Health) and HAS, who have been involved in the elaboration of clinical guidelines, to determine the level of consensus between experts about the strength of a given recommendation[4]. These experts rated the strength of 37 prototypical recommendations extracted from several hundreds recommendations occurring in recent clinical guidelines published by the HAS. They ranked the strength of each recommendation according to a predefined 6-point scale defined as follows:

```
CAT1- well-identified best practice, which is compulsory
CAT2- practice well adapted to the clinical situation that
presents demonstrable benefits
CAT3- accepted practice which can be advised, or to be considered
CAT4- practice left to the discretion of the physician
CAT5- statement explaining or justifying a course of action
CAT6- a useful information item
```

Fig. 3. Categories for evaluating the strength of recommendations

For each deontic verb, used in recommendations, we were subsequently able to associate a numerical score quantifying its rhetoric strength depending on the previous analysis of Guidelines corpora[5] and on explicit rules on guidelines vocabulary and terminology (including verbs for recommendations) mentioned in HAS internal documents on Guidelines' authoring methodology. This will serve as a starting point

---

[4] This represents a very significant sample, considering the total number of such experts within any given European country.
[5] Guidelines may contain explicit gradings on the level of certainty and strength for their recommendations. These make possible to analyse statistically the occurrence of deontic verbs as a function of these gradings.

to map the rhetorical strength of deontic expressions onto the emotional categories of Greta. It is an empirical solution, based on existing corpora, to the problem of relating grades of recommendations (as described by the experts) to the linguistic formulation of the strength of a recommendation. It also supports the extension of the set of performatives used by Greta as described in the next section.

5 Mapping Rhetorical Structures onto Multimodal Communicative Acts

The process by which the rhetorical strength of textual recommendations will be visualized rests on a mapping from deontic operators onto multimodal communicative acts. These can be described as the dynamic expression of traditional communicative acts (order, advice, propose, etc.), using communicative parameters and dynamic animation of non-verbal behavior, in particular facial expressions. The rationale for such a mapping derives from the pre-existing commonality between certain deontic operators used in the description of recommendations and the set of primitive communicative acts originally embedded in the APML control language (which contains communicative acts such as *advice*). This mapping attempts to generalize these commonalities by relating deontic operators to communicative acts but also their perceived strength to the *rheme* part of APML expressions. The *Rheme* corresponds to novel information that is brought into the conversation, and most of the nonverbal behavior occurs within the *rheme* [25].

5.1 Definition of Specific Expressions

As introduced previously (Section 2.2), performatives are described along three dimensions. We have elaborated the mapping between the six categories of the recommendations' strength scale and the performatives by looking at the common values for these 3 dimensions[6]. In particular, communicative acts are defined as a pair whose first element corresponds to the meaning of the communicative act. It is specified by APML tags, while the second element represents the signals that convey this meaning. Let us see the mapping for each category:

- CAT1: the practice is compulsory. It corresponds to an order, a request of carrying out an action. In APML, the performative "order" is described by a frown (sign of anger), head up and look down. Recommendations may not encapsulate social relationship thus any behavior linked to this aspect have been eliminated (in this case, head up and look down). To highlight the importance of the recommendation, the emphasis tag is added. It is shown through head nods.
- CAT2: the practice is a strong recommendation but not as much as CAT1. As for CAT1, no notion regarding social relationship is needed here. Thus,

---

[6] In the current version of our system, the mapping is done with the performatives that are already defined within the APML language. However, we aim to extend the list of performative to improve coverage.

CAT1 is represented by a less intense frown. The APML tag is "order" but with lesser intensity expression. No emphasis tag is added.
- CAT3: This type of recommendation can be considered as an advice. It is displayed using the eyebrow shape to convey the performative 'advice': slight rising of the eyebrows.
- CAT4: The recommendation mentions a possible course of action as a suggestion. Suggestion is characterized by raised eyebrows and tilted head.
- CAT5: the recommendation is used to inform physicians but with a certain emphasis. It is translated by looking at one's addressee and performing a head nod on the emphasized word. Two tags are used, one performative "inform" and the emphasis tag.
- CAT6: the physician is simply informed: that is displayed through gaze behavior, namely looking at the addressee.

The APML tags are added to the text automatically using the following rules:
- Performative tags: they mark the whole sentence.
- Emphasis tags: emphasis is linked to the intonational structure of the sentence. As new information should receive the most attention, emphasis is marked. We use the *rheme*/theme structure. New information is part of the *rheme*. The emphasis tag is set around the deontic verb while the *rheme* tag goes around the sentence (as the performative).
- Certainty tags: to mark negation contained in deontic verb (e.g. French *interdire* (to forbid) of CAT1; *ne pas prescrire* (not to prescribe) of CAT2), we use the tag "certainly-not". It spans the same text as the performative. Certainly-not is shown by a frown.

Let us see in the next two sections how this automatic transformation between G-DEE and APML happens.

5.2 XSL-Based Transformations and APML Generation

Because Greta's input format for multimodal communicative acts (APML) is XML-based, it is a natural choice, from a technical standpoint, to use XSLT transformations to generate APML from the deontic operator structure based itself on XML. The XSLT processor already integrated in G-DEE has been extended to support the generation of APML formulas. XSL style sheets need to define the mapping between various categories, in particular the communicative acts defined in the APML DTD. This includes the transfer of sentence fragments belonging to the recommendation from the marked-up guideline to Greta's TTS system. Firstly, the XML file generated by the text analysis module contains marked-up recommendations which are the starting point for the further XSL transformations. An example of such a marked-up recommendation is presented in Figure 2. We thus defined XSL style sheets to transform automatically this file into an APML file controlling Greta. In this example, the presence of a deontic operator of the type CAT1 justifies the presence of an <emphasis> mark-up in APML. In turn, the type of deontic operator determines the type of communicative act. This transformation is based on the conceptual mapping between the

```
<apml>
   <performative type="order"> <rheme>
   A treatment with morphine
   <emphasis> imposes
   </emphasis>
   a clinical supervision and a reassessment of pain levels
   </rheme> </performative>
</apml>
```

Fig. 4. APML expression resulting from the XSL transformation to map recommendations to speech acts

deontic operators recognized by G-DEE and the set of previously predefined communicative acts in Greta (see Fig. 4 for an example of transformation).

This conceptual mapping essentially establishes a correspondence between the expressivity of the written text and that of the recommendation pronounced by Greta. XSL style sheets are specific to each category of the recommendations strength. G-DEE characterizes which kind of deontic verb the recommendation contains using the mapping described in the section above. In some cases there exists a direct, one-to-one mapping such as with the "advice" operator, or, to some extent the "propose" one (as they exist in both mark-up systems). The deontic operators which are meant to "suggest" (e.g. French *être laissé à* (to be left to) / *pourrait* (may) of CAT4) can be mapped to the *suggest* communicative act. Strong negative recommendations (such as the French "*déconseiller*" (Advice not to, discourage)) have been mapped to the *refuse* communicative act in APML, while for mild negative recommendations we use the disagree communicative act.

To intensify negative deontic verbs the communicative function Certainly-not is added to the text. Certainly-not is part of the cluster 'certainty' and is marked by a frown. Frown is attached to a negative signal as it is often linked to goal obstruction [10]. Finally, for CAT1 and CAT5 cases, the emphasis tag is added around the deontic verb.

Table 1. Excerpt of the mapping table between deontic verbs and APML performative types

Deontic verb	APML
CAT1 – APML: order	
ordonner (to order) / *impose* (to impose) / *devra associer* (will have to associate)	Performative "order"+emphasis/*rheme*
interdire (to forbid)	Performative "order" + certainty "certainly_not" +emphasis/*rheme*
CAT2 – APML: recommend	
recommander (to recommend) / *prescrire* (prescribe) / *contre-indiquer* (to counterindicate)	Performative "recommend"
déconseiller (to advise not to) / *ne pas recommander* (not to recommend) / *ne pas prescrire* (not to prescribe)	Performative "recommend" + certainty "certainly_not"

Overall, the existing APML performative set of the communicative acts can support a consistent mapping: the only limitation lies in the lack of explicit nuances between some forms of positive recommendations, which should be the object of further work can be compensated in part using redundancy in behaviors, that is where meanings are conveyed over different modalities (e.g. raised eyebrow and head nod). Next section illustrates this mapping through an example.

6 Example Results

The XSLT transformations generate the most appropriate communicative act for the deontic verb considered in the recommendation. The following screenshots of Greta enable to visualize the differences between expressions according to performative type which are mainly focus on the eyebrow and the head nod (although differences can only be really seen on the dynamic animation).

The following examples correspond to two of the six categories defined for the strength scale. For the category 2, the dedicated style sheet enables to transform a marked-up recommendation to an APML format (Fig. 5) that supports the mapping of the French "*il est recommandé*" ("it is recommended") deontic verb to the *recommend* performative type, defined in section 5.1.

```
<apml>
    <performative type="recommend"> <rheme>
    <emphasis> It is recommended
    </emphasis> to perform a venous Doppler examination as
    part of the management of all patients with ulcers of the
    lower limbs. </rheme> </performative>
</apml>
```

Fig. 5. The resulting APML file corresponds to a recommend performative type

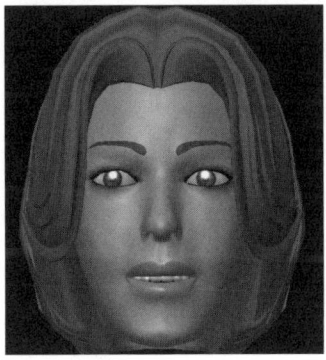

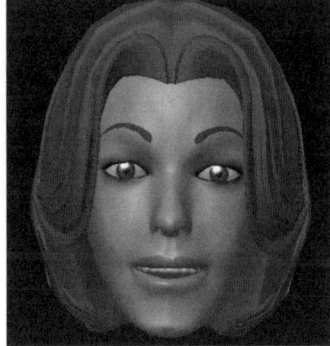

Fig. 6. The resulting expression of Greta corresponding to a *recommend* performative type

Fig. 7. The resulting APML file corresponding to a *suggest* performative type

The corresponding expression of Greta consists of a recommendation with an emphasis on the deontic verb "*il est recommandé*" (it is recommended) and a frown with a head nod (Fig. 6), while the suggest conversational act (Fig. 7) is associated to a slight raising of the eyebrows and a head tilt.

7 An Application to Consensus Judgments of Recommendations' Strength

We conducted an evaluation of the system with six medical experts drawn from the group of the fourteen experts that have participated in the definition of recommendations' strengths. For this evaluation, we devised a test suite of nine prototypical recommendations, representative of the whole spectrum of recommendations' strength. The main objective of this evaluation consists of determining whether Greta improves the perception of recommendations strengths, for instance by generating a stronger consensus or helping to disambiguate between neighboring categories. Since standard deviation is a simple and well described measure of consensus [26], we analyzed its value throughout our experiments. Each of the six medical experts rated the recommendations strength first from reading them and secondly from seeing them presented by Greta. The average strength score as well as the standard deviation were calculated for each recommendation, without and with Greta (Fig. 8).

The effect of Greta appears to vary greatly depending on the category in which a recommendation has been indexed. However, one of the main problem in guideline authoring is the level of consensus. In that sense, figures obtained from isolated users do not reflect the actual dynamics of a working group. A lack of consensus (measured e.g. through a high value for standard deviation) can have significant implications during a face-to-face consensus meeting and this is why improving consensus is a major objective in the process of Guidelines' elaboration.

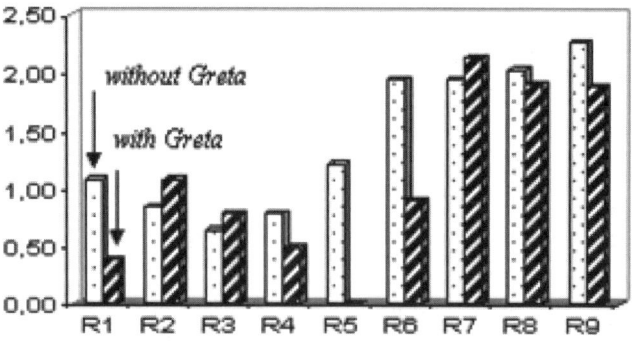

Fig. 8. Impact of Greta on the standard deviation of experts' judgments of recommendations' strength: this impact is stronger for "borderline" categories (R4, R5 and R6), where consensus is most difficult to reach

Most importantly, we observed a very significant effect of Greta on the standard deviation of recommendations' strength, and that effect is more pronounced, and highly significant, for intermediate categories, such as CAT3 (R4), CAT4 (R5) and CAT5 (R6), which are known to be the object of significant debate in working groups. This effect is also remarkable for the strongest type of recommendations CAT1 (R1), for which difficulties in reaching consensus have been often reported. The decrease in standard deviation can be interpreted as a better consensus between experts: this suggests that using Greta would potentially improve the efficiency of a working group. To a large extent, the system presented here can restore the link between the wording of a recommendation and its intended impact on the reader. As a tool to assist the authoring of guidelines, it should help selecting the appropriate level of emphasis required as well as balancing the importance of recommendations across the document as a whole.

8 Conclusions

ECAs have been mostly described in dialogue and interface applications, with little work on their use in the visualization of textual properties. The most natural applications in that area would be to dramatize the affective aspects of the underlying text. Yet, we suggest that dramatization, as provided by the non-verbal behavior of ECAs, can also be of use to visualize the rhetorical content of texts, and that this can have practical applications as well. The principle behind this approach is that communicative acts which are used to define ECA non-verbal behavior naturally overlap with some of the rhetorical intentions embedded in texts. The mapping between these two aspects may not be trivial, and in these first experiments we had to provide an empirical solution based on domain expertise.

These first results are very encouraging and future work will extend this approach using more sophisticated expressive mechanisms such as gestures, possibly also relating non-verbal behavior to further contents of the deontic operators, such as the recommended course of action.

Acknowledgments. Gersende Georg is partly funded through a post-doctoral fellowship from "Region Ile-de-France". We thank all the medical experts from the French National Health Authority (HAS) and Inserm (French National Institute of Health) for their participation in data collection and in evaluation experiments.

References

1. Hoorn, J., Konijn, E.: Personification: Crossover between Metaphor and Fictional Character in Computer Mediated Communication. In: The annual meeting of the International Communication Association, San Diego, CA (2003)
2. Nass, C., Steuer, J., Tauber, E.: Computers are Social Actors. In: Proceedings of the SIGCHI conference on Human factors in computing systems: celebrating interdependence, Boston, Massachusetts, United States, pp. 72–78 (1994)

3. Abbattista, F., Lops, P., Semeraro, G., Andersen, V., Andersen, H.: Evaluating virtual agents for e-commerce. In: Falcone, R., Barber, S., Korba, L., Singh, M.P. (eds.) AAMAS 2002. LNCS (LNAI), vol. 2631. Springer, Heidelberg (2003)
4. Allbeck, J., Badler, N.: Toward Representing Agent Behaviors Modified by Personality and Emotion. In: Falcone, R., Barber, S., Korba, L., Singh, M.P. (eds.) AAMAS 2002. LNCS (LNAI), vol. 2631. Springer, Heidelberg (2003)
5. Sackett, D., Rosenberg, W., Gray, J., Haynes, R., Richardson, W.: Evidence-based medicine: what it is and what it isn't. BMJ 312(7023), 71–72
6. Georg, G., Jaulent, M.-C.: A Document Engineering Environment for Clinical Guidelines. In: Proceedings of the 2007 ACM Symposium on Document Engineering, Winnipeg, Manitoba, Canada, pp. 69–78. ACM Press, New York (2007)
7. Pelachaud, C.: Multimodal expressive embodied conversational agent. In: ACM Multimedia, Brave New Topics session, Singapore, pp. 683–689 (2005)
8. De Carolis, B., Pelachaud, C., Poggi, I., Steedman, M.: APML, a Markup Language for Believable Behavior Generation. In: Prendinger, H., Ishizuka, M. (eds.) Life-like Characters. Tools, Affective Functions and Applications, pp. 65–86. Springer, Heidelberg (2003)
9. Moulin, B., Rousseau, D.: Knowledge acquisition from prescriptive texts. In: Proceedings of the 3rd international conference on Industrial and engineering applications of artificial intelligence and expert systems, Charleston, South Carolina, United States, pp. 1112–1121 (1990)
10. Ekman, P.: About brows: Emotional and conversational signals. In: von Cranach, M., Foppa, K., Lepenies, W., Ploog, D. (eds.) Human ethology: Claims and limits of a new discipline: contributions to the Colloquium, pp. 169–248. Cambridge University Press, Cambridge (1979)
11. Chovil, N.: Discourse-oriented facial displays in conversation. Research on Language and Social Interaction 25, 163–194 (1991)
12. Poggi, I.: Mind Markers. In: Rector, M., Poggi, I., Trigo, N. (eds.) Gestures, Meaning and use, pp. 203–207. University Fernando Pessoa Press, Oporto (2003)
13. Bolinger, D.: Intonation and its Part. Stanford University Press (1996)
14. Silverman, K., Beckman, M., Pitrelli, J., Ostendorf, M., Wightman, C., Price, P., Pierrehumbert, J., Hirschberg, J.: ToBI: A Standard for Labeling English Prosody. In: Proceedings of the International Conference on Spoken Language Processing, Banff, Alberta, pp. 867–870 (1992)
15. Poggi, I., Pelachaud, C.: Performative faces. Speech Communication 26, 5–21 (1998)
16. Darwin, C.R.: The expression of emotions in man and animals. Murray, London (1872)
17. Piwek, P., Hernault, H., Prendinger, H., Ishizuka, M.: T2D: Generating Dialogues Between Vir-tual Agents Automatically from Text. In: Intelligent Virtual Agents 2007, Paris, pp. 161–174 (2007)
18. Charon, R.: Narrative Medicine - A Model for Empathy, Reflection, Profession, and Trust. The Journal of the American Medical Association 286(15), 1897–1902 (2001)
19. Marsella, S., Gratch, J., Rickel, J.: Expressive behaviors for virtual worlds. In: Prendinger, H., Ishizuka, M. (eds.) Life-like Characters. Tools, Affective Functions and Applications, pp. 317–376. Springer, Heidelberg (2003)
20. De Carolis, B., De Rosis, F., Carofiglio, V., Pelachaud, C., Poggi, I.: Interactive Information Presentation by Embodied Animated Agent. In: International Workshop on IPNMD, Verona, Italy (2001)
21. De Rosis, F., De Carolis, B., Carofiglio, V., Pizzutilo, S.: Shallow and Inner Forms of Emotional Intelligence in Advisory Dialog Simulation. In: Prendinger, H., Ishizuka, M. (eds.) Life-like Characters. Tools, Affective Functions and Applications, pp. 271–294. Springer, Heidelberg (2003)

22. De Rosis, F., Novielli, N., Carofiglio, V., Cavalluzzi, A., De Carolis, B.: User modeling and adaptation in health promotion dialogs with an animated character. Journal of Biomedical Informatics 39(5), 514–531 (2006)
23. Bickmore, T., Caruso, L., Clough-Gorr, K.: Acceptance and usability of a relational agent interface by urban older adults. In: CHI 2005 extended abstracts on Human factors in computing systems, pp. 1212–1215 (2005)
24. Bickmore, T., Pfeifer, L., Paasche-Orlow, M.: Health Document Explanation by Virtual Agents. In: Intelligent Virtual Agents 2007, Paris, pp. 183–196 (2007)
25. Cassell, J., Torres, O., Prevost, S.: Turn Taking vs Discourse Strcuture: How Best to Model Multimodal Conversation. In: Wilks, Y. (ed.) Machine Conversations. Kluwer, Dordrecht (1999)
26. Burke, M., Dunlap, W.: Estimating interrater agreement with the average deviation index: a user's guide. Organizational Research Methods 5(2), 159–172 (2002)

Evaluation of Justina: A Virtual Patient with PTSD

Patrick Kenny, Thomas D. Parsons, Jonathan Gratch, and Albert A. Rizzo

Institute for Creative Technologies,
University of Southern California
13274 Fiji Way Marina Del Rey, CA 90292, USA
{kenny,tparsons,gratch,rizzo}@ict.usc.edu

Abstract. Recent research has established the potential for virtual characters to act as virtual standardized patients VP for the assessment and training of novice clinicians. We hypothesize that the responses of a VP simulating Post Traumatic Stress Disorder (PTSD) in an adolescent female could elicit a number of diagnostic mental health specific questions (from novice clinicians) that are necessary for differential diagnosis of the condition. Composites were developed to reflect the relation between novice clinician questions and VP responses. The primary goal in this study was evaluative: can a VP generate responses that elicit user questions relevant for PTSD categorization? A secondary goal was to investigate the impact of psychological variables upon the resulting VP Question/Response composites and the overall believability of the system.

Keywords: Virtual Humans, Virtual Patients, Psychopathology.

1 Introduction and Background

The potential of using virtual humans as virtual standardized patients (VP) for use in clinical assessments, interviewing and diagnosis training is becoming recognized as the technology advances [2,3]. These VPs are embodied interactive agents [4,6,11,24,26] who are designed to simulate a particular clinical presentation of a patient with a high degree of consistency and realism [15, 28]. VPs have commonly been used to teach bedside competencies of bioethics, basic patient communication, interactive conversations, history taking, and clinical decision making. [5,18,26] VPs can provide valid, reliable, and applicable representations of live patients [33]. Research into the use of VPs in psychotherapy training is in its nascent stages [8,14,20]. Since virtual humans and virtual environments can allow for precise presentation and control of dynamic perceptual stimuli (visual, auditory, olfactory, gustatory, ambulatory, and haptic conditions), conversations and interactions, they can provide ecologically valid assessments that combine the control and rigor of laboratory measures with a verisimilitude that reflects real life situations [1,18,23,30]. Although progress has been made toward establishing systems that are sensitive to component psychological processes, more studies are required to understand the effectiveness of these systems for training and education, to measure the believability of the characters with respect to their verbal and non-verbal behavior and how different genders, races and personality or interview styles interact with the characters.

This current project builds on our previous work in building a VP for conduct disorder [15] and aims to improve child and adolescent psychiatry residents, and medical students' interview skills and diagnostic acumen for a difficult subject through practice with a female adolescent virtual human with post-traumatic stress disorder (PTSD). This interaction with a VP provides a context where immediate feedback can be provided regarding trainees' interviewing skills in terms of psychiatric knowledge, sensitivity, and effectiveness. Use of an embodied natural language-capable virtual character is beneficial in providing trainees with exposure to psychiatric diagnoses such as PTSD that is prevalent in their live patient populations and believed to be under-diagnosed due to difficulty in eliciting pertinent information. Virtual reality patient paradigms, therefore, will provide a unique and important format in which to teach and refine trainees' interview skills and psychiatric knowledge.

In this paper we describe a series of subject tests of a virtual patient system performed with medical students to evaluate its usefulness and effectiveness as a medium to communicate with the students. The evaluation consisted of an assessment of the system as a whole through questionnaires and data collection of the questions and responses in the interview. Additionally we investigate the relationship of the questions with a number of psychological variables such as openness to interaction with the VP and willingness to be immersed in the virtual environment.

2 Designing a Patient with Post Traumatic Stress Disorder (PTSD)

2.1 Virtual Justina

One of the challenges of building complex interactive VPs that can act as simulated patients has been in enabling the characters to act and carry on a dialog like a real patient that has the specific mental condition in the domain of interest. Additional issues involve the breadth and depth of expertise required in the psychological domain to generate the relevant material for the character and dialog. In our first attempt to design a VP 'Justin'[15] we choose a domain, conduct disorder, that was more forgiving of inappropriate responses to user questions and where the patient would be somewhat resistant to answering questions. Inappropriate or out of domain responses

Fig. 1. Justina Virtual Patient

were seen as part of the disorder and this did not negatively impact the interview process. The current domain of PTSD is less forgiving and requires the system to respond appropriately based on certain criteria for PTSD as described in the Diagnostic and Statistical Manual of mental disorders (DSM-IV) category (309.81) [9]. For the PTSD domain we built an adolescent girl character called Justina, see Figure 1. Justina has been the victim of an assault and shows signs of PTSD. The technology used for the system is based on the virtual human technology developed at USC [16,29] and is the same as what was used with the previous VP 'Justin'. The system uses speech recognition, question / response and a procedural animation system to control the character.

2.2 PTSD Domain

The experience of victimization is a relatively common occurrence for both adolescents and adults. However, victimization is more widespread among adolescents, and its relationship to various problem outcomes tends to be stronger among adolescent victims than adult victims. Whilst much of the early research on the psychological sequelae of victimization focused on general distress or fear rather than specific symptoms of PTSD, anxiety, or depression, studies have consistently found significant positive correlations between PTSD and sexual assault, and victimization in general and violent victimization in particular [22]. Although there are a number of perspectives on what constitutes trauma exposure in children and adolescents, there is a general consensus amongst clinicians and researchers that this is a substantial social problem [25]. The effects of trauma exposure manifest themselves in a wide range of symptoms: anxiety, post-trauma stress, fear, and various behavior problems. New clinicians need to come up to speed on how to interact, diagnose and treat this trauma.

According to the most recent revision to the American Psychiatric Association's DSM Disorders, PTSD is divided into six major categories; refer to the DSM-IV category 309.81 [9] for a full description and subcategories;

A. Past experience of a traumatic event and the response to the event.
B. Re-experiencing of the event with dreams, flashbacks and exposure to cues.
C. Persistent avoidance of trauma-related stimuli: thoughts, feelings, activities or places, and general numbing such as low affect and no sense of a future.
D. Persistent symptoms of anxiety or increased arousal such as hyper vigilance or jumpy, irritability, sleep difficulties or can't concentrate.
E. Duration of the disturbance, how long have they been experiencing this.
F. Effects on their life such as clinically significant distress or impairment in social or educational functioning or changes in mental states.

Diagnostic criteria for PTSD includes a history of exposure to a traumatic event in category A and meeting two criteria and symptoms from each B, C, and D. The duration of E is usually greater than one month and the effects on F can vary based on severity of the trauma. Effective interviewing skills are a core competency for the clinicians, residents and developing psychotherapists who will be working with children and adolescents exposed to trauma. A clinician needs to ask questions in each of these categories to properly assess the patient's condition.

2.3 Question / Response Categorization

Domain building for the VP consisted of role-playing sessions to gather the verbal and non-verbal behavior for the patient along with the set of questions typically asked by a clinician. Additionally, iterative discussions with psychiatry faculty from the Keck School of medicine at USC were performed to enhance the corpus of questions and responses. The goal was to build enough of the domain to cover the six categories in the PTSD DSM criteria and cover the kinds of questions people would ask a patient. The corpus was used for the statistically natural language question/response system [17, 32]. The natural language system works by selecting responses based on input questions. The set of questions and responses are manually mapped by a domain expert. For this application domain there were a total of 459 questions that mapped roughly 4 to 1 to a set of 116 responses. The aim was to build the domain corpus with what we could anticipate and then elicit questions from the user that s/he may ask of the VP for the specific traumatic experience and use those questions in an iterative process to further build the corpus. Since PTSD falls in the diagnostic category of anxiety disorders, rather than assessing for all of the specific criteria, we initially focused at a high level upon the six major clusters of symptoms following a traumatic event. While this did not give the character depth but breadth, for initial testing this seemed prudent. Next, we developed two additional categories that we felt would aid in assessing user questions and VP responses that are not included in the DSM;

G. A general category meant to cover questions regarding establishing rapport, establishing relations, clarifications, opening and closing dialog.
H. A category to cover accidental mouse presses or miscellaneous items. Users interact with the system with speech, however they need to press the mouse button while talking. Sometimes when people are thinking they have a tendency to press the button then release without saying anything, this causes the system to respond with an off topic response, and can confuse new users.

Table 1 is an example of some questions and responses from Justina for each of the six categories. Once all of the responses were established a voice actor was used to record the voice for Justina to be used by the system.

Table 1. Question / Response Categorization

Category	User Question	Justina Response
1(A) Trauma	So, what happened to you that night?	Something really bad happened.
2(B) Re-experience	Do you still think about what happened?	Sometimes I feel like the attack is happening all over again
3(C) Avoidance	Do you go out with your friends?	I just stay away from everyone now.
4(D) Arousal	Do you feel jumpy?	I feel like I have to watch my back all the time.
5(E) Duration	How long has this been going on?	A few months
6(F) Life Effect	Are you upset?	Sometimes I don't do anything but stay in my room and cry.
7(G) Communication	Hi Justina, I'm Doctor..	Hello
8(H) Other	Button Press	I don't get what you mean.

3 Method

Although our primary goal in this study was evaluative: to assess the effectiveness of a virtual standardized patient to generate responses that elicit user questions relevant for a virtual character that has PTSD, a secondary goal was to investigate the relationship between a number of psychological variables and the resulting VP Question/Response composite. An important issue in the study of intelligent virtual agents is to identify under what circumstances a person interacting with the virtual agent is open to the interaction. We were interested in the psychological variables of hypnotizability and absorption [31] as well as immersiveness [37] and presence [10,21] in relation to a person's experience of an interaction with the VP. Although these variables have been little explored, results from a recent study reveal that physiological arousal appeared to be moderated by participant hypnotizability and absorption levels [19,36]. High-absorption individuals may be more capable of imagining that the VP has PTSD when it is suggested. It was hypothesized that participants' scores on measures of absorption, immersion, and presence would be positively and significantly correlated with a measure of their VP Question/Response composite.

3.1 Participants

Participants were asked to take part in a study of novice clinicians interacting with a VP system. They were not told what kind of condition the VP had if any. Two recruitment methods were used: poster advertisements on the university medical campus; and email advertisement and classroom recruitment to students and staff. A total of 15 people (6 females, 9 males; mean age = 29.80, SD 3.67) took part in the study. Ethnicity distribution was as follows: Caucasian = 67%; Indian = 13%; and Asian = 20%. The subject pool was made up of three groups: 1) Medical students (N=7); 2) Psychiatry Residents (N=4); 3) Psychiatry Fellows (N=4). For participation in the study, students were able to forgo certain medical round time with the time spent in the interview and questionnaires, which took approximately 45 minutes.

3.2 Setup

The VP system consisted of the virtual character Justina, as seen in Figure 1 and 2, along with a headset for speech input and mouse button which was required to be

Fig. 2. Testing setup and interaction

pressed while speaking. A control station was adjacent to the subject to run the system and log the data. Cameras were setup to record the subjects face and the interaction with the VP from the side for later post processing analysis and review.

3.3 Process and Procedure

Medical students currently perform interview training with human actors acting as standardized patients. The actors portray some clinical problem in what is called an Objective Structured Clinical Examination (OSCE) [13,34,35]. These tests typically take from 20-30 minutes, a faculty member watches the student perform and students are videotaped. The evaluation consists of self assessment rating along with faculty assessment and a review of the videotape. This practice is common, although varies based on the actors, available faculty members and space and time at the university. Although schools commonly make use of standardized patients to teach interview skills, the diversity of the scenarios that standardized patients can characterize is limited by the availability of human actors and their skills at portraying the condition. Additionally the actors most likely vary their performance from subject to subject and location to location. This is an even greater problem when the actor needs to be an adolescent, elder or portray a difficult condition. Our process is similar to an OSCE, but the actor is replaced with a virtual patient and an observer is replaced by video recording. Using virtual patients will allow standard performances for all subjects.

The subject testing was divided into three phases, a pre-test and pre-questionnaire, the interview and a post-questionnaire. The pre-questionnaire was performed in a separate room from the interview and took about 10 minutes. For the interview the participants were asked to perform a 15 minute interaction with the VP and assess any history or initial diagnosis of a condition of the character. The participants were asked to talk normally as they would to a standardized patient, but were informed that the system uses speech recognition and was a research prototype. They were free to ask any kind of question and the system would try to respond appropriately. At the end of the 15 minute exchange they would be sent to another room to take the post-questionnaire. Data was logged during the interview for later processing. The video recordings and system logs of the interaction could be re-played for review, critique and commentary by child and adolescent psychiatry attendings, as a teaching tool for residents, or for groups of medical students learning about PTSD, or even for students using distance learning that don't have access to this technology.

The data in the system was logged from various modules. Figure 2 is a diagram of how the user interacts with the VP system and the logging and annotation pipeline. First the user speech is recorded from the automated speech recognition (ASR) engine and later transcribed, this is the actual text said by the user. Next the output text from the ASR is logged, this text is usually not 100% accurate due to speech models, accents, or voice types. This output when compared with the actual text will give the accuracy of the speech engine, due to time constraints we were not able to compute the accuracy of the ASR, but know that it could be improved. The ASR output is sent to the natural language (NL) statistical question/response system. The NL system records a transcript of the entire dialog session, this is used later to help analyze the interaction. System messages are logged and can be used to reconstruct what was happening in all parts of the system as needed. Cameras record participant's facial expressions and system interaction with the patient for analysis at a later time.

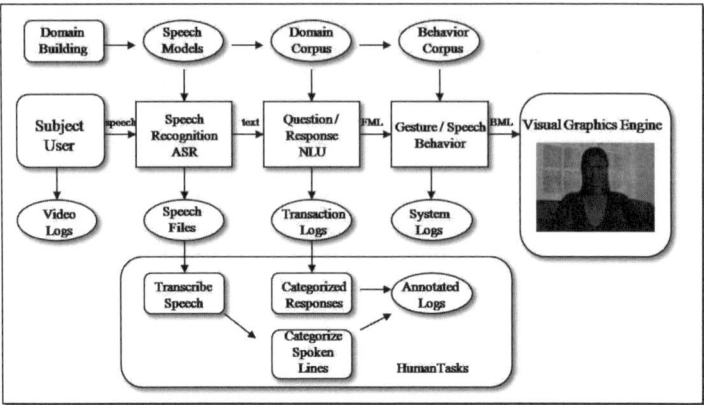

Fig. 3. Interaction and Data Logging Pipeline

3.4 Measures

As mentioned above, an important issue in the study of intelligent virtual agents is to identify under what circumstances a person interacting with the agent is open to the interaction. By this we mean how open is the person interacting with the VP open to new experiences and interacting with novel technologies. Again, results from a recent study that we conducted reveal that physiological arousal appeared to be moderated by participant openness to such interactions [19,36]. High-absorption individuals may be more capable of imagining that the VP has PTSD when it is suggested.

The following standardized and unstandardized measures were used to assess the impact of absorption and immersiveness upon the "believability" of the system. Prior to the experiment itself, the subjects were required to fill in the following standardized questionnaires: 1) Tellegen Absorption Scale (TAS). The TAS questionnaire aims to measure the subject's openness to absorbing and self-altering experiences. The TAS is a 34-item measure of absorption, and is a widely used questionnaire with well established reliability and validity [31]. 2) Immersive tendencies questionnaire (ITQ). The ITQ measure individual differences in the tendencies of persons to experience "presence" in an immersive VE. The majority of the items relate to a person's involvement in common activities. While some items measure immersive tendencies directly, others assess respondents' current fitness or alertness, and others emphasize the user's ability to focus or redirect his or her attention. The ITQ is comprised of 18 items, and each is rated on a 7-point scale and is a widely used questionnaire with well established reliability and validity [37]. Subjects also completed unstandardized measures that were developed specifically for this protocol: 1) Virtual Patient Pre-Questionnaire (VPQ1). This scale was developed to establish basic competence for interaction with a virtual character that is intended to be presented as one with PTSD, although no mention of PTSD is on the test. 2) Justina Pre-questionnaire (JPQ1). We developed this scale to gather basic demographics and ask questions related to the user's openness to the environment and virtual reality user's perception of the technology and how well they think the performance will be. There were 5 questions regarding the technology and how well they thought they might perform with the agent.

After the experiment the subjects were instructed to fill in the following standardized questionnaire: 1) Presence questionnaire (PQ). The Presence Questionnaire is a common measure of presence in immersive virtual reality. Presence has been described of as comprising three particular characteristics: sense of being within the VE; extent that the VE becomes the dominant reality for users; and extent to which users view the VE as a place they experienced rather than simply images they observed. The PQ is a widely used questionnaire with well established reliability and validity [37] Subjects were also asked to complete unstandardized questionnaires developed specifically for this protocol: 1) Justina Post-questionnaire (JPQ2). We developed this scale to survey the user's perceptions related to their experience of the virtual environment in general and experience interacting with the virtual character in particular the patient in terms of its condition, verbal and non-verbal behavior and how well the system understood them and if they could express what they wanted to the patient. Additionally there were questions on the interaction and if they found it frustrating or satisfying. There were 25 questions for this form. 2) Virtual Patient Post-questionnaire (VPQ2). This scale was exactly the same as the Virtual Patient Pre-questionnaire and will be used in the future for norming of a pre-post assessment of learning across multiple interactions with the VP. In the future we will also include social presence and rapport scales and include a control set that will just go thru a fixed script with the interview.

3.5 Data Analytics

Participants completed the VPQ1; JPQ1; TAS; and ITQ prior to the VP trial. Following this, participants received instructions on how to interact with the patient, then the trial started. After 15 minutes the trial was completed, and participants then completed a PQ; JPQ2; and VPQ2. When all the trials were completed the speech for each participant was transcribed and annotated with one of the categories in Table 1.

Here we focused on effective interview skills—a core competency for psychiatry residents and developing psychotherapists. The keys aspects of the interview that we looked at were: interpersonal interaction; attention to the VP's vocal communications, as well as verbal and non-verbal behavior. Specifically, we wanted to assess whether the clinician established and maintained rapport, as well as ask questions related to the reason for referral. We also wanted to assess whether the user (clinician in training) made attempts to gather information about the VP's problems. Finally, we wanted to see if the user would attempt detailed inquiry to gain specific and detailed information from the VP, separating relevant from irrelevant information.

VP Question/Response Composite

Question/response composites were developed to reflect the shared variance existing between the responses of a VP simulating PTSD in an adolescent female and of DSM IV TR-specific Questions (from novice clinicians) that are necessary for differential diagnosis. The question/response composites drawn from novice clinician questions and VP responses were referred to as VP Question/Response composites or (VP_QR'). Again, the primary goal in this study was evaluative and the VP_QR' scores were calculated to assess whether a virtual standardized patient could generate responses that elicit user questions relevant for PTSD categorization. For the VP_QR' scores, we first calculated eigenvalues via least squares procedures and separate composite measures were created for each observation. The resulting weights were used in conjunction with the original variable values to calculate each observation's score. The VP_QR' scores were standardized according to a z-score.

Primary and Secondary Analyses

To assess whether the responses of a VP simulating PTSD in an adolescent female could elicit a number of DSM IV TR-specific questions (from novice clinicians) that are necessary for differential diagnosis, our data analysis was completed in two stages. In the first stage, the reference distribution is a correlation of each cluster of questions (from the novice clinicians) making up a particular DSM PTSD Category with each (corresponding) cluster of responses from the VP representing the same DSM PTSD Category. In the second stage, variance from each individual's psychological distributions is controlled. Herein, the reference distribution reflects a semipartial correlation controlling for the psychological factors that may be impacting the relation between each cluster of questions (from the novice clinicians) making up a particular DSM PTSD Category with each (corresponding) cluster of responses from the VP representing the same DSM PTSD Category. We also assessed the impact of absorption and immersiveness upon the "believability" of the system.

4 Results and Evaluation

4.1 Assessment of the System

Assessment of the system was completed with the data gathered from the log files in addition to the questionnaires. The log files were used to evaluate the number and types of questions that the subjects were asking, along with a measure to see if the system was responding appropriately to the questions. For a 15 minute interview the participants asked on average, 68.6 questions with the minimum being 45 and the maximum being 91. Figure 4 is a graph showing the average number of questions, asked by the subjects, lighter color, and responses by the system, darker color for each of the 8 DSM categories. It is interesting to note that most of the questions asked were either general questions (Category #G, Average 24 questions) or questions about the Trauma (Category #A, 13 questions), followed by category #C and #B, 8. The larger number of questions asked in #G was partially due to clarification questions, however we did not break down the category further to try to classify this. The distribution of questions in each category for each participant was roughly equivalent, which meant in general people asked the same kinds of questions

There are several areas in the system that can be problematic due to technological issues which would cause the system to mis-recognize the question as out of domain, something the natural language system did not know about, and generate an inappropriate response. One area was the speech recognition system. We used a speaker independent speech recognizer that did not contain all of the words or phrases asked by the subjects, as it was not known all the questions they would ask. Additionally the system did not perform as well for women voices as with men. The natural language system deals with out of domain questions by responding with an off topic response, in our case the phrase 'I don't get what you mean'. This was a particular issue, based on the questionnaires, where the subjects got frustrated, as the system responded with this phrase too many times and there was not enough variability with out of domain responses. This response was said in total 411 times across all subjects, comparing that to the total responses of, 1066, the ratio was one in every 2.5 responses. While there is no standard for a reasonable set of questions to out of domain responses, this ratio at least gives us a measure as to how well the system was performing. While this value may seem high and did frustrate some subjects, most subjects were able to continue with questioning and get appropirate responses to perform a diagnosis. Future analysis on the speech recognition word error rate and accuracy will yield data as to what words and questions are needed to improve the speech models. It is clear from the transcriptions that the domain we built was not sufficient to capture all of the questions people were asking, the results from this study will be added to the domain for future testing. The interviewing method that people used to ask questions varied by individual; there were many different styles and personality factors that influenced the length and type of question, for example some people asked multiple segment questions, like 'hi how are you, why did you come here today?'. There are many novice assumptions by the subjects in how well this technology performs. Natural language and speech recognition is still a hard problem.

From the post questionnaires on a 7 point likert scale, the average value subjects rated the believability of the system to be 4.5. Subjects were also able to understand the patient, 5.1. People rated the system at 5.3 as frustrating to talk to, due to speech recognition problems, out of domain questions or inappropriate responses. However most of the participants left favorable comments that they thought this technology will be useful, they enjoyed the experience and trying different ways to talk to the character and also trying to get an emotional response for a difficult question. When the patient responded back appropriately to a question they found that very satisfying.

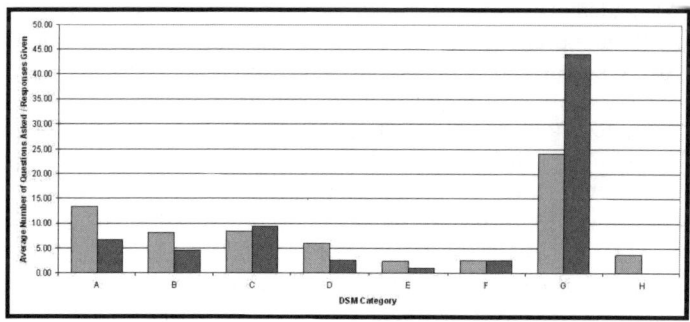

Fig. 4. Average number of questions asked, lighter, and responses given

4.2 Assessment of Student Questions and the Students

For this phase of the analysis, we aimed at investigating the relationship between a number of psychological variables and the resulting VP Responses. A summary of relations (measures as effect sizes "r") between each 1) DSM PTSD Category cluster of user questions; and 2) each (corresponding) cluster of responses from the VP representing the same DSM PTSD Category. Please note that these are "clusters" of Question/Response pairs that reflect different diagnostic categories used for differential diagnosis.

The present focus is on effect sizes indicating strength of correlation, that is, effect sizes that describe the strength of association between question and response pairs for a given diagnostic category. Given our small sample size, we wanted a more conservative estimate of effect. Hence, an effect size (herein we use "r" as a standard of effect size) of 0.20 was regarded as a small effect, 0.50 as a moderate effect, and 0.80 as a large effect. Moderate effects existed between User Questions and VP Response pairs for Category A ($r = 0.45$), Category B ($r = 0.55$), Category C ($r = 0.35$), and Category G ($r = 0.56$), but only small effects were found for Category D ($r = 0.13$) and Category F ($r = 0.13$). After controlling for the effects of the Tellegen Absorption Scale, increased effects were found for Category A ($r = 0.48$), Category C ($r = 0.37$), Category D ($r = 0.15$), and Category F ($r = 0.24$).

We also assessed impact of psychological characteristics such as absorption and immersiveness upon the "believability" of the VP and Student interaction. To assess this relation we created a composite variable that included scores from the TAS and the ITQ (Trait Composite). Strong effects existed between the Trait Composite and the Presence Questionnaire ($r = 0.78$), and moderate effects existed between the Trait Composite and the Justina Post-questionnaire ($r = 0.40$).

5 Discussion of Results

The primary goal in this study was evaluative: can a virtual standardized patient generate responses that elicit user questions relevant for PTSD categorization? Findings suggest that the interactions between novice clinicians and the VP resulted in a compatible dialectic in terms of rapport (Category G), discussion of the traumatic event (Category A), and the experience of intrusive recollections (Category B). Further, there appears to be a pretty good amount of discussion related to the issue of avoidance (Category C). These results comport well with what one may expect from the VP (Justina) system. Much of the focus was upon developing a lexicon that, at minimum, emphasized a VP that had recently experienced a traumatic event (Category A) and was attempting to avoid (Category B) experienced that my lead to intrusive recollections (Category C). However, the interaction is not very strong when one turns to the issue of hyper-arousal (Category D) and impact on social life (Category F). While the issue of impact on social life (Category F) may simply reflect that we wanted to limit each question/response relation to only one category (hence, it may have been assigned to avoidance instead of social functioning), the lack of questions and responses related to hyper-arousal and duration of the illness (Category E) reflects a potential limitation in the system lexicon. These areas are not necessarily negatives for the system as a whole. Instead, they should be viewed as potential deficits in the systems lexicon.

A secondary goal was to investigate the impact of psychological variables upon the VP Question/Response composites and the general believability of the system. After controlling for the effects of these psychological variables, increased effects were found for discussion of the traumatic event (Category A), avoidance (Category C), hyper-arousal (Category D), and impact on social life (Category F). Further, the impact of psychological characteristics revealed strong effects upon presence and believability. These findings are consistent with other findings suggesting that hypnotizability, as defined by the applied measures, appears moderate user reaction. Future studies should make use of physiological data correlated with measures of immersion to augment and quantify the effects of virtual human scenarios.

6 Future Work

We presented an approach that allows novice mental health clinicians to conduct an interview with a virtual character that emulates an adolescent female with trauma exposure. The work presented here builds on previous initial pilot testing of virtual patients and is a more rigorous attempt to understand how to build and use virtual humans as virtual patients along with the issues involved in building domains, the speech and language models and working with domain experts. The lessons learned here can be applied across any domain that needs to build large integrated systems for virtual humans. We believe this is a desirable application area and a small enough domain that we can perform meaningful evaluations on using VPs in real settings.

We will continue to perform more rigors subject testing with both professional medical students and with non experts to evaluate how well the different populations perform and further studies in comparing real OSCE's with real actors to the virtual patient. Investigate how incorporation of rapport [7,12] using facial analysis will further enhance the virtual patient interaction.

Additional analysis of the data includes: comparing the questions and the responses to assess how many were on and off-topic; Compute the word error rate for the speech recognition engine to assess its performance; Investigate tools to help build question/response sets for different domains; Explore ways to automate the process of classifying subject questions with the DSM categories; Build an agent framework that will be able to recognize conversation attributes such as; opening or closing statements, empathy, topic areas, follow-up and clarification questions along with more autonomous behavior, like asking questions to the clinician, assertiveness and, initiative levels, saving the conversation history and topic recognition and tracking.

People have many different interviewing and personality styles, some people are more direct, while others more empathetic. The system needs to be able to recognize these and adjust its responses. Studies are needs on incorporating learning objectives into the interview session and investigating if the virtual patient system has a learning impact is something that is valuable and will be the focus of future subject testing.

It is our belief that with adding more questions and responses the accuracy of the system will rise along with the depth of the conversions the clinician can have with the VP. In order to be effective VPs must be able to interact in a 3D virtual world, must have the ability to react to dialogues with human-like emotions, and be able to converse in a realistic manner with behaviors and facial expressions that match the clinical condition of interest. The combination of these capabilities allows them to serve as unique training and learning tools whose special knowledge and reactions can be continually fed back to trainees. Our initial goal of this study was to focus on a VP with PTSD, but a similar strategy could be applied to teaching a broad variety of psychiatric diagnoses to trainees at every level from medical students, to psychiatry residents, to child and adolescent psychiatry residents.

Acknowledgments. This work was sponsored by the U.S. Army Research, Development, and Engineering Command (RDECOM), and the content does not necessarily reflect the position or the policy of the Government, and no official endorsement should be inferred. We wish to thank the Keck School of Medicine at USC, Caroly Pataki, Michele Pato, Cheryl StGeorge and Jeffrey Sugar and special thanks to Mary Slater-Kenny as the voice of Justina. This work was sponsored in part by a USC Provost grant for Teaching with Technology and the V-Humans Project.

References

1. Andrew, R., Johnsen, K., Dickerson, R., Lok, B., Cohen, M., Stevens, A., Bernard, T., Oxendine, C., Wagner, P., Lind, S.: Comparing Interpersonal Interactions with a Virtual Human to those with a Real Human. IEEE Transactions on Visualization and Computer Graphics (2006)
2. Bernard, T., Stevens, A., Wagner, P., Bernard, N., Schumacher, L., Johnsen, K., Dickerson, R., Raij, A., Lok, B., Duerson, M., Cohen, M., Lind, D.S.: A Multi-Institutional Pilot Study to Evaluate the Use of Virtual Patients to Teach Health Professions Students History-Taking and Communication Skills. In: Proceedings of the Society of Medical Simulation Meeting (2006)
3. Bickmore, T., Pfeifer, L., Paasche-Orlow, M.: Health Document Explanation by Virtual Agents. In: Intelligent Virtual Agents 2007, Paris (2007)
4. Bickmore, T., Cassell, J.: Social Dialogue with Embodied Conversational Agents. In: van Kuppevelt, J., Dybkjaer, L., Bernsen, N. (eds.) Advances in Natural, Multimodal Dialogue Systems. Kluwer Academic, New York (2005)
5. Bickmore, T., Giorgino, T.: Health Dialog Systems for Patients and Consumers. Journal of Biomedical Informatics 39(5), 556–571 (2006)
6. Cassell, J., Bickmore, T., Billinghurst, M., Campbell, L., Chang, K., Vilhjálmsson, H., Yan, H.: An Architecture for Embodied Conversational Characters. In: Proceedings of the First Workshop on Embodied Conversational Characters, Tahoe City, California, October 12-15 (1998)
7. Cassell, J., Gill, A., Tepper, P.: Coordination in Conversation and Rapport. In: Proceedings of the Workshop on Embodied Natural Language, Association for Computational Linguistics, Prague, CZ, June 24-29 (2007)
8. Deladisma, A., Johnsen, K., Raij, A., Rossen, B., Kotranza, A., Kalapurakal, M., Szlam, S., Bittner, J., Sinwson, D., Lok, B., Lind, D.: Medical student satisfaction using a virtual patient system to learn history-taking and communication skills. Medicine Meets Virtual Reality (MMVR) 16 (2008)
9. DSM, American Psychiatric Association 2000 (DSM-IV-TR) Diagnostic and statistical manual of mental disorders, 4th edn, text revision. American Psychiatric Press, Inc. Washington (2000)
10. Gerhard, M., Moore, D., Hobbs, D.: Continuous presence in collaborative virtual environments: Towards the evaluation of a hybrid avatar-agent model for user representation. In: de Antonio, A., Aylett, R., Ballin, D. (eds.) Proceedings of the International Conference on Intelligent Virtual Agents, Madrid, Spain, pp. 137–153 (2001)

11. Gratch, J., Rickel, J., André, E., Badler, N., Cassell, J., Petajan, E.: Creating Interactive Virtual Humans: Some Assembly Required. IEEE Intelligent Systems, 54–63 (July/August, 2002)
12. Gratch, J., Ning, W., Jillian, G., Edward, F., Robin, D.: Creating Rapport with Virtual Agents. In: 7th International Conference on Intelligent Virtual Agents, Paris, France (2007)
13. Hardin, R.M., Stevenson, M., Downie, W.W., Wilson, G.M.: Assessment of clinical competence using objective structured examination. British Medical Journal 1, 447–451 (1975)
14. Johnsen, K., Raij, A., Stevens, A., Lind, D., Lok, B.: The Validity of a Virtual Human Experience for Interpersonal Skills Education. In: Proceedings of the SIGCHI conference on Human Factors in Computing Systems, pp. 1049–1058. ACM Press, New York (2007)
15. Kenny, P., Parsons, T.D., Gratch, J., Leuski, A., Rizzo, A.A.: Virtual Patients for Clinical Therapist Skills Training. In: Pélachaud, C., Martin, J.-C., André, E., Chollet, G., Karpouzis, K., Pelé, D. (eds.) IVA 2007. LNCS (LNAI), vol. 4722, pp. 197–210. Springer, Heidelberg (2007)
16. Kenny, P., Hartholt, A., Gratch, J., Swartout, W., Traum, D., Marsella, S., Piepol, D.: Building Interactive Virtual Humans for Training Environments. In: Proceedings of I/ITSEC, November 2007, Best Paper Nominee (2007)
17. Leuski, A., Patel, R., Traum, D., Kennedy, B.: Building effective question answering characters. In: Proceedings of the 7th SIGdial Workshop on Discourse and Dialogue, Sydney, Australia (2006)
18. Lok, B., Rick, F., Andrew, R., Kyle, J., Robert, D., Jade, C., Stevens, A., Lind, D.S.: Applying Virtual Reality in Medical Communication Education: Current Findings and Potential Teaching and Learning Benefits of Immersive Virtual Patients. Journal of Virtual Reality (to appear, 2006)
19. Macedonio, M., Parsons, T.D., Rizzo, A.A.: Immersiveness and Physiological Arousal within Panoramic Video-based Virtual Reality. Cyberpsychology and Behavior 10, 508–516 (2007)
20. McGee, J.B., Neill, J., Goldman, L., Casey, E.: Using multimedia virtual patients to enhance the clinical curriculum for medical students. Medinfo 9(Part 2), 732–735 (1998)
21. McQuiggan, S., Rowe, J., Lester, J.: The Effects of Empathetic Virtual Characters on Presence in Narrative-Centered Learning Environments. In: Proceedings of the 2008 SIGCHI Conference on Human Factors in Computing Systems, Florence, Italy (to appear, 2008)
22. Norris, F.H., Kaniasty, K., Thompson, M.P.: The psychological consequences of crime: Findings from a longitudinal population-based study. In: Davis, R.C., Lurigio, A.J., Skogan, W.G. (eds.) Victims of Crime, 2nd edn., pp. 146–166. Sage Publications, Inc., Thousand Oaks (1997)
23. Parsons, T.D., Bowerly, T., Buckwalter, J.G., Rizzo, A.A.: A controlled clinical comparison of attention performance in children with ADHD in a virtual reality classroom compared to standard neuropsychological methods. Child Neuropsychology (2007)
24. Prendinger, H., Ishizuka, M.: Life-Like Characters – Tools, Affective Functions, and Applications. Springer, Heidelberg (2004)
25. Resick, P.A., Nishith, P.: Sexual assault. In: Davis, R.C., Lurigio, A.J., Skogan, W.G. (eds.) Victims of Crime, 2nd edn., pp. 27–52. Sage Publications, Inc., Thousand Oaks (1997)
26. Rickel, J., Johnson, W.: Animated agents for procedural training in virtual reality: Perception, cognition, and motor control. Applied Artificial Intelligence 13(4-5), 343–382 (1999)

27. Rizzo, A.A., Pair, J., Graap, K., Treskunov, A., Parsons, T.D.: User-Centered Design Driven Development of a VR Therapy Application for Iraq War Combat-Related Post Traumatic Stress Disorder. In: Proceedings of the 2006 International Conference on Disability, Virtual Reality and Associated Technology, pp. 113–122 (2006)
28. Stevens, A., Hernandex, J., Johnsen, K., et al.: The use of virtual patients to teach medical students communication skills. The Association for Surgical Education Annual Meeting, New York, April 7–10 (2005)
29. Swartout, W., Gratch, J., Hill, R., Hovy, E., Marsella, S., Rickel, J., Traum, D.: Toward Virtual Humans. AI Magazine 27(1) (2006)
30. Tartaro, A., Cassell, J.: Playing with Virtual Peers: Bootstrapping Contingent Discourse in Children with Autism. In: Proceedings of International Conference of the Learning Sciences (ICLS), Utrecht, Netherlands, June 24-28 (2008)
31. Tellegen, A., Atkinson, G.: Openness to absorbing and self-altering experiences ("absorption"), a trait related to hypnotic susceptibility. Journal of Abnormal Psychology 83, 268–277 (1974)
32. Traum, D., Roque, A., Leuski, A., Georgiou, P., Gerten, J., Martinovski, B., Narayanan, S., Robinson, S., Vaswani, A.: Hassan: A virtual human for tactical questioning. In: Proceedings of the 8th SIGdial Workshop on Discourse and Dialogue, Antwerp, Belgium, September 2007, pp. 71–74 (2007)
33. Triola, M., Feldman, H., Kalet, A.L., Zabar, S., Kachur, E.K., Gillespie, C., et al.: A randomized trial of teaching clinical skills using virtual and live standardized patients. Journal of General Internal Medicine 21(5), 424–429 (2006)
34. Walters, K., Osborn, D., Raven, P.: The development, validity and reliability of a multi-modality objective structure clinical examination in psychiatry. Medical Education 39, 292–298 (2005)
35. Wessel, J., Williams, R., Finch, E., Gémus, M.: Reliability and validity of an objective structured clinical examination for physical therapy students. Journal of Allied Health 32(4), 266–269 (2003)
36. Wiederhold, B.K., Dong, P.J., Kaneda, M., Cabral, I., Lurie, Y., May, et al.: An investigation into physiological responses in virtual environments: an objective measurement of presence. In: Riva, G., Calimberti, C. (eds.) Toward cyberpsychology: Mind, cognition and society in the internet age. IOS Press, Amsterdam (2001)
37. Witmer, B., Singer, M.: Measuring presence in virtual environments: a presence questionnaire. Presence: Teleoperators and Virtual Environments 7(3), 225–240 (1998)

Elbows Higher!
Performing, Observing and Correcting Exercises by a Virtual Trainer

Zsófia Ruttkay and Herwin van Welbergen

HMI, Dept. of CS, University of Twente,
P.O. Box 217, 7500AE Enschede, The Netherlands
{z.m.ruttkay,H.vanWelbergen}@ewi.utwente.nl
http://hmi.ewi.utwente.nl

Abstract. In the framework of our Reactive Virtual Trainer (RVT) project, we are developing an Intelligent Virtual Agent (IVA) capable to act similarly to a real trainer. Besides presenting the physical exercises to be performed, she keeps an eye on the user. She provides feedback whenever appropriate, to introduce and structure the exercises, to make sure that the exercises are performed correctly, and also to motivate the user. In this paper we talk about the corpora we collected, serving a basis to model repetitive exercises on high level. Then we discuss in detail how the actual performance of the user is compared to what he should be doing, what strategy is used to provide feedback, and how it is generated. We provide preliminary feedback from users and outline further work.

1 Introduction

In the framework on our ongoing project [11], we aim at developing a Reactive Virtual Trainer (RVT), an intelligent virtual agent (IVA) with the rich functionality of human trainers. We are aiming at a RVT who can act as a stand-in for a real trainer, being capable of:

1. 'Understanding' and *intelligently dealing with a big variety of exercise* tasks, specified by a human expert in a near-natural high-level language.
2. Introducing and *explaining the exercise* to be performed.
3. *Presenting the exercises* in different tempi, accompanied by
    a. rhythmic counting
    b. and/or other acoustics such as music.
4. *Monitoring the user's motion*, and *correcting inaccuracies* in tempo or formation of the exercises.
5. *Monitoring the user's general physical and emotional state,* acknowledge it and possibly adjust the exercises to be followed.
6. *Providing emphatic or other feedback* to keep the user motivated.
7. Keeping track of past sessions of the user, and *evaluating current performance* in that context.

8. *Adjusting his/her own appearance* and, if appropriate, *biomechanical characteristics* to those of the user, in order to get closer to him psychologically and in the intensity and quality of the motions presented.
9. Use *conversation style and feedback strategy* adjusted to the age and personality of the user.

Our envisioned RVT may be used in different scenarios, where a real trainer cannot be present because of financial or other reasons: at home to motivate and supervise regular fitness or rehabilitation training, at work to insist on exercises best suited for people to prevent work-related injuries, see [14] for more.

The above applications are very timely, seeing the increase in age and the 'white collar' jobs in Western societies [5, 18]. A Tai Chi exercise performer was developed to demonstrate exercises using unaltered samples from a motion capture samples DB [4]. Medical and psychological consultancy applications with an empathic IVA have proven to be a success [3, 8]. Different sensors and portable devices [2] are in the focus of industry to be used to augment physical training. Also, currently Wii has been used as instrument to do sport game or other physical training [21].

Babu et al [1] presented the Virtual Human Physiotherapist Framework. In their short account they concentrate on exercise-specific monitoring, based on 3d colour markers placed on the body of the trainee. They provided a proof-of-concept demo of the monitoring mechanism, but have not covered the related issues in the breadths as we do.

The virtual Fitness Trainer of Philips, a companion projected onto an immersive screen in front of the trainee exercising on a home-trainer, evaluates his performance based on heart rate feedback [6, 7, 20]. We expect that in our case, where the RVT has richer (as of types of feedbacks, see above list) and more natural contact (based on visual perception input and subtle feedback strategy) with the trainee, the positive effect will be more significant than what the Philips researchers have reported so far [20].

The idea of vision-based physiotherapy was raised [15]. Sony's successful Eye-Toy: Kinetic 'game' [16] offers personalised as well as ready-made sequences of fitness exercises, commented on by one of two virtual trainers. In our application we provide more active and situated feedback based on more subtle monitoring of the user.

In a previous paper [14] we explained the above goals and the related technological challenges in depth, and gave account on the first chunk of work we accomplished. Namely, we showed how our animation engine could be used, together with music beat detection, for the major task 3, and 3b. We also described a low-level XML-based scripting language which can serve as a bridge between the authoring language and the animation engine.

In this paper we report on new work since then, contributing to goals 1, 2, 3a and particularly, 4. In the next section, we discuss our two different corpora we are building up as a reference for designing the motion and communication repertoire for the RVT. Then we outline the architecture of the RVT, and go into details of monitoring the performance of the user and providing feedback. Finally, we account on feedback from users, and outline further work.

2 Modeling Physical Exercises

In order to design and implement a RVT who resembles in his motion and interaction to a real expert trainer, we need to study practices from real life. For our work, we have established two kinds of corpora. The first one contains exercises performed as a trainee would do, while the second one contains recordings of sessions with a trainer and a trainee. The two corpora are complementary, providing information on motion characteristics and on training sessions and interaction strategy, respectively.

2.1 The Exercise Recordings Corpus

The Exercise Recordings (ER) corpus contains motion-capture + video recording of physical exercises, performed by healthy people or by professional fitness trainers. We selected the persons with the goal of getting the exercises performed correctly. A selected exercise is recorded in different tempi, some by several different performers. In case of expert trainers performing an exercise, we also recorded 'typically wrong' performances of the exercise in question.

We made recordings by using passive optical mocap method in the Vicon 460 optical motion capture laboratory of our university. 41 small white marker balls about 1 cm of diameter are fixed to the body of the user, basically to track the motion of the limb joints, hands, torso and head (See Figure 1.a). Six infra-red cameras trace the light reflected from the markers, and 3d coordinates of each marker are provided automatically, with a 120 Hz frequency, and written to a file for further (off-line) processing.

We have a body of 25 minutes of recordings, of 10 exercises, performed, altogether 101 times in different variations, by 3 subjects. Cardiovascular exercises are to improve general (heart) condition, strengthening ones to develop certain muscles, and stretching ones to warm up and relax before/ after sessions

This corpus is used to analyze the motion characteristics of exercises, individual variations and correlations between parameters like time and intensity, see further in [17]. We are currently working on making parameterized models for the motion of the 'major joints involved', based on the mocap samples. We wish to use a more advanced, multi-layered animation engine where parameterized procedural motion is layered with physical balancing to improve on quality of the motion performed by the RVT, see [19].

The other type of information on the motion characteristics is to gather knowledge for modeling the 'natural', 'fast' and 'slow' tempi for different exercises. Finally, by analyzing 'wrong' performances, we plan to improve our monitoring and diagnosis process (to be discussed in Section 3).

2.2 The Training Recordings Corpus

The Training Recordings (TR) contains (audio-) video recordings of training sessions. We aim at collecting a big body of recordings systematically, where the trainer, the client and the type of exercises are carefully chosen to end up with a rich collection with samples for all aspects. We also have some exercises from the TR session which were 'performed' (albeit by different people) in the ER corpus.

Currently we have 75 minutes of recordings of 4 complete exercise sequences covering a variety of goal, performed in different postures (standing, seated, ...), with the same trainee and a professional physiotherapist trainer.

We annotate the videos by using the ELAN tool [22]. We analyze the video and audio with two goals:

- to learn about the structure of training sessions and the individual exercises, and
- to spot and categorize different types of feedbacks.

See [9] for more on preliminary results.

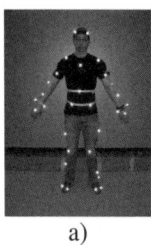
a)

b)

Fig. 1. a) A frame from the MR mocap corpus, with the subject wearing reflective markers in the Vicon lab. b) A frame from the TR video corpus, taken in a real gym hall. The trainer on the right addresses the subject pointing out a correction, while performing the exercise too.

3 The Architecture of the RVT

The RVT serves two functions:

- to *author* an exercise sequence;
- to *'make the trainee perform'* this sequence.

The first function is assured by a simple editing tool, which allows compilation of the exercise to be performed Also, the tempo is specified.

The second task is achieved by introducing the exercise, starting it once the user is ready, and then performing a monitoring-performing loop, where the 'performance' involves both spoken feedback (if appropriate) and the presentation of the exercise by the RVT. Figure 2 shows the major modules of the architecture of the RVT.

3.1 Monitoring the User

We designed a single-camera vision system based on ParleVision components [10], which can identify 'natural' colour-coded markers the user is wearing (gloves and socks). The *Feature Extractor* uses traditional image processing techniques to compute centre of mass for each marker as current location. Left and right limbs are identified also on the basis of the model of the human body and a short history of the tracking. The current location of the markers is normalised, with respect to the body geometry of the virtual trainer. The *Feature Interpreter* makes the diagnosis about the

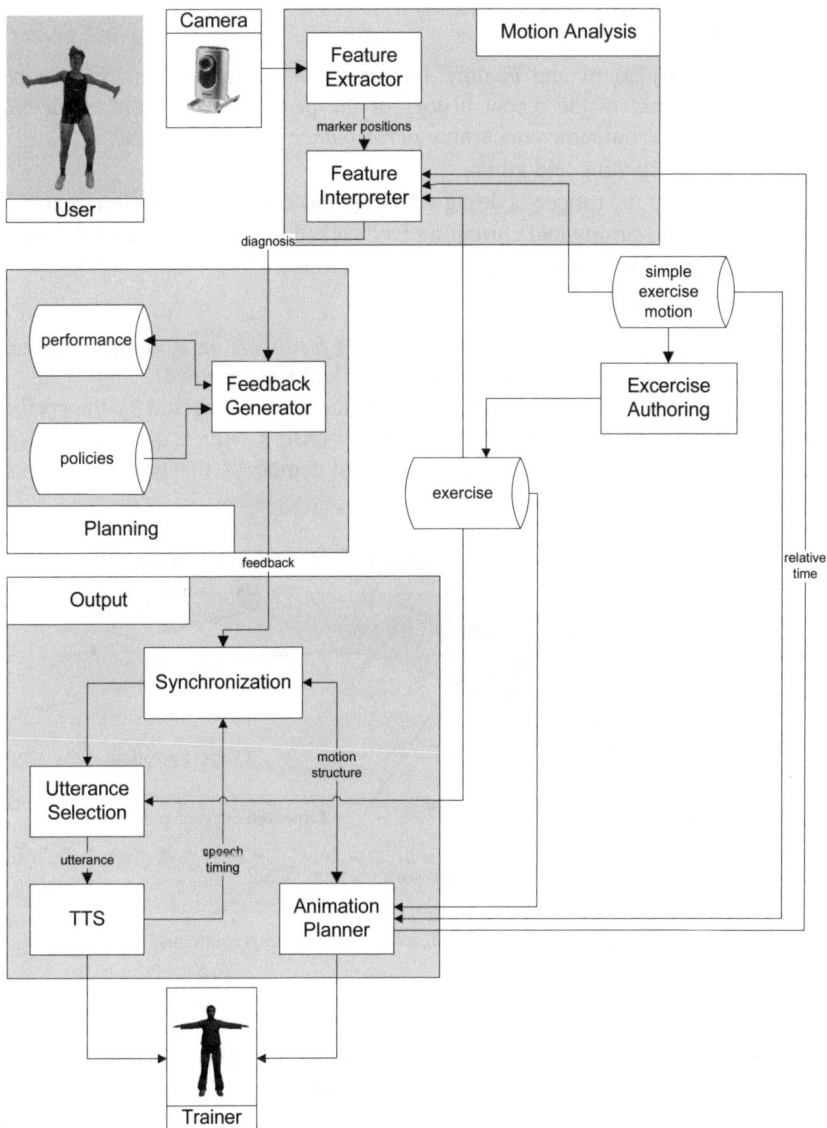

Fig. 2. The architecture of the RVT, with major modules and data transfer

user's performance, knowing of the (normalized) marker positions. Based on the relative time in performing the exercise, and the structure and tempo of the exercise to be performed, and having access to the motion definition of the simple exercises, the Feature Interpreter generates an *allowed location* for each marker (see Figure 3). Based on comparing the actual and allowed location of the marker, the motion of the limb of the trainee is diagnosed as 'correct', 'slow' or 'fast' or 'wrong'.

3.2 Providing Feedback

The diagnosis provided by the Feature Interpreter about the current movement is evaluated with respect to the recent history of the performance. Based on declared feedback policies, the outcome concerning *performance feedback* can be:
- not to say anything and go on,
- confirm that the trainee is doing well (e.g. has corrected a previous error),
- to provide warning and correcting feedback, concerning tempo or formation;
- readjust own tempo, to be in sync with the trainee;
- to abandon the exercise.

Another source of feedback is the *exercise-related feedback*, on structuring the exercises (e.g. last time an exercise is to be performed, half time reached).

The actual feedback is decided on the basis of the events triggered by the performance of the 4 limbs and the relative time of the exercise. *Conflicts* are solved by giving higher priority to recently realized change of tempo of the trainer, and novel anomalies diagnosed with limbs.

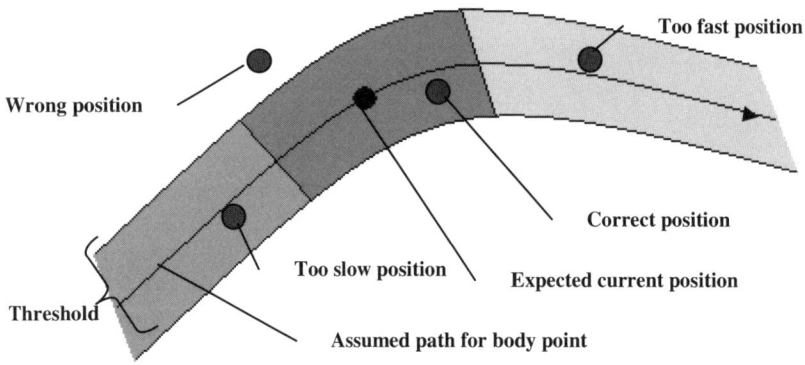

Fig. 3. Assumed motion path of a body point, with expected, early, late and wrong positions

3.3 Presenting on Time, in Sync

The *Synchronization* module receives eventual tempo adjustment instruction for the virtual trainer, and a list of prioritized and time-stamped messages to be conveyed for the user. For different language variants one is selected randomly. The utterances are aligned to beat, or adjusted also to motion tempo (e.g. counting down to indicate the end of the exercise), and the motion of the virtual trainer is generated [13].

4 Preliminary Evaluation

Our current implementation (see Figure 4) was pre-tested with some 10 'casual users' belonging to two major groups: adults around 50, and students around 23, native Dutch speakers. They had to perform a short session (3-5 minutes) introduced and guided by the VT. Afterwards, they were interviewed. The actions of the VT were logged and analyzed after the session.

Most of our users got engaged with the VT, she made several of them sweat. Users saw the VT as someone they would welcome, should she offer more 'serious' exercises, tailored to their needs. This, in general, is a promising feedback, especially that the pre-test had the clear features of a try-out, not of a real application setting.

The critical feedbacks addressed the following issues:

- Too general/misleading 'wrong' feedback.
- Speech quality (Loquendoe TTS) was not engaging.
- Trainer embodiment should be improved.

Fig. 4. Screen snapshot showing the Virtual Trainer in action

Acknowledgement

We thank B. Varga and G. Nagy for their work on recording the corpora, A. Zanderink, F. Zijlstra, M. Klaas, R. Rookhuiszen and R. Leemkuil for implementing some modules of the RVT, and the physiotherapists Dr. M. Horváth and N. van Regteren for giving insight to and contribution with recordings of training practices.

This research has been supported by the GATE project, funded by the Netherlands Organization for Scientific Research (NWO) and the Netherlands ICT Research and Innovation Authority (ICT Regie). The motion capture recordings of the ER corpus were carried out in the framework of a Short Mission supported by the EU Action COST2102.

References

1. Babu, S., Zanbaka, C., Jackson, J., Chung, T.-O., Lok, B., Shin, M.C., Hodges, L.F.: Virtual Human Physiotherapist Framework for Personalized Training and Rehabilitation, Short Paper Graphics Interface 2005, Victoria, British Columbia, Canada (2005)

2. Buttussi, F., Chittaro, L., Nadalutti, D.: Bringing mobile guides and fitness activities together: a solution based on an embodied virtual trainer. In: Proceedings of MOBILE HCI 2006: 8th International Conference on Human-Computer Interaction with Mobile Devices and Services, September 2006, pp. 29–36. ACM Press, New York (2006)
3. Bickmore, T., Picard, R.: Towards caring machines. In: Proc. of CHI (2004)
4. Chao, S.-P., Chiu, C.-Y., Yang, S.-N., Lin, T.-G.: Tai Chi synthesizer: a motion synthesis framework based on key-postures and motion instructions. Computer Animation and Virtual Worlds 15, 259–268 (2004)
5. Cybertherapy, http://www.cybertherapy.info/pages/main.htm
6. HomeLab Philips, http://www.research.philips.com/technologies/misc/ homelab/
7. IJsselsteijn, W., de Kort, Y., Westerink, J., De Jager, M., Bonants, R.: Fun and Sports: Enhancing the Home Fitness Experience. In: Rauterberg, M. (ed.) ICEC 2004. LNCS, vol. 3166. Springer, Heidelberg (2004)
8. Marsella, S., Johnson, L., LaBore, C.: Interactive Pedagogical Drama for Health Interventions. In: AIED 2003, 11th International Conference on Artificial Intelligence in Education, Australia (2003)
9. Nagy, G.: Modelling Physical Exercises, Engineering Design Report. Pazmany Peter Catholic University, Dept. of Information Technology, Budapest (2007) (in Hungarian)
10. Parlevision showcase at HMI, University of Twente, http://hmi.ewi.utwente. nl/showcases/parlevision
11. Reactive Virtual Trainer showcase at HMI, University of Twente, http://hmi.ewi. utwente.nl/showcase/The%20Reactive%20Virtual%20Trainer
12. Reichel, H.S., Groza-Nolte, R.: Fizioterápia. Medicina Kiadó, Budapest (2001)
13. Reidsma, D., van Welbergen, H., Poppe, R.W., Bos, P., Nijholt, A.: Towards Bidirectional Dancing Interaction. In: Harper, R., Rauterberg, M., Combetto, M. (eds.) ICEC 2006. LNCS, vol. 4161, pp. 1–12. Springer, Heidelberg (2006)
14. Ruttkay, Z.M., Zwiers, J., van Welbergen, H., Reidsma, D.: Towards a Reactive Virtual Trainer. In: Gratch, J., Young, M., Aylett, R.S., Ballin, D., Olivier, P. (eds.) IVA 2006. LNCS (LNAI), vol. 4133, pp. 292–303. Springer, Heidelberg (2006)
15. Shafaei, S., Rahmati, M.: Physiotherapy Virtual Training by Computer Vision Approach. In: Third International Workshop on Virtual Rehabilitation (2004), http://www. iwvr.org/2004/
16. Sony: EyeToy: Kinetic, http://www.us.playstation.com/Content/OGS/ SCUS-97478/Site/, http://www.eyetoykinetic.com
17. Varga, B.: Movement modeling and control of a Reactive Virtual Trainer, Master Thesis, Pazmany Peter Catholic University, Dept. of Information Technology, Budapest (2008)
18. Virtual Rehabilitation (2007), http://www.aristea.com/iwvr2007/
19. Welbergen, H., Ruttkay, Zs.: Multilayered Motion Modelling for Physical Exercises, MiG, Utrecht, The Netherlands (submitted, 2008)
20. Westerink, J., de Jager, M., de Kort, M.Y., IJsselsteijn, W., Bonants, R., Vermeulen, J., van Herk, J., Roersma, M.: Raising Motivation in Home Fitnessing: Effects of a Virtual Landscape and a Virtual Coach with Various Coaching Styles. In: ISSP 11th World Congress of Sport Psychology, Sydney, Australia, August 15 - 19 (2005)
21. Wii Sports, http://nlbe.wii.com/software/02/
22. Wittenburg, P., Brugman, H., Russel, A., Klassmann, A., Sloetje, H.: Elan: a professional framework for multimodality research. In: Proc. of LREC 2006 (2006)

A Virtual Therapist That Responds Empathically to Your Answers

Matthijs Pontier[1,2] and Ghazanfar F. Siddiqui[1,2]

[1] Vrije Universiteit Amsterdam, Department of Artificial Intelligence,
De Boelelaan 1081a, 1081 HV Amsterdam, The Netherlands
[2] Center for Advanced Media Research Amsterdam
Buitenveldertselaan 3, 1082 VA Amsterdam, The Netherlands
{mpontier,ghazanfa}@few.vu.nl
http://www.few.vu.nl/~{mpontier,ghazanfa}

Abstract. Previous research indicates that self-help therapy is an effective method to prevent and treat unipolar depression. While web-based self-help therapy has many advantages, there are also disadvantages to self-help therapy, such as that it misses the possibility to regard the body language of the user, and the lack of personal feedback on the user responses. This study presents a virtual agent that guides the user through the Beck Depression Inventory (BDI) questionnaire, which is used to measure the severity of depression. The agent responds empathically to the answers given by the user, by changing its facial expression. This resembles face to face therapy more than existing web-based self-help therapies. A pilot experiment indicates that the virtual agent has added value for this application.

Keywords: Virtual agent, Self-help therapy, Emotion modeling.

1 Introduction

Self-help therapies have been investigated for several decades. Self-help therapy started with bibliotherapy, in which clients follow a therapy from a book. Previous research indicates that this is a very effective form of therapy; e.g., a meta-analysis by Cuijpers [7] concluded that bibliotherapy in unipolar depression is an effective treatment modality, which is no less effective than traditional individual or group therapy.

The advent of new communication technologies, like internet and videoconferencing, can also assist in the field of mental healthcare. Since the last decade, a lot of self-help programs have been delivered through the internet [5], [6], [12]. Several previous studies concluded that self-help therapies are useful and efficient in reducing mental health problems convincingly (e. g., [7], [12]). Compared to traditional therapy methods, web-based self-help may be more efficient and less expensive [4], [9].

Web-based self-help therapy can also be a solution for people who would otherwise not seek help, wishing to avoid the stigma of psychiatric referral or to protect their privacy [13]. The majority of persons with a mental disorder in the general population do not receive any professional mental health services (an estimated 65%) [4].

In many occupations, such as the police force, the fire service and farming, there is much stigma attached to receiving psychological treatment, and the anonymity of web-based self-help therapy would help to overcome this [11]. Also many other people feel a barrier to seek help for their problems through regular health-care systems; e.g., in a study by Spek et al. [12] about internet-based cognitive behavioural therapy for subtreshold depression for people over 50 years old, many participants reported not seeking help through regular health-care systems because they were very concerned about being stigmatised. Patients may be attracted to the idea of working on their own to deal with their problems, thereby avoiding the potential embarrassment of formal psychotherapy [13]. Self-help therapy can also be offered to patients while they are on a waiting list, with the option to receive face to face therapy later, if required [11].

However, there is also critique on internet-based self-help therapy. Drop-out rates from self-help therapy can be high, especially when the use of self-help is unmonitored by a health care practitioner [13]. A wide range of drop-out rates for bibliotherapy have been estimated: from about 7% [7] up to 51.7% [12]. People may miss personal feedback when performing self-help therapy, which might decrease their motivation. By making self-help therapy more similar to face to face therapy, it can become a more personal and entertaining experience, which might decrease drop-out.

Several self-help therapy programs are already available on the internet. Two well-known examples of CBT (Cognitive Behavioural Therapy) programs are 'BluePages' and 'MoodGYM' [5]. BluePages gives information about the symptoms of depression whereas MoodGYM is designed to prevent depression [5], [6]. However, none of the existing online self-help therapies include a virtual agent that provides a kind of face to face assistance.

There have already been developed several agents in the health-supporting domain. For example, [3] describes a virtual agent that explains health documents to patients.

This study presents an application for performing the Beck Depression Inventory questionnaire [2]. The application is equipped with a virtual agent that responds empathically to the responses of the user. As the virtual agent is emotionally responsive to the answers given by the user throughout the questionnaire, the experience should resemble face to face therapy more than a similar application without a virtual agent.

2 The Application

In the application, the user performs the BDI questionnaire [2]. The main goal of the BDI is to measure the characteristic attitudes and symptoms of depression. The BDI is a self-report inventory that consists of 21 multiple-choice questions, and is generally used for measuring the severity of depression. Every question has at least four answer options ranging in intensity from 0 to 3.

The virtual agent asks the questions to the user, and the user selects the appropriate answer from a given drop-down box. This virtual agent has a certain emotional state, consisting of two emotions: happiness and empathy. According to Eisenberg [8], empathy is *"An affective response that stems from the apprehension or comprehension of another's emotional state or condition, and that is similar to what the other person is feeling or would be expected to feel."* Because in this application a depression questionnaire is conducted, which means empathy concerns rather sad things, showing empathy consists of showing sadness. If during the questionnaire the user appears to be more depressed the virtual agent will show more sadness, expressed by a relatively sad facial expression. On the other hand, if the user appears to be completely fine, the agent will show a relatively happy facial expression. When the webpage is loaded for the first time, the original emotional state of the virtual agent is loaded, which is a calm emotional state, with very little sadness and an average level of happiness.

2.1 The Emotion Model of the Agent

The virtual agent responds empathically towards the user, by showing the right facial expressions on the answers given by the user. In consultation with clinical psychologists, we defined an impact of these answers as a real number in the domain [-1, 1] on the emotions of the virtual agent, represented by a real number in the domain [0, 1]. Further we defined in consultation with the clinical psychologists how the agent should behave towards the users using these impacts. The impacts are used to detect how the user is feeling. When the user gives a lot of answers that indicate he or she is not feeling well, the agent should show empathy, by showing a sad facial expression, without showing any happiness. When the user is feeling fine, the agent should show a neutral, calm facial expression, with some happiness and no sadness. Because it would be undesirable if the emotions of the agent suddenly shift from very sad to very happy or vice versa, with any change in emotions, the old levels of the emotions are taken into account. If the answers of the user have no impact on the emotions of the agent, its facial expression should slowly return to the original emotional state it had at the start of the application. We have developed the following formula that meets the requirements as described above:

$$\text{New\_emotion} = \text{Old\_emotion} + \text{Decay} + \text{Change}$$
$$\text{Decay} = (\text{Original\_emotion} - \text{Old\_emotion}) * \text{Decay\_factor}$$
$$\text{Change} = \zeta * \text{Impact} / (1 + (\text{Original\_emotion} - \text{Old\_emotion}) * \text{Impact})$$

New_emotion can be calculated by taking the old emotion, and adding decay and change. Here Old_emotion is the emotion of the virtual agent before the formula is applied. Decay is the size of the decay effect (i.e., how quickly the emotion will move towards the original emotion if the user response has no impact). Change is the change of the emotion of the virtual agent, according to the impact of the answer given by the user.

Decay is calculated by subtracting Old_emotion from Original_emotion, and multiplying the result with the Decay_factor. In this formula, Decay_factor is a variable that determines the size of the decay effect, which is taken 0.1 in this paper. Original_emotion is the emotion of the virtual agent at the start of the application.

Change is calculated by multiplying the impact of the answer given by the user with ζ and dividing the result by 1 + (Original_emotion - Old_emotion) * Impact. In this formula, ζ is a variable that determines the speed with which the answers given by the user can modify the emotions of the virtual agent. Dividing Impact by 1 + (Original_emotion - Old_emotion) * Impact manages that when the current emotion of the agent deviates more from the original emotion the agent had at the start of the application, and the answer given by the user pushes the emotion of the agent even further away from the original emotion, the change will be relatively smaller as when the user's answer would push the agent's emotion back towards the original emotion with the same impact.

The emotions of the agent during two scenarios are shown in Figure 1. In this figure, along the x-axis the time is given and along the y-axis the levels of emotions are given. The pink line shows happiness and the blue line shows sadness. In scenario 1, the agent interacts with a severely depressed user, while in scenario 2 the agent interacts with a user who scores average on feelings of depression.

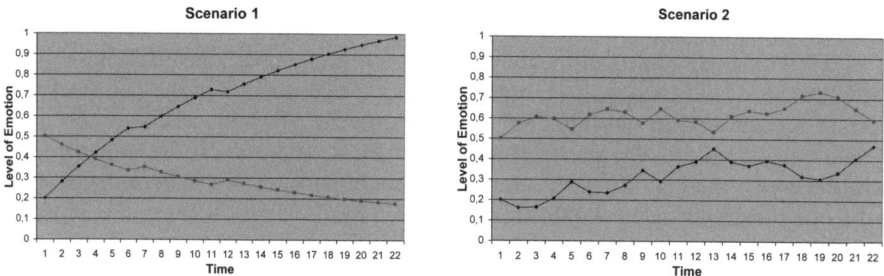

Fig. 1. The emotions of the agent during scenario 1 and scenario 2

As can be seen in Figure 1, initially the virtual agent has a very small level of sadness (0.2) and an average level of happiness (0.5). Each time point the virtual therapist asks a question to the user and gets an answer. This answer affects the emotions of the agent. The user in scenario 1 is severely depressed, coming to a final score of 61 on the BDI questionnaire. During the questionnaire, the agent notices this, which increases her level of sadness and decreases her level of happiness. This results in the agent showing empathy towards the user by means of a sad facial expression.

In scenario 2, the user got an average score on the BDI questionnaire (33). This means the user scores average on feelings of depression. In Figure 1, the different emotional reactions of the agent on the answers of the user in scenario 2 can clearly be seen. On answers that indicate the user is depressed, such as the answer just before time point 5, the agent's level of sadness increases, and the agent's level of happiness decreases. On the other hand, on answers that indicate the user is not depressed, such as the answer just before time point 6, the agent's level of sadness decreases, and the agent's level of happiness increases. At the end of the questionnaire, the agent shows a facial expression with an average level of both sadness and happiness.

2.2 The Resulting Website

For creating the virtual agent, we used Haptek's peopleputty software [10]. Through this program we created the face of the virtual agent. Further, we created nine different emotional states using the sliders 'happy' and 'sad' given by the software; one for each possible combination of the three levels of the emotions happiness and sadness (showing empathy). We created a webpage for the BDI questionnaire, on which the virtual agent was embedded as a Haptek player. We used JavaScript [1], a scripting language, in combination with scripting commands provided by the Haptek software [10], to control the Haptek player within a web browser.

We performed a pilot experiment to test whether the virtual agent has added value for the application. To recruit participants, we invited people by sending an e-mail with a link to the website. The first page of this website does not show the virtual agent, and contains 8 demographic questions. The next page contains the 21 questions of the BDI questionnaire. The virtual agent is shown on top of this page. Before the first question, the agent introduces itself, and states that it will guide the user through the questionnaire. Instead of showing all the questions in a form, only one question at a time was shown, to let the application more resemble face to face therapy. Each question is shown in a text area below the virtual agent. Below the question, there is a dropdown-box from which the user can select an answer.

Because some of the possible participants were Dutch, we created a Dutch, as well as an English version of the website. In the English version, the virtual agent asks the question using speech. Because at the moment of this study, we did not have a Dutch speech synthesis engine available that included lip-syncing, in the Dutch version the question was only shown in the text area below the virtual agent. We used this shortcoming of the Dutch version to investigate the added value of speech in this application.

Each time the user selected an answer from the drop-down box, the virtual agent changed its emotional state depending on the calculated values of the emotions, as described in section 3.1. Each answer has a score, as described in Section 2, and during the questionnaire these scores are accumulated to calculate the final score. When the user presses the submit button, it proceeds towards the next page, where the virtual agent gives feedback about the final score of the user on the questionnaire, showing an appropriate facial expression. When the final score was below 16, the virtual agent indicates that the user is less depressed as average and shows a facial expression with a low level of sadness and a medium level of happiness. When the user scores between 16-41, the virtual agent indicates that the user scores average on feelings of depression and shows a facial expression with a medium level of sadness and a low level of happiness. When the user scores above 41, the virtual agent indicates that the user scores high on feelings of depression and shows a facial expression with a high level of sadness and a low level of happiness. If the user responded that he or she considers committing suicide, the agent stringently advises the user to contact his or her general practitioner.

After receiving the feedback, the user clicks a button to proceed to the next page, which contains the evaluation form. This page consists of 5 questions about the virtual agent, such as whether the user prefers performing the BDI questionnaire with or without the virtual agent. The virtual agent itself is not shown on this page, to prevent the user from giving socially desirable answers towards the agent.

3 Experiment

We have performed a pilot experiment to test whether the virtual agent has added value for this application.

Participants. The participants were recruited by sending an e-mail with an invitation to participate in the experiment. The participants could choose between a Dutch and an English version of the questionnaire. 28 participants completed the experiment, of which 16 the English, and 12 the Dutch version.

Procedure. First the participants entered some demographic information in a web-form, without the virtual agent. Next the application with the virtual agent was loaded, and the participants performed the BDI questionnaire. When the questionnaire was finished, the participants received feedback from the virtual agent about their result. In the English version, the question was shown in a text area below the agent, and the agent additionally asked the questions to the participants using speech. In the Dutch version however, the virtual agent could not speak, and the text was only shown in the text area below the agent. Finally, the participants filled in an evaluation questionnaire, without the virtual agent present to prevent socially desirable answers towards the agent. The complete procedure can still be performed at [14].

Results. The participants evaluated on an eight-point scale whether they thought the virtual agent was friendly, interested, trustworthy and kind. In both the English and the Dutch version, for all properties, the participants scored the agent just above moderate, as can be seen in Table 1. No statistical differences were found between the English version with voice, and the Dutch version without voice.

Table 1. The score of the virtual agent on several properties on an eight-point scale

	English		Dutch	
	M	SD	M	SD
Friendly	4.75	1.98	4.91	1.30
Interested	4.50	2.13	4.00	1.95
Trustworthy	4.31	1.66	4.27	1.56
Kind	4.75	1.69	4.55	1.44

Further, the participants answered the question "If you were to administer the same questionnaire, would you rather do this with or without virtual interviewer?" on the evaluation form. For the English version, with speech, 81% of the participants preferred to perform the questionnaire with the virtual agent (sign test, $p = .021$). However, for the Dutch version, without speech, only 64% of the participants preferred to fill in the questionnaire with the virtual agent (sign test, $p = .55$).

The participants were asked to explain their answer in an open question. On this question people gave various responses, but a response that came back several times was that with the virtual agent, it "feels more personal" and that it "feels friendlier". Participants also indicated that it was more fun to perform the questionnaire with the virtual agent. This indicates that the virtual agent makes it more attractive and entertaining to perform the questionnaire, and adding the virtual agent to a self-help application might decrease drop-out of the self-help therapy.

Participants that preferred to perform the questionnaire without a virtual agent gave as reasons for this that the agent was still too "cold and computer-like", and with the Dutch version, without speech, that the agent did not have any added value.

The responses on the open question "How do you think the virtual interviewer should be improved?" indicated that there is still a lot of work to do. In the Dutch version, a lot of participants responded that the agent should speak, while in the English version many participants responded that the voice of the agent should be friendlier. In both versions many participants responded the agent should give feedback on each answered question.

4 Discussion

This study presents a virtual agent that guides the user through a questionnaire about depression. The agent responds empathically to the answers given by the user, by changing its facial expression.

A pilot experiment has been performed to test the applicability of a virtual agent in this application. Due to time limitations, the way of recruitment of the participants was not ideal, and the participants are probably not a very good representation of the target group of online self-help applications. When the application has been improved, an extensive validation will need to be performed before it can be used in practice. However, the experiment has led to some interesting results that can be used to determine a direction for further research.

In both the English version, with speech, and the Dutch version, without speech, the participants found the agent moderately friendly, interested, trustworthy and kind. Further, although there were not many participants, an interesting statistical significant result was found. For the English version, the amount of participants who preferred to perform the questionnaire with the virtual agent was significantly bigger than the amount of participants who preferred to perform the questionnaire without the virtual agent. For the Dutch version also more participants preferred to perform the questionnaire with the agent than without, but this result was not significant. However, none of the participants appeared to actually be depressed, and the agent thus will not have shown much obvious empathic expressive behaviour. In the Dutch version of the application, without speech, this means the participants just saw a rather passive face above the questions. Given this information, and that there were only 12 participants, it is not very surprising that the for the Dutch questionnaire, the amount of participants that preferred performing the questionnaire with the agent was not significantly bigger than the amount of participants that preferred performing the questionnaire without the agent.

Taking into account many improvements can still be made to the application, the results described above are very promising, and motivate further research in this direction. The response of a participant that he did not feel shy of the virtual interviewer as he would with a real one further indicates nicely the use of this kind of applications. For people who feel uncomfortable with undergoing face-to-face therapy and therefore choose not to seek help, an application like this can be a nice solution.

As also indicated by the responses in the experiment, many improvements can still be made to the application. As pointed out by many participants in the open questions, the agent should provide appropriate feedback after each answer on a question. This should increase the humanness of the agent, enhancing the feelings of a personal, realistic experience during the questionnaire.

Another possible point of improvement is the voice of the agent. The fact that with the Dutch version, without speech, the amount of participants that preferred to perform the questionnaire with the virtual agent was not significantly bigger than the amount of participants that preferred to perform the questionnaire without the virtual agent indicates that this is an important issue. Moreover, in the open questions, many participants gave responses that indicated that speech should be added (in the Dutch version) or improved (in the English version). Since the application should ultimately also be available for Dutch speaking users, possibilities for adding Dutch speech synthesis including lip-syncing should seriously be considered. Also possibilities to create a friendlier voice that is able to show emotions should be considered.

Acknowledgements

We kindly want to thank Annemieke van Straten and Tara Donker for their input to this paper, and Jan Treur, Tibor Bosse and the anonymous reviewers for their comments on earlier drafts of this paper.

References

1. About JavaScript – MDC,
2. http://developer mozilla.org/en/docs/About_JavaScript
3. Beck, A.T.: Depression: Causes and Treatment. University of Pennsylvania Press, Philadelphia (1972)
4. Bickmore, T.W., Pfeifer, L.M., Paasche-Orlow, M.K.: Health Document Explanation by Virtual Agents. In: Pélachaud, C., Martin, J.-C., André, E., Chollet, G., Karpouzis, K., Pelé, D. (eds.) IVA 2007. LNCS (LNAI), vol. 4722, pp. 183–196. Springer, Heidelberg (2007)
5. Bijl, R.V., Ravelli, A.: Psychiatric morbidity, service use, and need for care in the general population: results of The Netherlands Mental Health Survey and Incidence Study. American Journal of Public Health 90(4), 602–607 (2000)
6. Christensen, H., Griffiths, K.M., Jorm, A.F.: Delivering interventions for depression by using the internet: randomised controlled trial. BMJ 328, 265–269 (2004)
7. Christensen, H., Griffiths, K.M., Korten, A.: Web-based cognitive behavior therapy: analysis of site usage and changes in depression and anxiety scores. Journal of Medical Internet Research 4(1), 3 (2002)
8. Cuijpers, P.: Bibliotherapy in unipolar depression, a meta-analysis. Journal of Behavior Therapy & Experimental Psychiatry 28(2), 139–147 (1997)
9. Eisenberg, N.: Empathy-related emotional responses, altruism, and their socialization. In: Davidson, R.J., Harrington, A. (eds.) Visions of compassion: Western scientists and Tibetan Buddhists examine human nature, pp. 131–164. Oxford University Press, London (2002)
10. Griffiths, F., Lindenmeyer, A., Powell, J., Lowe, P., Thorogood, M.: Why Are Health Care Interventions Delivered Over the Internet? A Systematic Review of the Published Literature. Journal of Medical Internet Research 8(2), 10 (2006)
11. Haptek, Inc., http://www.haptek.com
12. Peck, D.: Computer-guided cognitive–behavioural therapy for anxiety states. Emerging areas in Anxiety 6(4), 166–169 (2007)

13. Spek, V., Cuijpers, P., Nyklíĉek, I., Riper, H., Keyzer, J., Pop, V.: Internet-based cognitive behavior therapy for emotion and anxiety disorders: a meta-analysis. Psychological Medicine 37, 1–10 (2007)
14. Williams, C.: Use of Written Cognitive-Behavioural Therapy Self-Help Materials to treat depression. Advances in Psychiatric Treatment 7, 233–240 (2001)
15. http://www.few.vu.nl/~ghazanfa/welcome.php

IDEAS4Games: Building Expressive Virtual Characters for Computer Games

Patrick Gebhard[1], Marc Schröder[1], Marcela Charfuelan[1], Christoph Endres[1], Michael Kipp[1], Sathish Pammi[1], Martin Rumpler[2], and Oytun Türk[1]

[1] DFKI, Saarbrücken and Berlin, Germany
firstname.lastname@dfki.de
[2] FH Trier, Umwelt-Campus Birkenfeld, Germany
m.rumpler@umwelt-campus.de

Abstract. In this paper we present two virtual characters in an interactive poker game using RFID-tagged poker cards for the interaction. To support the game creation process, we have combined models, methods, and technology that are currently investigated in the ECA research field in a unique way. A powerful and easy-to-use multimodal dialog authoring tool is used for the modeling of game content and interaction. The poker characters rely on a sophisticated model of affect and a state-of-the art speech synthesizer. During the game, the characters show a consistent expressive behavior that reflects the individually simulated affect in speech and animations. As a result, users are provided with an engaging interactive poker experience.

1 Motivation

Virtual characters are widely used in a variety of applications, including computer games, where they are notably used for non-player characters, i.e. characters controlled by the computer. In general, virtual characters have the purpose to enrich the game play experience by showing an engaging and consistent interactive behavior. Because this issue influences acceptance in general, computer games, and virtual characters in particular, have to be designed carefully according to Loyall's *suspension of disbelief principle* [1]:

> ... a character is considered to be believable if it allows the audience to suspend their disbelief ...

The creation of interactive expressive characters with a consistent behavior comes with a whole range of challenges, such as interaction design, emotion modeling, figure animation, and speech synthesis [2]. The interactive drama game Façade [3] or the Mission Rehearsal Exercise Project [4] explicitly address these problems in combination, in an integrated application. However, relevant research is also being carried out in a range of relevant individual disciplines, including believable facial, gesture and body animation of virtual characters [5] [6], modeling of personality and emotion [7] [8], expressive speech synthesis [9] and control mechanisms for story as well as high-level interaction [10].

In the project IDEAS4Games, we investigate how modern ECA technologies can help to improve the process of creating computer games with interactive expressive virtual characters. Based on the experience of a computer game company (RadonLabs, Berlin), we identified four main challenges in creating computer games:

- *Fast creation of game demonstrators.* In order to compete with other game companies the implementation of demonstrators has to be fast and reliable.
- *Localization of game content.* To sell games in other countries content has to be translated into the respective language. The more dialogs a game contains, the higher the costs for the translation.
- *Easy interaction.* The success of a game is tremendously related to an easy interaction concept.
- *Consistent quality.* The quality of audio and visual presentation should be consistent for the whole game. Every exception lowers its acceptance.

Based on our experience in the research area of ECAs, we rely on the following models and technologies for addressing these challenges.

- *Flexible multimodal dialog and story modeling.* Other than the approach of Bosser et al. [11] that is based on an autonomous agent framework for computer games, we rely on the authoring tool SceneMaker [12] for modeling game interaction and dialog content. SceneMaker is able to create a Java executable that flexibly integrates other modules.
- *Sophisticated simulation of affect in real time.* Affect is key to an advanced behavior modeling of virtual characters [21]. Based on a simulation of affect, one can transfer aspects of human affective behavior onto virtual characters to enhance their believability.
- *Expressive speech synthesis.* To enhance the consistent affective behavior of characters, utterances should reflect their affective state. In addition, if it was possible to use synthesized speech rather than recording every utterance needed in the dialog, this would help to control the costs for content localization. It would also enable new game features such as the integration of a player's name in dialogs.
- *3D virtual characters.* Our characters provide lip-sync facial expressions, body expressions such as breathing, and gestural animations.

We focus on the employment of virtual characters in interactive social games, like board or card games. They provide a huge field of application of virtual characters that are in the role of competitors and enrich the overall gaming experience by an engaging performance.

2 Poker Demonstrator

In order to demonstrate that it is possible to meet the requirements stated above, we created a poker computer game and presented it at the CeBIT 2008 fair to a large number of people to collect initial feedback.

Fig. 1. Poker Demonstrator at the CeBIT 2008 fair

By using real poker cards with unique RFID tags, a user can play draw poker [13] against the two 3d virtual characters Sam and Max (see Fig. 1). Sam is a cartoon-like, friendly looking character, whereas Max is a mean, terminator-like robot character. Both are rendered by the open-source 3d visualisation engine Horde3D [14]. The human user acts as the card dealer and also participates as a regular player.

As shown in Fig. 1, we use a poker table which shows three areas for poker cards: one for the user and one for each virtual character. These areas of the table are instrumented with RFID sensor hardware, so that the game logic can detect which card is actually placed at each specific position. A screen at the back of the poker table displays an interface which allows users to select their actions during the game, using a computer mouse. These include general actions, such as playing or quitting a game, and poker game actions: bet a certain amount of money, call, raise, or fold. This screen also shows the content relevant for the poker game: the face of the user's cards, the number of Sam's and Max's cards respectively, all bets, and the actual money pot. The two virtual characters are shown above this interaction screen on a second 42" monitor.

When a user approaches the poker table and initiates a game, Sam and Max explain the game setup and the general rules. In a next step the user has to deal the cards. During the game the virtual characters Sam and Max react to events, notably when the user deals or exchanges their cards. Time is also considered – for example, Sam and Max start complaining if the user deals the cards too slowly, or they express their surprise about erratic bets.

In order to support Sam's and Max's individual character style, different poker algorithms are used. Sam, who represents a human-like poker player, uses a rule based algorithm, whereas Max, the robot poker player, relies on a brute-force algorithm that estimates a value for each of the 2.58 million possible combinations of five poker cards.

Based on game events, the affect of each character is computed in real-time and expressed through the character's speech and body. The richly modeled characters, as well as the easy interaction with the game using real poker cards, have stirred up a lot of interest in the CeBIT 2008 audience (see Fig. 1).

2.1 Poker Algorithms

Brute Force Approach. The brute force strategy starts by calculating an evaluation function, which maps a hand (5 cards) to a numeric value. The higher the value, the better are the chances for its owner to score. Our approach is to put all 2,598,960 possible hands in its *natural* order, which is given by the rules of the game and the fact of one hand winning against another in direct comparison. After generating and sorting that list, we assign values to the hands in an inductive manner:

- The lowest hand in the list is assigned a value of zero.
- For each following hand we assign values as follows:
  1. If the comparison of this hand with the previous hand would result in a draw game, the same value is assigned.
  2. If this hand would win against the previous hand, then its value is the one of the previous hand plus the amount of hands that had the same value as the previous hand.

The main purpose of the algorithm is to determine an action to be performed by the player. When playing Draw Poker, the player has a choice at one point to exchange up to three cards. There are 26 options to do so: One option not to exchange cards at all, five to exchange one card, ten to exchange two cards, and ten to exchange three cards. For each option the expected value of the outcome can be calculated. When no card is exchanged, it is simply the evaluation value of the hand. In case of exchanging cards, it is the average value of all possible combinations of the remaining cards with cards from the remaining 47 cards of the deck. In order to determine the best possible action, we have to check $1 + 5 * 47 + 10 * 1,081 + 10 * 16,215 = 173,196$ resulting hands and their value.

Rule-based Approach. Our rule-based approach was implemented in *JESS* (*Java Expert Systems Shell*) [15], a fast Java-based expert system. The main advantages of a rule based approach are: (a) short development time, (b) readability of code, and (c) ease of implementing variations in terms of "playing style". The latter is of high importance when an AI module is to be used in conjunction with virtual character that should display a personal style. The

```
(defrule predict-full-house-1m-a
    "When having two pairs, add prediction full house"
    (not (hand (type full_house)))
    (hand (type two_pair) (indices $?i))
 ⇒ (assert (maybe (type full_house) (missing 1) (have ?i))))
```

Fig. 2. An example of a JESS rule

drawback of rule-based systems are usually (a) performance and (b) maintainability of large code bases. However, for our application performance time of this module was negligible and the code base was quite small. The JESS module is called in the *draw* phase of the game. It handles the recognition of the system's current hand, prediction of possible hands after the draw and the decision which cards to exchange (see Fig. 2 for a sample rule).

This is done using 36 rules, organized in 4 modules. The system's strategy is to keep the highest current hand and exchange all others (up to 3). To determine which cards to exchange the system consults its computed predictions, usually aiming for the highest potential hand achievable with the fewest new cards. Here, heuristics can be easily changed, e.g. to model risky players, beginners or cautious players, due to the small and readable code base.

3 Modeling of Interaction and Story

The behavior of the two virtual players has been modeled with the authoring tool SceneMaker. Our authoring approach relies on the separation of dialog content and narrative structure, which we have introduced in [12,16].

3.1 Authoring Dialog Content

Dialog content is organized in *scenes* - pieces of contiguous dialog. Scenes are defined in a multimodal script that specifies the text to be spoken as well as the agents' verbal and nonverbal behavior. The utterances can be annotated with dialog act tags that influence the computation of affect (see Section 4 for details). In addition, system commands (e.g., for changing the camera position) can be specified. Scenes (see Fig. 3 for an example) are created by an author with standard text processing software. The major challenge when using scripted dialog is variation. The characters must not repeat themselves because this would severely impact their believability. For this purpose we use blacklisting: once a scene is played, it is blocked for a certain period of time (e.g., five minutes), and variations of this scene are selected instead. For each scene, several variations

```
Scene_de: Welcome(3)
Sam:   [camera 1] Hi. It's great to have [point] you here for a game. [GoodEvent 0.7].
Max:   My pal [S look-to-other] and I [nod] were a bit bored the last seconds.
Sam:   [look-to-user] Poker isn't a game that can be played by only two ... [M mod]
```

Fig. 3. A Welcome scene from the multimodal script

can be provided that make up a *scene group*. In our poker scenario there are 335 scenes organized in 73 scene groups. The number of scenes in a scene group varies between two and eight scenes, depending on how much variation is needed.

The name of the scene group is followed by the variation number in parentheses. Each dialog contribution starts with the characters name. At the beginning of this scene the camera is moved to position 1. Sam's first utterance is annotated with the appraisal tag [**GoodEvent**] because the arrival of the user is appraised positively by him. This tag will elicit the emotion *joy* with a certain intensity. The gesture specification in Max's utterance [**S nod**] shows that it is possible to specify gestures of other characters, e.g. for back-channeling behavior. The multimodal dialog script is parsed by the scene compiler which generates a single Java class file for each scene.

3.2 Authoring Narrative Structure

The narrative structure – the order in which the individual scenes are played – is defined by the *sceneflow*. Technically, the sceneflow is modeled as a hypergraph that consists of nodes and edges (transitions). *Supernodes* contain subgraphs. Each node can be linked to one or more scenes and scene groups. Different branching strategies (e.g. logical and temporal conditions as well as randomization) can be used by specifying different edge types [12]. The sceneflow is modeled using the sceneflow editor, our graphical authoring tool that supports authors with drag'n'drop facilities to *draw* the sceneflow by creating nodes and edges (see Fig. 4).

Fig. 4 shows the graph structure for the supernode *On* representing the situation that a human user is interacting with the installation. After an initial welcome phase (represented by the supernode *Welcome*), he or she is either prompted to remove the cards from the table (*ClearHands*) or to deal out the cards to start a new game (*Deal*), etc.

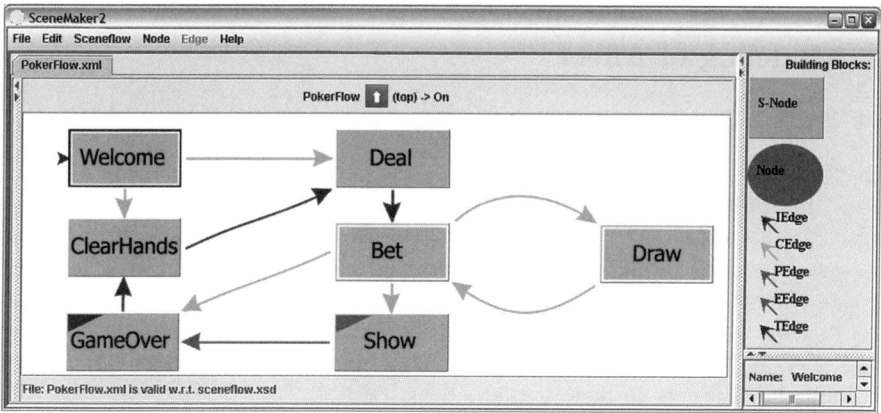

Fig. 4. Sceneflow editor showing the supernode *On*

At runtime the sceneflow graph is traversed by selecting nodes and edges based on the current game state and the actions of the three players. The scenes that are selected and executed during such a traversal control the multimodal behavior of the virtual agents. Transitions in the sceneflow are triggered either by the players' actions or as a result of context queries. In both cases sceneflow variables used for conditional branching are updated as described in the following section. The sceneflow is stored as an XML document and parsed by the sceneflow compiler which generates Java source code that can be executed by the sceneflow interpreter. After the compilation process both scenes and sceneflow are deployed as a single Java application.

3.3 Game Control

Each time the user places or removes a card, an event is generated and sent to the poker event handler. The same happens when the user chooses an action (bet, call, raise, or fold) by pressing the respective button on the graphical user interface (see Fig. 1).

The poker event handler receives these low-level events and updates the data model that represents the game state as well as the graphical user interface that visualizes it. It also analyses the situation and generates higher level events, e.g., that all cards have been removed from the table or that the user has changed the cards of a player in the drawing phase. At the end of this process, it updates the respective sceneflow variables, which may enable transitions in the sceneflow and trigger the next scenes.

Apart from scenes and scene groups the author can also attach commands to a node. These commands are executed by the sceneflow interpreter each time the node is visited. There are commands that modify the game state (e.g. selecting the next player after a scene has been played in which one of the players announces that he drops out) and commands that access the poker logic to suggest the next action (e.g. deciding which cards should be changed in the drawing phase and which action the virtual players should perform in the betting phase).

4 Modeling of Affect

4.1 Affect Computation

For the affect computation in real-time, we rely on ALMA, a computational model of affect [17]. It provides three affect types as they occur in human beings: *Emotions* (1) reflect short-term affect that decays after a short period of time. Emotions influence e.g. facial expressions, and conversational gestures. *Moods* (2) reflect medium-term affect, which is generally not related to a concrete event, action or object. Moods are longer lasting affective states, which have a great influence on humans' cognitive functions. *Personality* (3) reflects individual differences in mental characteristics.

ALMA implements the cognitive model of emotions developed by Ortony, Clores, and Collins (OCC model of emotions) [18] combined with the *BigFive*

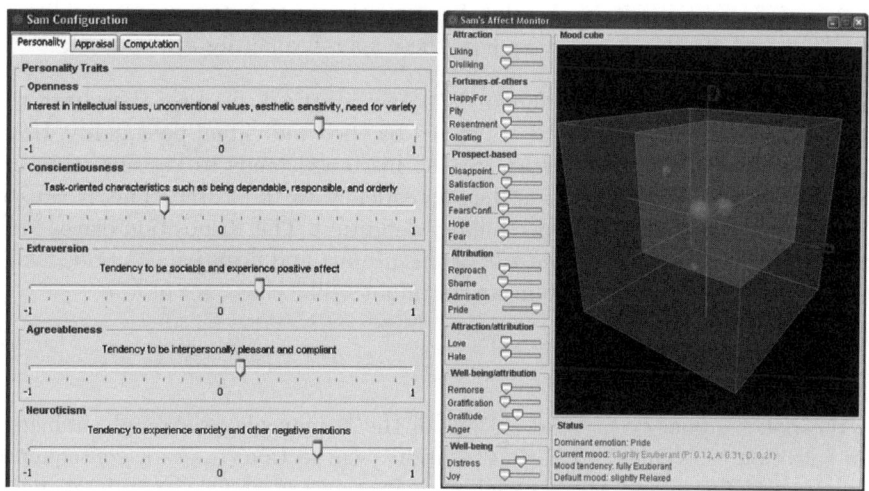

Fig. 5. ALMA Personality Configuration and Affect Monitor dialogs

model of personality [19] and a simulation of mood based on the PAD model [20]. The emotion simulation considers the impact of mood and personality on emotion intensities. Personality traits can be configured for each character (see Fig. 5 - left side). Once configured, a *default mood* is computed.

ALMA enables the computation of 24 OCC emotions with *appraisal tags* [21] as input. Elicited emotions influence an individual's mood. The higher the intensity of an emotion's, the higher the particular mood change. A unique feature is that the current mood also influences the intensity of emotions. This simulates, for example, the intensity increase of *joy* and the intensity decrease of *distress*, when a individual is in an *exuberant* mood. The PAD model of mood distinguish eight different mood classes: *exuberant, bored, dependent, disdainful, relaxed, anxious, docile, hostile*. Generally, a mood is represented by a triple of the mood traits pleasure (P), arousal (A), and dominance (D). The mood's trait values define the mood class. If, for example, every trait value is positive (+P,+A,+D), the mood is *exuberant* (see Fig. 5 - right side, highlighted area). An AffectMonitor (see Fig. 5 - right side) visualizes in real time emotions and their intensities, as well as the current mood.

Mood and Emotion Control Affective Behavior Aspects. According to the current mood and emotions elicited during the game play, the following aspects of the virtual poker characters' affective behavior are changed:

- *Breathing* is related to the mood's arousal and pleasure values. For positive values, a character shows fast distinct breathing. The breathing is slow and faint for negative values.
- *Speech quality* is related to the current mood or an elicited emotion. In *relaxed* mood, a character's speech quality is *neutral*. In *hostile* or *disdainful* mood or during negative emotions, it is *aggressive*, in *exuberant* mood or during

positive emotions it is *cheerful*. In any other mood, the speech quality is *depressed*. Section 5 describes how these speech qualities are realised.

Initially, Sam's and Max's behavior is nearly identical, because they have the same default *relaxed* mood that is defined by their individual personality. However, Sam is slightly more *extroverted*, so his mood tends to become more *exuberant*. Max, on the other side, has a tendency to be *hostile*. This is caused by his negative *agreeableness* personality definition. This disposition drives Max towards negative mood (e.g. *hostile* or *disdainful*) and negative emotions faster than Sam, which supports the mean, terminator-like character of Max.

4.2 Authoring Affect

In the poker game, events and actions of the human player strongly influence the selection of scenes and their execution (see Section 3). As a consequence the actions of Sam and Max as well as their appraisal of the situation changes. For example, if Sam loses the game because he has bad cards, a related scene will be selected in which he verbally complains about that. An appraisal tag used in the scene lets Sam appraise the situation as a bad event during the scene execution. The corresponding appraisal tag is [**BadEvent**]. When this is used as input for the affect computation, it will elicit the emotion *Distress*, and if such events occur often, they will lead to a *disdainful* or *hostile* mood. In general, appraisal tags can be seen as a comfortable method for dialog authors to define on an abstract level input for a affect computation module. For doing this, no conceptual knowledge about emotions or mood is needed.

Affect in the Poker Game. The poker game scenario covers the simulation of all 24 OCC emotions. The following overview gives a brief explanation how some of them are computed based on appraisal tags. They are grouped according to the OCC classification of emotions.

- *Prospect-based emotions*: *hope, satisfaction*, or *disappointment*. During the card change phases, Sam and Max randomly utter their expectation of getting good new cards. Those utterances also contain the appraisal tag [**GoodLikelyFutureEvent**]. As a consequence *hope* is elicited. After all cards have been dealt out, Sam's and Max's poker engine compares the current cards with the previous ones. Depending on the result either the [**EventConfirmed**] or the [**EventDisconfirmed**] appraisal tag is passed to the affect computation. In the first case (cards are better), *satisfaction* is elicited, otherwise *disappointment*.
- *Fortunes-of-Others*: *gloating* or pity. If Sam loses the game, Max (who does not like Sam) always comments the situation with a smirk: "With such cards you could never win!". The appraisal tag [**BadEventForBadOther**] elicits *gloating* during the utterance. Sam always expresses his sorrow if a user loses a game (Sam likes users). During that, *pity* is elicited by the appraisal tag [**BadEventForGoodOther**].

The influence of emotions and mood on the behavior (speech, breathing) can be observed during the game play and gives some hints on the current situation of Sam and Max.

5 Expressive Synthetic Voices with Reliable Quality

Convincing speech generation is a necessary precondition for an ECA to be believable. That is true especially for a system featuring emotional expressivity. To address this issue, we have investigated the two currently most influential state-of-the-art speech synthesis technologies: unit selection synthesis and statistical-parametric synthesis.

Two criteria are traditionally used for assessing the quality of synthetic speech: intelligibility and naturalness. In the context of this paper, these criteria must be rephrased in line with the aim of supporting the user in *suspending his disbelief*. We address intelligibility in terms of a *reliable quality* of the generated speech output, and naturalness in terms of a *natural expressivity*.

Unfortunately, in state-of-the-art speech synthesis technology, the two criteria are difficult to reconcile. The most natural-sounding synthesis technology, unit selection [22], can reach close-to-human naturalness in limited domains, such as speaking clocks or weather forecasts [23], but it suffers from unpredictable quality in unrestricted domains – when suitable units are not available, the quality can drop dramatically. On the other hand, statistical-parametric synthesis, based on Hidden Markov Models (HMMs) [24], has a very stable synthesis quality, but the naturalness is limited because of the excessive smoothing involved in training the statistical models. Insofar, there is currently no technology that simply fulfils both the reliability and the naturalness criterion.

In IDEAS4Games, we investigated two methods for approximating the criteria formulated above. For our humanoid character, Sam, we created a custom unit selection voice with a "cool" speaking style, featuring high quality in the poker domain and some emotional expressivity; for our robot-like character, Max, we used an HMM-based voice, and applied audio effects to modify the sound to some extent in order to express emotions.

5.1 Expressive Domain-Oriented Unit Selection

Sam's voice was carefully designed as a domain-oriented unit selection voice [25]. Domain-oriented voices sound highly natural within a given domain; they can also speak arbitrary text, but the quality outside the domain will be seriously reduced. We designed a recording script for the synthesis voice, consisting of a generic and a domain-specific part. We used a small set of 400 sentences selected from the German Wikipedia to cover the most important diphones in German [26]. This is a small number – usually, between 1,500 and 10,000 sentences are recorded for a synthesis voice of standard quality. Consequently, it is clear that the general-domain performance of the voice must be expected to be quite restrained. In addition, about 200 sentences from the poker domain were recorded, i.e. sentences related to poker cards, dealing, betting, etc. – sentences of the kind that are needed for the game scenario. The 600 sentences of the recording script were produced by a professional actor in a recording studio. The speaker was instructed to utter both kinds of sentences in the same "cool" tone of voice.

In a similar way, the same speaker produced domain-oriented voice databases consisting of approximately 600 sentences each for a *cheerful*, an *aggressive* and a *depressed* voice. We built a separate unit selection synthesis voice for each of the four databases, using the open-source voice import toolkit of the MARY text-to-speech synthesis platform [27].

To generate speech from text, the MARY TTS system converts any given text into a linguistic *target*, an abstract description of the ideal units needed to generate the speech. The unit selection component identifies an optimal sequence of *units*, i.e. short audio-snippets from a given voice database, that best fit to the given target as well as to their respective neighboring units, and concatenates the selected units like pearls on a string, with only minimal signal processing involved. If – and only if – suitable candidates are found, the resulting speech will be of high quality. Naturally, the speech output will have the same speaking style as the original recordings. Changing expressive style, in this method, can only be done by switching from one voice database to another.

In our application, most of Sam's utterances are spoken with the neutral poker voice. As the voice database contains many suitable units from the domain-oriented part of the recordings, the poker sentences generally sound highly natural. Their expressivity corresponds to the "cool" speaking style realised by our actor. Selected utterances are realised using the *cheerful*, *aggressive*, and *depressed* voices. For example, when Sam loses a round of poker, he may utter a frustrated remark in the *depressed* voice; if the affect model predicts a positive mood, he may greet a new user in a *cheerful* voice, etc. Note that, as for the neutral poker voice, these emotional voices can potentially speak any text; however, only within a relatively small domain of poker-related emotional utterances, the quality will be optimal.

5.2 Robust HMM-Based Synthesis

The voice of the robotic character, Max, uses statistical-parametric synthesis based on Hidden Markov Models. When the voice is built, the statistical models are trained on a speech database. After training, the original data is not needed anymore; at runtime, speech is generated from the statistical models by means of a vocoder. As a result of the training process, the speech generated with an HMM-based voice has a certain degree of resemblance to the speaker and speaking style of the database used for training; however, the speech sounds muffled due to the averaging in the statistical models as well as the vocoding. While their speech is less natural, HMM-based synthesis voices have one great advantage, however: the quality is largely stable, independently of the text spoken.

We started from the open-source system HTS released by the Nagoya Institute of Technology [28], ported the code to Java, and incorporated it into our MARY TTS platform [29]. The HMM-based voice for our robotic character Max was created from recordings of a male speaker in the BITS speech synthesis corpus [30], a phonetically balanced German speech corpus containing about 1700 sentences per speaker. The resulting voice sounds rather monotonous, but intelligible and of stable quality throughout all utterances.

Expressivity for Max' voice is performed differently from Sam's voice. Here, we use audio effects to modify the generated speech. At the level of the parametric input to the vocoder, we can modify the pitch level, pitch range and speaking rate. In addition, we apply audio signal processing algorithms that modify the generated speech signal, using linear predictive coding (LPC) techniques. In this framework, our code provides a vocal tract scaler, which can simulate a longer or shorter vocal tract; a whisper component, adding whisper to the voice; a robot effect, etc. We can control the degree to which these effects are applied. Through a trial-and-error procedure, we determined coarse settings for expressive speech: *aggressive* speaking style is simulated by a lengthened vocal tract, lowered pitch, increased pitch range, faster speech rate, and slightly whisperised speech; *cheerful* speech is simulated by a shortened vocal tract, higher pitch level and range, and faster speech rate; and *depressed* speech is simulated by a slightly lengthened vocal tract, lower pitch level and much narrower pitch range, and a slower speech rate. A lot could be optimised about the application of these effects, e.g. the extent to which they are applied could be linked to the emotional intensity in a gradual way. Here, we merely intended to illustrate that audio effects can be applied to shape an HMM-based voice towards a certain expression.

6 Discussion and Conclusion

We have presented an ECA-based computer game employing RFID-tagged poker cards as a novel interaction paradigm. It is built around the integration of three components: a powerful and easy-to-use multimodal dialog authoring tool, a sophisticated model of affect, and a state-of-the-art speech synthesizer.

In our authoring approach, the separation of dialog content and narrative structure makes it possible to modify these two aspects independently. The script can be edited using standard text processing software – no programming skills are required. When the structure and the script are compiled into Java code, they are checked for integrity. This approach makes it extremely easy to rapidly develop and fine-tune an interactive application.

The flexible dialog script is most effective when it is supported by an equally flexible high-quality speech synthesizer. We have shown that the state of the art is still limited in that respect. On the one hand, the HMM-based voice used for our character Max is appropriately flexible, so that any changes in the dialog script can easily be rendered with his voice, in a limited but reliable quality. On the other hand, we have created a very high quality unit selection voice for our character Sam, but the quality comes at the price of high effort and limited flexibility. Extending the domain, or adding other expressions, would require additional recordings. Also, with unit selection technology it is not yet possible to express degrees of a certain expression: Sam's voice is either cheerful or not, but it is not possible to make him sound "slightly cheerful", or to make him sound "increasingly aggressive" as the game unfolds. We are working on voice interpolation methods that should make this possible in the future [31,32].

Our computational model of affect plays a major role in modeling the expressive behavior of the poker characters Sam and Max. This is supported by two aspects: 1) the use of appraisal tags, which simplifies the generation of affect so that authoring affect does not require any deep knowledge about a model of emotion or mood; and 2) the real-time computation of different affect types as they occur in humans. Especially the continuous computation of short-term emotions and medium-term moods allow for a smooth blending of different aspects of affective behavior that help to increase the believability and the expressiveness of virtual characters.

Overall, we have shown how state-of-the-art research approaches can be combined in an interactive poker game to show new possibilities for the modeling of expressive virtual characters and for future computer game development.

Acknowledgments. The work reported here was supported by the EU ProFIT project IDEAS4Games (EFRE program). Part of this research has been carried out within the framework of the Excellence Cluster Multimodal Computing and Interaction (MMCI), sponsored by the German Research Foundation (DFG). The authors would like to thank Michael Schneider for supporting us with his RFID-technology expertise. Also, we would like to thank Nicolas Schulz for supporting us with the Horde 3D visualisation engine and extensive support.

References

1. Loyall, A.B.: Believable Agents: Building Interactive Personalities. PhD thesis, School of Computer Science, Carnegie Mellon University, Pittsburgh, PA (1997)
2. Gratch, J., Rickel, J., André, E., Cassell, J., Petajan, E., Badler, N.I.: Creating interactive virtual humans: Some assembly required. IEEE Intelligent Systems 17, 54–63 (2002)
3. Mateas, M., Stern, A.: Façade: An experiment in building a fully-realized interactive drama. In: Game Developers Conference, Game Design Track (2003)
4. Swartout, W., Gratch, J., Hill, R., Hovy, E., Marsella, S., Rickel, J., Traum, D.: Toward virtual humans. AI Magazine 27, 96–108 (2006)
5. Martin, J.C., Niewiadomski, R., Devillers, L., Buisine, S., Pelachaud, C.: Multimodal complex emotions: Gesture expressivity and blended facial expressions. International Journal of Humanoid Robotics, Special Edition Achieving Human-Like Qualities in Interactive Virtual and Physical Humanoids (2006)
6. Kipp, M., Neff, M., Kipp, K.H., Albrecht, I.: Toward natural gesture synthesis: Evaluating gesture units in a data-driven approach. In: IVA 2007. LNCS (LNAI), vol. 4722, pp. 15–28. Springer, Heidelberg (2007)
7. de Rosis, F., Pelachaud, C., Poggi, I., Carofiglio, V., de Carolis, B.: From greta's mind to her face: Modelling the dynamics of affective states in a conversational embodied agent. Int. Journal of Human Computer Studies 59, 81–118 (2003)
8. Marsella, S., Gratch, J.: Ema: A computational model of appraisal dynamics. In: Agent Construction and Emotions (2006)

9. Schröder, M.: Emotional speech synthesis: A review. In: Proceedings of Eurospeech 2001, Aalborg, Denmark, vol. 1, pp. 561–564 (2001)
10. Prendinger, H., Saeyor, S., Ishizuk, M.: Mpml and scream: Scripting the bodies and minds of life-like characters. In: Life-like Characters – Tools, Affective Functions, and Applications, pp. 213–242. Springer, Heidelberg (2004)
11. Bosser, A.G., Levieux, G., Sehaba, K., Bundia, A., Corruble, V., de Fondaumière, G., Gal, V., Natkin, S., Sabouret, N.: Dialogs taking into account experience, emotions and personality. In: Ma, L., Rauterberg, M., Nakatsu, R. (eds.) ICEC 2007. LNCS, vol. 4740, pp. 356–362. Springer, Heidelberg (2007)
12. Gebhard, P., Kipp, M., Klesen, M., Rist, T.: Authoring scenes for adaptive, interactive performances. In: Proc. of the 2nd Int. Joint Conference on Autonomous Agents and Multi-Agent Systems, pp. 725–732. ACM, New York (2003)
13. Wikipedia: Draw Poker (2008), http://en.wikipedia.org/wiki/Draw_poker
14. Schulz, M.: Horde3D – Next-Generation Graphics Engine. Horde 3D Team (2006–2008), http://www.nextgen-engine.net/home.html
15. Hill, E.: Jess in Action: Java Rule-Based Systems, Manning, Greenwich, CT, USA (2003)
16. Klesen, M., Kipp, M., Gebhard, P., Rist, T.: Staging exhibitions: methods and tools for modelling narrative structure to produce interactive performances with virtual actors (2003). Virtual Reality 7, 17–29 (2003)
17. Gebhard, P.: Alma - a layered model of affect. In: Proc. of the 4th Int. Joint Conference on Autonomous Agents and Multiagent Systems, pp. 29–36. ACM, New York (2005)
18. Ortony, A., Clore, G.L., Collins, A.: The Cognitive Structure of Emotions. Cambridge University Press, Cambridge (1988)
19. McCrae, R., John, O.: An introduction to the five-factor model and its applications. Journal of Personality 60, 175–215 (1992)
20. Mehrabian, A.: Pleasure-arousal-dominance: A general framework for describing and measuring individual differences in temperament. Current Psychology: Developmental, Learning, Personality, Social 14, 261–292 (1996)
21. Gebhard, P., Kipp, K.H.: Are computer-generated emotions and moods plausible to humans? In: Gratch, J., Young, M., Aylett, R.S., Ballin, D., Olivier, P. (eds.) IVA 2006. LNCS (LNAI), vol. 4133, pp. 343–356. Springer, Heidelberg (2006)
22. Hunt, A., Black, A.W.: Unit selection in a concatenative speech synthesis system using a large speech database. In: Proceedings of ICASSP 1996, Atlanta, Georgia, vol. 1, pp. 373–376 (1996)
23. Black, A.W., Lenzo, K.A.: Limited domain synthesis. In: Proceedings of the 6th International Conference on Spoken Language Processing, Beijing, China (2000)
24. Yoshimura, T., Tokuda, K., Masuko, T., Kobayashi, T., Kitamura, T.: Simultaneous modeling of spectrum, pitch and duration in HMM-based speech synthesis. In: Proceedings of Eurospeech 1999, Budapest, Hungary (1999)
25. Schweitzer, A., Braunschweiler, N., Klankert, T., Möbius, B., Säuberlich, B.: Restricted unlimited domain synthesis. In: Proc. Eurospeech 2003, Geneva, Switzerland (2003)
26. Hunecke, A.: Optimal design of a speech database for unit selection synthesis. Master's thesis, Universität des Saarlandes, Saarbrücken, Germany (2007)
27. Schröder, M., Hunecke, A.: Creating German unit selection voices for the MARY TTS platform from the BITS corpora. In: Proc. SSW6, Bonn, Germany (2007)
28. Zen, H., Nose, T., Yamagishi, J., Sako, S., Masuko, T., Black, A., Tokuda, K.: The HMM-based speech synthesis system version 2.0. In: Proc. of ISCA SSW6, Bonn, Germany (2007)

29. Charfuelan, M., Schröder, M., Türk, O., Pammi, S.C.: Open source HMM-based synthesisser for the MARY TTS platform. In: Proceedings of the 16th European Signal Processing Conference (EUSIPCO 2008) (submitted, 2008)
30. Ellbogen, T., Schiel, F., Steffen, A.: The BITS speech synthesis corpus for German. In: Proc. 4th Conference on Language Resources and Evaluation (LREC), Lisbon, Portugal, pp. 2091–2094 (2004)
31. Turk, O., Schröder, M., Bozkurt, B., Arslan, L.: Voice quality interpolation for emotional text-to-speech synthesis. In: Proc. Interspeech 2005, Lisbon, Portugal, pp. 797–800 (2005)
32. Schröder, M.: Interpolating expressions in unit selection. In: Paiva, A.C.R., Prada, R., Picard, R.W. (eds.) ACII 2007. LNCS, vol. 4738, Springer, Heidelberg (2007)

Context-Aware Agents to Guide Visitors in Museums

Ichiro Satoh

National Institute of Informatics
2-1-2 Hitotsubashi, Chiyoda-ku, Tokyo 101-8430, Japan
ichiro@nii.ac.jp

Abstract. This paper presents an agent-based system for building and operating context-aware services in public museums. Using RFID-tags or location-sensors, the system detects the locations of users and deploys user-assistant agents at computers near the their current locations. When users move between exhibits in a museum, this enables agents to follow them to annotate the exhibits in personalized form and navigate them to the next exhibits along their routes. It provides users with intelligent virtual agents in the real-world and enables them to easily interact with their agents though user movement between physical places. To demonstrate the utility and effectiveness of the system, we constructed and operated location/user-aware visitor-guide services in a science museum as a case study in our development of agent-based ambient computing in wide public spaces.

1 Introduction

The use of user/location-aware user-assistant services in public spaces, including museums, has attracted a great deal of attention from researchers over the past few years. This is because most visitors to museums lack sufficient knowledge about the exhibits and they need supplementalary annotations on these. However, as their knowledge and experiences are varied, they may become puzzled (or bored) if the annotations provided to them are beyond (or beneath) their knowledge or interest. In this paper, we explore a user/location-aware system for assisting visitors in museums by using intelligent virtual agent (IVA) technology to solve this problem.

However, the requirements of the real world, including public museums, particularly science museums, are completely different to those of IVAs in virtual environments and experimental spaces, e.g., laboratories. For example, one of the goals of science museums is to provide experiences to visitors that enhance their knowledge of science from exhibitions. Attractive characters and visual effects from multimedia terminals in museums may please visitors, particularly children with pleasant feelings, but they often prevent visitors from viewing the exhibits. Interactions between visitors and agents may often lower the extent of the visitors' learning experiences in museums, because they tend to be preoccupied with the interactions. It may also be difficult for visitors, particularly children, elderly, and handicapped people, to interact with IVAs through the buttons or touch panels of portable computers, cellular phones, and stationary terminals.

Nevertheless, our system provided agents to visitors in a science museum to study potential problems with agents, in particular IVAs, used within the real world. Our agents were designed to play supporting or extra roles, because exhibits meant to play the leading roles in the museums. The system detected the presence of visitors standing in front of specified exhibits and it deployed their agents at stationary computers close to their current positions. They recorded and maintained the experiences of visitors, e.g., a record of the exhibits that the visitors had previously viewed. In addition, they recorded and maintained profile information on the visitors and provided them with personalized annotations about the exhibits that they were standing in front of. Also, to free visitors from the burden of complex operations of interacting with their agents and configuring their annotations, we used their movements between exhibits as implicit operations to select the annotations that they wanted to view/hear and to evaluate what they had learned from the exhibits, because visitor movement is one of the most basic and natural behaviors in museums.

Our final goal was to offer context-aware services in large public spaces, e.g., cities. Nevertheless, we designed, implemented, and operated several context-aware services in a science museum as a case study before we provide ambient computing services through the wide public spaces of cities. This paper addresses the design, implementation, and application of the agent-based visitor-guide system. However, we intend to present our experiences, e.g., usability and human aspects, in future papers.

2 Basic Approach

This paper addresses the issue of location/user-aware services from IVAs running on stationary computing devices located at museums instead of on portable devices carried by visitors. Our system was inspired by the real requirements of museums rather than mere academic interest.

2.1 Background

There have been many academic and commercial attempts to provide context-aware services to visitors in public museums. A typical approach has been to provide visitors with audio annotations from portable audio players. These have required end-users to carry players and explicitly input numbers attached to exhibits in front of them if they wanted to listen to audio annotations about the exhibits. Many academic projects have provided portable multimedia terminals or PDAs to visitors. These have enabled visitors to interactively view and operate annotated information displayed on the screens of their terminals, e.g, the Electronic Guidebook [3] and Museum Project [2]. They have assumed that visitors are carrying portable terminals, e.g., PDAs and smart phones and they have explicitly input the identifiers of their positions or nearby exhibits by using user interface devices, e.g., buttons, mice, or the touch panels of terminals. However, such operations are difficult for visitors, particularly children, the elderly, and handicapped people, and tend to prevent them from viewing the exhibits to their maximum extent.

To solve this problem, several academic projects have provided visitors with portable terminals equipped with location-sensing systems to locate visitors, e.g., Hippie [7], Imogl [6], and Rememberer [3]. However, most museums have avoided using such devices because they are too expensive to lend to visitors and they also require regular maintenance, e.g., replacing or recharging the batteries every day. In addition, one of the most serious problems associated with portable smart terminals and AR systems is that they prevent visitors from focusing on the exhibits because they tend to become interested in the device rather than the exhibitions themselves and therefore concentrate on operating the PDA buttons or touch panel instead of looking at the exhibits. There have been several academic attempts to use AR systems in museums, but the equipment for these systems tends to be expensive and fragile so that they cannot be operated without professional operators. In fact, cost issues represent one of the most serious problems in deploying context-aware services in the real world.

2.2 Design Principles

To solve these problems, visitor-guide services for exhibits should be provided from stationary computers located close to the exhibits. Such services should be selected and provided to visitors in personalized form according to their knowledge and experience, e.g., the exhibits that the visitors previous viewed. We introduce the notion of agents, including IVAs, to context-aware service providers with user profiles. Visitors move between exhibits in museums. Agents as visitor guides are responsible for recording and maintaining user preferences and experiences and providing visitors with annotation services adapted to them from stationary devices. However, the requirements of agents in museums are different to the requirements of IVAs in virtual environments and entertainment.

- Agents provide annotations about exhibits to assist visitors to learn about or be impressed by them. When a visitor is in front of an exhibit, his/her agent selects/configures annotations and provides him/her with the annotations according to his/her preferences and the exhibits that he/she previous viewed, without his/her explicit operations.
- Visitors move from exhibit to exhibit in a museum. Therefore, when they move to another exhibit, their agents should be deployed at computing devices close to their destination exhibits. That is, our virtual agents should accompany their visitors and annotate exhibits in front of the visitors in the real-world on behalf of museum guides or curators.
- Many museums assign explicit or implicit routes to visitors, because the order in which they view exhibits often affects what they learn about the exhibits. Agents should navigate their visitors to exhibits along routes assigned to the visitors.
- User-manipulation must be simple and natural, because it is difficult for visitors to explicitly operate devices, including buttons, mice, touch panels, and smart cards. We introduce visitor movements as a user-friendly interaction with agents, because visitor movement is one of the most basic and natural behaviors in museums.
- Although there are many visitors in museums, existing IVA systems cannot cope with conflicts caused by multiple users in the real world. For example, more than two visitors may simultaneously view and hear at most one annotation provided from a stationary device at an exhibition space under the impression that the annotation is for them. To solve this problem, we use the visual representation of agents, e.g. characters, as a method of assisting visitors to know who the annotation is for.

- Annotations about exhibits may often be changed and modified, because real museums frequently replace and relocate displays in their exhibition spaces. Non-professional administrator, e.g., curators, should be able to easily define and customize visitor-guide agents.

Unlike agents in a virtual environment, entertainment is not alway needed, although it may be useful for children who want to experience annotation services. This is because, children under school age are liable to be more impressed by the entertainment features of annotation services in exhibits rather than the exhibits themselves and their annotative content. In fact, most exhibits in museum are visual objects. Therefore, if agents have overly attractive visual features, e.g., animation and 3D modeling, these actually prevent visitors from fully viewing the exhibits. Agents should be as invisible and inconspicuous as possible.

3 Deployable Virtual Agent Platform

Our user/location-aware visitor-guide system consists of three subsystems: 1) an agent host, 2) context-aware directory servers, called CDSs, and 3) context-aware virtual agents (Fig. 1). The first can execute context-aware virtual agents, where we assume that the computing devices are located at specified spots in public spaces. The second is an autonomous entity that defines application-specific services for visitors. The third is responsible for reflecting changes in the real world and the location of users when services are deployed at appropriate computers. User/location-aware visitor-guide services are encapsulated within the third subsystem so that the first and second subsystems are independent of any application specific services or other agents, which are simultaneously running to provide different services.

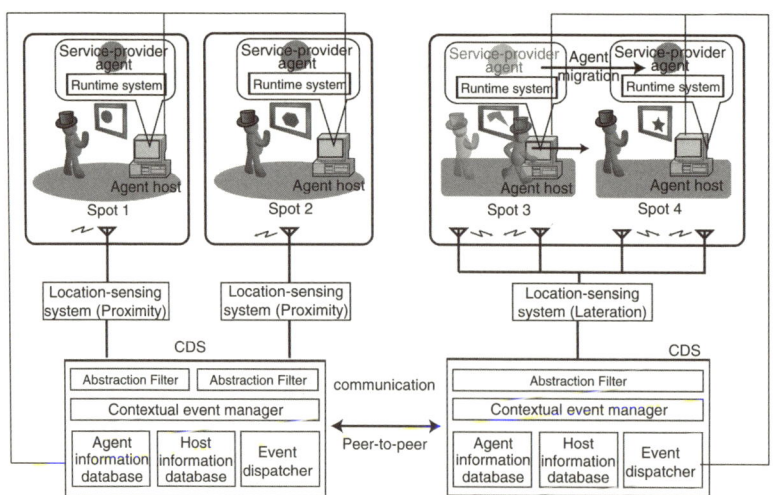

Fig. 1. Architecture of Context-aware Virtual Agent System

3.1 Agent Host

Each agent host is a computer that can provide visitor-guide services through user-interface devices, e.g., display screens and loudspeakers. It provides a runtime system for executing and migrating agents to other hosts. Each runtime system is built on the Java virtual machine (Java VM), which conceals differences between the platform architectures of the source and destination hosts.

Agent execution management: Each runtime system governs all the agents inside it and maintains the life-cycle state of each agent. When the life-cycle state of an agent changes, e.g., when it is created, terminates, or migrates to another host, the runtime system issues specific events to the agent. Some navigation or annotation content, e.g., audio-annotation, should be played without any interruptions.

Agent migration management: Each agent host can exchange agents with another host through a TCP channel using mobile-agent technology. When an agent is transferred over the network, not only the code of the agent but also its state is transformed into a bitstream by using Java's object serialization package and then the bit stream is transferred to the destination. The agent host on the receiving side receives and unmarshals the bit stream. Agents may have to acquire various resources, e.g., video and sound, or release previously acquired resources.

3.2 Context-Aware Agent Deployment

Each CDS is responsible for monitoring location-sensing systems and spatially binding more than one virtual agent to each user. It maintains two databases. The first stores information about each of the agent hosts and the second stores each of the agents attached to users. It can exchange this information with other CDSs in a peer-to-peer manner.

Location-sensing management: Tracking systems can be classified into two types: proximity and lateration. The first approach detects the presence of objects within known spots or close to known points, and the second estimates the positions of objects from multiple measurements of the distance between known points. The current implementation assumes that museums have provided visitors with tags. These tags are small RF transmitters that periodically broadcast beacons, including the identifiers of the tags, to receivers located in exhibition spaces. The receivers locate the presence or position of the tags. To abstract away differences between the underlying location-sensing systems, the CDSs map geometric information measured by the sensing systems to specified areas, called *spots*, where the exhibits and the computing devices that play the annotations are located.

Discovery and deployment of agents: When the underlying sensing system detects the presence (or absence) of a tag in a spot, the CDS that manages the system attempts to query the locations of the agent tied to the tag from its database. If the database does not contain any information about the identifier of the tag, it multicasts a query message that contains the identity of the new tag to other CDSs. It then waits for reply messages from other CDSs. Next, if the CDS knows the location of the agent tied to the newly visiting tag, it instructs the agent to migrate to a computing device.

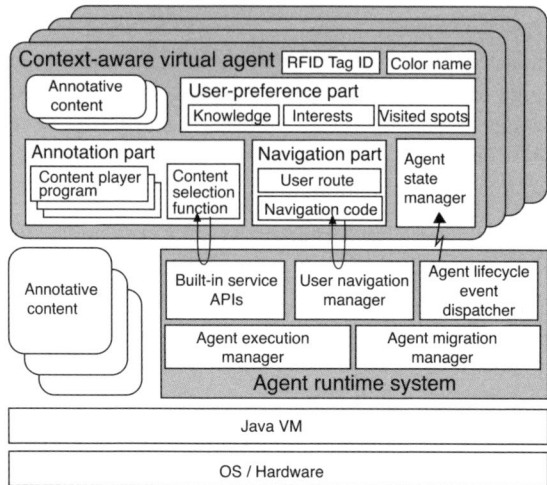

Fig. 2. Architecture of agent host

4 Context-Aware Virtual Agent

Each agent is attached to at most one visitor and maintains his/her preference information and programs that provide annotation and navigation to him/her. To enable agents to be easily developed and configured without any professional administrators, we divided each agent into three parts:

- **The user-preference part** maintains and records information about visitors, e.g., their knowledge, interests, routes, name, and the durations and times they spend at exhibits they visit.
- **The annotation part** defines a task for playing annotations about exhibits or interacting with visitors.
- **The navigation part** defines a task for navigating visitors to their destinations.

When an agent is deployed at another computer, the runtime system invokes a specified callback method defined in the annotation part and then one defined in the navigation part. Although these parts are implemented as Java objects, they are loosely connected with one another through data attributes by using Java's introspection mechanism so that they can be replaced without any compilations or linkages for their programs. The current implementation uses the standard JAR file format for archiving these parts because this format can support digital signatures, enabling authentication. Each agent keeps the identifier of the tag attached to its visitor. Each agent can specify the requirements that its destination hosts must satisfy in CC/PP form [12] and the runtime system can select an appropriate destination among multiple destination candidates through comparing between the capabilities required by agents and the capabilities of the candidates.

4.1 User-Preference Part

This is responsible for maintaining information about a visitor. In fact, it is almost impossible to accurately infer what a visitor knows or is interested in from data that are measured by sensing systems. Instead, the current implementation assumes that administrators will explicitly ask visitors about their knowledge and interests and manually input the information into this part. Nevertheless, it is still possible to make an educated guess with some probability as to what a visitor may be interested in, if we know which spots he/she has visited, how many he/she has visited, and how long he/she was visited. Each agent has a mechanism to automatically record the identifiers, the number of visits to, and the length of stays at spots by visitors. This part is implemented as a hashtable to maintain the collection of data entries. Each entry is a pair of a name and a value, where the former is string data and the latter has an arbitrary data structure represented as Java objects. The second and third parts can access entries with key names so that these parts can be combined loosely and replaced by compatible parts.

4.2 Annotation Part

Each agent is required to select annotations according to the current spot and route in addition to the information stored in the user-preference part and play the content in the personalized form of its user. This part defines the content-selection function and the set of programs for playing the selected content. The function maps more than one argument, e.g., the current spot, the user's selected route, and the number of times he/she has visited the spot into a URL referring to the annotative content. The content can be stored in the agent, the current agent host, or external http servers. That is, each agent can carry a set of its content, play the selected content at its destinations, directly play the content stored at its destinations, or download and play the content stored in Web servers on the Internet. Such content is provided in a variety of multimedia representations, e.g., text, image, video, and sound. The annotation part defines programs for playing this content. The current implementation supports (rich) text data, html, image data, e.g, JPEG and GIF, video data, e.g., animation GIF and MPEG, and sound data, e.g., WAV and MP3. The format for content is specified in an MIME-based attribute description. Since the annotation part is defined as Java-based general-purpose programs, we can easily define interactions between visitors and agents. The current implementation can divide this part into three sub-parts: opening, annotation, and closing, which are played in turn.

4.3 Navigation Part

Our agents are required to navigate visitors to their destinations along routes recommended by museums or visitors. After executing their annotation part, the navigation part is invoked by the runtime system to provide visual (or audio) information on the screens of the displays (or from loudspeakers) of the current agent host. For example, the agents display directions to exhibits that their visitors should next see. We also introduced visitor movements between exhibits as an implicit operation for selecting the routes that they wanted and evaluating what they had learned from the exhibits, because visitor movement is one of the most basic and natural behaviors in museums. This part provides the four navigation patterns, outlined in Fig. 3.

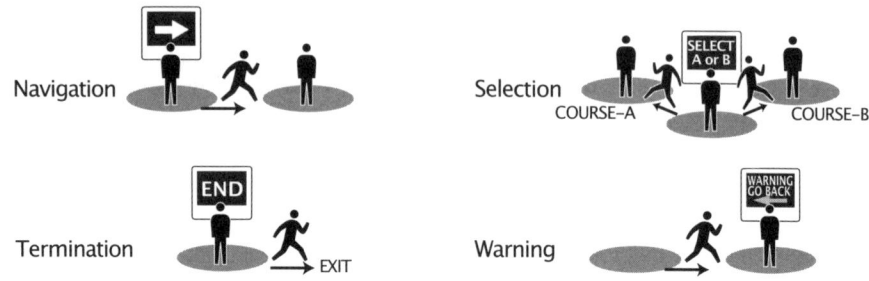

Fig. 3. User-navigation patterns

- *Navigation* instructs users to move to at least one specified destination spot.
- *Selection* enables users to explicitly or implicitly select one spot or route from one or more spots or routes close to their current spots by moving to the selected spot or one spot along the selected route.
- *Termination* informs users that they have arrived at the final destination spot.
- *Warning* informs users that they had missed their destination exhibit or their routes.

The user's route is described as a sequences of primitives corresponding to the above free patterns with our language for specifying the itineraries of mobile agents for network management [11] and they are stored in the user-preference part. No agent knows the spatial directions to the destinations because the directions themselves depend on the spatial relationships between the locations of the current agent host and the locations of the destinations, as well as the direction to the current host's screen. The current implementation permits administrators to manually input the directions of possible destinations and the direction to the screen. Agent hosts provide built-in APIs to their visiting agents. For example, if an agent has at least one destination, it invokes a specified API corresponding to the first pattern with the name of the destination; its current host returns the direction to the destination to it or displays the direction on the screen on its behalf.

5 Current Status

This section describes the current implementation of our system. It was implemented using Sun's Java Developer Kit version 1.5 or later versions.

Support for location-sensing systems: The current implementation supports two commercial tracking systems. The first is the Spider active RFID tag system, which is a typical example of proximity-based tracking. It provides active RF-tags for users. Each tag has a unique identifier that periodically emits an RF-beacon (every second) that conveys an identifier within a range of 1-20 meters. The second system is the Aeroscout positioning system, which consists of four or more readers located in a room. These readers can measure differences in the arrival timings of WiFi-based RF-pulses emitted from tags and estimate the positions of the tags from multiple measurements of the distance between the readers and tags; these measurement units correspond to about two meters.

Context-Aware Agents to Guide Visitors in Museums   449

Performance evaluation: Although the current implementation was not built for performance, we measured the cost of migrating a null agent (a 5-KB agent, zip-compressed) and an annotation agent (1.2-MB agent, zip-compressed) from a source host to a destination host that was recommended by the CDSs. The latency of discovering and instructing an agent attached to a tag after the CDS had detected the presence of the tag was 420 ms and the respective cost of migrating the null and annotation agent between two hosts over a TCP connection was 38 ms and 480 ms. This evaluation was operated with three computers (Intel Core 2 Duo 2 GHz with Windows XP Professional and JDK 1.5) connected via a Fast Ethernet. This cost is reasonable for migrating agents between computers to follow that visitors moving between exhibits.

Security mechanism: Security is essential in deployable software. The current implementation can directly use the security mechanisms provided in the Java language environment. The Java VM explicitly restricts agents so that they can only access specified resources to protect hosts from malicious agents. To protect against malicious agents being passed between agent hosts, each runtime system supports a Kerberos-based authentication mechanism for agent migration. It authenticates users without exposing their passwords on the network and generates secret encryption keys that can be shared selectively between parties that are mutually suspicious parties.

Remarks: Each runtime system has a queuing mechanism for exclusively executing agents for multiple simultaneous users. When two users enter the same spot, the CDS sends two notification messages to one of the agent hosts in the spot in the order in which they entered. The agent host can send events to the agents bound to the two users in that order, or can explicitly send an event to one of the agents. After the first has handled the event, it sends the same event to the second one. The current implementation provides Web-based APIs with AJAX technology to control agents. Therefore, an operator can create and customize agents through a Web browser running on his/her (portable) computer.

6 Early Experience

We constructed and operated an experiment at the Museum of Nature and Human Activities in Hyogo, Japan, using the proposed system. Figure 4 has a sketch that maps the spots located in the museum. The experiment was carried out at four spots in front of specimens of stuffed animals, i.e., a bear, deer, racoon dog, and wild boar. Each spot could provide five different pieces of animation-based annotative content about the animals, e.g., its ethology, footprints, feeding, habitats, and features, and had a display and Spider's active RFID reader with a coverage range that almost corresponded to the space, as shown in Fig. 5. When a visitor first participated in the experiment, an operator input his/her point of interest and the route for the new visitor and created his/her virtual agent by using a Web browser running on a portable terminal (Apple iPod Touch) equipped with a WiFi interface. Most curators have a positive impression of such portable management terminals rather than stationary management terminals, because stationary terminals require museums to provide spaces to house them. They also provide each visitor with a colored pendant including RFID tags, where these pendants are green, orange,

Fig. 4. Experiment at Museum of Nature and Human Activities in Hyogo

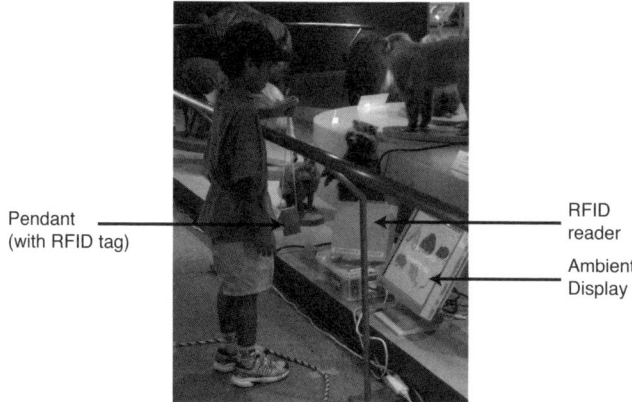

Fig. 5. Spot at Museum of Nature and Human Activities in Hyogo

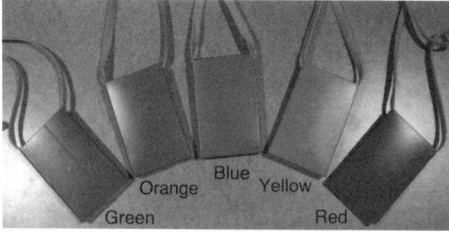

Fig. 6. Five colorful pendants

blue, yellow, or red (Fig. 6). A newly created agent is provided with its own color corresponding to the color of the pendant attached to the agent, because visitors could distinguish between their agents and others' agents through their pendants' colors. This may seem to be naive but it effectively enables visitors to readily identify their agents.

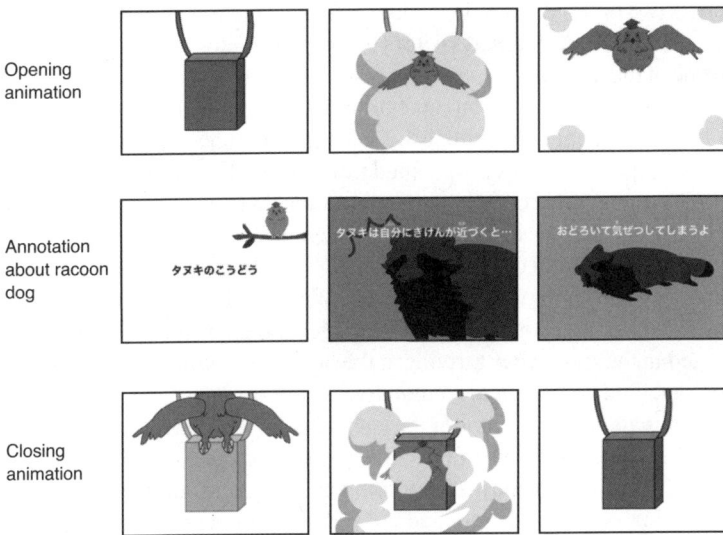

Fig. 7. Opening animation, annotation animation, and closing animation for orange pendant

The experiment was designed to enable visitors to imagine that their agents, which were virtual owls, were within their pendant. When a visitor entered a spot with the specimen of a racoon dog, his/her agent migrated from his/her pendant to the display located in the spot. As shown in Fig. 7, an agent tied to the orange pendant plays the opening animation to inform that its target is a visitor with an orange pendant, where the animation shows the agent's character appearing as an orange pendant. It next plays the annotation and then the closing animation. The durations of the opening animation, annotation, and closing are 7 sec, shorter than 40 sec, and 5 sec.

We simultaneously provided two kinds of routes for visitors to evaluate the utility of our user-navigation support. Both routes navigated visitors to destination spots along the way (Fig. 8). They enabled all visitors to walk around an exhibition both consisting of four spots two or three times, as shown on the right of Fig. 4. That is, a visitor might visit

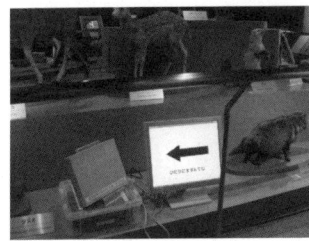
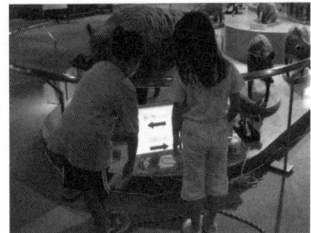

Navigation to one destination   Selection from two destinations

Fig. 8. Navigation patterns for user navigation at Museum of Nature and Human Activities in Hyogo

the same spots two or three times depending on the navigation providing by his/her agent. In addition, the first route enabled visitors to explicitly select subjects they preferred by moving to one of the neighboring spots corresponding to the subjects selected in specified spots at specified times. The second route provided visitors with several quizzes to review what they had learnt about the animals by selecting neighboring spots corresponding to their answers in specified spots at specified times. Both the experiments offered visitors animation-based annotative content about the animal in front of them so they could learn about it while observing the corresponding specimen.

The experimental system consisted of one CDS and four agent hosts. It provided GUI-based monitoring and configuration for agents. When the CDS detected the presence of a tag bound to a visitor at a spot, it instructed the agent bound to the user to migrate an agent host contained in the spot. After arriving at the host, the runtime system invoked a specified callback method defined in the annotative part of the agent. The method first played the opening animation for the color of the agent and then called a content-selection function with his/her route, the name of the current spot, and the number of times that he/she had visited the spot. The latency of migrating an agent and starting its opening animation at the destination after visitors had arrived at a spot was within 2 seconds, so that visitors could view the opening animation soon after they began standing in front of the exhibits. The method next played the selected content and then played the closing animation. After that, the runtime system invoked a specified callback method defined in the navigation part. An agent bound to a user could recommend two or more destination spots by using the *selection* pattern provided on its current agent host. When a visitor moved to one of the spots, his/her agent could record their selection. If the selection corresponded to a quiz choice, when a user moved to a spot corresponding to a correct or incorrect answer, their agent modified the visitor's profile that was maintained within it. Furthermore, if a user left out his/her route, the navigation part invoked a method to play warning content for him/her to return to his/her previous spot.

We conducted the experiment over two weeks. Each day, more than 60 individuals or groups took part in the experiment. Most of the participants were groups of families or friends aged from 7 to 16. Most visitors answered questionnaires about their answers to the quizzes and their feedback on the system in addition to their gender and age. Almost all the participants (more than 95 percent) provided positive feedback on the system. Typical feedback was "We were very interested in or enjoyed the system", "We could easily answer the quizzes by moving between the spots", and "We gained detail knowledge about the animals by watching them in front of where we were standing positions." Nevertheless, in the first three days of the experience, about 10 percent visitors lost interest in the annotations among their courses, while viewing closing animation on the disappearance of their IVAs for 5 sec in the spots. Later in the experiment, we must have abandoned the animations and instead displayed the images of the IVAs for 5 sec. Consequently, visitors could continue to view the annotations. They appeared to be interested at watching the specimens of animals instead of IVAs. In fact, most visitors only paid attentions to the colors of the IVAs instead of the characters and visual effects.

As application-specific services could be defined and encapsulated within the agents, we were able to easily change the services provided by modifying the corresponding agents while the entire system was running. More than two different visitor-guide services could also be simultaneously supported for visitors. Even while visitors were participating, curators with no knowledge of context-aware systems were able to configure the annotative content by doing drag-and-drop manipulations using the GUI-based configuration system. Such dynamic configuration is useful, because museums need to continuously provide and configure services for visitors.

7 Related Work

There have been many attempts to provide museum visitor-guide systems. Most existing works has assumed that visitors carry portable devices, e.g., the Electronic Guidebook, [3], Museum Project [2], Hippie system [7], ImogI [6], and Rememberer [3]. A few researchers have attempted approaches to support users from stationary sensors, actuators, and computers. However, most of these systems have stayed at the prototype- or laboratory-level and have not been operated and evaluated in real museums. Therefore, the results obtained may not be applicable in practical applications. Of these, the PEACH project [8] has developed a visitor-guide system for use in museums and has evaluated it in a museum. The system supported PDAs in addition to ambient displays and estimated the location of visitors by using infrared and computer-vision approaches. The project proposed a system for enabling agents to migrate between computing devices [5] by displaying an image of an avatar or character corresponding to the agent on remote computing devices, but could not migrate agents themselves to computing devices. The Virtual Anatomy Assistant Ritchie [13] was an attempt to seamlessly integrate characters into user actions by using 3D-optical tracking systems but it focused on conversation between individual users and agents and did not support any multiple user settings.

8 Conclusion

We designed and implemented an agent-based system for building and operating context-aware visitor-guide services in public museums. When a visitor moves from exhibit to exhibit, his/her agent can be dynamically deployed at a computer close to the current exhibit to accompany him/her and play annotations about the exhibit according to his/her knowledge, interest, and the exhibits that he/she is watching. His/her agent can also navigate him/her to exhibits along his/her route. That is, our system provides a spatial link between users and their assistant agent. To support large-scale context-aware systems, the system is managed in a non-centralized manner. Using the system, we constructed and operated location/user-aware visitor-guide services at a museum as case studies in our development of ambient computing services in public spaces. Unfortunately, our experiment revealed several counterproductive effects of the IVAs themselves when they were used in the real world. As a result, we reduced the appearance of the characters of the IVAs because they were more of a hindrance than a help to visitors who were studying the exhibits. We believe that IVAs in the real world should play a supporting role for visitors to achieve their purposes, e.g., viewing exhibits, instead of any leading role.

Finally, we would like to identify further issues that need to be resolved. Some may consider that non-centralized management architecture we used needed to operate the context-aware visitor-guide services in museums, as described in this paper. However, our final goal is to provide these large-scale context-aware services in large spaces, e.g., cities. We therefore need to demonstrate the scalability of the framework for large-scale context-aware services. We have developed an approach to testing context-aware applications on mobile/ubiquitous computers [9] and we plan to apply this to the development of virtual agents. This paper only described the design and implementation of agents and the effectiveness and utility of the model; however, the experiments provided us with other interesting experiences, e.g., usability, human aspects, and user modelling. We intend to present these experiences in our future papers. Although we experienced counterproductive results form IVA technology in our experiment, we need further evaluation with other applications in the real world to minimize these effects.

References

1. Brumitt, B.L., Meyers, B., Krumm, J., Kern, A., Shafer, S.: EasyLiving: Technologies for Intelligent Environments. In: Proceedings of International Symposium on Handheld and Ubiquitous Computing, pp. 12–27 (2000)
2. Ciavarella, C., Paterno, F.: The Design of a Handheld, Location-aware Guide for Indoor Environments. Personal and Ubiquitous Computing 8(2), 82–91 (2004)
3. Fleck, M., Frid, M., Kindberg, T., Rajani, R., O'BrienStrain, E.,, E., Spasojevic, M.: From Informing to Remembering: Deploying a Ubiquitous System in an Interactive Science Museum. IEEE Pervasive Computing 1(2), 13–21 (2002)
4. Kruppa, M., Spassova, L., Schmitz, M.: The Virtual Room Inhabitant - Intuitive Interaction with Intelligent Environments. In: Zhang, S., Jarvis, R. (eds.) AI 2005. LNCS (LNAI), vol. 3809, pp. 225–234. Springer, Heidelberg (2005)
5. Kruppa, M., Kruger, A.: Performing Physical Object References with Migrating Virtual Characters. In: Maybury, M., Stock, O., Wahlster, W. (eds.) INTETAIN 2005. LNCS (LNAI), vol. 3814, pp. 64–73. Springer, Heidelberg (2005)
6. Luyten, K., Coninx, K.: ImogI: Take Control over a Context-Aware Electronic Mobile Guide for Museums. In: Workshop on HCI in Mobile Guides, in conjunction with 6th International Conference on Human Computer Interaction with Mobile Devices and Services (2004)
7. Oppermann, R., Specht, M.: A Context-Sensitive Nomadic Exhibition Guide. In: Thomas, P., Gellersen, H.-W. (eds.) HUC 2000. LNCS, vol. 1927, pp. 127–142. Springer, Heidelberg (2000)
8. Rocchi, C., Stock, O., Zancanaro, M., Kruppa, M., Kruger, A.: The Museum Visit: Generating Seamless Personalized Presentations on Multiple Devices. In: Proceedings of 9th international conference on Intelligent User Interface, pp. 316–318. ACM Press, New York (2004)
9. Satoh, I.: A Testing Framework for Mobile Computing Software. IEEE Transactions on Software Engineering 29(12), 1112–1121 (2003)
10. Satoh, I.: A Location Model for Pervasive Computing Environments. In: Proceedings of IEEE 3rd International Conference on Pervasive Computing and Communications (PerCom 2005), pp. 215–224. IEEE Computer Society, Los Alamitos (2005)
11. Satoh, I.: Building and Selecting Mobile Agents for Network Management. Journal of Network and Systems Management 14(1), 147–169 (2006)

12. World Wide Web Consortium (W3C). Composite Capability/Preference Profiles (CC/PP) (1999), http://www.w3.org/TR/NOTE-CCPP
13. Wiendl, V., Dorfmuller-Ulhaas, K., Schulz, N., Andre, E.: Integrating a Virtual Agent into the Real World: The Virtual Anatomy Assistant Ritchie. In: Pélachaud, C., Martin, J.-C., André, E., Chollet, G., Karpouzis, K., Pelé, D. (eds.) IVA 2007. LNCS (LNAI), vol. 4722, pp. 211–224. Springer, Heidelberg (2007)

Virtual Institutions: Normative Environments Facilitating Imitation Learning in Virtual Agents

Anton Bogdanovych[1], Simeon Simoff[1], and Marc Esteva[2]

[1] School of Computing and Mathematics,
University of Western Sydney
Sydney, NSW, Australia
{a.bogdanovych,s.simoff}@uws.edu.au

[2] Artificial Intelligence Research Institute (IIIA, CSIC)
Campus UAB, Barcelona, Catalonia, Spain
esteva@iiia.csic.es

Abstract. The most popular two methods of extending the intelligence of virtual agents are explicit programming of the agents' decision making apparatus and learning agent behaviors from humans or other agents. The major obstacles of the existing approaches are making the agent understand the environment it is situated in and interpreting the actions and goals of other participants. Instead of trying to solve these problems we propose to formalize the environment in a way that these problems are minimized. The proposed solution, called Virtual Institutions, facilitates formalization of participants' interactions inside Virtual Worlds, helping the agent to interpret the actions of other participants, understand its options and determine the goals of the principal that is conducting the training of the agent. Such formalization creates facilities to express the principal's goals during training, as well as establishes a way to communicate desires of the human to the agent once the training is completed.

1 Introduction

Non-gaming Virtual Worlds like Second Life (http://secondlife.com) or Active Worlds (http://activeworlds.com) constantly grow in popularity. Their significance was highlighted by many researchers (i.e. [1], [2]). A report by *Gartner predicts that 80% of the Internet users will be actively participating in non-gaming Virtual Worlds by the end of 2011* [1].

The popularity of such Virtual Worlds creates a demand for intelligent autonomous agents operating within these virtual environments. The need for human-like sales assistants in E-Commerce environments, computer controlled teachers in virtual classrooms, or smart guides and travel agents in tourism systems stimulates researchers to look for more and more complex software architectures controlling the behavior of the autonomous agents.

The behavior of the majority of such virtual characters today is often controlled using preprogrammed scripts, finite state machines, or tree searches. None of these methods is well known for generalization capabilities. Consequently, common approaches for such virtual characters lead to ennui and frustration of the humans interacting with them. After a short period of interaction, the actions of computer-controlled characters tend to appear artificial and lack the element of surprise human participants would provide. Moreover, if a human acts in a way not envisaged by the programmers of the characters, such characters simply appear to behave "dumb" [3].

Another serious problem is making virtual agents appear believable. Carnegie-Mellon set of requirements for believable agents include personality, social role awareness, self-motivation, change, social relationships, and "illusion of life". Integrating these believability characteristics into virtual environments is associated with computational and architectural complexity; is platform and problem dependent, and is essentially far from achieving a high level of believability [4]. No existing virtual agent was yet able to pass the Turing test [4], adaptations of which are the only known research method of believability assessment [3].

Instead of explicitly programming various believability characteristics some researchers rely on the simulation theory. The key hypothesis behind this theory can be best summarized by the cliché "to know a man is to walk a mile in his shoes" [5]. It is assumed that simulation and imitation are the key technologies for achieving believability. In particular, using these techniques to produce more human-like behavior is quite popular in cognitive systems research [6].

Applying simulation theory to the development of autonomous agents is known as *imitation learning*. This approach is not new but it is not as popular as other types of learning and, most importantly, it has not been very well developed. Most of the imitation learning research is focused on robots intended for deployment in physical world [3]. This focus led to a situation where research aimed at behavior representation and learning struggles with issues arising from embodiment dissimilarities [7], uncontrollable environmental dynamics [8], perception and recognition problems ([6], [7]) and noisy sensors [6].

The aforementioned problems do not exist in Virtual Worlds. The sensors available there are not noisy, all participants normally share similar embodiment (in terms of avatars) and the environment is controllable and easily observable. Thus, using imitation learning for virtual agents represented as avatars within Virtual Worlds ought to be more successful than applying it to robots situated in the physical world. Despite this fact, only a few scholars have taken this direction and most of them are concerned with gaming environments, where virtual agents are used as computer controlled enemies fighting with human players [9], [10].

Focusing on video games makes possible to introduce a number of limitations and simplifications, which are not acceptable in non-gaming Virtual Worlds. The algorithms described in [9] seem to be quite successful in teaching the agent reactive behaviors, where next state an agent should switch into is predicted on the basis of the previous state and a set of environment observed parameters. These algorithms also prove to be quite useful in learning strategic behavior inside a particular video game (Quake II). The main limitation we see in this approach is that players' long term goals are assumed to be quite simple, namely to collect as many items as possible and to defeat their opponents [11].

In non-gaming Virtual Worlds, the situation is not that simple, as goals are more complex, and there is also a need to recognize the goals, desires and intentions of the human. For understanding the goals it is required to be able to assign the context to the training data and sort it into different logical clusters. Recognizing the desires and intentions is necessary when the agent is to replace the human in doing a particular task. For example, a principle may wish to train a virtual agent to participate in an auction on human's behalf. One of the reasons why such tasks cannot be achieved by the algorithms presented in [9] and [10] is that there is no mechanism provided there to communicate human requests, and there is no method for the agent to infer human's desires and intentions.

In respect to making agents understand the desires and intentions of the humans, existing approaches fall under one of the following extreme cases. First case is to purely rely on explicit communication between agents and humans, when every goal, belief, desire, intention and action the human trains the agent to perform is formalized for the agent. Another case is the fully implicit communication between humans and agents, when any explicit form of communication is considered unacceptable. As a result in the first case it often becomes easier to program the agents than to train them and in the second case only simple reactive behaviors can be learned and more complex behaviors are mostly left out (as the agent can not recognize complex human desires or intentions).

In this paper we explore the following two research hypotheses in relation to using imitation learning for virtual agents in non-gaming Virtual Worlds:

Hypothesis 1: It is impossible for the agent to implicitly recognize all the desires and intentions of the human and, therefore, a high-level communication language is required in some cases for the human to make the agent aware of those.

Hypothesis 2: For the agent to be able to handle the complexity of the human actions and goals, it should not purely rely on its own intelligence but should expect some help from the environment it is situated in.

The first hypothesis is an attempt to find a happy medium between the previously described extreme cases. In non-gaming Virtual Worlds participants' goals are not as trivial as in video games. On the one hand, always providing the agent with detailed formalization of human's actions, goals, intentions and desires is even less effective here. On the other hand, the goals and intentions of the humans are too complex for the agent to be able to infer them implicitly.

The second hypothesis is based on the suggestion made by Russel and Norvig that the agent's ability to successfully participate in some environment and extend its intelligence there is highly dependent on the complexity of this environment [12]. Authors mention that having the agent situated in a fully observable, deterministic and discrete environment significantly simplifies agent programming and valuably facilitates agent learning. As an example of a fully observable, deterministic and discrete environment, we consider Virtual Institutions, which are Virtual Worlds with normative regulation of participants' interactions. Through introducing the normative regulation of the interactions Virtual

Institutions help to interpret human actions, identify logical states of the agent and let humans communicate with an agent using a high level language in situations where it is impossible for the agent to recognize humans desires and intentions.

The key message this paper is trying to communicate is that *shifting some efforts into formalizing the environment may prove being more beneficial than spending them on improving agents' intelligence.* To support this message we show how the application of Normative Multiagent Systems can help in formalizing the interactions of humans and agents participating in a Virtual World.

The remainder of the paper is structured as follows. Section 2 provides a description of the Virtual Institutions concept. In Section 3 it is shown how using Virtual Institutions can facilitate imitation learning in virtual agents. Finally, Section 4 summarizes the contribution and outlines the directions of future work.

2 Virtual Institutions as Normative Virtual Worlds That Enable Learning from the Behavior of the Inhabitants

We consider Virtual Institutions [13] being a new class of normative Virtual Worlds, that combine the strengths of 3D Virtual Worlds and Normative Multiagent Systems, in particular, Electronic Institutions [14]. In this "symbiosis" the 3D Virtual World component spans the space for visual and audio presence, and the electronic institution component takes care of enabling the formal rules of interactions among participants. The normative system of the Virtual Institutions provides context and background knowledge for learning, helping to explain the tactical behavior and goals of the humans. The 3D representation provides the necessary environment to observe the behaviour of the humans. It assumes similar embodiment for all participants, including humans and autonomous agents, so every action that a human performs can be observed and, if necessary, reproduced by an autonomous agent.

One of the initial stages in the development of Virtual Institutions is formal specification of institutional rules. The specification defines which actions require institutional verification, assuming that the rest are safe and can be instantly performed. On the one hand, the specification plays the key role in the environment formalization and eventually helps the agent to put its actions into context. On the other hand, due to its formal nature and the available formal verification mechanisms, rules specification is a powerful way to ensure the validity of the participants' interactions and provide guarantees of correct rule enforcement. The specification is expressed through three types of conventions and their corresponding dimensions (for detailed explanation see [14]):

Conventions on language form the *Dialogical Framework* dimension. It determines language ontology and illocutionary particles that agents should use, roles they can play and the relationships or incompatibilities among the roles.

Conventions on activities form the *Performative Structure* dimension. It establishes the different interations agents can engage in, and the role flow policy

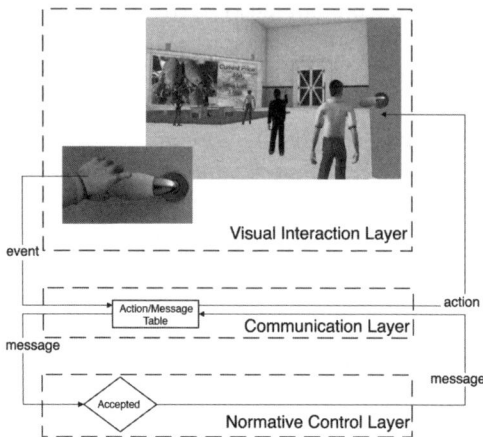

Fig. 1. The operation of the three-layered architecture of the virtual institution

among them. Each interaction protocol is specified in the so-called scenes, which define the possible interactions agents may have.

Conventions on behaviour form the *Norms* dimension. It captures the consequences of agents' actions within the institution. Such consequences are modeled as commitments (obligations) that agents acquire as consequence of some performed actions and that they must fulfill later on.

A virtual institution is enabled by a three-layered architecture presented by three conceptually (and technologically) independent layers, as shown in Fig 1.

Normative Control Layer. Its task is to regulate the interactions between participants by enforcing the institutional rules.

Communication Layer. It causally connects the above discussed institutional dimensions with the virtual world. It transforms the actions of the virtual world into messages, understandable by the institutional infrastructure and vice versa, using the Action/Message table created by the system designers.

Visual Interaction Layer supports the immersive interaction space of a virtual institution. Technologically, this layer includes a 3D virtual world and the interface that converts communication messages from the Communication Layer.

Fig 1 outlines the interaction between all these layers on an example of the agent requesting to enter a room inside the Virtual World. The human moves the mouse pointer over the door handle and clicks the mouse button (requesting the avatar in the Virtual World to open a door by pushing it handle). With the help of the Action/Message table this event is then translated into a message understandable by the Normative Control Layer. In case such a message is accepted in the Normative Control Layer the response message is sent back to the Communication Layer. The Communication Layer again consults the Action/Message table and transforms this response into the corresponding action which is executed in the Visual Interaction Layer. In the given example this action will result in opening the door and moving the avatar through it.

3 Enabling Imitation Learning with Virtual Institutions

An important feature of Virtual Institutions is that every human participant (principal) is always supplied with a corresponding software agent. The *couple agent/principal* is represented by an avatar. Each avatar is controlled by either a human or the autonomous agent. The agent is always active, and when the human is driving the avatar and acts in the Virtual World, the agent observes those actions. This allows the deployment of learning algorithms in order for the agent to learn how to make the decisions on behalf of the human.

The formal specification of a Virtual Institution contributes to learning the human-like behavior through providing the autonomous agents with a way of translating the actions performed by the human into the formal context of the institution. The dimensions of the institutional specification contribute to the quality of learning in the following way:

- *Dialogical Framework*: the roles of the agents enable the separation of the actions of the human into different logical patterns. The message types specified in the ontology help to create a connection between the objects present in the Virtual Worlds, their behaviors and the actions executed by the avatars.
- *Performative Structure*: Enables grouping of human behavior patterns into actions relevant for each room.

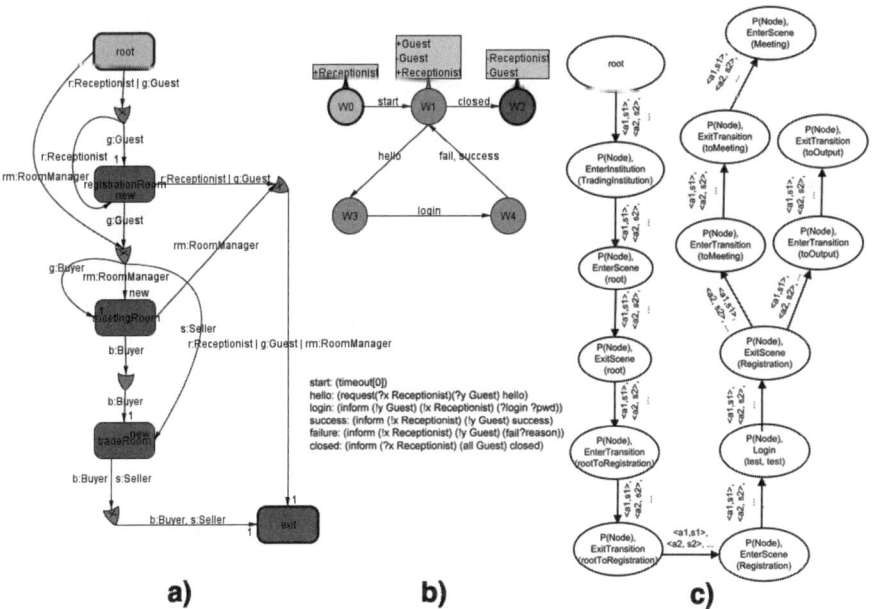

Fig. 2. Institutional Specification and Agent Training

– *Scene Protocols*: Enable the creation of logical landmarks within human action patterns in every room.

Fig. 2 illustrates how the institutional specification influences the imitation learning of autonomous agents in Virtual Institutions.

Fig. 2 a) outlines a performative structure of a prototypical Virtual Institution containing three rooms (registrationRoom, meetingRoom and tradeRoom). In the performative structure graph these rooms are connected through transitions (corridors). The arcs in the graph (visualized as doors) are marked with the permissions of the agents playing particular roles to enter certain rooms or corridors. Each of the rooms is associated with its interaction protocol. To determine the entrance into the institution and its exit, invisible rooms "root" and "exit" are included into every Performative Structure.

A protocol of each scene is specified by a finite state machine establishing the possible interactions agents may have. Fig. 2 b) shows the specification of the registrationRoom scene. The upper part is the scene protocol, while the lower part outlines the institutional level messages that can change the scene state.

Fig. 2 c) illustrates how the institutional specification can be used to simplify the imitation learning of virtual agents. It represents agent's decision graph created with the help of the institutional specification.

The decision graph is created and modified while the principal is acting in the Virtual Institution. These actions can be of two types. The actions that require institutional verification are the institutional level actions, while the rest are the actions of the visual level. The training of the virtual agents for believable behavior in Virtual Institutions happens on both visual and institutional levels. The actions of the visual level are important for capturing, for example, human-like control of avatar movement. The actions of the institutional level, on the one hand, allow the autonomous agent to make decisions about when to start and stop collecting data about the actions of the visual level and which context to assign to the collected sequences. Analyzing the sequence of institutional level actions on its own allows to understand how to reach different rooms and separate the sequences of actions there into meaningful logical states of the agent.

Through the three-layered architecture presented in Fig. 1 each institutional level action is transformed into an institutional message. The nodes of the agent's decision graph correspond to these institutional messages. Each of the nodes is associated with two variables: the message name and the probability of this message to be executed. The arcs connecting the nodes are associated with a set of parameters the avatar in question was able to sense in the environment and the recorded sequences of the visual level actions that represent believable avatar transitions between the institutional level actions.

In order to communicate human's desires to the agent we have defined a list of textual commands. Each command starts with the special keyword "Do:" and the rest of the command is an institutional message. When the agent receives such an instruction it searches its current decision graph for the node with the corresponding institutional message, and backtracks through it to the current node. Hence, it executes the sequence of the most probable actions (with the highest probability) that lead from the current node to the target node. More details about the learning algorithm can be found in [13].

4 Conclusions and Future Work

We have presented the concept of Virtual Institutions as the facilitator of imitation learning and a mechanism of high level communication of human desires to a virtual agent. Future work includes applying Virtual Institutions to the domain of electronic markets, using the described learning mechanisms within this domain and conducting experiments on the evaluation of believability.

Acknowledgments

This research is partially supported by an ARC Discovery Grant DP0879789, the e-Markets Research Program (http://e-markets.org.au), projects AT (CONSOLIDER CSD2007-0022), IEA (TIN2006-15662-C02-01), EU-FEDER funds, the Sabatical Programme of the Spanish Ministerio de Educacion y Ciencia Grant SAB2006-0001 and by the Generalitat de Catalunya under the grant 2005-SGR-00093. Marc Esteva wishes to expresses his gratitude to the Spanish Government for the Ramon y Cajal contract sponsoring his research.

References

1. Gartner, Inc.: Gartner Says 80 Percent of Active Internet Users Will Have A "Second Life" in the Virtual World by the End of 2011 (2007), http://www.gartner.com/it/page.jsp?id=503861
2. Cascio, J., Paffendorf, J., Smart, J., Bridges, C., Hummel, J., Hurtsthouse, J., Moss, R.: Metaverse Roadmap: Pathways to the 3D Web. Report on the cross-industry public foresight project, July 4 (2007)
3. Bauckhage, C., Gorman, B., Thurau, C., Humphrys, M.: Learning Human Behavior from Analyzing Activities in Virtual Environments. MMI-Interaktiv 12, 3–17 (2007)
4. Livingstone, D.: Turing's test and believable AI in games. Computers in Entertainment 4(1), 6–18 (2006)
5. Breazeal, C.: Imitation as social exchange between humans and robots. In: Proc. of the AISB Symposium on Imitation in Animals and Artifacts, pp. 96–104 (1999)
6. Schaal, S.: Is imitation learning the route to humanoid robots? Trends in cognitive sciences 3(6), 233–242 (1999)
7. Ekvall, S., Kragic, D.: Grasp recognition for programming by demonstration. In: Proceedings of the 2005 IEEE International Conference on Robotics and Automation, pp. 748–753 (2005)
8. Aleotti, J., Caselli, S., Reggiani, M.: Toward Programming of Assembly Tasks by Demonstration in Virtual Environments. In: 12th IEEE Workshop Robot and Human Interactive Communication, pp. 309–314, October 31 - November 2 (2003)

9. Gorman, B., Thurau, C., Bauckhage, C., Humphrys, M.: Believability Testing and Bayesian Imitation in Interactive Computer Games. In: Nolfi, S., Baldassarre, G., Calabretta, R., Hallam, J.C.T., Marocco, D., Meyer, J.-A., Miglino, O., Parisi, D. (eds.) SAB 2006. LNCS (LNAI), vol. 4095, pp. 655–666. Springer, Heidelberg (2006)
10. Le Hy, R., Arrigony, A., Bessiere, P., Lebeltel, O.: Teaching bayesian behaviors to video game characters. Robotics and Autonomous Systems 47, 177–185 (2004)
11. Thurau, C., Bauckhage, C., Sagerer, G.: Learning Human-Like Movement Behavior for Computer Games. In: Proceedings of the 8-th Simulation of Adaptive Behavior Conference (SAB 2004), pp. 315–323. MIT Press, Cambridge (2004)
12. Russel, S., Norvig, P.: Artificial Intelligence a Modern Approach. Prentice Hall, Pearson Education International, Upper Saddle River (2003)
13. Bogdanovych, A.: Virtual Institutions. PhD thesis, University of Technology, Sydney, Australia (2007)
14. Esteva, M.: Electronic Institutions: From Specification to Development. PhD thesis, Institut d'Investigació en Intelligència Artificial (IIIA), Spain (2003)

Enculturating Conversational Agents Based on a Comparative Corpus Study

Afia Akhter Lipi[1], Yuji Yamaoka[1], Matthias Rehm[2], and Yukiko I. Nakano[3]

[1] Dept. of Computer and Information Sciences, Tokyo University of Agriculture and Technology, Japan
{50007646211,50007646208}@st.tuat.ac.jp
[2] Institute of Computer Science, Augsburg University, Germany
rehm@informatik.uni-augsburg.de
[3] Dept. of Computer and Information Science, Seikei University, Japan
y.nakano@st.seikei.ac.jp

1 Introduction

When encountering people who have a different cultural background from our own, many of us feel uncomfortable because gestures and facial expressions may not be familiar to us. Thus, to enhance the believability of conversational agents, culture-specific nonverbal behaviors should be implemented into the agents. In our previous study [1], with the goal of building a user interface that incorporates a user's cultural background, we have collected comparative conversation corpus in Germany and Japan, and investigated the differences in gestures and posture shifts between these two countries. Based on [1], this paper reports a more detailed analysis about posture shifts, and proposes a chat system with an embodied conversational agent (ECA) that can act as a language trainer.

2 Analysis of Comparative Corpus

For the empirical basis of our ECA, we calculated statistics for speakers' posture shift frequency and distribution in our comparative corpus. The results are shown in Table 1. We use Bull's posture coding scheme in [2], but change the abbreviations for better readability. Each speaker-turn is divided into three sections: first (F), middle (M), and end (E). The "Ratio" column shows the distribution of posture shift occurrences.

For example, head postures most frequently occurred at (F) in both the Japanese and German data (36% in the German data and 31% in the Japanese data). Leg and Head postures were not very different in each culture. Both German and Japanese rested on the right or left leg throughout the conversation, and the most frequent head posture was "Turn head away". This head posture was reported as a typical turn-taking signal in Duncan's analysis of American people [3]. Significant difference was found in arm postures. German people frequently put their hands in their pockets or folded their arms. On the contrary, Japanese people put one hand on their face at the beginning of a turn, and joined their hands or put one hand on top of the other with fingers touching the wrist at the middle and end of a turn. These results are useful in deciding when the agent should change her/his posture, and what types of posture are appropriate according to a given culture.

Table 1. Frequently occurring postures in each section of a turn

			Germany			Japan	
	Section	Ratio	Frequency # 1	Frequency # 2	Ratio	Frequency # 1	Frequency # 2
Leg	F	39%	Weight on left leg (33%)	Weight on right leg (24%)	28%	Weight on right leg (45%)	Weight on left leg (25%)
	M	22%	Weight on right leg (42%)	Weight on left leg (42%)	26%	Weight on right leg (43%)	Stop leaning sideways (21%)
	E	34%	Weight on left leg (38%)	Weight on right leg (25%)	31%	Weight on right leg (37%)	Stop leaning sideways (26%)
Head	F	36%	Turn head away (40%)	Straighten head (32%)	31%	Straighten head (39%)	Turn head away (29%)
	M	29%	Straighten head (31%)	Turn head away (25%)	23%	Straighten head (42%)	Turn head away (24%)
	E	28%	Straighten head (43%)	Turn head away (20%)	25%	Turn head away (50%)	Straighten head (32%)
Arm	F	33%	Put hands in pockets (56%)	Put hand on elbow (13%)	27%	Puts hand on face (30%)	JoinHands (24%)
	M	30%	Put hands in pockets (57%)	Fold arms (14%)	23%	Puts hand on face (28%)	JoinHands (28%)
	E	32%	Put hands in pockets (34%)	Fold arms (25%)	32%	JoinHands (35%)	Puts hand on wrist (19%)

3 Enculturated Language-Trainer Agents

Based on our corpus study, we came up with the idea of implementing such culture-specific behaviors in ECAs. In Fig. 1 a student first chooses which language s/he wants to learn. When s/he chooses Japanese, a human Japanese teacher types in Japanese text. The text is sent to TTS and appropriate postures are determined based on

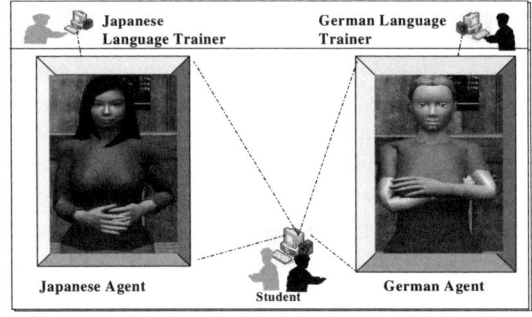

Fig. 1. Language- trainer Agents

the empirical data in Table 1. Finally, the speech sound and agent-posture animations are produced on the student's computer display. Likewise, if the student chooses German, then German text typed by a real German teacher is converted into speech accompanied by typical German posture animations. Thus, the system not only teaches language, but also makes the learner familiar with the culture-specific non-verbal behaviors. In the future, we hope that this system can be used as a distance-learning system by which a user can train her/himself how to smoothly communicate with people from other cultures.

Acknowledgment. This work is funded by the German Research Foundation (DFG) under research grant RE 2619/2-1 (CUBE-G) and the Japan Society for the Promotion of Science (JSPS) under a Grant-in-Aid for Scientific Research (C) (19500104).

References

1. Rehm, M., et al.: Creating a Standardized Corpus of Multimodal Interactions for Enculturating Conversational Interfaces. In: Proceedings of Workshop on Enculturating Conversational Interfaces by Socio-cultural Aspects of Communication, 2008 International Conference on Intelligent User Interfaces (IUI 2008) (2008)
2. Bull, P.E.: Posture and Gesture. Pergamon Press, Oxford (1987)
3. Duncan, S.: On the structure of speaker-auditor interaction during speaking turns. Language in Society 3, 161–180 (1974)

The Reactive-Causal Architecture: Towards Development of Believable Agents*

Ali Orhan Aydın, Mehmet Ali Orgun, and Abhaya Nayak

Macquarie University, Department of Computing, NSW, Australia
{aaydin,mehmet,abhaya}@ics.mq.edu.au

Abstract. To develop believable agents, agents should be highly autonomous, situated, flexible, and emotional. By this study, our aim is to present an agent architecture called Reactive-Causal Architecture that supports development of believable agents.

Keywords: Intelligent Agent Architecture, Believable Agents.

1 Introduction

The most commonly accepted attributes of intelligent agents are autonomy, situatedness, and flexibility [1]. Believable agents should have strong autonomy; since, they are interacting in real time with human actors. To have stronger autonomy agents should be capable of learning and have motivations to generate goals [3,4]. In addition, for believable agents researchers have stressed that an additional core attribute is affect display [5].

In the literature, there have been many proposed agent architectures. However, to the best of our knowledge the existing architectures do not satisfy these properties all together. Reactive-Causal Architecture (ReCau) is proposed to fill this gap. By extending the boundaries of theories of needs, ReCau employs an emotion model for affect display. In this paper, we present an overview of ReCau.

2 The Reactive-Causal Architecture

ReCau consists of reactive, deliberative and causal layers. The reactive and deliberative layers are in their classical form. The reactive layer is meant to interface with the environment. The middle layer has deliberative capabilities such as learning and planning. The decision-making and emotion generation occurs in the causal layer.

ReCau adopts Existence, Reletadness and Growth theory to support motives which enables an agent to generate goals [6]. Whenever a ReCau agent receives a condition related with its needs, it generates a goal to satisfy the corresponding

---

* This work is supported in part by Australian Research Council under Discovery Project No. DP0452628.

need. The goals of an agent are held in a queue in a hierarchical order. A ReCau agent pursues the lowest level needs first.

If a goal of a ReCau agent is to learn, it starts learning in accordance with social learning theory. Learning is realised through a reinforcement learner. Whenever the deliberative layer receives a condition and a goal, it develops plans to meet the goal. Each plan corresponds to a mean value of satisfaction degree and a variance value. The manipulator provides the means to collaborate with other agents and evaluate plan alternatives. These functions are performed by utilizing pro-attitudes each of which has a certain impact over the mean values.

By using mean values and variance values, the decision-making mechanism generates satisfaction degrees for every plan alternative. Then a ReCau agent selects the most satisfactory plan alternative as its intention. If the satisfaction obtained from the intention is above or below certain limits, the emotion generation mechanism generates corresponding positive or negative emotion. Emotions are classified as strong and regular emotions. If a generated emotion is a strong emotion, then it changes the order of the corresponding need in the hierarchy.

In accordance with the intention and emotion, tasks are dispatched by the deliberative layer. Task execution is guided by the reactive layer. When a plan is completely executed, a ReCau agent starts pursuing the next goal in the queue.

3 Conclusion

Reactive-Causal Architecture (ReCau) can be employed to develop believable agents. The emotion model, integrated with theories of needs, enables a ReCau agent to be highly autonomous and affective at the same time. Moreover, the employment of satisfaction degrees as a metric to select among plan alternatives provides a higher degree of flexibility to a ReCau agent.

References

1. Jennings, N.R., Sycara, K., Wooldridge, M.: A Roadmap of Agent Research and Development. Autonomous Agents and Multi-Agent Systems 1(1), 7–38 (1998)
2. Mateas, M., Stern, A.: Towards Integrating Plot and Character for Interactive Drama. In: Proceedings of the AAAI Fall Symposium on Socially Intelligent Agents, pp. 113–118. AAAI Press, Menlo Park (2000)
3. Russell, S., Norvig, P.: Artificial Intelligence: A Modern Approach, 2nd edn. Prentice Hall, Englewood Cliffs (2002)
4. Luck, M., D'Inverno, M.: A Formal Framework for Agency and Autonomy. In: First International Conference on Multi-Agent Systems, pp. 254–260. AAAI Press / MIT Press (1995)
5. Bates, J.: The Role of Emotion in Believable Agents. Communications of the ACM 37(7), 122–125 (1994)
6. Alderfer, C.P.: Existence, Relatedness, and Growth: Human Needs in Organizational Settings. Free Press (1972)
7. Rao, A.S., Georgeff, M.P.: Modeling Rational Agents within a BDI-Architecture. In: Knowledge Representation and Reasoning, pp. 473–484. Morgan Kaufmann Publishers, San Mateo (1991)

Gesture Recognition in Flow in the Context of Virtual Theater

Ronan Billon, Alexis Nédélec, and Jacques Tisseau

CERV: Centre Européen de Réalité Virtuelle
Laboratoire d'Informatique des SYstèmes Complexes
{billon,nedelec,tisseau}@enib.fr
http://www.cerv.fr/

Keywords: Motion-capture, gesture recognition, virtual theatre, synthetic actor.

Introduction. Our aim is to put on a short play featuring a real actor and a virtual actor, who will communicate through movements and choreography with mutual synchronization. Although the theatrical context is a good ground for multimodal communication between human and virtual actors, this paper will mainly deal with the ability to perceive the gestures of a real actor. During the performance, our goal is to match real-time observation with recorded examples. The recognition event will be sent to a virtual actor which reply with its own movements. We follow an approach similar to that proposed by [1] or [2]. They made the assumption that the first phase is to summarize the variation of the movement into a symbolic description. Then, the recognition phase is performed when a new trajectory is consistent with this description. The main section will describe our method of creating a signature from gestural data and the recognition system in real-time flow. To conclude, we shall present some results.

Gesture Representation and Perception. First, we must store an example of the gesture to be recognized. Our idea is that the data for movements from any motion-capture system can be reduced to a single artificial signature. Motion-capture systems generate a large amount of data and each type of sensor has its own characteristics, so we have decided to move away from those technologies and turn our system into a more generic recognition system. We decided to work on the variation during the gesture. For that purpose, we use the well-known PCA[1], that preserves variance, whether it is an angle or acceleration variation. Although we use PCA, our approach is quite different compared to [3] or [4]. Our gestures are all independent from one other. Instead of trying to find one gesture within a number of gestures, it is the gesture itself that signals when it recognizes itself in the stream. We can find similarities with multiagent systems where each agent is a gesture (perception of the real-time flow, decision of the similarity recognition and action of sending an event). During the performance we have to compare the real-time observation with the stored examples. Two main problems are raised

---

[1] Principal Component Analysis.

by recognition: firstly, segmentation, e.g., finding the beginning and the end; then the fidelity computation between the recorded gesture and the observed one. We use a forward spotting scheme that executes gesture segmentation and recognition simultaneously as stated by [5].

Results. We evaluated our system with laboratory experiments and a real demonstration in front of an audience. Laboratory experiments: we measured a rate of 100% recognition under good conditions and more than 80% under poor conditions. Furthermore, each gesture was recognized 0.7 second before the effective end of gesture. Real demonstration: We did a short demonstration of about five minutes which was performed in front of a live audience at a contemporary dance festival[2] for seven days. The real actor played capoeira according to the scenario and the reactions of the opponent.

Conclusion. This paper has presented a new method of perceiving gestures made by a real actor in real-time. These gestures are recorded during rehearsals and pre-processed to generate a signature. During the performance, the recognition system uses this signature and compares it with its observation. The basics of the gestural recognition system are achieved at this point. We have a versatile system that can record, load or unload gestures during execution time. It must now be included in the overall process of the play.

Acknowledgements. The research described here was supported in part by France Telecom R&D and in part by the Regional Council of Brittany.

References

1. Campbell, L.W., Bobick, A.: Recognition of human body motion using phase space constraints. In: Fifth International Conference on Computer Vision, 1995, pp. 624–630 (1995)
2. Bobick, A., Wilson, A.: A state-based technique for the summarization and recognition of gesture. ICCV 00, 382 (1995)
3. Vasilescu, M.A.O.: Human motion signatures: Analysis, synthesis, recognition. In: ICPR 2002: Proceedings of the 16 th International Conference on Pattern Recognition (ICPR 2002), Washington, DC, USA, vol. 3, p. 30456. IEEE Computer Society, Los Alamitos (2002)
4. Forbes, K., Fiume, E.: An efficient search algorithm for motion data using weighted pca. In: SCA 2005: Proceedings of the 2005 ACM SIGGRAPH/Eurographics symposium on Computer animation, pp. 67–76. ACM, New York (2005)
5. Kim, D., Song, J., Kim, D.: Simultaneous gesture segmentation and recognition based on forward spotting accumulative hmms. Pattern Recogn. 40(11), 3012–3026 (2007)

---

[2] http://festivals.lequartz.com/antipodes/2008/journal.php3

… Automatic Generation of Conversational Behavior for
Multiple Embodied Virtual Characters:
The Rules and Models behind Our System

Werner Breitfuss[1], Helmut Prendinger[2], and Mitsuru Ishizuka[1]

[1] Graduate School of Information Science and Technology, University of Tokyo
7-3-1 Hongo, Bunkyo-ku
Tokyo, Japan
werner@mi.ci.i.u-tokyo.ac.jp
[2] National Institute of Informatics 2-1-2 Hitotsubashi, Chiyoda-ku Tokyo, Japan
helmut@nii.ac.jp

Abstract. In this paper we presented the rules and algorithms we use to automatically generate non-verbal behavior like gestures and gaze for two embodied virtual agents. They allow us to transform a dialogue in text format into an agent behavior script enriched by eye gaze and conversational gesture behavior. The agents' gaze behavior is informed by theories of human face-to-face gaze behavior. Gestures are generated based on the analysis of linguistic and contextual information of the input text. Since all behaviors are generated automatically, our system offers content creators a convenient method to compose multimodal presentations, a task that would otherwise be very cumbersome and time consuming.

1 Introduction and Motivation

Combining synthetic speech and human-like conversational behavior like gaze and gestures for virtual characters is a challenging and tedious task for human animators. As virtual characters are used in an increasing number of applications, such as computer games, online chats or virtual worlds like Second Life, the need for automatic behavior generation becomes more pressing. Thus, there have been some attempts to generate non-verbal behavior for embodied agents automatically by some researchers ([3],[4]). A drawback of most current systems and tools, however, is that they consider only one agent, a salient feature of our system is that we generate the behavior not only for the speaker agent, but also for the listener agent, who might use backchannel behavior in response to the speaker agent. A previous version of this system has been presented in [1] in this paper we focus mainly on novel the rules that are used to plan and generate the behavior and are were refined after conducting a study which was described in [2].

2 Behavior Generation

Behavior generation in our system operates on the utterance level, for which rules are defined. The input we use thus consists of a text line spoken by one of our two agents.

Based on contextual and linguistic information of the text, the behavior for the speaking and the listening agent is suggested. (For further information on the input, see [1]). Three levels of rules are applied. The first set of rules traverses a tree like structure and suggests certain behavior based on the nodes in the tree. In the next level, we identify words and phrases that might be connected to a specific gesture, using WordNet. In the third level we iterate and recheck the suggested gestures and gaze patterns and align them to each other. One advantage of this approach is that the rule sets can be easily extended, and other non-verbal and verbal behavior could be included easily, such as emotion expression and verbal back channeling.

```
GAZE                                      GESTURE
FOR each THEMA node in the tree           FOR each OBJECT node in the tree
  IF at the beginning of the utterance      IF contains NEW node
  Or 85% of the time                        Or 70% of the time
  Look at speaker                           Single beat
FOR each RHEMA node in the tree           SPECIFIC
  IF at the end of the utterance            IF word == (big, huge, wide)
  Or 95% of the time                        IF word form == (superlative)
  Look at speaker                           set p = 2
                                            Show vertical size (p)
```

Fig. 1. Different Generation Rules

After this iteration, which is the last step of our behavior module, the generated behavior trees are forwarded to our output module that actually produces the final script in form of an MPML3D File ([5]). that can be displayed on the local computer or in the Virtual World of Second Life

References

1. Breitfuss, W., Prendinger, H., Ishizuka, M.: Automated Generation of Non-verbal Behavior for Virtual Embodied Characters. In: Proc. of the Int'l Conf. on Multimodal Interfaces (ICMI 2007), pp. 319–322. ACM Press, New York (2007)
2. Breitfuss, W., Prendinger, H., Ishizuka, M.: Automatic Generation of Gaze and Gestures for Dialogues between Embodied Conversational Agents: System Description and Study on Gaze Behavior. In: Proc. of the AISB 2008 Symposium on Multimodal Output Generation (MOG 2008), pp. 18–25 (2008)
3. Cassell, J., Vilhjálmsson, H., Bickmore, T.: BEAT: the Behavior Expression Animation Toolkit. In: Proceedings of SIGGRAPH 2001, pp. 477–486 (2001)
4. Kipp, M.: Creativity meets automation: Combining nonverbal action authoring with rules and machine learning. In: Gratch, J., Young, M., Aylett, R.S., Ballin, D., Olivier, P. (eds.) IVA 2006. LNCS (LNAI), vol. 4133, pp. 230–242. Springer, Heidelberg (2006)
5. Nischt, M., Prendinger, H., André, E., Ishizuka, M.: MPML3D: a reactive framework for the Multimodal Presentation Markup Language. In: Gratch, J., Young, M., Aylett, R.S., Ballin, D., Olivier, P. (eds.) IVA 2006. LNCS (LNAI), vol. 4133, pp. 218–229. Springer, Heidelberg (2006)

Implementing Social Filter Rules in a Dialogue Manager Using Statecharts

Jenny Brusk

Dept. of Game Design, Narrative and Time-based Media, Gotland University
621 67 Visby, Sweden
jenny.brusk@hgo.se

1 Introduction

This paper presents an implementation of Prendinger and Ishizuka's [3,4] social filter rules using statecharts. Following their example, we have implemented a waiter character in a coffee-shop that can interact with a user-controlled customer and a system-controlled boss. Due to space limitations, all further references to their work will be implicit, instead we refer to the literature.

The aim with this paper is to show the potential of using Harel statecharts [2] for modelling socially equipped game characters. The work is based on the assumption that statecharts successfully can be used for designing (game) dialogue managers (see e.g. [1]). There are also other advantages in using statecharts, e.g. (1) the fact that the world wide web consortium (W3C) has introduced a new standard for describing (dialogue) flow, StateChartXML (SCXML)[1], that combines the semantics of Harel statecharts with XML syntax, (2) statechart theory is an extension to ordinary finite-state machines (commonly used in games), featuring parallel and hierarchical states as well as broadcast communication, and (3) statecharts is a method that we think can support treating dialogue flow as part of the overall game flow.

2 Implementation

The waiter character statechart (see fig 1) consists of three independent states; Dialogue_manager, Actions_manager and Emotions_manager. The infinity-symbol (∞) represents non-atomic states, i.e. complex states that contain contracted (or unspecified) sub-states. The communicative acts the waiter can perform at a certain point will be determined by the following factors:

- Personality type: *Extraversion* (outgoing, neutral and introverted), affects the choice of communicative act (i.e. target state), and *Agreeableness* (friendly, indifferent and unfriendly), is expressed in the linguistic style.
- Current emotion state: Joy, Anger, or Neutral.
- Threat: social distance between speaker $SD(S, H)$ + the power the hearer has over the speaker $P(H, S)$.

---

[1] http://www.w3.org/TR/scxml/

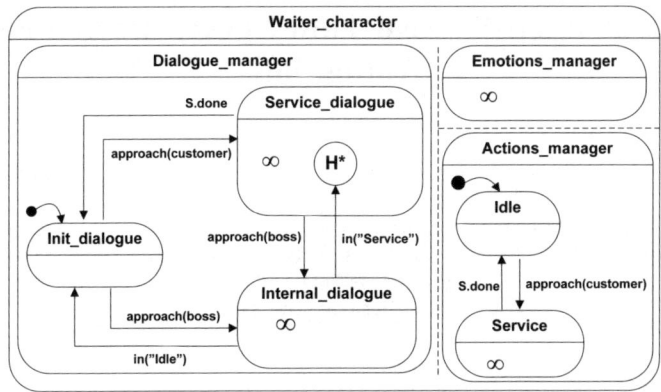

Fig. 1. Waiter character statechart

– Attitude towards respecting or violating conventional practises associated with the social role.

An example of a transition based on these parameters can be formulated as in the SCXML code below, interpreted as "if the waiter is in the state Anger and the threat is below 4, then go to state Take_order, otherwise go to the state Greet":

```
<transition cond="In('Anger') && threat < 4" target="Take_order"/>
<transition target="Greet"/>
```

It is also possible for the character to initiate a parallel dialogue session with the boss character, which could interrupt a waiter-customer dialogue. The waiter will adjust its behaviour towards the new social interplay according to the factors above. After this interruption, the waiter-customer dialogue will be resumed, enabled through the history function in statecharts (H*), as will the waiter's behaviour.

References

1. Brusk, J., Lager, T., Hjalmarsson, A., Wik, P.: DEAL - Dialogue Management in SCXML for Believable Game Characters. In: Proceedings of ACM Future Play, Toronto, Canada, November 14-18, pp. 137–144 (2007)
2. Harel, D.S.: A Visual Formalism for Complex Systems. Science of Computer Programming 8, 231–274 (1987)
3. Prendinger, H., Ishizuka, M.: Let's talk! Socially intelligent agents for language conversation training. Dautenhahn, K. (ed.) IEEE Trans on Systems, Man, and Cybernetics - Part A: Systems and Humans. Special Issue on "Socially Intelligent Agents - The Human in the Loop" 31, 465–471 (2001)
4. Prendinger, H., Ishizuka, M.: Simulating affective communication with animated agents. In: Proc 8th IFIP TC.13 Conf on Human-Computer Interaction (INTERACT 2001), Tokyo, Japan, pp. 182–189 (2001)

Towards Realistic Real Time Speech-Driven Facial Animation

Aleksandra Cerekovic[1], Goranka Zoric[1], Karlo Smid[2], and Igor S. Pandzic[1]

[1] Faculty of Electrical Engineering and Computing, University of Zagreb
{Aleksandra.Cerekovic, Igor.Pandzic, Goranka.Zoric}@fer.hr
[2] Ericsson Nikola Tesla, Karlo
smid@ericsson.com

Abstract. In this work we concentrate on finding correlation between speech signal and occurrence of facial gestures with the goal of creating believable virtual humans. We propose a method to implement facial gestures as a valuable part of human behavior and communication. Information needed for the generation of the facial gestures is extracted from speech prosody by analyzing natural speech in real time. This work is based on the previously developed HUGE architecture for statistically based facial gesturing, and extends our previous work on automatic real time lip sync.

A facial gesture is a form of non-verbal communication made with the face or head, used continuously instead of or in combination with verbal communication. There are a number of different facial gestures that humans use in everyday life, such as different head and eyebrow movements, blinking, eye gaze, frowning etc. We are interested in speech driven facial gestures. In the state of the art literature there is no system which would cover complete set of speech driven facial gestures in the real time. Albrecht et al. in [1] introduce a method for automatic generation of head and eyebrow raising and lowering, gaze direction, blinking, random eye movement from the speech. However, this system needs a preprocessing step. Real time speech driven facial animation is addressed in [2]. Speech energy is calculated and used as a variable parameter to control eyebrows frowning or forehead wrinkling. In our work we include wider set of speech-driven facial gestures generated in the real time.

In the previous work an automatic lip sync system is implemented [3]. Our current work aims to develop an automatic system for full facial animation driven by speech in the real time using universal architecture for statistically based human gesturing, HUGE [4]. It is capable of producing and using statistical models for facial gestures based on any kind of inducement, in our case the speech signal (Figure 1). In the first stage we extract unvoiced segments from the input speech signal, identify it as a pause and use as an audio state to produce a statistical model. However, to fine tune the output, we additionally add rules which correspond to the rules that facial gestures have as a communication channel. If identified pause is not longer than four frames, we consider it as a punctuator whereas longer pause is considered as thinking and word search pause. In addition to voluntary eye blinks, we have implemented periodic

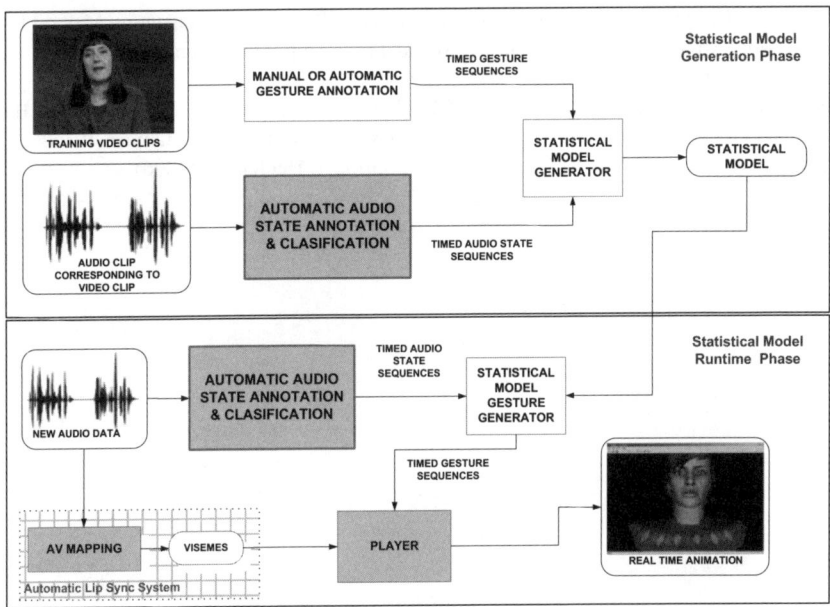

Fig. 1. Universal architecture of HUGE system adapted to audio data as inducement

eye blinks (i.e. manipulators). Therefore, we currently distinguish head and eyebrow movements and blinking as punctuators, head and eyebrow movements during thinking and word search pauses and blinking as manipulator. Once we know a gesture type, we know amplitude and duration of the specific gesture since they are also obtained from the statistical model. Having timed gesture sequences and also correct lip movements, we are able to create facial animation (visage|SDK is used).

Even by incorporating such a small number of speech features as well as communication rules, facial animation looks far more natural than when we had only lip sync. Next we are planning to add head and eyebrow movements correlated with pitch and gaze corresponding to communication rules found in the literature.

Acknowledgments

The work was partly carried out within the research project ``Embodied Conversational Agents as interface for networked and mobile services" supported by the Ministry of Science, Education and Sports of the Republic of Croatia.

References

1. Albrecht, I., Haber, J., Seidel, H.: Automatic Generation of Non-Verbal Facial Expressions from Speech. In: Proceedings of CGI (2002)
2. Malcangi, M., de Tintis, R.: Audio Based Real-Time Speech Animation of Embodied Conversational Agents. LNCS. Springer, Heidelberg (2004)

3. Zoric, G., Pandzic, I.: Real-Time Language Independent Lip Synchronization Method Using a Genetic Algorithm, special issue of Signal Processing Journal on Multimodal Human-Computer Interfaces (2006)
4. Smid, K., Zoric, G., Pandzic, I.P.: [HUGE]: Universal Architecture for Statistically Based HUman GEsturing. In: Gratch, J., Young, M., Aylett, R.S., Ballin, D., Olivier, P. (eds.) IVA 2006. LNCS (LNAI), vol. 4133, pp. 256–269. Springer, Heidelberg (2006)

Avatar Customization and Emotions in MMORPGs

Shun-an Chung and Jim Jiunde Lee

Institute of Communication Studies,
No.1001, University Road, Hsinchu, Taiwan 300, R.O.C.
pikatori@gmail.com

Abstract. This paper looks at how avatar customization details affect players' emotions in MMORPGs, which aspects they prefer in the environment of creating an avatar.

Keywords: Avatar Customization, emotion, avatar's characteristics.

1 Introduction

With the progressing of aesthetic technology, game-production industries put more emphasis on the artificial aspect of games. In MMORPGs, players pay attention to deliberate avatar customization because it offers specific signals for individual differences, privacy, emotions, and even construct their own identities.

2 Relevant Research

2.1 Avatar

Avatar is a "person" represents the user in the virtual space. In game environments, avatars are both a concrete portrait of a player and a protagonist in the narrative[1]. More humanized avatars can offer players more pleasure and entertainment value [2].

2.2 Emotions and Avatars

Emotion is a kind of strong feeling, which is excited by surroundings, and which causes internal or external responses. Izard(According to [1]) argues that emotion is a common experience for human beings, via expressing emotions, people's life experiences could be enriched. As Louchart and Aylett [3] said, the avatar which player operates represents him/herself as it experiences the online environment, faces others, is involved in activities, and inspires emotions.

3 Method

After comparing online role-playing games in Taiwan, we adopted "Perfect World" to execute the experiment because it not only offers general characters to choose from but allows players to elaborate many aspects in detail. Thirty participants were divided into two teams: team A could change all of the avatar's alterable parts; team B had to keep the original settings. Then the researcher measured these participants' emotions after finishing this stage by filling out an emotion-testing questionnaire.

4 Results

This research analyzed three parts of emotions: pleasure, arousal, and affected degree differences between teams A and B.

After the statistical analysis, the research found that in terms of pleasure, the average score of team A was 6.21, and team B was 4.47; thus team A's scores were higher than team B's. In terms of emotional arousal, the average score of team A was 5.25, and team B was 5.04; although the average score of team A was higher than that of team B, there was no obvious diversity. In terms of emotional arousal, the average score of team A was 5.16, and team B was 4.93, showing that the emotions of team A were more affected than those of team B. It seems that there are significant differences between players who have the freedom to control the avatar customization stage and those who do not. In other words, players who can freely change their avatars have higher pleasure, higher emotional arousal, and more emotional effects than players who can only use the original characters.

5 Discussion

According to the results, players' emotions will be more pleasant, experience higher emotional arousal and be more affected if their avatars are more close to their imagination, even the sense of being an entire "person". This research implies that players need more deliberate design of their avatars' in order to increase their preference and for the game. And designers could think more about the avatars details to improve game design development in the future.

References

1. Burn, A.: Computer games: Text, narrative and play, ch. 6, pp. 72–87 (2006)
2. Salem, B., Earle, N.: Designing a non-verbal language for expressive avatars. In: Proceedings of the third international conference on Collaborative virtual environments (2000)
3. Louchart, R., Aylett, S.: Towards a narrative theory of virtual reality. Springer, London (2003)

Impact of the Agent's Localization in Human-Computer Conversational Interaction

Aurélie Cousseau and François Le Pichon

VirtuOz,
75009 Paris, France
http://www.virtuoz.com

Abstract. How can we deliver more credible and effective conversational interactions between human and machine? The goal of this study is to define the needs and key characteristics of a dialogue before setting these factors against reality and technological advancements in the field of Intelligent Virtual Agents. This is the first step of our research, where we study the importance of the localization of the agent in the context of a human-machine dialogue.

Keywords: conversational interaction, human-machine.

1 Introduction

This study examines the influence of localization, the graphic type, and sex of an Intelligent Virtual Agent (IVA) in acts of dialogue. All conversational exchanges with a VirtuOz agent [1] are saved and tagged. Therefore it's possible for us to rank conversations undertaken by VirtuOz agents according to specfic criteria.

2 Method

Our study relies on the analysis of a broad corpus of conversations from the VirtuOz database. We have chosen 18 VirtuOz agents, each a different type, with a different mission, and with web locations which generated 3 millions conversations.

For each agent, we calculated the percentage of conversations where certain situations displaying social criteria arose. We measured the importance of sympathy in user interactions by determining the level of salutations, compliments and thanks in the first and last retorts of the user. To understand the level of respect between interlocutor and agent, we calculated the level of insults and the number of words with sexual orientation. To test the stratum of user speculation that their interlocutor is a computer program, we measured the number of conversations, where we observed the use of the agent's name and plotted them against the frequency where the user asks if the agent is a robot. We then delved deeper into the results to investigate the influence of factors such as the agents' broadcast channel, the presence of its name and its sex.

3 Impact of the Agent's Localization

Over the course of the analyses performed bt our VirtuOz agents, we studied the importance of the agent's localization, its' avatar design and sex. Our observations are recorded in the results table below.

Localization of the agent	"Are you real ?"	First interaction			Last interaction		
		Hello	Insults	Flirt	Goodbye	Insults	Flirt
Instant Messaging	2,06	33,62	4,96	1,88	3,81	5,00	2,13
Dedicated page	0,92	19,25	2,98	1,46	2,08	1,81	1,26
Home page	0,44	14,14	2,06	1,26	0,85	1,38	1,13
Type of avatar							
Human video	0,54	18,58	2,16	1,31	1,29	1,50	1,37
Illustration	0,81	17,02	2,84	1,47	1,86	2,00	1,24
Sex of the avatar							
Female	0,70	17,87	2,62	1,43	1,87	1,66	1,34
Male	0,68	20,23	2,36	1,31	2,12	2,03	1,49

Fig. 1. Distribution of the dialogue acts ("Are you real ?", Salutations, Insults and Flirt) according to the localization of the agent, the type and the sex of its' avatar. (% on 3 millions conversations)

First of all, the analysis of the results in this table (Fig. 1) shows that conversations performed over an instant messaging platform generated a higher level of social interactions than conversations that took place over a platform on a web site. Social interactions are defined a themes of dialogue that are more often used in interaction between humans. Thus, the instant messaging software presents the agent in a strong context of use, encouraging the humanization of the agent.

According to Richard Bourrelly [2], depicting an agent with a welcoming and communicative avatar, absorbed in a graphical environment suited to its role, and combining it with a rich dialogue interface, will allow the user to understand who and to what he can speak to. According to the compairison table (Fig. 1), in the framework of an internet site, it is important to place a specialized agent on the relevant page corresponding to his field of expertise. Immersed in the universe of the agent, the dialogue will be able to start from intelligent grounds favorable to a humanized exchange.

For Peirce [3], a sign is a perceptible element which finds its meaning in a specific context. Here, the interface of a conversational agent comprises two major signs, the avatar and the name of the agent. Their primary function is to personify the agent. By studying the data from (Fig. 1), we do not observe any particular trend in the influence of the graphic type or sex of the avatar.

While the datas (Fig. 1) show a real difference in the social approach of the agent depending on where the agent is deployed, they also lead us to the conclusion that the type of avatar is not the only component in the humanization of the agent.

Future Work. This preliminary study reveals that the localization of the agent is at least as important as the graphic representation of the avatar. Future studies will examine in more detail what is termed as "Immersion", the first contact between the human and the IVA.

References

1. VirtuOz, S.A.: French software company specialized in conversational IVA since 2002 (2002), http://www.virtuoz.com
2. Bourrelly, R.: Mieux négocier, ed. Eyrolles,éditions d'organisation, Paris (2007)
3. Peirce, C.S.: Illustrations of the Logic of Science. In: Dans The Essential Peirce, vol. 1, pp. 109–199. Indiana University Press (1992)

Evolving Expression of Emotions in Virtual Humans Using Lights and Pixels

Celso de Melo and Jonathan Gratch

Institute for Creative Technologies
University of Southern California
13274 Fiji Way, Marina Del Rey, CA 90292, USA
{demelo,gratch}@ict.usc.edu

An intelligent virtual human should not be limited to gesture, face and voice for the expression of emotion. The arts have shown us that complex affective states can be expressed resorting to lights, shadows, sounds, music, shapes, colors, motion, among many others [1]. Thus, we previously proposed that lighting and the pixels in the screen could be used to express emotions in virtual humans [2]. Lighting expression inspires in the principles of lighting, which are regularly explored in theatre or film production [3]. Screen expression acknowledges that, at the meta level, virtual humans are no more than pixels in the screen which can be manipulated to convey emotions, in a way akin to the visual arts [4]. In particular, we explore the *filtering* technique where the scene is rendered to a temporary texture, modified using shaders and, then, presented to the user. Now, having defined the expression channels, how should we express emotions through them? We are presently exploring an evolutionary approach which relies on genetic algorithms (GAs) to learn mappings between emotions and lighting and screen expression. The GAs' clear separation between generation and evaluation of alternatives is convenient for this problem. Alternatives can be generated using biologically inspired operators – selection, mutation, crossover, etc. Evaluation, in turn, can rely on artificial critics, which define fitness functions from art theory, or on human critics. Humans can be used to fill in the gaps in the literature as well as accommodate the individual, social and cultural values with respect to the expression of emotion in art [1].

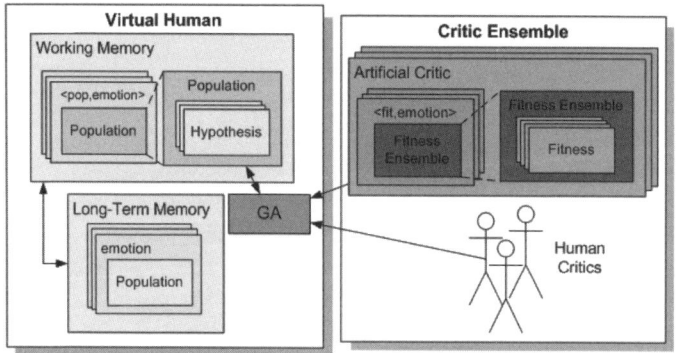

Fig. 1. The evolutionary model

Building on the expression model, the evolutionary model learns mappings between emotions and lighting and screen expression, see Fig. 1. The model revolves around two key entities: the *virtual human* and the *critic ensemble*. The virtual human tries to evolve the best way to express some emotion. For every possible emotion, it begins by generating a random set of *hypotheses*, which constitute a *population*. A hypothesis reflects some configuration of lighting and screen expression. The population evolves resorting to a *genetic algorithm* under the influence of feedback from the critic ensemble. The ensemble is composed of *artificial critics*, which classify hypotheses according to guidelines from art theory, and *human critics*, which classify hypotheses according to subjective criteria. The set of populations (one per emotion) is kept in the *working memory* while being evolved but, can be saved persistently in the *long-term memory*.

A study was conducted where subjects were asked to evolve mappings of joy and sadness into lighting and screen expression. No artificial critics were used in this study. Subjects were asked to evolve five generations for each emotion. No gesture, facial or vocal expression was used throughout the whole experiment. The results were promising and showed that subjects classified each succeeding generation with increasing fitness and that they were comfortable interpreting emotions in virtual humans even though expression was exclusively based on lights and pixels. Interestingly, however, the notion of joy and sadness seemed to vary among subjects, see Fig. 2. We are presently in the process of analyzing the collected data and what we seek to understand is what are the subjective fitness functions humans are using and how general these are. It would also be interesting to expand the mappings to the six basic emotions – anger, disgust, fear, joy, sadness and surprise – and, furthermore, explore more expression modalities such as the camera and sound of which much knowledge already exists in the arts [1].

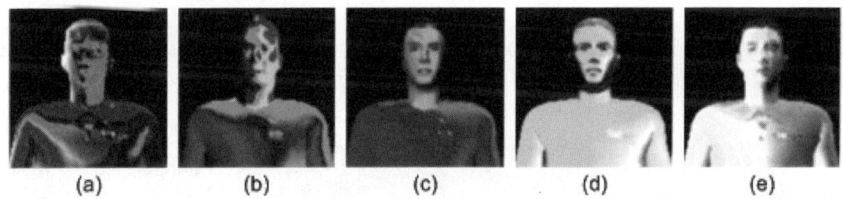

Fig. 2. Five top hypotheses for joy from different subjects in the study. What are the subjective fitness functions? What is common among them?

References

1. Sayre, H.: A World of Art, 5th edn. Prentice-Hall, New Jersey (2007)
2. de Melo, C., Paiva, A.: The Expression of Emotions in Virtual Humans using Lights, Shadows, Composition and Filters. In: ACII 2007, pp. 546–557. Springer, Heidelberg (2007)
3. Millerson, G.: Lighting for Television and Film, 3rd edn. Focal Press, Oxford (1999)
4. Gross, L., Ward, L.: Digital Moviemaking, 6th edn. Thomson/Wadsworth, Belmont (2007)

Motivations and Personality Traits in Decision-Making

Etienne de Sevin

University of Paris 8 / INRIA
INRIA Paris-Rocquencourt, Mirages, BP 105
78153 Le Chesnay Cedex, France
etienne.de_sevin@inria.fr

Abstract. By modifying the intensity of the motivations to test the adaptability of our real-time model of action selection, we put in evidence a relation between motivations and personality traits. We can give to the virtual human some corresponding personality traits such as greedy, lazy or dirty in order to obtain more interesting and believable virtual humans.

Keywords: motivations, personality traits, real-time decision making.

1 Introduction

A large number of personality characteristics are related to motivations [1]. In this paper, we test some personality traits [2] for the virtual human in a simulated environment in real-time [3] by modifying motivational parameters.

2 Motivations and Personality Traits

The personality traits in this test correspond to a specific set of motivation intensities. We defines that the virtual human behaves with greedy, lazy and dirty tendencies in

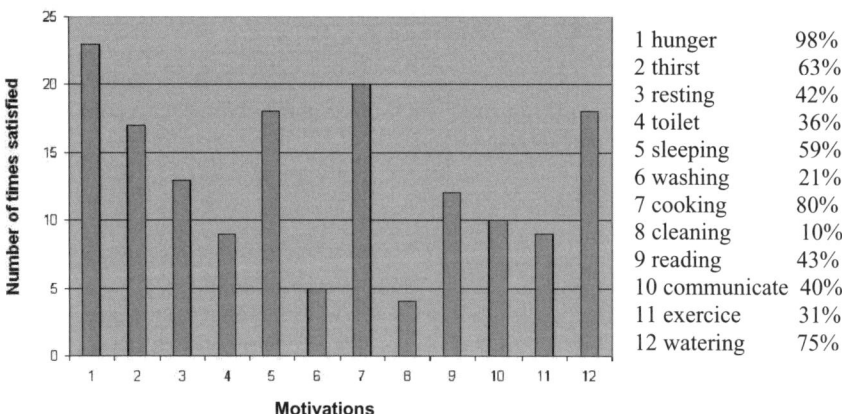

Fig. 1. Number of times the motivations are satisfied by the action selection model during the 65000 iterations and the parameterization of motivations

its behaviors. Thirst, hunger, cooking (greedy) and sleeping (lazy) are high whereas doing exercise (lazy), washing and cleaning (dirty) are low. Figure 1 shows that the action selection model satisfies the motivations accordingly and doesn't neglect any of them.

3 Conclusion

The motivational variations give to the virtual human distinctiveness in its behaviors since it does not react in the same way to the same situation. It depends also of the context of the simulation. In this test, the possible personality traits are linked to the twelve motivations defined in the apartment scenario [4]. However, as the number of motivations is not limited, we can add some new personality traits to the virtual human. It could be an easy way to test personality traits by tweaking motivational parameters.

As we can reuse the model easily, we plan to study the link between motivations intensity and personality traits with Embodied Conversational Agents (ECAs) [5]. We will test the model with users to have more experimentations and validations about the influence of motivation intensity on personality traits.

Acknowledgments

This research was supported partly by the Network of Excellence HUMAINE IST-507422 and partly by the STREP SEMAINE IST-211486.

The author would like to thank designers for their implication in the design of the simulation and Catherine Pelachaud for the helpful advices.

References

1. Deci, E.L., Ryan, R.M.: Intrinsic motivation and self-determination in human behavior. Plenum, New York (1985)
2. Allbeck, J., Badler, N.: Toward representing agent behaviors modified by personality and emotion. In: The 1st International Joint Conference on Autonomous Agents and Multi-Agent Systems, Bologna, Italy (2002)
3. de Sevin, E., Thalmann, D.: A motivational Model of Action Selection for Virtual Humans. In: Computer Graphics International (CGI). IEEE Computer SocietyPress, New York (2005)
4. de Sevin E.: An Action Selection Architecture for Autonomous Virtual Humans in Persistent Worlds, PhD. Thesis, VRLab EPFL (2006)
5. André, E., Pelachaud, C.: Interacting with Embodied Conversational Agents. In: Chen, F., Jokinen, K. (eds.) New Trends in Speech Based Interactive Systems. Springer, Heidelberg (2008)

A Flexible Behavioral Planner in Real-Time

Etienne de Sevin

University of Paris 8 / INRIA
INRIA Paris-Rocquencourt, Mirages, BP 105
78153 Le Chesnay Cedex, France
etienne.de_sevin@inria.fr

Abstract. Often real-time behavioral planner can not adapt to the change of the environments and be interrupted if it is necessary. Although Hierarchy is needed to obtain complex and proactive behaviors, some functionality should give flexibility to these hierarchies. This paper describes our motivational model of action which integrates a hierarchical and flexible behavioral planner.

Keywords: behavioral planning, flexibility, hierarchy, real-time.

1 Introduction

To bridge the gap between reflective and embodied cognition and action, efficient action selection architectures should associate hierarchical and reactive systems. Our action selection model for autonomous virtual humans [1] is based on hierarchical classifier systems [2] which can respond to environmental changes rapidly with its external rules and generate situated and coherent behavior plans with its internal rules. Therefore the virtual humans could satisfy their motivations wherever it is.

Our model is also based on free flow hierarchy [3] to allow compromise and opportunist behaviors increasing the flexibility of hierarchical systems. The activity is propagated through the hierarchy and no choices are made before the lowest level: the action one. The model has to choose the appropriate behavior at each point in time according to the motivations and environmental information.

2 Behavioral Planner

To reach specific locations where the virtual human can satisfy its motivations, behavioral sequences of locomotion actions need to be generated, according to environmental information and internal context of the hierarchical classifier system [1]. However these sequences can be interrupted if one motivation is more urgent to satisfy or by opportunism according to perceptions. In spite of high priority motivations, low priority ones are satisfied according to the parameters of the model. In this test (see Figure 1), the "cleaning" motivation parameter was the highest and the "watering" one was the lowest.

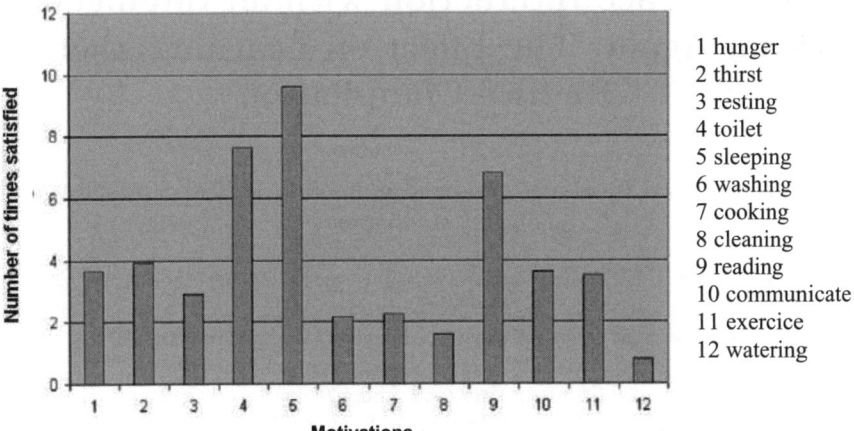

Fig. 1. Time-sharing of the twelve motivations over 65000 iterations

3 Conclusion

The results prove that our behavior planner is robust and flexible because it chooses the good sequence of actions to go to the locations where the virtual human can satisfy its motivations according to the distance, the opportunistic and compromise behaviors [1]. None of the motivations are neglected according to the parameters. We will reuse the model for Embodied Conversational Agents (ECAs) [4].

Acknowledgments

This research was supported partly by the Network of Excellence HUMAINE IST-507422 and partly by the STREP SEMAINE IST-211486.
The author would like to thank designers for their implication in the design of the simulation and Catherine Pelachaud for the helpful advices.

References

1. de Sevin, E.: An Action Selection Architecture for Autonomous Virtual Humans in Persistent Worlds, PhD. Thesis, VRLab EPFL (2006)
2. Donnart, J.Y., Meyer, J.A.: A hierarchical classifier system implementing a motivationally autonomous animat. In: The 3rd Int. Conf. on Simulation of Adaptive Behavior. The MIT Press/Bradford Books (1994)
3. Tyrrell, T.: Computational Mechanisms for Action Selection, in Centre for Cognitive Science. Phd. thesis, University of Edinburgh (1993)
4. André, E., Pelachaud, C.: Interacting with Embodied Conversational Agents. In: Chen, F., Jokinen, K. (eds.) New Trends in Speech Based Interactive Systems. Springer, Heidelberg (2008)

Face to Face Interaction with an Intelligent Virtual Agent: The Effect on Learning Tactical Picture Compilation

Willem A. van Doesburg[1], Rosemarijn Looije[1], Willem A. Melder[1], and Mark A. Neerincx[1,2]

[1] TNO Defence, Security & Safety, Kampweg 5 3796 DE, Soesterberg, The Netherlands
willem.vandoesburg@tno.nl, rosemarijn.looije@tno.nl,
willem.melder@tno.nl, mark.neerincx@tno.nl,
http://www.tno.nl
[2] Delft University of Technology, Mekelweg 4, 2628 CD Delft, The Netherlands

Abstract. Learning a process control task, such as tactical picture compilation in the Navy, is difficult, because the students have to spend their limited cognitive resources both on the complex task itself and the act of learning. In addition to the resource limits, motivation can be reduced when learning progress is slow. Intelligent Virtual Agents may help to improve tutoring systems by offering a student feedback on their performance via speech and facial expression, which imposes limited "communication load" and motivating ways of interaction. We present our Intelligent Virtual Agent, called Ashley, and an experiment in which we examined whether learning of tactical picture compilation is better when the feedback is provided by Ashley compared to a standard text-box interface. This first experiment did not show user interface effects, but provided new requirements for the next version of the feedback system and Ashley: The task to learn was yet too simple for substantial feedback effects, and the timing of Ashley's feedback should be improved, and Ashley's presence should be better scheduled.

1 Introduction

We have developed a tutoring system that can provide feedback on a task while aiming at minimal cognitive load of the feedback and maximal student motivation. In this study we compare the way that our tutoring system presents feedback. Feedback is presented in a classical way via a text-based user interface and via a new type of user interface that uses an Intelligent Virtual Agent for both verbal and nonverbal communication. This new interface is called Ashley. We have evaluated both manners of feedback and their impact on learning and motivation.

2 Experiment

During the evaluation we have tested the following hypotheses. (H1) Participants that receive feedback from Ashley have slower responses than participants that receive feedback in text. (H2) Participants learn faster during the training with Ashley than with text based feedback. (H3) Participants like the Ashley feedback better than text feedback. (H4) Participants who learn with Ashley make more accurate decisions after learning than students that receive text based training.

The experimental task is a radar task from the naval warfare domain. The task of the subject is to learn, of all ships currently on radar, to identify the ship that is most threatening.

20 participants took part in the experiment; 10 male and 10 female, aged 19-40 $\underline{M}$ age = 24.5, SD = 4.6.

The participants were divided into two groups. Each group started a training with the same instruction, but during the training they received feedback differently. One group received feedback by **text**, and the other group received feedback from **Ashley**. After training subjects received two tests. One of which was a speeded test where the subjects had to perform under time pressure. We have collected data on the task performance, response times, learning and subjective experience. For the performance data we compared the difference between the answers from our tutoring system with trainee answers. We distinguish between the ordinal difference (we call this ranking disagreement) and linear difference (called threat disagreement). For the the subjective experience data we used a questionnaire with a seven point Likert scale.

3 Results

Participants who received feedback from Ashley did not have significantly longer response times than participants that received text feedback. The two sided t-tests applied on this data had the following results; trainingset ($t=-0.4$, $df=18$, $p=0.70$), test ($t=-1.05$, $df=18$, $p=0.31$) and speeded test ($t=-0.96$, $df=18$, $p=0.35$).

We have examined the responses to the questionnaire, the questions in the motivational category especially, to see if participants liked Ashley better than the text feedback. We used a Wilcoxon rank-sum test to test the difference. We found no significant difference.

There were no significant differences in task performance. There was no significant two sided difference in the ranking disagreement measure between Ashley and the text condition in the standard test ($t=-1.37$, $df=18$, $p=0.41$) and the speeded test ($t=-1.55$, $df=18$, $p=0.14$). Additionally, there was no significant two sided difference between the mean threat disagreement measure between Ashley and the text condition in standard test ($t=-1.32$, $df=18$, $p=0.20$) and speeded test ($t=-1.49$, $df=18$, $p=0.15$).

We have examined the learning curves of the subjects in both groups. We found no significant difference in learning between the two conditions.

We believe that in this experiment learning the task was too easy, the subjects we're able gain expertise even without much feedback.

Creating and Scripting Second Life Bots Using MPML3D

Birgit Endrass[1,2], Helmut Prendinger[1], Elisabeth André[2], and Mitsuru Ishizuka[3]

[1] National Institute of Informatics, Tokyo, Japan
[2] University of Augsburg, Germany
[3] University of Tokyo, Japan

1 Motivation

The use of virtual characters becomes invariably common in computer games and other applications. So-called "bots" (computer-controlled virtual characters), may explain virtual settings in a natural way. We describe an example scenario in the online world Second Life [1] in order to explain to content creators how to create bots and script their behavior using the well established authoring language MPML3D [2]. MPML3D does not assume knowledge of a high-level programming language. In Second Life users are represented by their avatars (virtual characters controlled by humans) and its environment provides character models, animations and a graphics engine.

2 Example Scenario in SL

Our example scenario contains a poster session where users can be informed about NII (National Institute of Informatics) projects. It involves two bots (see Fig. 1) that explain the setting. When a user, i.e., his or her avatar, approaches the scene, the bots will perceive the avatar, adjust their position and start the presentation.

The virtual character's appearance can be chosen from several predefined templates. To create individual avatars Second Life provides character modifications as well as the possibility to upload user-created objects (like clothes or body parts). Additionally, appearances or items can be bought in online marketplaces and "attached" to the character.

To control the bots' behavior, we use the authoring language MPML3D. The head section of the MPML3D script provides information about the entities. For the example scenario, two bots (Yuki and Ken) are loaded into the scene.

```
<Entities>
<Entity type="human" name="yuki"resourcePath="girl"> </Entity>
<Entity type="human" name="ken" resourcePath="boy"> </Entity>
</Entities>
```

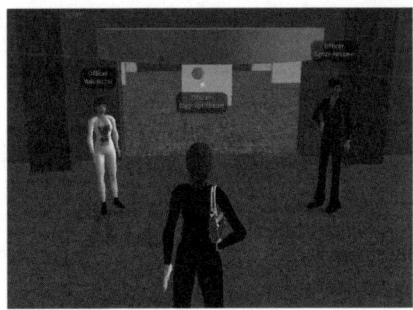

Fig. 1. Screenshot of example scenario in Second Life with two bots (left & right) and an user avatar (center)

The body part of the MPML3D script defines the bots' actions (verbal and nonverbal behavior. In the example below the female bot (Yuki) performs a sentence accompanied by a pointing gesture (while speaking the 6th word of the sentence (i.e., "this")). Behavior of different bots does not have to be sequential. In the example Kenzo will perform a head nod while he is introduced by Yuki.

```
<Parallel>
<Action name="yukiSpeak">yuki.speak("My name is Yuki. And this is my coworker Kenzo.")</Action>
<Action startOn="yukiSpeak[6].begin">yuki.gesture("POINT_LEFT")</Action>
<Action startOn="yukiSpeak[10].begin">ken.gesture("AGREE")</Action>
</Parallel>
```

SL provides a large set of gestures. Additional animations can be uploaded to Second Life in the commonly used BVH motion file format.

Our approach combines the following features: (1) Simplicity: no previous knowledge in modeling or in high-level programming languages is required, (2) Reusability: existing animations can be easily converted and reused in SL, (3) Low cost: all tools used in this approach are available for free, (4) Community: using Second Life as a platform a large community of users can be reached.

Other approaches in this area do either not provide an authoring tool for easy scripting of agent behavior or do not use free online worlds as a platform.

Acknowledgements. The first author was supported by an International Internship Grant from NII under a Memorandum of Understanding with the University of Augsburg and a Grant from the Elitenetzwerk Bayern (Elite Network Bavaria).

References

1. http://secondlife.com/ (last viewed: 07.04.2008)
2. Nischt, M., Prendinger, H., André, E., Ishizuka, M.: MPML3D: a reactive framework for the multimodal presentation markup language. In: Procedings of 6th International Conference on Intelligent Virtual Agents, pp. 218–229 (2006)

Piavca: A Framework for Heterogeneous Interactions with Virtual Characters

Marco Gillies, Xueni Pan, and Mel Slater

University College London Department of Computer Science
{m.gillies,s.pan,m.slater}@cs.ucl.ac.uk

Animated virtual humans are a vital part of many virtual environments today. Of particular interest are virtual human that we can interact with in some approximation of social interaction. However, creating characters that we can interact with believably is an extremely complex problem. Part of this difficulty is that human interaction is highly multi-modal. This multi-modality implies that there will be a great variety of ways of interacting with a virtual character. The diverse styles of interaction also imply diverse methods of generating behavior. For example animation can be played back from motion capture, generated procedurally, or generated by transforming existing motion capture

Handling and combining many diverse methods of animation on a single character requires a single representation for all of them. We use an abstract functional representation of animation. different types of animation can be represented in this way, and using a single representation makes it possible to use the diverse methods interchangeably and transparently. Functions can also represent methods for transforming and composing animations. This means that complex animations can easily be built from a few basic primitives through composition and other manipulations.

Figure 1(a) shows the type of heterogeneous interaction that is possible with our characters. The human participant's behavior is input with typical sensors for an immersive VR system, a microphone and head tracker. However, these two inputs are used in a variety of different ways by different behaviors. The head tracker is used to obtain the position of the participant in order to maintain an appropriate conversational distance, realistic gaze, and to respond to the participants posture shifts. The audio from the microphone is used for speech interaction. Speech interaction is either controlled by a human controller, if the character is an avatar, or by a dialogue engine. When the character speaks a number of other behaviors are triggered. The character's lip movements will be synchronized to the speech and the character will gesture. When the participant is speaking the character also gives head nods and other feedback signals and the gaze behavior will look at the participant more.

Figure 1(b) shows how a heterogeneous character can be implemented using our functional model. The example we give is of a user controlled character, whose speech is controlled by a human being (the controller) selecting speech sequences from a library of possible utterances, while the character interacts with another person (the participant). This system has 3 inputs: the position of the participant, an audio signal of the voice of the participant and input from the person controlling the character, specifying speech utterances. The position input is a 3-vector whose value is obtained every frame from a head tracker on the participant. The voice signal is obtained from an ordinary

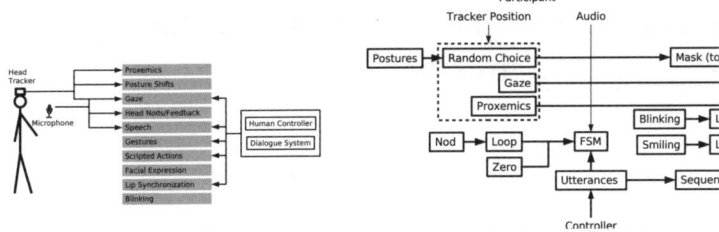

Fig. 1. (a) Heterogeneous Interactions (b) Implementing the Interaction

Fig. 2. A real and virtual human interacting in an immersive virtual environment

microphone. For this application we simply threshold the audio value to detect whether the participant is speaking. The controller has a user interface with a number of buttons used to trigger speech utterances.

Figure 2 shows some still frames from an interaction with our virtual character. The framework has been release as part of the open source project Piavca (available at http://piavca.sourceforge.net/), we encourage readers to try out the functionality.

Acknowledgements

The authors wish to thank the members of the University College London Department of Computer Science Virtual Environments and Computer Graphics group. This work was supported in part by BT plc. and a grant from the UK Engineering and Physical Sciences Research Council.

Interpersonal Impressions of Agents for Developing Intelligent Systems

Kaoru Sumi[1] and Mizue Nagata[2]

[1] National Institute of Information and Communications Technology,
Interactive Communication Media and Contents Group
3-5 Hikaridai, Seika-cho, Soraku-gun, Kyoto 619-0289, Japan
kaoru@nict.go.jp
http://www2.nict.go.jp/x/x164/people/kaoru/index-e.html
[2] The Department of Early Childhood Care and Education,
The Faculty of Human Life, Jumonji University
Sugasawa, Niiza-shi, Saitama, Japan

Abstract. This paper reports an experiment on the interpersonal impressions given by a character agent.

Keywords: Facial expression, Interpersonal impression of agents.

1 Introduction

When one is developing intelligent systems, it is important to consider how provide this feeling of affinity with the system and the presence for communicating with the system that has human-like intelligent functions such as recommendation or persuasion [1]. According to Media Equation [2], people treat computers, television, and new media as real people and places, so if an agent behaves in a disagreeable manner to users, it makes the users uncomfortable.

We planned an experiment to examine how the impression that the user gets from the agent's answer is affected by the combination of facial and word expressions. It was intended to clarify the impression that the agent gave the user by answering when interacting with the user in an emotion-arousing scenario.

2 Experiment on Impressions of Replies from the Agent

We chose six kinds of feelings ("Joy", "anger", "sadness", "antipathy", "fright", and "surprise"). From the total of 216 combinations, covering multiple feelings that the user felt (6) and the facial expressions for the agent's interaction with the user (6) and word expressions[1] used by the agent (6), we selected 96 patterns in this experiment. These covered the 6 feelings and 16 patterns: empathetic words and consistent facial expressions, nonempa-

---

[1] At first, as empathetic dialogue, "I think so, too" or "I don't think so" as nonempathetic dialog were spoken. Then, emotionally, the dialogue of "that's nice", "that's aggravating", "that's sad", "that's disgusting", "that's scary", and "that's a big surprise" were spoken.

thetic words and consistent facial expressions, word consistent and facial inconsistent, and word inconsistent and facial consistent. This is because the conditions of nonempathetic and both inconsistent word and facial expressions are nonsensical in normal communication. This is the condition where the word and facial expressions are consistent, which is the condition for double bind communication [3], but it can be considered as either word or facial being empathetic to the user. The case of nonempathetic condition and inconsistent word and facial expressions can be considered as pathological. In this experiment, we use nine factors: three factors for interpersonal impression evaluation were "affable–inaffable", "serious–unserious", and "conversable–inconversable" and original six factors "reliable–unreliable", "gentle–bitter", "pretentious–humble", "empathetic–unempathetic", "authoritative–unauthoritative", and "offensive–inoffensive". The examination was conducted in the form of a questionnaire on the Web. The content was displayed on user's own PC monitor after the user accessed the target URL.

A total of 1236 people, 568 male and 668 female (AV. 38.0, SD 11.5), were assigned 96 contents. More than ten users were assigned to each content.

3 Discussion

First, we predicted that an agent having the same facial and word expressions as the emotions aroused by the scenario would led to the most favorable impression, so we set these data as the control group. In fact, there were more favorable impressions than those obtained for the control group. For example, the facial and word expressions were "joy" when the user's emotion was "joy" for the control group. It is very interesting that when the user's emotion was "joy", the agent's words for "joy" with facial expressions of "surprise", "sadness", or "fright" were most favorable. On the other hand, when the user's emotion was "fright", the agent's words for "fright" with facial expressions of "antipathy" or "sadness" were the most favorable. These facial expressions were recognized as the emotion conveyed by the words and were more empathetic and somewhat meaningful emotions.

References

1. Fogg, B.J.: Persuasive Technology –Using Computers to Change What We Think and Do. Elsevier, Amsterdam (2003)
2. Reeves, B., Nass, C.: The Media Equation: How People Treat Computers, Television, and New Media Like Real People and Places. Cambridge University Press, Cambridge (1996)
3. Bateson, G., Jackson, D.D., Haley, J., Weakland, J.: Toward a Theory of Schizophrenia, Behavioral Science, vol. 1, pp. 251–264 (1956)

Comparing an On-Screen Agent with a Robotic Agent in Non-Face-to-Face Interactions

Takanori Komatsu[1] and Yukari Abe[2]

[1] Shinshu University, International Young Researchers Empowerment Center,
Tokida 3-15-1, Ueda, Nagano 3868567, Japan
tkomat@shinshu-u.ac.jp
[2] Future University-Hakodate, Graduate School of Systems Information Science,
Kamedanakano 116-2, Hakodate, Hokkaido 0418655, Japan
g2108003@gmail.com

Abstract. We investigated the effects of an on-screen agent and a robotic agent on users' behaviors in a non-face-to-face interaction, which is a much more realistic style than the face-to-face interaction on which most studies have focused. The results showed that the participants had more positive attitudes or behaviors toward the robotic agents than the on-screen agent. We discuss the results of this investigation and focus in particular on how to create comfortable interactions between users and various interactive agents appearing in different media.

Keywords: Human-agent interaction, robotic and on-screen agents, Shiritori game.

1 Introduction

Communication media terminals, such as PDAs, cell phones, and mobile PCs, are devices that are used globally and are a part of our daily lives. Most people have access to such media terminals. Various interactive agents, such as robotic agents [1,2] and embodied conversational agents (ECA) appearing on a computer display [3,4], are now being developed to assist us with our daily tasks. The technologies that these interactive agents can provide will soon be applied to these widespread media terminals.

Some researchers have started considering the effects of these different agents appearing on these media terminals on users' behaviors and impressions, especially for comparisons of on-screen agents appearing in a computer display with robotic agents [5,6,7]. Most of these studies reported that most users stated that they felt much more comfortable with the robotic agents and that these agents were much more believable interactive partners compared to on-screen agents. In these studies, the participants were asked to directly face the agents during certain experimental tasks. However, this "face-to-face interaction" does not really represent a realistic interaction style with the interactive agents in our daily lives. Imagine that these interactive agents were basically developed to assist us with our daily tasks. We can assume that the users are engaged in tasks when they need some help from the agents. Thus, these users do not look at the agents much but mainly focus on what they are doing. This interaction style is called a "non-face-to-face interaction." We assumed that this style is much more realistic than the "face-to-face interaction" on which most former studies focused.

We experimentally investigated the effects of an on-screen agent and a robotic agent on users' behaviors in non-face-to-face interactions. We discuss the results of this investigation and focus in particular on how to create comfortable interactions between users and various interactive agents appearing in different media.

2 Experiment

2.1 Experimental Setting

We set up a "non-face-to-face interaction" between users and agents in this experiment by introducing a dual task setting; that is, one task was obviously assigned to participants as a dummy task, and the experimenter observed other aspects of the participants' behaviors or reactions, of which the participants were completely unaware.

First, the participants were told that the purpose of this experiment was to investigate the computer mouse trajectory while they played the puzzle video game "picross[1]" by Nintendo Co., Ltd. (shown on the right of Fig. 1). The picross game was actually a dummy task for participants. While they were playing the picross game, an on-screen or robotic agent placed in front of and to the right of the participants talked to them and encouraged them to play another game with it. The actual purpose of this experiment was to observe the participants' behavioral reactions when the agent talked to them.

2.2 An On-Screen Agent and a Robotic Agent

The on-screen agent we used was the CG robot software "RoboStudio[2]" by NEC Corporation (the left part of Fig. 1), and the robotic agent was the "PaPeRo robot[3]" by NEC Corporation (the center of Fig 1). RoboStudio was developed as simulation

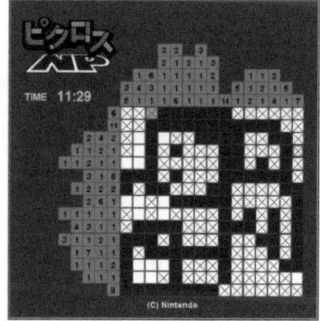

Fig. 1. On-screen agent "RoboStudio" by NEC Corporation (left), robotic agent "PaPeRo" by NEC Corporation (center), and "picross" puzzle game by Nintendo Co., Ltd (right).

---

[1] http://tgcontent.nintendo-europe.com/enGB/games_DS_TGP/picross/overview_picross.php
[2] http://www.incx.nec.co.jp/robot/english/cgpapero/index.html
[3] http://www.incx.nec.co.jp/robot/english/papero2005/index.html

Fig. 2. Experimental Scene: participants in Screen group (left) and in Robotic group (right)

software of the PaPeRo, and it was equipped with the same controller used with PaPeRo. Therefore, both the on-screen and the robotic agents could express the same behaviors.

2.3 Participants

The participants were 20 Japanese university students (14 men and 6 women; 19-23 years old). Before the experiment, we ensured that they did not know about the PaPeRo robot and RoboStudio. They were randomly divided into the following two experimental groups.

- **Screen Group (10 participants):** The on-screen agent appeared on a 17-inch flat display (The agent on the screen was about 15 cm tall) and talked to participants. The agent's voice was played by a loudspeaker placed beside the display.
- **Robotic Group (10 participants):** The robotic agent (It was about 40 cm tall) talked to participants.

Both the robotic agent and the computer display (on-screen agent) were placed in front of and to the right of the participants, and the distance between the participants and the agents was approximately 50 cm. The sound pressure of the on-screen and robotic agent's voice at the participants' head level was set at 50 dB (FAST, A). Pictures showing both experimental groups are shown in Fig. 2.

2.4 Procedure

First, the dummy purpose of the experiment was explained to the participants, and they were asked to play the picross game for about 20 minutes after receiving simple instructions on how to play it. The game was a web-based application, so the participants used the web browser to play it on a laptop PC (Toshiba Dynabook CX1/212CE, 12.1 inch display).

The experimenter then gave the instruction that "This experiment will be conducted by this agent. The agent will give you the starting and ending signal." After these instructions, the experimenter exited the room, and the agent started talking to the participants, "Hello, my name is PaPeRo! Now, it is time to start the experiment. Please tell me when you are ready." Then, when the participants replied "ready" or "yes," the agent said, "Please start playing the picross game," and the experimental session started. The agent was located as described earlier so that the participants could not look at the agent and the picross game simultaneously.

One minute after starting the experiment, the agent said to them "Umm...I'm getting bored... Would you play Shiritori (see Fig. 3 about the rules of this game) with me?" Shiritori is a Japanese word game where you have to use the last syllable of the word spoken by your opponent for the first syllable of the next word you use. Most Japanese have a lot of experience playing this game, especially when they are children. If the participants acknowledged this invitation, i.e., said "OK" or "Yes," then the Shiritori game was started. If not, the agent repeated this invitation every minute until the game was terminated (20 minutes). After 20 minutes, the agent said "20 minutes have passed. Please stop playing the game," and the experiment was finished.

The agent's behaviors (announcing the starting and ending signals and playing the last and first game) were remotely controlled by the experimenter in the next room by means of the wizard of oz (WOZ) manner.

Japanese Last and First Game (*Shiritori*)
Rule:
- Two or more people take turns to play.
- Only nouns are permitted.
- A player who plays a word ending in the mora "*N*" loses the game, as no word begins with that character.
- Words may not be repeated.

Example:
 Sakura (cherry blossom)-> *rajio* (radio)-> *onigiri* (rice ball)-> *risu* (squirrel) -> *sumou* (sumo wrestling) -> *udon* (Japanese noodle)
  Note: The player who played the word *udon* lost this game.

Fig. 3. Rules of Shiritori from Wikipedia[4]

2.5 Results

In this experiment, we assumed that the effects of different agents on the participants' impressions would directly reflect on their behaviors. We then focused on the following behaviors: 1) whether the participants acknowledged the agent's invitation and actually played the Shiritori game, 2) whether the participants looked at the agent or the picross game during the game, 3) how many puzzles the participants succeeded in solving.

1) Whether the participants acknowledged the agent's invitation and actually played the Shiritori game. In the robotic group, eight out of the 10 participants acknowledged the agent's invitation and actually played the Shiritori game with the agent. However, in the screen group, only four out of the 10 participants did so (Fig. 4). Statistical analysis showed a significant tendency between these two experimental groups (p=0.067, p<.1 (+)).

---

[4] http://en.wikipedia.org/wiki/Shiritori

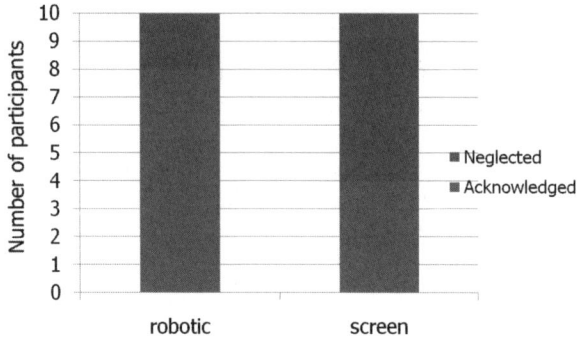

Fig. 4. Rate of participants acknowledging or ignoring the agent's invitation to play Shiritori

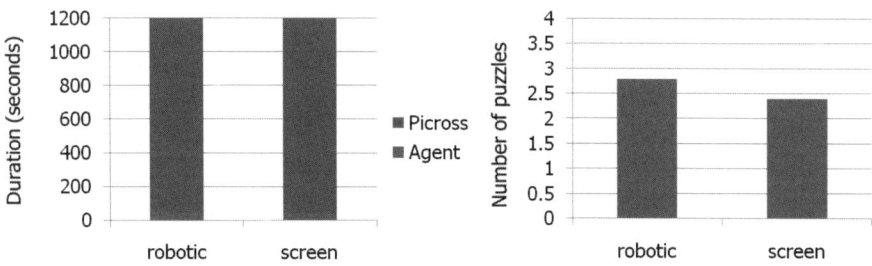

Fig. 5. (left) Duration of participants looking at picross game or agent. (right) How many puzzles they succeeded in solving.

2) Where the participants looked (agent or picross game). In the robotic group, the participants' average duration of looking at the robotic agent was 46.3 seconds. In the screen group, the average duration was 40.5 seconds (Fig. 5, left). These results revealed that most participants in both groups concentrated on looking at the picross game during this 20-minute (1,200 seconds) game. The results of an ANOVA showed no significant differences between the two groups on this issue ($F(1,18)=0.17$, n.s.).

3) How many puzzles the participants succeeded in solving. In the robotic group, the participants' average number of puzzles solved was 2.8, and in the screen group, it was 2.4 (Fig. 5, right). The results of an ANOVA showed no significant differences between these two groups ($F(1,18)=2.4$, n.s.).

The results of this experiment are summarized in the following:

- **Screen group:** The participants in the screen group showed the same achievement level on the picross game with the robotic Group (Fig. 5, right). They did not look at the on-screen agent much during the experiment (Fig. 5, left), and they also did not acknowledge the invitation from the on-screen agent (Fig. 4).
- **Robotic group:** The participants in the robotic group showed the same achievement level on the picross game with the Screen group (Fig. 5, right). They also did not look at the robotic agent much (Fig. 5, left). However, most of them acknowledged the robotic agent's invitation and actually played the Shiritori game (Fig. 4).

3 Discussion and Conclusions

The results of this experiment showed that most participants acknowledged the robotic agent's invitation for the Shiritori game, while many neglected the on-screen agent's invitation. The participants in both groups showed nearly the same attitudes toward the picross game; that is, they did not look at the agent much but concentrated on the task, and they achieved nearly the same level on the picross game. The participants in the robotic group interacted with the robotic agent (playing the Shiritori game) without neglecting their given tasks (playing the picross game). Therefore, the robotic agent was appropriate for interacting with users in a non-face-to-face interaction, which is much more similar to the interaction we encounter in our daily lives compared with the style observed in a typical face-to-face interaction setting. Actually, these results are similar to those of former studies that focused on face-to-face interaction; that is, these studies argued that the robotic agents are much more comfortable and believable interactive partners than on-screen agents.

Let us consider why the participants acknowledged the robotic agent's invitation even though they were not really looking at the robotic agent. Beforehand, we reviewed Kidd and Breazeal's [8] investigation. They conducted an experiment comparing a physically present robot with a robot appearing on television as live TV output. The results were that the participants did not show different behaviors and impressions of these different robots. They then concluded that *"it is not the presence of the robot that makes a difference, rather it is the fact that the robot is a real, physical thing, as opposed to a fictional animated character on screen, that leads people to respond differently."* Therefore, the participants' beliefs or mental models about a "robot" based on their expectation or stereotypes (such as "A robot would be nice to talk to") would affect their attitudes toward an interaction with a robotic agent. More specifically, these beliefs and mental models would lead to making the participants assign certain types of personality or characteristics to this robotic agent and would then cause the participants to acknowledge the robotic agent's invitation, even though they did not look at the robotic agent much.

In any case, the results of this experiment were similar to the ones obtained in former studies regarding face-to-face interaction. We will apply these results to create an appropriate interaction between users and on-screen agents. Specifically, we will conduct a follow-up experiment to observe how the participants who acknowledged the PaPeRo robot's invitation will react to the on-screen agent CG PaPeRo. If these participants actually assign certain types of personality or characteristics to the robotic agent, they should regard an on-screen agent having a similar appearance to a robotic agent as being the same person or character, and they would then acknowledge the on-screen agent's invitation as if acknowledging the robotic agents' invitation. We believe that this follow-up experiment will lead to a novel interactive methodology for inducing participants to act toward an on-screen agent just like he or she would toward a robotic agent.

Acknowledgments. The part of this study was performed through Special Coordination Funds for Promoting Science and Technology of the Ministry of Education, Culture, Sports, Science and Technology, the Japanese Government.

References

1. Gravot, F., Haneda, A., Okada, K., Inaba, M.: Cooking for a humanoid robot, a task that needs symbolic and geometric reasoning. In: The 2006 IEEE International Conference on Robotics and Automation, pp. 462–467 (2006)
2. Imai, M., Ono, T., Ishiguro, H.: Robovie: Communication Technologies for a Social Robot. International Journal of Artificial Life and Robotics 6, 73–77 (2003)
3. Cassell, J., Stocky, T., Bickmore, T., Gao, Y., Nakano, Y., Ryokai, K., Tversky, D., Vaucelle, C., Vilhjalmsson, H.: MACK: Media lab Autonomous Conversational Kiosk. In: Imagina 2002 (2002)
4. Prendinger, H., Ishizuka, M. (eds.): Life-Like Characters. Springer, Heidelberg (2004)
5. Shinozawa, K., Naya, F., Yamato, J., Kogure, K.: Differences in effects of robot and screen agent recommendations on human decision-making. International Journal of Human-Computer Studies 62, 267–279 (2004)
6. Powers, A., Kiesler, S., Fussel, S., Torrey, C.: Comparing a computer agent with a humanoid robot. In: The 2nd ACM/IEEE /International Conference on Human-robot /Interactions, pp. 145–152 (2007)
7. Wainer, J., Feil-Seifer, D.J., Sell, D.A., Mataric, M.J.: Embodiment and Human-Robot Interaction: A Task-Based Perspective. In: 16th IEEE International Conference on Robot & Human Interactive Communication, pp. 872–877 (2006)
8. Kidd, C., Breazeal, C.: Effect of a robot on user perceptions. In: 2004 IEEE/RSJ International Conference on Intelligent Robots and Systems, pp. 3559–3564 (2004)

Sustainability and Predictability in a Lasting Human–Agent Interaction

Toshiyuki Kondo, Daisuke Hirakawa, and Takayuki Nozawa

Dept. of Computer and Information Sciences,
Tokyo University of Agriculture and Technology, Japan
t_kondo@cc.tuat.ac.jp
http://www.livingsys.lab.tuat.ac.jp/

Abstract. Sustainability is one of important features to be considered in a human–agent interaction (HAI) process, because humans are unavoidably accustomed to the agents and it seems to become boredom as the agents' behaviors come to be predictable. To be clear the relationship between the sustainability and the predictability, the interaction processes between human subjects and a virtual entertainment robot with either of three different interaction models are evaluated.

1 Introduction

Recently there have been gradually increasing the works dealing with human–agent interaction in AI and robotics fields. Most of these agents are widely acknowledged, since they are carefully designed to behave amusingly. However it is reasonable that these pre-designed scenarios make us dissatisfied feelings as time goes by. According to the concept of *spontaneous motivation* in psychology, we essentially have an intellectual curiosity with respect to unfamiliar agents. Thus it seems natural that active interaction is gradually getting stagnant as the agent behavior becomes predictable. While complicated, unpredictable, or inconsistent behaviors (e.g. random) would probably decrease our interests. Based on this, we assumed that maintaining the behavior predictability in a moderate level contributes to the sustainable HAI. To be clear this, we prepared three *interaction models* (i.e. strategies for behavior selection) with different predictabilities. Using the models, we experimentally evaluate the lasting interaction processes among human subjects and the virtual entertainment robot.

2 Methods

A virtual Sony ERS-7 robot simulator was developed. It equipped with virtual touch sensors for interpreting human mouse operation as touch interaction (i.e. beat/stroke). In response to the interaction, it displays one of eight predetermined motions. As has been noted, we prepared following three interaction models; FSM (finite state machine), RND (random), and EMT (emotion). Here, EMT model has two hidden (i.e. unobservable) variables named *arousal* and *pleasure*, it referred to Russell's circumplex model for facial expression [1].

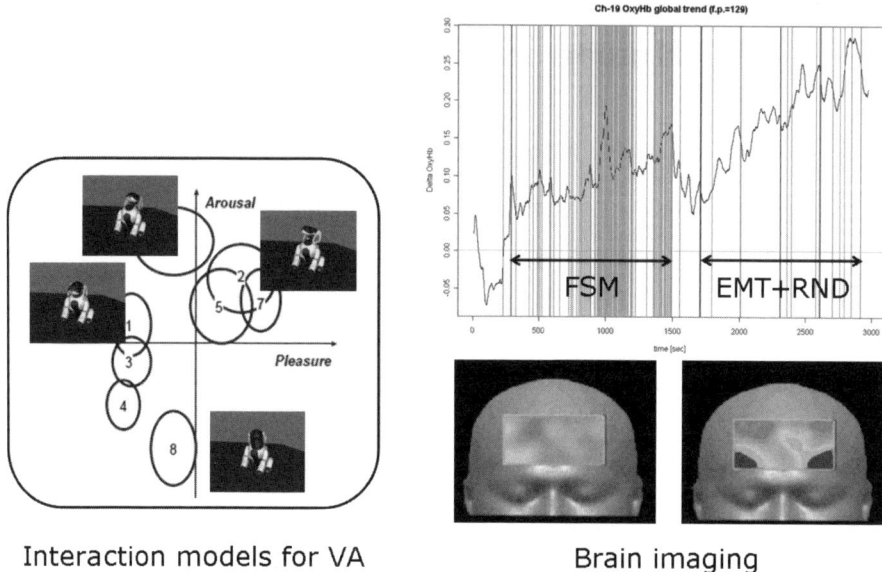

Fig. 1. Interaction models for VA and Brain imaging

3 Human–Agent Interaction Experiments

In the experiments, we asked the subjects to elicit one of target behaviors as many as they can, so as to constrain the purpose of the interaction for evaluation reason. Each subject interacted with the virtual robot with one of the interaction models for 20 minutes. After each duration, they were asked to fill a questionnaire that is aiming at revealing the behavior predictability, animacy, and sustainability impressed by the virtual robot just interacted. Experimental results show that maintaining the predictability in a moderate level contributes to the sustainability of the HAI process. Also brain imaging results suggest that not only the predictability but also awareness of intentionality play important role for realizing sustainable HAI.

Acknowledgment

This research was partially supported by a Grant-in-Aid for Scientific Research on Priority Areas (No.454) from the MEXT, and "Symbiotic Information Technology Research Project" of Tokyo University of Agriculture and Technology.

Reference

1. Russell, J.A.: Reading emotions from and into faces: Resurrecting a dimensional-contextual perspective. In: Russell, J.A., Fernández-Dols, J.-M. (eds.) The Psychology of Facial Expression, pp. 295–320. Cambridge University Press, Cambridge (1997)

Social Effects of Virtual Assistants. A Review of Empirical Results with Regard to Communication

Nicole C. Krämer

University Duisburg-Essen, Forsthausweg 2,
47057 Duisburg, Germany
nicole.kraemer@uni-due.de

Abstract. Early as well as recent evaluation studies indicated that embodied conversational agents induce social reactions on the part of the user. When confronted with virtual agents, human participants show communication behaviors that are similar to those shown in human-to-human interaction. The paper gives a review on relevant research.

Keywords: social effects, evaluation studies.

There is ample empirical evidence that embodied conversational agents can induce social-emotional effects similar to those in human-to-human interactions. In contrast to dealing with GUIs, people conversing with ECAs show for example social reactions such as social facilitation or inhibition, a tendency to socially desirable behavior or increased cooperation. Machines with basic social cues are able to affect the human user in similar ways as a human interlocutor would do. Here, the effects on the users´ communication behavior are presented. Although early studies on dialogue systems indicate that „Talking to a computer is not like talking to your best friend" (see the title of Jönsson & Dahlbäck´s paper in 1988 [1]), recent studies using interfaces incorporating social cues hint to astonishing effects on human´s communication behavior.

Quantitative aspects. Concerning quantitative aspects, several studies consistently show that human-like appearances on the screen may lead to a distinct increase of natural speech utterances on the part of the user: When a TV/VCR system is represented by an anthropomorphic figure instead of a merely text- or audio-based interface, human users are more inclined to use natural language (instead of e.g. a remote control) when interacting with the system [2]. Virtual faces thus seem to "invite intuitive interaction": A virtual person is perceived at least so social that potentially existing restraints to talk to a machine might be overcome. Additionally, within the GrandChair project empirical data indicate that elderly participants tell longer stories when they are confronted with an artificial child than when there was no addressee [3].

Qualitative aspects. Qualitative content analyses of users´ utterances while interacting with the agent Max within the Heinz Nixdorf Museum in Paderborn, Germany, show that utterances resemble those of human to human communication [4]: 57% of users formally greeted the agent (probably since triggered to do so by the agent) and, more surprisingly, at least 30% bade farewell – the latter being especially astonishing as users simply had to step back from screen and keyboard. More than one third of questions addressed to the agent implied human-likeness („Do you have a girlfriend?", „Can you dance?") or tested this very assumption. A more controlled experimental study, which by means of a between subjects design systematically compared a text based interface with a speech based and an agent based interface, supported and augmented these findings [2]: When a TV/VCR system was represented by a virtual face, the users showed more polite phrases ("thank you" or "please") and used personal pronouns such as "you" more often. Furthermore, orders were given in a more personalized way ("Could you record James Bond" instead of "James Bond should be recorded"). There was also a striking tendency to more often repeat sentences in exactly the same or slightly altered way – that did not occur when interacting with merely text- or speech-based systems. This might be attributed to the fact that the virtual face was not able to give immediate feedback as to whether the utterance had been understood. People thus seemed to come to the conclusion that no immediate reaction from the face meant that the system did not understand, while they patiently waited for an answer when no face was displayed. This suggests that the statement "One cannot not communicate" [5] might also be true for agents: as soon as a face is displayed on the screen, humans will take every behavior – even a blank face – as communication. Other studies moreover show that self-disclosing utterances are more frequent when the system is represented by a talking head than compared to a merely speech-based system [6]. In sum, it can be stated that there is a relation between anthropomorphized embodiment and communicative reactions that are similar to human to human interaction.

References

1. Jönsson, A., Dahlbäck, N.: Talking to a computer is not like talking to your best friend. In: Proceedings of the First Scandinavian Conference on Artificial Intelligence, Tromsø, Norway, March 9-11 (1988)
2. Krämer, N.C.: Social communicative effects of a virtual program guide. In: Panayiotopoulos, T., Gratch, J., Aylett, R.S., Ballin, D., Olivier, P., Rist, T. (eds.) IVA 2005. LNCS (LNAI), vol. 3661, pp. 442–543. Springer, Heidelberg (2005)
3. Smith, J.: GrandChair: Conversational Collection of Family Stories. MIT Press, Media Lab, Cambridge (2000)
4. Kopp, S., Gesellensetter, L., Krämer, N.C., Wachsmuth, I.: A conversational agent as a museum guide. Design and evaluation of a real-world application. In: Panayiotopoulos, T., Gratch, J., Aylett, R.S., Ballin, D., Olivier, P., Rist, T. (eds.) IVA 2005. LNCS (LNAI), vol. 3661, pp. 329–343. Springer, Heidelberg (2005)
5. Watzlawick, P., Beavin, J.H., Jackson, D.D.: Pragmatics of Human Communication. A Study of Interactional Patterns, Pathologies, and Paradoxes. W. W. Norton & Co, New York (1967)
6. Oviatt, S., Darves, C., Coulston, R.: Toward adaptive conversational interfaces: Modeling speech convergence with animated personas. ACM Transactions on Computer-Human Interaction 11(3), 300–328 (2004)

SoNa: A Multi-agent System to Support Human Navigation in a Community, Based on Social Network Analysis

Shizuka Kumokawa[1], Victor V. Kryssanov[2], and Hitoshi Ogawa[2]

[1] Graduate School of Science and Engineering, Ritsumeikan University
[2] Faculty of Information Science and Engineering, Ritsumeikan University
1-1-1 Noji-Higashi, Kusatsu, Shiga 525-8577, Japan
shizuka@airlab.ics.ritsumei.ac.jp,
{kvvictor,ogawa}@is.ritsumei.ac.jp

1 Motivation

When one joins an existing community, she or he may have little-to-no knowledge of what other members already know. "Traditional" and electronic media are useful to collect information, but when the request is related to social experience, it is hard to locate and extract social knowledge from essentially subjective individual accounts. Asking somebody would then be a better way to search information, but it has disadvantages, too: the practicable area of potential contacts is limited by the range of one's personal network, the contacted individuals' knowledge, time limitations, etc. Communication on a personal basis in a new community is always trial-and-error driven, and it does not guarantee obtaining the information of interest even when such information is freely available from some of the community members. As a result of this, newcomers usually experience significant difficulties in adaptation to the community. In an attempt to help overcome this problem, the presented study proposes a system to assist personalized knowledge sharing. The system allows the user to navigate a social space of a community and to locate members who would address the user's needs.

2 Proposed System

Fig.1 shows the architecture of the system developed in this study. The agent manager is to create a multi-agent network reflecting the social network of a community in focus. Each member of the community is represented as an agent. When two members have a relationship in the real world, their agents are connected. There is a "trust value" parameter characterizing the link between two agents, and the larger this value, the more (potentially) influential the agents' recommendations when advising each other. Each agent has static parameters, such as name, age, gender, and personal characteristics (sociability, friendliness, etc.), and a dynamic profile with information about the current location of the member and activities she or he is being involved in. The agent also has knowledge about other members' fields of expertise, which is based on the member's personal experience. By inputting a query, which is then sent to the corresponding agent in the agent network, the system's user can search the community members for a specific expertise and/or personal characteristics. If the user is a newcomer and has not received her/his own agent, an agent, whose parameters are most similar to the user's personal characteristics, is selected as a proxy to process the query. The query is propagated over the network, and each agent sends responses if it has knowledge about members who would be potentially useful for the user. The information-gathering algorithm, which is based on the work of Walter with co-authors [1], thus produces a set of responses from the community members. The next step is selecting members, who meet the user's criteria, from the available set. The system filters the responses to remove any "useless" recommendations (e.g. about currently unavailable members), using the agents' dynamic parameters. Finally, the recommendation is delivered to the user via the multimodal interface. The community's physical environment and the agents are visualized in a 3D graphics virtual reality, reflecting the members' current locations in the physical world. The recommended members are indicated with color, and the user can "browse" the community with a haptic device "PHANToM" to get a force feedback that is adjusted to represent personal characteristics of the members (for example, if a member is generally not friendly, there is a high friction when "touching" the corresponding agent.)

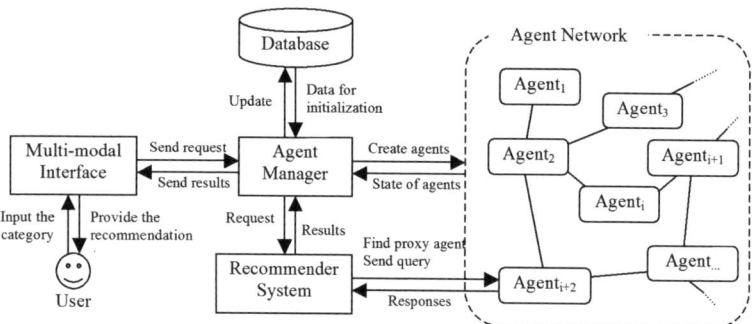

Fig. 1. System architecture

3 Summary

A prototype of the proposed system has been developed and is used at a laboratory of Ritsumeikan University to assist students, both members and not members of the laboratory, in their daily university life. The prototype utilizes data of the laboratory's social network, which is represented as an agent network, where each agent has knowledge about expertise and personal characteristics of other members. An agent gathers information, using its own social "connections" to find community members most suitable to communicate to in a given situation. The system's multi-modal interface then allows the user to explore the community and its members' expertise and, to an extent, personal characteristics. Over the time of its deployment, the developed system has proven a useful tool for assisting human navigation in a social space.

Reference

1. Walter, F.E., Battiston, S., Schweitzer, F.: A Model of a Trust-based Recommender System on a Social Network. In: Autonomous Agents and Multi-Agent Systems, pp. 57–74 (2008)

Animating Unstructured 3D Hand Models

Jituo Li[1], Li Bai[2], and Yangsheng Wang[1]

[1] Digital Interactive Media Lab, Institute of Automation, Chinese Academy of Sciences,
No 95, Zhongguancuan East Road, Beijing, P.R. China 100080
{jituo.li, yangsheng.wang@ia.ac.cn}
[2] School of Computer Science, University of Nottingham, UK
bai@cs.nott.ac.uk

Abstract. We present a novel approach for automatically animating unstructured hand models. Skeletons of the hand models are automatically extracted. Local frames on hand skeletons are created to skin and animate hand models. Self-intersection of hand surfaces is avoided. Our method produces realistic hand animation efficiently.

Keywords: hand animation, skinning, skeleton, local frame.

1 Introduction

In general hand animation methods [1,2,3], the hand surface is linearly bound onto the skeleton by handcrafting, which can be rather time-consuming, and is prone to surface intersection in hand animation. In our work, we can automatically extract the skeleton of a hand model. We skin and animate the hand model with local frames which bypasses surface intersection adaptively and produces realistic hand animation results.

2 Hand Skeleton Extraction and Hand Segmentation

As shown in Fig.1, We extract out the finger tips on the points with local maximal geodesic distances from the wrist boundary, and extract out the valley points on the valleys of hand contour. With the Gaussian curvature variation on the hand surface, we locate more feature points. Geodesic curves linking proper feature points segment the hand model into fingers, a palm and a wrist. We position the joints as the mid

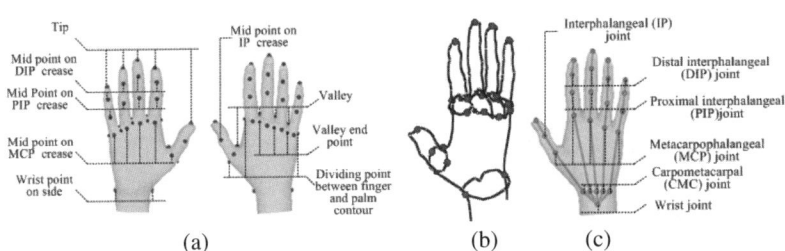

Fig. 1. Skeleton extraction: (a) Feature points (b) Hand segmentation (c) Joints on skeleton

points of the two intersections between the rays from the features against the hand model. The hand skeleton is represented as a hierarchical set of joints.

3 Local Frame Based Hand Skinning and Animation

We define the local frame $u_i v_i w_i$ on joint J_i as: u_i is in the direction of tangent of the skeleton on J_i; if J_i is on the fingers, v_i parallels with the plane fitting the contour of the corresponding knuckle; otherwise, v_i parallels with the plane fitting the palm contour. The intermediate local frames on a bone are linearly interpolated with the local frames on joints. In each hand part, we cluster the mapping points of the vertices on their corresponding bone, and present the vertices only with the local frames on these clustered points. To a point P_i, we have $\mathbf{X}_i = \sum \lambda_j \mathbf{M}_j \mathbf{x}_{i,j}$, where $\mathbf{X}_i$ is P_i's world coordinate ; $\mathbf{x}_{i,j}$ is the local coordinate of P_i in local frame F_j; $\mathbf{M}_j$ is the transformation matrix between F_j and the world coordinate, λ_j is the factor with $\sum \lambda_j = 1$. If P_i is on the fingers, P_i is controlled with only one local frame. If P_i is on the palm, P_i is controlled with two local frames, as shown in Fig. 2(a), $\lambda_j = (\beta - \alpha)/\beta$, $\lambda_{j+1} = \alpha/\beta$.

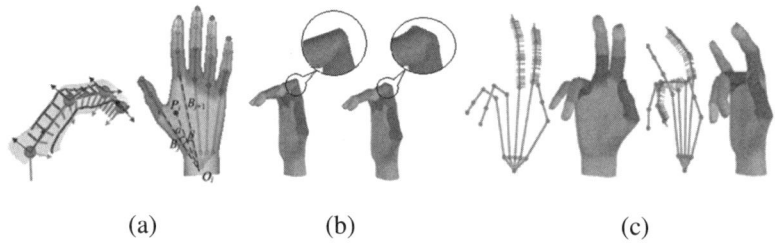

(a) (b) (c)

Fig. 2. Skinning and animation: (a) Skinning with local frames (b) Comparison between our method (left) and typical animation method (right) (c) More animation results

As shown in Fig.2(b),(c), our method can avoid surface self-interaction between adjacent phalanxes in animation, and can produce realistic hand animation results(to make it clear, not all local frames are displayed). Our system is able to animate a hand with 11744 triangles at 45 frames per second on a Pentium IV 1.5GHZ, 512RAM PC.

4 Conclusions

We have presented an approach for generating hand skeletons from hand surface automatically, and then skin and animate hand models with local frames on the hand skeletons. The whole process can be completed automatically which greatly reduces the pretreatment process of hand animation. Our local frame based hand skinning and animation method can avoid surface interactions, and deform the hand model realistically and efficiently.

Acknowledgement. The authors would like to acknowledge the support from China Hi-Tech Research and Development Programme (2007AA01Z341) and the UK Engineering and Physical Sciences Research Council (EP/F013477/1).

References

1. Albrecht, I., Haber, J., Seidel, H.P.: Construction and Animation of Anatomically Based Human Hand Models. In: Proc. SIGGRAPH/Eurographics Symp.on Computer Animation, pp. 98–109. ACM Press, New York (2003)
2. Rhee, T., Neumann, U., Lewis, J.P.: Human Hand Modeling from Surface Anatomy. In: SI3D 2006, pp. 27–34. ACM Press, New York (2006)
3. Lee, J., Yoon, S.H., Kim, M.S.: Realistic Human Hand Deformation. Comp. Anim. and Virtual Worlds. 17, 479–489 (2006)

Verification of Expressiveness of Procedural Parameters for Generating Emotional Motions

Yueh-Hung Lin[1], Chia-Yang Liu[2], Hung-Wei Lee[3], Shwu-Lih Huang[2], and Tsai-Yen Li[1]

[1] Department of Computer Science, National Chengchi University, Taiwan
{g9339,li}@nccu.edu.tw
[2] Department of Psychology, Research Center for Mind, Brain, and Learning, National Chengchi University, Taiwan
{95752003,slh}@nccu.edu.tw
[3] Department of Applied Psychology, Hsuan Chuang University, Taiwan
spoon@hcu.edu.tw

1 Motivation and System Overview

Body movements are crucial for emotion expression of a virtual agent. However, the perception of expressiveness of an animation has always been a subjective matter. Research in psychology has asked professional actors to perform emotional motions to study how body gestures deliver emotions [2]. In this work, we aim to design an animation system that can generate human body motions in a more systematic manner and then study the expressiveness of the generated animation as a preceding step for the study of the relation between motion and emotion.

We propose to stratify the variables relevant to describing motion and emotion into four layers (from upper to lower): *emotional*, *style*, *motion*, and *procedural* layers with their own respective sets of parameters. In the emotional layer, emotions can be modeled with either the basic emotions approach [1] or the dimensional approach. The style layer serves as an intermediate layer for describing expressiveness of an animation. In the motion layer, the parameters specific to a target type of motion are defined. In the procedural layer, generic animation procedures are used to generate parameterized motions. In this paper, we focus on designing generic animation procedures for the generation of human body motions and studying how the parameters in these procedures are mapped onto the parameters in the upper layers through the example of walking motion.

In procedural animation, a motion is specified by defining appropriate keyframes and interpolations. The interpolations are performed on the procedural parameters such as joint angles or points in the 3D space. If the points in the Cartesian space are specified as a curve, the curve is re-parameterized by arc length. Given these procedural parameters, the motion parameters for interpolation are defined as the locations of the control points that are used to specify a spline curve (such as a Bezier curve) in the parameter-time space. By designing appropriate mapping between the motion parameters and the style parameters, we hope to produce expressive walking motions.

We have adapted the style attributes defined in [2] for our style parameters which include *smooth-jerky*, *stiff-loose*, *slow-fast*, *soft-hard*, and *expanded-contracted*. For example, in our implementation, we assume that each joint is equipped with a virtual

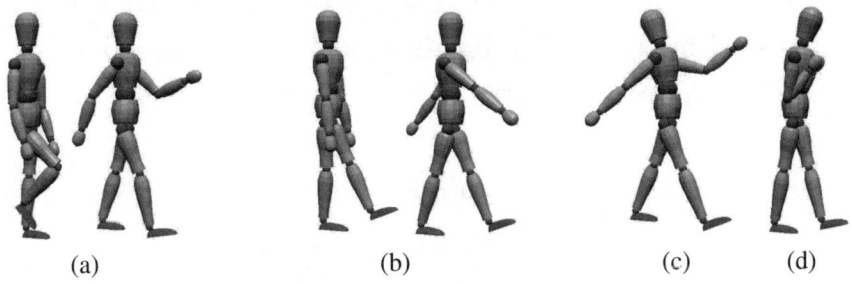

Fig. 1. Examples of (a) loose (b) stiff (c) expanded (d) contracted walking motions

spring of constant stiffness that determines how displacement affects spring force. Typical snapshots for the stiff-loose parameter at extreme values are shown in Fig. 1(a) and (b). The parameter of expanded-contracted is realized by changing the expansiveness of keyframe definitions as illustrated in Fig. 1(c) and (d).

2 Preliminary Experimental Results

We have conducted two psychological experiments to verify the expressiveness of the generated animations. In both experiments, thirty participants were invited to compare two side-by-side videos with a target and a standard stimulus, respectively, and then rate the target stimulus for the five style parameters. The standard stimulus is set to mid-point values on all of the five style parameters while the target stimulus is one of the 32 combinations of the five dimensions with either a high or a low value. We use the point-biserial correlation to reveal the correspondence between the manipulation of animations and the participants' subjective experiences. The correlation coefficients are as follows: smooth-jerky (0.65), stiff-loose (0.56), slow-fast (0.92), soft-hard (-0.04, not significant) and expanded-contracted (0.95). In order to investigate why the manipulation of the soft-hard parameter is not effective, we conducted another experiment in which only one style parameter is changed at a time but with finer granularity. The F-values of the linear trend analysis on the result are as follows: smooth-jerky (79.59), stiff-loose (67.07), slow-fast (573.54), soft-hard (0.31, not significant) and expanded-contracted (952.70). From these results, we learn that four out of the five parameters have been implemented with satisfactory expressiveness. The only one exception (soft-hard) probably is due to the fact that it is not easy for a human to discern second-order changes (i.e. the acceleration of object motion) given that the lower order changes remain fixed.

References

1. Ekman, P., Davidson, R.J.: The Nature of Emotion. Oxford University Press, New York (1994)
2. Montepare, J., Koff, E., Zaitchik, D., Albert, M.: The Use of Body Movements and Gestures as Cues to Emotions in Younger and Older Adults. Journal of Nonverbal Behavior, 133–152 (1999)

Individualised Product Portrayals in the Usability of a 3D Embodied Conversational Agent in an eBanking Scenario

Alexandra Matthews, Nicholas Anderson, James Anderson, and Mervyn Jack

Centre for Communication Interface Research
University of Edinburgh, Scotland, UK

Abstract. A cohort of 65 participants interacted with an extrovert male Embodied Conversational Agent (ECA) in four individualised product portrayals. Results suggest that customers respond more positively to portrayals by ECAs of savings account product offers in contrast to investment account product offers, both in terms of usability and relevance.

1 Introduction

This paper presents results of an empirical evaluation of an Embodied Conversational Agent (ECA) within an eBanking scenario where an extrovert male agent offers eBanking services and individualised product portrayals. It is widely accepted that humans react and behave toward computers in a similar way as they do with other humans, as depicted by the Computers as Social Actors theorem (CASA [1]). The social rules that govern human-to-human interaction, stereotypes and even personality attribution also apply to Human-Computer Interaction (HCI) (for example, [2,3]). ECAs have the ability to assist in predicting, influencing and determining customers' buying behaviour by gathering information as well as target through individualised product portrayals.

2 Experiment Design and Procedure

The application is an eBanking service set in a virtual branch of the Case Bank, with an ECA playing the role of financial advisor. While the ECA was carrying out one of the tasks, they informed the customer that they were looking at the recent transactions on their account and believed that they had an (individualised) product that would be of benefit to them.

The agent recommended one of two product types, in one of two possible ways; a savings account with a high (HPP) or a low planning propensity (LPP) portrayal and an investment account with a high (HRA) or a low risk aversion (LRA) portrayal. Each portrayal highlighted different features; HPP offered flexibility, LPP offered the best interest rate whilst saving regularly, HRA offered no risk and a guaranteed 15% gross minimum return and LRA benefits from 75% of stock market growth.

A repeated-measures within-participants design was used, where participant age and gender were balanced together with the presentation of the four scenarios. Participants were asked to complete a usability questionnaire, a product uptake questionnaire and asked to rate the relevance of each product.

3 Results

The usability data were analysed by a series of repeated measures ANOVAs. There was a significant within-participants effect of product type on the mean usability scores ($p = 0.006$, $F = 4.687$). Overall, participants rated the four interactions extremely similarly. There was a significant between-participants effect of age ($p = 0.041$, $F = 4.404$) where older participants rated the overall usability slightly lower ($M = 5.26$) than the younger participants ($M = 5.6$).

The product uptake data ANOVAs show one significant within-participant effect of product ($p = 0.002$, $F = 6.436$) on the mean product uptake scores. Pairwise comparisons revealed the mean for the HPP product portrayal ($M = 4.93$) and the LPP ($M = 5.00$) were rated significantly higher in terms of likelihood of product uptake, indicating that participants felt that they would be more likely to take up those products, compared to the LRA product portrayal ($M = 4.52$). Overall however participants rated the four product portrayals similarly.

The relevance ratings for the four products showed significant differences ($p < 0.000$, $F = 26.036$). Participants rated (30-point scale) both the LPP product ($M = 22.23$) and the HPP product ($M = 20.04$) as significantly more relevant than the HRA product ($M = 14.99$) and the LRA product ($M = 12.37$).

4 Conclusions

All products achieved positive reactions in terms of both presentation and possible future purchasing behaviour. In relation to the type of product offered, significant differences were observed which suggest that the best types of products to offer via an ECA that will appeal to the majority of customers are savings accounts. This indicates that the Bank should choose carefully when selecting products to recommend through an ECA, as customers receive an overall more positive impression if they feel they have recieved an individualised product offer.

References

1. Reeves, B., Nass, C.: The Media Equation: How People Treat Computers, Television and New Media Like Real People and Places. Cambridge University Press, Cambridge (1996)
2. Cassell, J., Sullivan, J., Prevost, S., Churchill, E. (eds.): Embodied Conversational Agents. MIT Press, Cambridge (2000)
3. Prendinger, H., Ma, C., Yingzi, J., Nakasone, A., Ishizuka, M.: Understanding the effect of life-like interface agents through users' eye movements. In: Proceedings of the 7th International Conference on Multimodal Interfaces (ICMI 2005), Trento, Italy, pp. 108–115. ACM Press, New York (2005)

Multi-agent Negotiation System in Electronic Environments

Dorin Militaru

Amiens School of Management, 18 place Saint Michel - 80 000 Amiens, France
GRID, 30 Avenue du Président Wilson, 94235 Cachan, France
dorin.militaru@supco-amiens.fr

Abstract. The main purpose of the study is to investigate how costly information could affect the performance of several types of searching agents in electronic commerce environments. Internet searching agents are tools allowing consumers to compare on-line Web-stores' prices for a product. The existing agents base their search on a predefined list of Web-stores and, as such, they can be qualified as fixed-sample size searching agents. However, with the implementation of new Internet pricing schemes, this search rule may evolve toward more flexible search methods allowing for an explicit trade-off between the communication costs and the product price. In this setting, the sequential optimal search rule is a possible alternative. However, its adoption depends on its expected performance. The present paper analyses the relative performances of two types of search agents on a virtual market with costly information. At the theoretical equilibrium of the market, we show that the sequential rule-based searching agents with a reservation price always allow consumers to pay lower total costs. We further test the robustness of this result by simulating a market where both sellers' and buyers' search agents have only partial information about the market structure.

Keywords: Electronic Commerce, Intelligent Agents, Game Theory.

JEL Classification: C68, C72, D4, D83.

1 Introduction

With very few exceptions [1], the research available on agent-based markets considered search costs as negligible in the analysis of market evolution. However, we believe that with the evolution of Internet pricing mechanisms and agent intermediation, the search costs will become an important aspect in modeling agent-based markets. Indeed, these two phenomena could change the basic rules of agent interactivity and market structure as we conceive of it today.

In the present study, we compare the FSS buying agents and RP buying agents performances in term of total costs (product price plus total searching cost) paid by the consumers. We try to determinate if it is profitable for the existing searching agents to adopt an optimal sequential searching rule. For this purpose, we analyze a market where FSS buying agents and RP buying agents coexist. In our model, we assume that the search costs correspond to a constant unit cost of communication paid by the agents to have access to the price information of a Website.

2 Conclusion

In this study, we analyzed the possible evolution of the fixed sample-size searching rule used by the existing agents towards an optimal sequential searching rule with reservation price. Indeed, a costly information search requires the agents to make a trade-off between communication costs and product prices. In order to study this possible evolution, we proposed a market model where FSS and RP buying agents coexist. A theoretical analysis of the market equilibrium showed that RP buying agents always allowed the consumers to pay lower total costs than FSS buying agents. Moreover, this result holds when some assumptions of the game theory model are relaxed by simulating the dynamics of a market where the selling agents use a dynamic pricing strategy and where the RP buying agents can fix their reservation price only an the basis of partial information about the prices charged on the market. We can thus conclude that in the future the sequential searching rule could be a good alternative for searching agents.

References

1. Kephart, J., Greenwald, A.: Shopbot And Pricebots. In: Proceedings Of The International Joint Conferences On Artificial Intelligence, Stockholm, Sweden (1999)
2. Mcknight, L.W., Bailey, J.P.: Internet Economics. MIT Press, Cambridge (1997)
3. Rothschild, M.: Searching For The Lowest Price When The Distribution Of Prices Is Unknown. The Journal Of Political Economy 82(4), 689–711 (1974)
4. Stahl, D.O.: Oligopolistic Pricing With Sequential Consumer Search. American Economic Review 79(4), 700–712 (1989)
5. Stiglitz, J.E.: Imperfect Information In The Product Market. In: Schmalensee, R., Willig, R.D. (eds.) Handbook Of Industrial Organization, pp. 769–847. North Holland, New-York (1989)
6. Varian, H.: A Model Of Sales. American Economic Review 70(4), 651–659 (1980)

A Study of the Use of a Virtual Agent in an Ambient Intelligence Environment

Germán Montoro[1], Pablo A. Haya[1], Sandra Baldassarri[2],
Eva Cerezo[2], and Francisco José Serón[2]

[1] Computer Science Department - Universidad Autónoma de Madrid
Francisco Tomás y Valiente 11. 28049 Madrid, Spain
{German.Montoro,Pablo.Haya}@uam.es
[2] Advanced Computer Graphics Group (GIGA) - University of Zaragoza
Engineering Research Institute of Aragon (I3A), Zaragoza, Spain
{sandra,ecerezo,seron}@unizar.es

Abstract. In this paper we present the results in the evaluation of the use of a virtual agent together with a spoken dialogue system in an ambient intelligence environment. To develop the study, 35 different subjects had to perform eight different tasks and fill in a questionnaire with their impressions. From the answers we can conclude that the use of the virtual agent does not provide an improvement in their appreciation of the conversation performance but it offers them a more human-like interaction. Only a minority of users preferred to maintain the conversation without the visual support of the virtual agent.

1 Introduction

In this paper we review a test case study of the use in a real ambient intelligence environment of a virtual agent to assist the spoken interaction. Users had to perform different eight tasks in the environment (such as switch on or off the lights or open the door) with and without the help of the virtual agent. The system was tested with 35 different users. From them, only eight subjects (23%) had previously interacted with an automatic response dialogue system. User's gender was balanced; while 45% of the subjects were women, the other 55% were men. Once users had finished with all the tasks they had to answer a questionnaire to measure their grade of satisfaction with the system.

2 Evaluation of the Use of the Virtual Agent

The questions were related to the employment of the virtual agent in the conversation. Most of the answers did not have a direct relation with the system performance and corresponded to the subjective impressions of the users:

- Almost half of the subjects (46%) felt more comfortable looking at the virtual agent while they interacted with the system. Only 26% of them felt some kind of discomfort and 28% of the subjects did not appreciate any difference. We can

notice that most of the users felt better having the visual support of the virtual agent but also that 1 out of 4 users considered this as a disadvantage.
- Most of the subjects (68%) did not feel disoriented when they did not use the virtual agent for the interaction. Only 11% of them felt lost. Therefore most of the users felt that they could continue with the dialogue in the same conditions even without the presence of the virtual agent.
- The same amount of subjects (37%) considered that the conversation was improved by the use of the virtual agent and that it was not affected. This shows that the employment of a virtual agent did not have a specific effect in the appreciation of the users of the conversation performance.
- Only 26% of subjects answered that looking at the agent did not provide them any stronger feeling of carrying out a conversation with a human being. This points out that the use of the virtual agent can result in a higher degree of naturalness and "humanity" to the interaction with the system.
- And perhaps the most important point was the user preference for the interaction with the environment. Almost half percent of the subjects (48%) preferred to interact with a virtual agent in their conversation with the ambient intelligence environment. Only 14% of them considered that they preferred to carry out the interaction without the support of the virtual agent. Finally, 38% of the subjects did not express any specific preference.

3 Conclusions

With these results we can see that the use of a virtual agent does not provide a strong improvement in the dialogue (as few as 11% of the subjects felt disoriented without it and only 1 out of 3 considered that the performance got better). This means that, for general users, the employment of a virtual agent does not become a required requisite to improve the performance of the interaction in an environment.

Nevertheless, half of the subjects preferred the interaction with the virtual agent to just the oral communication. Besides, most of the subjects answered that the virtual agent conferred more humanity to the interaction and the 46% of them felt more comfortable looking at it.

According to these results we can conclude that since the use of a virtual agent may not provide better results in the interaction, it can have a key role in offering the users a more natural, pleasant and human-like communication with the environment.

Acknowledgments

This work has been partly funded by project numbers TIN2007-64718 and TIN2007-63025. We would like to thank Álvaro Jiménez, the student who helped to develop this study, for his time and enthusiastic work.

Automatic Torso Engagement for Gesturing Characters

Michael Neff

University of California, Davis
Department of Computer Science and Program for Technocultural Studies
Davis, CA 95616, USA
neff@cs.ucdavis.edu

1 Introduction

Acting theorists such as Grotowski [1] have argued that full body engagement is a key principle in creating expressive movement. Unfortunately, torso use has received limited attention in virtual character research and torso involvement is often lacking. In procedural gesture animation, the character's gestural movements often begin at the shoulder and the character's torso is left relatively stiff. This reduces both the believability and expressiveness of the resulting motion.

We have developed a system for automatically varying a character's torso based on *drives* that must be present in a gesturing character, such as hand position. The system adds little overhead to the character system builder, while significantly improving the quality of the output motion. Different styles of torso engagement can be created to reflect character personality and/or mood. The system is straightforward and easy to implement, allowing easy integration into existing character systems while still adding effective movement detail to a character's torso. Various results have been generated for pre-existing gesture sequences in order to validate the approach.

2 Technical Overview

The key idea in our approach is to use body "drives" to automatically vary torso movement. *Drives* include hand position and gaze direction; parameters that define a character's gesturing behaviour and so should already be present in the animation system. Correlations are specified between these drives and the deformation of the torso. Thereby, when a character moves his arms, his torso will automatically be engaged in the movement. This work builds on previous work using correlations as an animation primitive [2,3].

The correlations can be easily layered together, scaled and shifted, allowing a high degree of customization to generate various effective movement styles for a character. An interface is provided which allows these styles to be interactively defined. The correlations can be set for an entire animation or varied over time by the character system for greater control. The net result is a very flexible

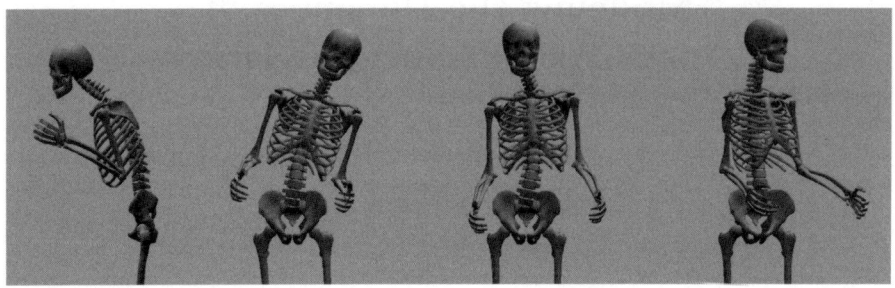

Fig. 1. Component posture deformations: crunch, side lean, beauty curve and twist

system for controlling torso engagement that should fit easily into existing character pipelines. An architecture has been developed that yields a straightforward implementation. The system is light weight and can be run in real-time.

An effective torso model, developed in [4,5], supports the definition of meaningful correlations. It provides a compact parameterization of the torso based on the performing arts literature. A wide range of potential torso deformations are possible, but in practice, four have proven particular useful for gesturing characters: *crunch*, *twist*, *beauty curve* and *side lean*. Examples of these deformations are shown in Figure 1 and may be layered together in any combination.

The drives are correlated with the pose deformations using linear maps. For example, the amount of vertical movement of a hand can be mapped to the amount of *crunch* torso deformation. These maps can be layered, scaled and translated, providing flexible control.

The correlations are used to update the parameters that define the character's posture. The positions of the arms are solved using IK routines, maintaining the original position of the gestures.

Results from applying the model to automatically add torso variation to preexisting gestural animation suggest that it produces motion that appears both more natural and more lively.

References

1. Grotowski, J.: Actor's training. In: Barba, E. (ed.) Towards a Poor Theatre. Routledge. A Theatre Arts Book, New York (1968)
2. Pullen, K., Bregler, C.: Motion capture assisted animation: Texturing and synthesis. ACM Transactions on Graphics 21(3), 501–508 (2002)
3. Neff, M., Albrecht, I., Seidel, H.P.: Layered performance animation with correlation maps. Computer Graphics Forum 26(3), 675–684 (2007)
4. Neff, M., Fiume, E.: Methods for exploring expressive stance. Graphical Models 68(2), 133–157 (2006)
5. Neff, M., Fiume, E.: Methods for exploring expressive stance. In: Proc. ACM SIGGRAPH / Eurographics Symposium on Computer Animation 2004, pp. 49–58 (2004)

Modeling the Dynamics of Virtual Agent's Social Relations

Magalie Ochs, Nicolas Sabouret, and Vincent Corruble

LIP6,Université Paris 6, France
{magalie.ochs, nicolas.sabouret, vincent.corruble}@lip6.fr

Introduction. To create believable virtual agents, both the social context and the emotions appearing during the interaction should be taken into account. Most existing models of social context are statically defined. However, research shows that the affective experience during an interaction can induce a modification of social context and more particularly of social relations. In this paper, we propose to explore the dynamics of virtual agent's social relations. The agent's social relations are updated according to emotions that appear during the interaction (both the agent's triggered emotions and those expressed by its interlocutor). The model proposed is qualitatively defined. In its implementation, numerical functions are proposed.

Virtual Agent's Social Relations. Based on [Bickmore and Cassell, 2001] and [Brown and Levinson, 1987], we propose a representation of social relations defined by four variables:

- $liking_{i,j}$ represents the degree of liking character i has for character j;
- $dominance_{i,j}$ corresponds to the power that i can exert on j;
- $solidarity_{i,j}$, a.k.a. social distance, is defined as the degree of similarity between i and j (e.g. similar religions, family, profession, gender, etc.);
- $familiarity_{i,j}$ represents the tendency that i has to transfer confidential information to j.

Note that social relations are defined from the point of view of character i. Thus, they are directed and not necessarily reciprocal.

The Emotion-based Dynamics of Social Relations. [Ortony, 1991] shows that, during an interaction, one's emotions and those of one's interlocutor may lead to a change in their social relations. In our model, we integrate the following dynamics of each variable of a social relation.

Liking. Based on [Ortony, 1991], in our model, character i's positive emotions triggered during an interaction with character j induces an increase in $liking_{i,j}$. Conversely, negative emotions *caused by* j decrease $liking_{i,j}$. Note that a negative emotion not caused by j has no impact on the degree of liking. For instance, let's imagine a situation in which Jack informs John about a very sad event, John will not necessarily like John less than before.

Dominance. Based on [Keltner and Haidt, 2001], in our model, an emotion of *pride* of i caused by j induces an increase in $dominance_{i,j}$ whereas *shame* infers a decrease. Based on [Knutson, 1996] which shows that some types of emotions expressed by someone affect the dominance value of the person who perceives it, in our model, *expression* of sadness or fear of j induces an increase in $dominance_{i,j}$, and the expression of anger, a decrease.

Solidarity. Based on [de Rivera and Grinkis, 1986], i's negative emotion caused by j induces a decrease in $solidarity_{i,j}$. However, [Keltner and Haidt, 2001] shows that expressed emotions reflect an individual's mental state (goals, beliefs, etc). Thus, in our model, the congruence between i's triggered emotion and the emotion expressed by his interlocutor j affects the solidarity variable: if the triggered emotion of i is *joy* or *hope* (and is not caused by j), and j expressed emotion of the same type, $solidarity_{i,j}$ increases; on the contrary, if j expresses an opposite type of emotion, $solidarity_{i,j}$ decreases.

Familiarity. In the literature, emotions do not seem to have a direct impact on familiarity (*i.e.* on the degree of confidentiality of the information transmitted by a person). However, research shows that one confides more in another when the former likes the latter [Collins and Miller, 1994]. Therefore, in our model, the value of $liking_{i,j}$ modifies the effect of $familiarity_{i,j}$ on the transfer of confidential information.

Ongoing work. Our model is currently being integrated into a virtual character's architecture for video games. Our aim is to allow game designers to programme characters whose behavior depends on its social relations and emotions during interactions.

References

[Bickmore and Cassell, 2001] Bickmore, T.W., Cassell, J.: Relational agents: a model and implementation of building user trust. In: CHI, pp. 396–403 (2001)
[Brown and Levinson, 1987] Brown, P., Levinson, S.: Politeness: Some Universals in Language Usage. Cambridge University Press, Cambridge (1987)
[Collins and Miller, 1994] Collins, N., Miller, L.: Self-disclosure and liking: A meta-analytic review. Psychological Bulletin 116, 457–475 (1994)
[de Rivera and Grinkis, 1986] de Rivera, J., Grinkis, C.: Emotion as relationships. Motivation and Emotion 10, 351–369 (1986)
[Keltner and Haidt, 2001] Keltner, D., Haidt, J.: Social functions of emotions. In: Mayne, T.J., Bonanno, G.A. (eds.) Emotions: Current issues and Future Directions, pp. 192–213. Guilford Press, New York (2001)
[Knutson, 1996] Knutson, B.: Facial expressions of emotion influence interpersonal trait inferences. Journal of Nonverbal Behavior 20, 165–182 (1996)
[Ortony, 1991] Ortony, A.: Value and emotion. In: Memories, thoughts, and emotions: Essays in honor of George Mandler. Erlbaum, Hillsdale (1991)

Proposal of an Artificial Emotion Expression System Based on External Stimulus and Emotional Elements

Seungwon Oh and Minsoo Hahn

ICU Digital Media Lab, 517-10 Dogok-dong, Gangnam-gu, Seoul, South Korea
{aegis901,mshahn}@icu.ac.kr

1 Introduction

Recently, for intelligent agents, the concept of emotion has become more and more important and many researchers think that emotion is one of the major factors for more natural and friendly interaction between human and agents [1]. There have been several researches to build more emotional creatures.

In robotics, many researchers developed their own emotion expression robots such as MIT's Kismet [2], Waseda's WE-4 [3]. Despite all these valuable attempts, it is believed that there are still a lot of things to explore when it comes to the affective capability enhancement of virtual agents. Human emotion is affected by the psychological internal state, the physical state, and the external environments. However, above studies might not be successful in handling various affective elements. In this paper, we consider various affective elements based on psychological studies and propose some human emotional elements and the external stimulus information-based artificial emotion expression system (AES) which can express emotion rather successfully.

2 Artificial Emotion Expression System (AES)

The reason why an agent needs an additional system to express emotion is that a virtual agent does not have emotion intrinsically. In order to make a rather natural and human-friendly interaction, an emotion expression system based on the human emotion-generation mechanism must be added. We explored some terms such as personality, tendency, machine-rhythm, mood, and emotion. And this paper defines internal affective elements based on the results of previous psychological studies at first. Then our emotion expression system for agents is described with the affective elements. Fig. 1 shows the AES architecture.

Personality is a lasting and consistent human behavioral feature. In our work, personality is defined by the Jung's Psychological Type Theory which was widely known as the Myers-Briggs Type Indicator (MBTI) test [4].

Tendency implies an organic trait in reacting constantly to external stimuli. According to Pierre Janet, there are various types of tendency from low to high level. It forms in a hierarchical order. Thus, the high level tendency is based on the low level tendency. An animal shows the low level tendency while a human exposes the high level tendency [5].

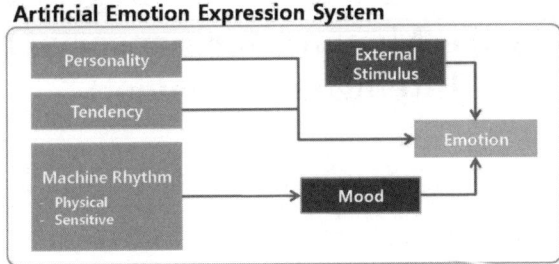

Fig. 1. Architecture of Artificial Emotion Expression System

The bio-rhythm represents variety periodic fluctuation arising from the human biotical activity. In this paper, the machine-rhythm plays the similar role to bio-rhythm for virtual agents.

The mood is a feeble and lasting affective state arising from obscure stimuli [6]. The mood is defined by the machine rhythm which changes day by day.

The emotion is a strong and instant affective state. The emotion is the expressed affective state to others. In this paper, four types of emotion, i.e., Fear, Anger, Depression, and Satisfaction are defined as the basic emotion types.

3 Conclusion

This paper defined an internal affective elements based on the results of previous psychological studies and proposed the artificial emotion expression system based on the human internal elements.

References

1. Norman, D.A.: Emotional Design, pp. 161–169. Basic Books (2004)
2. Breazeal, C.: Function meets style: insights from emotion theory applied to HRI. IEEE Transactions on Systems, Man and Cybernetics, Part C: Applications and Reviews 34(2), 187–194 (2004)
3. Miwa, H., Okuchi, T., Itoh, K., Takanobu, H., Takanishi, A.: A New Mental Model for Humanoid Robots for Human Friendly Communication. In: International Conference on Robotics & Automation (2003)
4. Myers-Briggs Type Indicator,
   http://en.wikipedia.org/wiki/Myers-Briggs_Type_Indicator
5. Sailot, I.: Pierre Janet's Hierarchy: Stages Or styles?
6. Plutchik, R.: Emotions and Life:Perspectives From Psychology, Biology, and Evolution. American Psychological Association (2003)

A Reactive Architecture Integrating an Associative Memory for Sensory-Driven Intelligent Behavior

David Panzoli, Hervé Luga, and Yves Duthen

Toulouse Research Institute for Informatics (IRIT)
Université Toulouse 1, 31042 Toulouse Cedex, France
{panzoli,luga,duthen}@irit.fr

1 Introduction

Realistic human-like agents are nowadays able to follow goals, plan actions, manipulate objects[1], show emotions and even converse with human people. Despite these agents being endowed with many abilities, the question of intelligence, even for simple animal agents, is still pending. Our research positions itself in an artificial life and virtual reality context. It corresponds to the direction of recent team research[2] suggesting that a creature's morphology and environment complexity are closely tied. The work presented in this paper intends to show in addition that human expert cognitive modeling is not a necessary prerequisite of behavioral intelligence.

The contribution of this work is dual. First, we provide a methodology for integrating a memory into a reactive connectionist agent (figure 1). Our approach is primarily concerned with avoiding the weaknesses of deliberative architectures. Namely, this memory must be integrated into the action-selection (i.e., does not

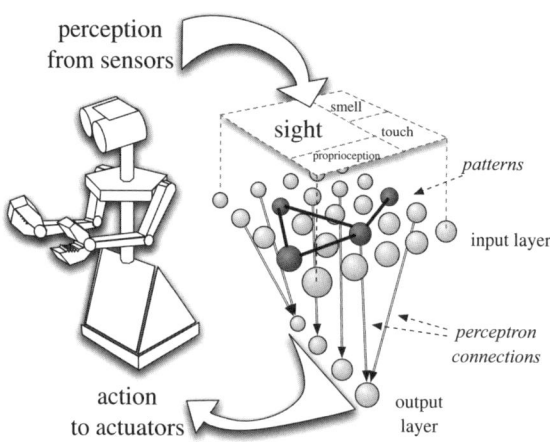

Fig. 1. The controller presented in this paper combines a traditional connectionist architecture with an associative network

require to be user defined) and grounded on perception (i.e., not symbolic). Then, we prove through experiments that memory is a bridge between complexity in the environment and intelligence in the behavior.

2 Experiments

Experiments[1] in a 3D environment (figure 2) show how building perception-based patterns of knowledge may help the agent enhance the action selection. A comparison with the behavior of a purely reactive agent stresses our agent's ability to generalize and anticipate situations.

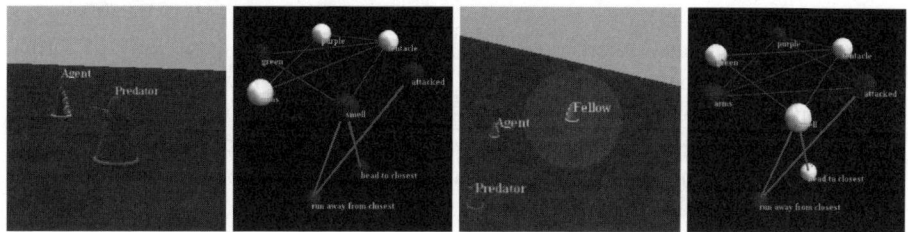

Fig. 2. In the environment, Agent builds patterns of knowledge in order to extend the action selection mechanism

3 Conclusion

In this paper, one original vision of intelligence is investigated. We believe the interest of a memory does not reside in the way representations can be manipulated but rather in the representations themselves, particularly how they can enhance the adaptation of the agent to the environment. Moreover, we postulate that the ability to create patterns of knowledge from the environment, and take advantage of them to improve the action selection may reveal intelligent behavior, insofar as the environment is complex enough.

References

1. Sanchez, S., Balet, O., Luga, H., Duthen, Y.: Autonomous Virtual Actors. In: 2nd International Conference on Technologies for Interactive Digital Storytelling and Entertainment, Darmstadt, Germany, June 2004. LNCS, pp. 68–78. Springer, Heidelberg (2004)
2. Lassabe, N., Luga, H., Duthen, Y.: A New Step for Evolving Creatures. In: IEEE-ALife 2007, Honolulu, Hawaii, pp. 243–251. IEEE, Los Alamitos (2007)

---

[1] All videos from which the pictures of this paper are extracted can be found at the following URL: http://www.irit.fr/~David.Panzoli/

Social Responses to Virtual Humans: Automatic Over-Reliance on the "Human" Category

Sung Park and Richard Catrambone

School of Psychology
Graphics, Visualization, and Usability Center (GVU)
Georgia Institute of Technology
gtg116s@mail.gatech.edu, rc7@prism.gatech.edu

In the human computer interaction (HCI) literature, responding socially to virtual humans means that people exhibit behavioral responses to virtual humans as if they are humans. Based partly on the work of Nass and his colleagues (for a review, see [1]), there is general agreement in the research community that people do respond socially to virtual humans. What seems to be vital, in terms of understanding why people respond socially to virtual humans, is how we respond socially to another person and how we know about others' temporary states (e.g., emotions, intentions) and enduring dispositions (e.g., beliefs, abilities).

Gilbert [2] suggested a unified framework for ordinary personology which attempts to incorporate attribution theories (i.e., how a stimulus engenders identification and dispositional inferences) and social cognitive theories (i.e., how responses engender impression). In his framework, appearance (e.g., a crucifix or a Mohawk) allows an actor to be classified (rightly or wrongly) as a member of a category (e.g., a priest or a punk), which then allows the observer to draw inferences about the actor's dispositions (he is religious or rebellious). Such a categorization process is rapid and automatic [3]. Automaticity is the idea that sufficient practice with tasks maps stimuli and responses consistently and therefore produces performance that is autonomous, involuntary, unconscious, and undemanding of cognitive resource (for a review, see [4]). This automatic nature of social responses can be equally observed when we interact with virtual humans (See Figure 1).

Human-like characteristics (the use of facial expression or voice implementation) act as a cue that leads a person to place the virtual human into the category "human" and thus, elicits social responses. These processes are practiced since birth and are strongly ingrained and have the characteristics (i.e., unconscious) of automaticity, such that a single cue may activate the human category (automatic over-reliance) and block other cues that would activate the virtual human category. Nass and his colleagues' finding--that participants denied that they thought the computer was a human but nevertheless responded socially--seems compatible with Bargh's [4] idea of "postconscious automaticity". In both cases the initial cues are conscious (i.e., human-like cues were clearly observable), yet their effects on category associations are not reportable (i.e., participants produce social behaviors when interacting with humans, yet explicitly deny considering the computer to be human-like).

Social behaviors are mediated by individual's cognitive and affective structures developed through experience [5]. Lack of knowledge regarding virtual humans might be one factor as to why we ignore asocial cues of virtual humans. This indicates that individuals do not have enough experience with virtual humans to form a rigid and concrete virtual human category. Experiencing a virtual human and realizing how the technology is still far from perfectly mimicking a human and identifying the asocial nature of virtual humans may contribute to forming a category of virtual human, where its disposition is non-human and non-social.

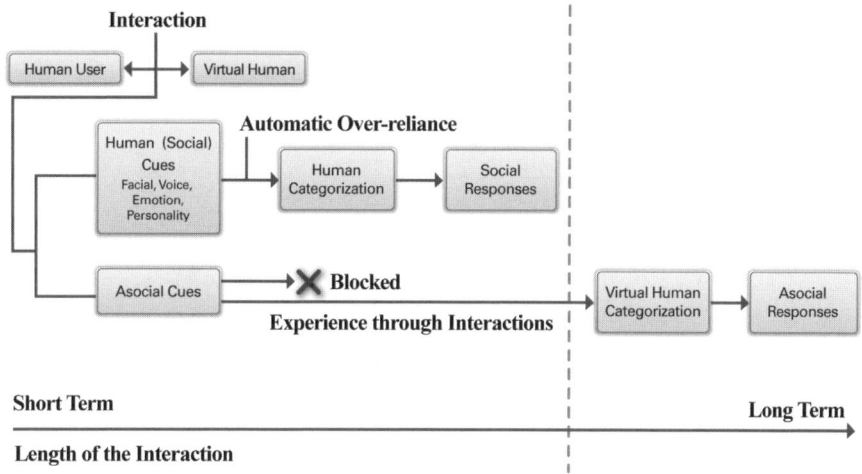

Fig. 1. How humans respond socially to virtual humans

Inevitability, length and time of the interaction become critical factors influencing social interaction with virtual humans. The more time spent with a virtual human, the greater chance there is to identify the asocial nature of a virtual human. One will be able to deliberately map the stimulus (virtual human) and the responses one would apply to a virtual human and not to a human. Uncertainty, ambiguity, or inconsistency of behavior will become a driving force (i.e., motivation), which will preclude categorization of the stimulus object or the production of behavior. These factors also prompt controlled attributional (as opposed to automatic) processing [5] and lead one to start isolating a special class of invariant properties that distinguish one (e.g., virtual humans in general) from the other (e.g., humans in general).

References

1. Moon, Y., Nass, C.: Machines and Mindlessness: Social Responses to Computers. Journal of Social Issues 56(1), 81–103 (2000)
2. Gilbert, D.: Ordinary Personology. In: Gilbert, D., Fiske, S., Lindzey, G. (eds.) The Handbook of Social Psychology, vol. 2, pp. 89–150. Oxford University Press, NY (1998)

3. Fiske, S.: Stereotyping, prejudice, and discrimination. In: Gilbert, D., Fiske, S., Lindzey, G. (eds.) The Handbook of Social Psychology, vol. 2, pp. 89–150. Oxford University Press, NY (1998)
4. Bargh, J.A.: Conditional automaticity: varieties of automatic influence in social perception and cognition. In: Uleman, J.S., Bargh, J.A. (eds.) Unintended thought, pp. 3–51. Guilford, NY (1989)
5. Feldman, J.M.: Four questions about human social behavior. In: Adamopoulos, J., Kashima, Y. (eds.) Social Psychology and Cultural Context: Essays in Honor of Harry C. Triandis. Sage, NY (1999)

Adaptive Self-feeding Natural Language Generator Engine

Jovan David Rebolledo Méndez and Kenji Nakayama

Graduate School of Natural Science and Technology,
Kanazawa University, Kanazawa-shi, 920-1192 Japan
{jovan,nakayama}@leo.ec.t.kanazawa-u.ac.jp
http://leo.ec.t.kanazawa-u.ac.jp

Abstract. A Natural Language Generation (NLG) engine is proposed based on the combination of NLG and Expert Systems. The combination of these techniques paves the way to employ user defined behaviors in virtual worlds as inputs to an expert system. Adaptive Algorithms can then be used to retrieve information from the Internet to give feedback to the user via the NLG engine. The combination of these AI techniques can bring about some benefits such as believability in the interaction between AI-driven and human-driven avatars in virtual worlds.

Keywords: NLG, Expert System, Machine Learning, Believable Agent.

1 Introduction

For years there have been attempts to incorporate Artificial Intelligence into games, educational systems, and some other different kinds of interactions [1] between a humans and computers. Those attempts shared a common goal: to develop *intelligence* that would give certain degree of independence, and decision-making to the computer system when interacting in an autonomously driven behavior with the user. There are also other systems that measure users' preferences by ranking interactions with links or applications. In order to get a better interface between users and systems, the concept of believability has been very important[2]. This paper explains a way for a user and an agent to interact by giving believability to the agent while keeping some common user's learned *traits*, by using a Natural Language Generation [3] and adaptive learning expert system [4] to achieve those objectives. Such a system could provide relevant categorized data from learnt criteria to its user, allowing it to preserve user patterns.

2 Natural Language Generation Engine Architecture

The proposed architecture for an *self-feeding NLG engine* is as follows: 1) The agent is initially fed through batteries of questions to the user, and an expert system will categorize the information provided by the user to fill the user's *knowledge base* of common patterns. Thus, the input data would serve as way to

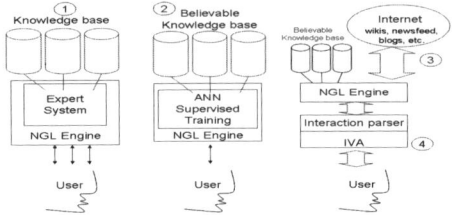

Fig. 1. Architecture of the proposed self-feeding believable NGL Engine

get to know the user's preferences; 2) Next, the system is trained by an Artificial Neural Network, where new information is presented, and by using a supervised learning method it would re-categorize new data. During this step, the system would gain some randomness, which would enable *believability* for the proposed agent; 3) Then, it would have a module for retrieving data that matches with its *knowledge*, from the web (wikis, newsfeed, blogs, etc.). By using machine the module would select in accordance with the previous *classified knowledge bases*. 4) Finally, by a dedicated module that interacts between the knowledge bases and the real time information, all of the system would be ported to an embodied IVA. By using a parser language that identifies the current subject, the system would present current themes, news, and comments about different topics, permitting a *conversational mode* with its original user, so it will help him in achieving certain tasks. The proposed architecture can be seen in Fig1.

3 Future Work

This proposed engine is the first stage of more complete AI interaction approach called *iTwinning System*, a dedicated personal agent that learns some user's behaviors, *generalizing* from their interaction, to assist him in achieving certain tasks in a virtual world or the internet, as well as being helpful, and *loadbalancing* work, and preserving the specific user's characteristics. The plan is to conduct a set of tests to users, providing a pool of question and answers, so that each user would select his preferences. By employing a generic algorithm it would separate the relevant information, and by a programmed robust feeder it would retrieve information from the Internet.

References

1. Schröder, M., Cowie, R.: Developing a consistent view on emotion-oriented computing Machine Learning for Multimodal Interaction Washinton, USA (2006)
2. Bates, J.: The role of emotion in believable agents. Communications of the ACM. 37(7) (1994)
3. Strong, C., Mishra, K., Mehta, M., Jones, A., Ram, A.: Emotionally Driven Natural Language Generation for Personality Rich Characters in Interactive Games. In: 3rd Conference on AI for Interactive Digital Entertainment CA, USA (2007)
4. Wiriyacoonkasem, S., Esterline, A.C.: Adaptive learning expert systems Southeastcon 2000. In: Proceedings of the IEEE Nashville, TN, USA (2005)

A Model of Motivation for Virtual-Worlds Avatars

Genaro Rebolledo-Mendez, David Burden, and Sara de Freitas

The Serious Games Institute, Coventry University
Coventry University Technology Park, Cheetah Road, Coventry CV1 2TL, UK
{GRebolledoMendez,DBurden,SFreitas}@cad.coventry.ac.uk

Abstract. This paper presents a model of motivation for AI-driven avatars in virtual worlds. This work aims at providing the avatar with motivational capabilities to detect and react to different motivational states aiming to enhance and maintain user's motivation. The algorithm and its associated rules are presented. Future work consists of assessing this model in virtual world applications in the context of a quiz.

1 Introduction

The introduction of the term 'affective computing' (Picard 1997) brought about an interest in understanding affective processes. The aim of this paper is to present a motivational model consisting of an algorithm and its associated behavioural rules for virtual world applications. The interest on motivation arises from its potential in helping users engage in tasks and activities framed in serious games for virtual worlds. The emergence of serious games has increased in the last five years opening up the potential for developing new technologies based upon immersive and interactive interfaces (de Freitas and Oliver, 2006).

2 Recognising and Reacting to Other Avatar's Motivation

Our research on motivation addresses motivation-recognition (de Vicente and Pain 2002; Lepper, Woolverton et al. 1993) and motivation-reaction (Ames 1990; Keller and Suzuki 1988). The model presented here involves an AI-driven avatar recognising an avatar's motivation and reacting appropriately following a set of theoretical constructs (Ames 1990; Keller and Suzuki 1988). To perpetually monitor the actor the AI-driven avatar follows the next steps: 1) the human-driven avatar is willing to answer questions posed by the AI- avatar in the virtual world; 2) the AI-driven collects contextual information such as number of correct/incorrect answers (effort, A in Table 1), processes the human-driven responses as either correct or incorrect (B in Table 1) and time taken (C in Table 1) to answer the current question; 3) based on a set of rules taken from Theory (Ames 1990; Keller and Suzuki 1988) see Table 1. By calculating the mean value of A, B and C the AI-avatar selects one rule from Table 1 and reacts aiming to sustain or enhance the current level of motivation.

Table 1. Motivation diagnosis based on effort (A), correctness of answer (B) and time taken (C)

A	B	C	Reaction
High	True	Long	Provide help, specify criteria for success and provide opportunity to problem-solving based on example.
High	True	Short	Provide help; use unexpected rewards and corrective feedback; praise if effort is high while considering the degree of task completion.
High	False	Long	Specify criteria for success.
High	False	Short	Use unexpected rewards and corrective feedback; praise when effort is high and based on task completion. Provide strategies to succeed on task.
Low	True	Long	Provide help; use novel, incongruous, conflictual and paradoxical events. Provide opportunities for success under moderate risk conditions.
Low	True	Short	Provide help; use novel, incongruous, conflictual and paradoxical events; use attributional feedback to connect success to personal effort.
Low	False	Long	Provide opportunities to achieve success under moderate risk conditions.
Low	False	Short	Use attributional feedback to connect success to personal effort. Praise considering degree of task completion. Provide strategies to succeed on task.

3 Conclusions and Future Work

The theoretical framework could enable the AI-driven avatar to have an understanding of users' avatars' motivation. The model consists of an algorithm to perpetually monitor the user and react considering a set of eight motivational rules. This work considers one-to-one interactions only. Work for the future includes studying the possibilities for one-to-many interactions and the evaluation of this model in the context of a quiz for higher education students. Possible uses of this technology include serious game adversaries (motivation is used *against* the gamer), personal tutors or learning companions. A longer-term goal consists on designing and evaluating a serious game for higher education where cognitive modelling and motivational modelling are integrated into a virtual learning situation.

References

Ames, C.A.: Motivation: What Teachers Need to Know. Teachers College Record 91(3), 409–421 (1990)

de Freitas, S., Oliver, M.: How can exploratory learning with games and simulations within the curriculum be most effectively evaluated? Computers and Education Special Issue 46, 249–264 (2006)

de Vicente, A., Pain, H.: Informing the Detection of the Student's Motivational State: An empirical Study. In: 6th International Conference on Intelligent Tutoring Systems, Biarritz, France, pp. 933–943. Springer, Heidelberg (2002)

Keller, J.M., Suzuki, K.: Use of ARCS motivation model in courseware design. In: Jonassen, D.H. (ed.) Instructional designs for microcomputer courseware, Lawrence Erlbaum Associates, Hillsdale (1988)

Lepper, M., Woolverton, M., et al.: Motivational Techniques of Expert Human Tutors: Lessons for the design of Computer-Based Tutors. In: Lajoie, S.P., Derry, S.J. (eds.) Computers as Cognitive Tools, pp. 75–105. Erlbaum, Hillsdale (1993)

Picard, R.W.: Affective Computing. MIT Press, Cambridge (1997)

Towards Virtual Emotions and Emergence of Social Behaviour

Dirk M. Reichardt

BA Stuttgart - University of Cooperative Education
D-70180 Stuttgart, Germany
reichardt@ba-stuttgart.de

In our approach we explore emotional effects of virtual agents *without* human interaction. The idea is the following: emotion is considered a valuable means to explain social phenomena like the emergence of costly punishment. Now, would it be possible to put together a population of agents which mimics human society? Would it even be possible to show that certain personality configurations and emotional behavior *emerge* from interactions in a virtual agent society experiment? In previous work we selected the public goods game with punishment option because of its intriguing property of affect-based decisions [1]. In this paper we present our first approach to virtual agent population modelling which adopts a setting of the public goods game as it is used for studies on cooperative behaviour (see [1][3]). The discrete version of the game distinguishes between the roles of the punisher, the cooperator and the defector. As shown in [3] the iterated public goods game leads to a predominance of the defectors. This is changed by giving the participants the freedom to choose whether to participate in the game or not which gives the punishers a chance to get back into the game and it is shown that eventually they become dominant (see [3]).

The basis for our agent is the model by Ortony, Clore and Collins (OCC) [2]. It defines an interface to the application by intensity variables which need to be adapted to the application scenario (see [4]). The expected payments and punishments are computed based upon a locally stored action history. This forms the rational basis of the agent. In our approach we introduce a concept of mood as a mid-term multidimensional accumulator which serves as an *affect filter*. In order to update mood we need to transform the OCC-emotions into the pleasure-arousal-dominance (PAD) space [5][7] in which mood is also described. The mood is shifted in the direction of the current emotion using the metaphor of the attraction of masses. The mood takes the role of the bigger mass which moves just a bit in direction of the smaller one, the emotion. The elicitation value of the current emotion is modified depending on the distance of the emotion to the mood coordinates in the PAD-space. Both, mood changes and emotion elicitation depend on the personality. The term personality describes the factors that determine the character of an individual and thus it is stable over time. We integrate the "big 5" model of personality which is the most popular model nowadays [6]. Since personality influences emotion elicitation, we integrate the personality traits directly in the elicitor functions of emotions. Personality also influences the *dynamics* of mood changes therefore we adopted the feature of a long term mood attractor which is defined by the personality [7].

Each agent in the simulated society is configured by setting the personality traits and two further parameters called *goalpreference* and *moraldisposition*. The latter determines whether and to what extent the agent shows shame or pride. The goal preference is necessary for decision making. There are two decisions to be made: contribution and punishment. In the simulation these are binary decisions (in contrast to earlier publications which use different settings). In case of the contribution decision we precalculate the potential action and reactions. For each potential future state, the system computes the emotional state. The decision is then made by comparing the alternatives with respect to the goal state of the agent's mood. The two goals - mood goal state and payoff - are compared using a *goalpreference* bias.

First simulations were done using personality parameters which we took from a personality test with human test persons who also played the game on our distributed public goods game test platform. The emotional agents replace cooperators *and* punishers in the simulation scenario since emotional agents can show both kinds of behavior. One observation is that in the presence of our agents the number of nonparticipants and defectors is almost stable.

Due to the high number of defectors the payoff is rather low, therefore the mood of the agents becomes negative. We identified small regions in which the average mood is $p<0, a>0, d>0$ which is called 'hostile'. This fits the parameters of the emotion 'anger' very well, therefore the highest elicitation values for anger can be seen in these regions. This leads to punishment decisions which make the defector less attractive and consequently their number decreases. Currently, the model is built up and is functionally tested and plausible reactions can be seen. Future experiments will include a setup with emotional agents only. This setup intends to analyse the evolution of strategies within the emotional agent framework.

References

[1] Fehr, E., Gächter, S.: Altruistic Punishment in Humans. Nature 415, 137–140 (2002)
[2] Ortony, A., Clore, G., Collins, A.: The Cognitive Structure of Emotions. Cambridge University Press, Cambridge (1988)
[3] Hauert, C., Traulsen, A., Brandt, H., Nowak, M., Sigmund, K.: Via Freedom to Coercion: The Emergence of Costly Punishment. Science 316, 1905–1907 (2007)
[4] Reichardt, D.: Interpretation of Intensity Variables for an Emotional Agent in the Public Goods Game. In: Reichardt, D., Levi, P. (eds.) Proceedings of the 2nd Workshop Emotion and Computing – Current Research an Future Impact, Osnabrück, Germany, September 10 (2007) ISSN 1865-6374
[5] Mehrabian, A.: Pleasure-arousal-dominance: A general framework for describing and measuring individual differences in temperament. Current Psychology 14, 261–292 (1996)
[6] Mehrabian, A.: Analysis of the Big-five Personality Factors in Terms of the PAD Temperament Model. Australian Journal of Psychology 48(2), 86–92 (1996)
[7] Gebhard, P., Kipp, K.H.: Are Computer-generated Emotions and Moods plausible to Humans? In: Gratch, J., Young, M., Aylett, R.S., Ballin, D., Olivier, P. (eds.) IVA 2006. LNCS (LNAI), vol. 4133, pp. 343–356. Springer, Heidelberg (2006)

Using Virtual Agents for the Teaching of Requirements Elicitation in GSD

Miguel Romero[1], Aurora Vizcaíno[2], and Mario Piattini[2]

[1] University of Bío-Bío, Department of Computer Science and Information Technologies, Chillán, Chile
mromero@pehuen.chillan.ubiobio.cl
[2] University of Castilla-La Mancha, Alarcos Research Group- Institute of Information Technologies & Systems, Dep. of Information Technologies & Systems - Escuela Superior de Informática, Ciudad Real, Spain
{Aurora.Vizcaino,Mario.Piattini}@uclm.es

Abstract. Requirements elicitation is particularly difficult in Global Software Development (GSD) environments owing principally to cultural differences and communication problems derived from the geographical distance that separates stakeholders. For this reason it is necessary to train professionals in the skills needed to confront a requirements elicitation in GSD. We have, therefore, designed a simulator which, by using virtual agents, will enable students and professionals to acquire a subset of the skills necessary for requirements elicitation in GSD.

Keywords: Requirements Elicitation training, Global Software Development, Educational Environment, virtual agents.

1 Introduction

Requirements elicitation is the first stage in the process of developing a software product and the most critical of all the phases in software development, because the mistakes made at this stage are more expensive and difficult to resolve owing to their impact upon the other stages. Unfortunately, professionals who have recently graduated from universities lack the skills and abilities necessary to carry out this task correctly since during their degree course little time is usually spent on training in this phase of software engineering, they often do not carry out professional practices, their teaching is centered on theory, and students rarely get involved in real projects. In addition, the current trends of software development and their effect upon requirements elicitation are not generally considered. Global Software Development (GSD) [1] is one of the current challenges of teaching and training in the requirements elicitation process. GSD is characterized by stakeholders who are geographically distributed around the World. Those GSD issues whitch affect a requirements elicitation process are: Cultural, language and time differences; inadequate communication, difficulties in knowledge management, and trust [2]. In order to confront the challenge of training professionals capable of developing a requirements elicitation process

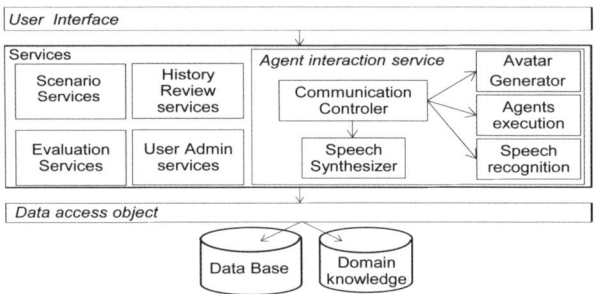

Fig. 1. Architecture of the Simulator

in GSD, we propose a simulator environment which, by using virtual agents, will enable students and professionals to acquire a subset of the skills necessary for requirements elicitation in GSD, such as: elicitation of real requirements based on stakeholder's need using an interview technique and computer mediated communications, or an understanding of the cultures and customs of other countries. The student interacts with various stakeholders which will be virtual humans and/or real humans, in order for them to obtain the functional and non-functional requirements of the software to be developed. Figure 1 shows the architecture of our simulator.

The Agent interaction service, which is the main component of our architecture, is where input from the user during conversations with a virtual stakeholder (virtual agents) is interpreted. This component is made up of the following component: the *Communication controller* is the link between the communication interface between the user and the virtual agent; *Agents Execution* provides a particular agent's answers to questions from the user, and the input and output is a text; the *Avatar Generator* generates a graphical representation of the virtual agent; The *Speech Recognition and Synthesizer* convert audio to text and text to audio, respectively.

This simulator may be an initial step towards students' participation in real projects developed between universities and the GSD industry, which would diminish the risk of non-qualified people being involved in real projects.

Acknowledgements. This work is partially supported by the MELISA project (PAC08-0142-3315), Junta de Comunidades de Castilla-La Mancha in Spain.

References

1. Herbsleb, J.D.: Global software engineering: The future of socio-technical coordination. In: FOSE 2007, Minneapolis, pp. 188–198. IEEE Computer Society, Los Alamitos (2007)
2. Damian, D.E., Zowghi, D.: The impact of stakeholders' geographical distribution on managing requirements in a multi-site organization. In: RE 2002, pp. 319–328 (2002)

A Virtual Agent's Behavior Selection by Using Actions for Focusing of Attention

Haris Supic

Faculty of Electrical Engineering, University of Sarajevo
Zmaja od Bosne bb, 71000 Sarajevo, Bosnia and Herzegovina
haris.supic@etf.unsa.ba

Abstract. In this paper we present an approach for focusing of IVA's attention in order to select appropriate behavior in continuously changing virtual environment. An action for focusing of attention represents the act of moving the attention to the currently relevant attributes of the local virtual environment.

Keywords: virtual agent, action for focusing of attention, behavior selection.

1 Action for Focusing of Attention

To move in continuously changing virtual environment, an intelligent virtual agent (IVA) must be able to react appropriately to changes and unexpected obstacles in the virtual environment. In such situations, it is necessary for an IVA to select appropriate action for focusing of attention, and if necessary, to change existing behavior. This paper presents an approach that use case-based reasoning to select appropriate action for focusing of attention. As a consequence of the selected action for focusing attention, an IVA receives synthetic perception stimulus [1]. An action for focusing of attention represents the act of moving the attention to the currently relevant attributes of the local virtual environment. Formally, an action for focusing of attention f is an n-tuple $f=(A_1, A_2, ...A_n)$, where A_i is a perception attribute, $i=1,2,...n$. A perception attribute is a relevant feature of the virtual environment that is important for the IVA's next behavior selection [2]. As an illustration, assume that the current IVA's intention is "*exit-from-room*" (see Figure 1). This intention directs the IVA's attention to the relevant perception attributes: the distance from obstacle to the left wall (L), the distance from obstacle to the right wall (R), the distance from the obstacle (D), the angle to the front-right corner of the obstacle (θ), and the distance from current position to the right wall (W) (see Figure 1). Illustrated perception attributes constitute the formal representation of the action for focusing of attention $f=(L,R,D,\theta,W)$. As a consequence of the selected action for focusing of attention an IVA receives synthetic perception stimulus. Thus, the new situation case is created and the most similar case is retrieved from the casebase. The appropriate behaviors are selected and adapted to the new relevant perception attributes based on retrieved case. We used an approach to synthetic perception similar to the one described by Tu and Terzopoulos [3].

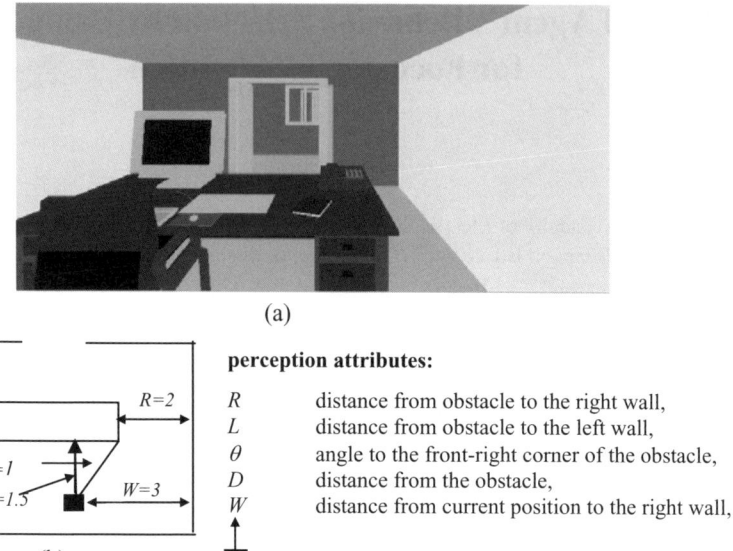

Fig. 1. An illustration of the relevant perception attributes for autonomous navigation in indoor virtual environments. (a) A simulated vision from the current position of the IVA (b) 2D representation of the relevant perception attributes for the current situation (adapted from [1]).

2 Conclusion

This paper has presented the role of an action for focusing of attention in IVA's moving through a virtual environment. An action for focusing of attention shifts an IVA's attention to the relevant aspects of a local virtual environment. An IVA determines next behavior based on the synthetic perception that is generated as a consequence of the previously selected action for focusing attention.

References

1. Supic, H., Ribaric, S.: Autonomous Creation of New Situation Cases in Structured Continuous Domains. In: Muñoz-Ávila, H., Ricci, F. (eds.) ICCBR 2005. LNCS (LNAI), vol. 3620, pp. 537–551. Springer, Heidelberg (2005)
2. Supic, H.: A Case-Based Approach to Intelligent Virtual Agent's Interaction Experience Representation. In: Pélachaud, C., Martin, J.-C., André, E., Chollet, G., Karpouzis, K., Pelé, D. (eds.) IVA 2007. LNCS (LNAI), vol. 4722, pp. 407–408. Springer, Heidelberg (2007)
3. Tu, X., Terzopouls, D.: Artificial fishes: Physics, locomotion, perception, behavior. In: Proc. of SIGGRAPH 1994, pp. 43–50 (1994)

Emergent Narrative and Late Commitment

Ivo Swartjes, Edze Kruizinga, Mariët Theune, and Dirk Heylen

Human Media Interaction group, Twente University
{swartjes,kruizingaee,theune,heylen}@cs.utwente.nl

Abstract. Emergent narrative is an approach to interactive storytelling in which stories result from local interactions of autonomous characters. We describe a technique for emergent narrative that enables the characters to fill in the story world during the simulation when this is useful for the developing story.

1 Introduction

The Virtual Storyteller [1] is a framework that can be used to author a story world, simulate virtual characters that live their lives in this story world, and tell stories about the resulting event sequence. The approach is that of emergent narrative [2], meaning that stories are not scripted in advance, but result from local interactions of virtual character agents. To facilitate the emergence of interesting stories, we give these character agents some responsibility for story progression, informed by techniques used by actors in improvisational theater [3]. One of these techniques is to fill in the story world in line with what the emerging story needs. Objects are added, character quirks and relationships defined, and the back stories of the characters are conveniently filled in, when this is useful for story progression. In this paper, we describe our exploration of this idea, and illustrate how we use it in the Virtual Storyteller.

2 Late Commitment

Emergence, as a guiding principle in both improvisational theater and emergent narrative, implies that there is no clear relationship for the author between the initial setup of the story world and what emerges. We would like her to be able to author parts of the story world in terms of *potentiality*, so that elements can be introduced whenever they become useful in the simulation.

For the Virtual Storyteller, we have enabled this by introducing a type of operators to the story domain that we call *framing operators*. The effects of a framing operator describe *commitments* to information about the story world, rather than *changes* (as with actions and events). Where a normal action is pursued *in character* (i.e., the agent's character does the action), a framing operator is an action *out of character* (i.e., the agent *itself* does the action, as if it says to the rest of the agents: "let's pretend my character hated your character, OK?"). The execution of a framing operator should create the illusion that its effects have been true since the start of the simulation. The idea builds on initial state revision in story planning [4].

3 Usage Examples

We have fully implemented support for framing operators in the Virtual Storyteller and have run experiments within the domain of pirate stories. To illustrate how late commitment facilitates the emergent narrative process, we sketch two examples of how we use it:

Finding motivations for the adoption of goals. In the Virtual Storyteller, character agents can adopt goals when the preconditions of these goals are met. For instance, adopting a goal of plundering another ship might require that there is another ship in sight, and that the character adopting the goal is a pirate captain. If these circumstances do not apply, the preconditions can still be achieved by introducing a ship in sight, and endowing one of the characters with the role of pirate captain.

Filling in the environment to support action selection. If the captain of the ship has adopted a goal to find out whether an approaching ship is friend or foe, he can make a plan involving looking through a binocular. This binocular can be framed to be in the captain's cabin, and the captain will plan to go to his cabin to get it.

We have found that the technique of late commitment increases the flexibility of character agents to improvise, not only by using framing operators to aid in the planning process, but also by being able to create the necessary context for the adoption of desired character goals. We plan to investigate how this technique can also be used for the agents to actively find reasons for emotional reactions and belief changes.

Acknowledgements

This research has been supported by the GATE project, funded by the Netherlands Organization for Scientific Research (NWO) and the Netherlands ICT Research and Innovation Authority (ICT Regie).

References

1. Theune, M., Rensen, S., op den Akker, R., Heylen, D., Nijholt, A.: Emotional characters for automatic plot creation. In: Technologies for Interactive Digital Storytelling and Entertainment (TIDSE 2004) (2004)
2. Aylett, R.: Emergent Narrative, Social Immersion and "Storification". In: Proceedings of the 1st International Workshop on Narrative Interaction for Learning Environments, Edinburgh (2000)
3. Swartjes, I., Vromen, J.: Emergent story generation: Lessons from improvisational theater. In: Intelligent Narrative Technologies: Papers from the AAAI Fall Symposium. AAAI Fall Symposium Series, vol. FS-07-05 (2007)
4. Riedl, M., Young, M.: Story planning as exploratory creativity: Techniques for expanding the narrative search space. In: Proceedings of the 2005 IJCAI Workshop on Computational Creativity (2005)

"I Would Like to Trust It but" Perceived Credibility of Embodied Social Agents: A Proposal for a Research Framework

Federico Tajariol, Valérie Maffiolo, and Gaspard Breton

OrangeLabs, France Telecom Group
2, av. Pierre Marzin 22300 Lannion - France
{federico.tajariol,valerie.maffiolo,
gaspard.breton}@orange-ftgroup.com

Abstract. We suggest a research framework to evaluate users' perceived credibility when they interact with an ESA. Our research framework only addresses ESA's nonverbal features, thus we do not take the content and the rhetorical arguments of the message delivered by ESAs into account.

Keywords: Nonverbal cues, perceived credibility, attractiveness, involvement.

1 Introduction

Embodied Social Agents (ESAs) act as assistants, guides, salespersons, entertainers, etc. ESAs are perceived by users as a sort of "social actors" [1], being neither simple mouse-and-screen interface nor human beings. In order to evaluate human-ESAs interaction, this "half-computer half-human" identity implies that we cannot apply usability criteria only, but we should also consider other relevant criteria [2]. Amongst these criteria, *credibility* is a fundamental concept to understand how human beings trust the ESA interacting with them [3]. In our view, *credibility* reflects the power of an ESA to act as a trusted (or believed) assistant in a fixed open-end human activity, by means of its moral qualities and its expertise.

Main experimental researches on ESAs did not clearly state the construct of perceived credibility, nor did operationalize it [4, 5, 6, 7]. Main limitations to the study of credibility concern both theoretical and methodological levels. Theoretical limits are: i) *credibility* is only considered either for moral or expertise dimension, rather than under both dimensions; ii) ESA's verbal cues and nonverbal behaviors are believed to have a direct influence on perceived *credibility*, without taking other factors into account (e.g., user's task involvement, ESA's attractiveness). Methodologically, the main weaknesses are the following: i) experimental studies only dealt with (cf. [8]) *presumed credibility* (a person believes a source because of general assumptions in the person's mind based on stereotypes: e.g., physicians are good expert, friends are trustworthy) and *surface credibility* (a person believes others through their surface traits, like dressing style or general look), whereas *earned credibility* (the strengthened type of credibility, based on the consistent performance of the ESA in accordance with users' expectations) has not been studied; ii) although perceived *credibility* would be a subjective measurement, no studies tried to calibrate participants' profile (personality, etc.), their involvement towards the task and participants' attractiveness and likeability towards the ESA; iii) generally, questionnaire are not statistically reliable, thus their construct and external validity is not satisfactory.

In light of these limits, we suggest a research framework to evaluate users' *perceived credibility* when they interact with an ESA. Our research framework only addresses ESA's nonverbal features, and more particularly, we do not take the content and the rhetorical arguments of the message delivered by ESAs into account. This framework owns two main characteristics:

i) We consider that *goodwill* is a required dimension to operationalize the construct of credibility. We think that this dimension could be relevant in specific types of interaction with ESAs, such as the submission of users' personal data or preferences in e-banking and commercial services. For example, by analyzing the behavior of visitors on a commercial Web site showing an ESA, Reeves [9] reported that the clients interacted with the ESA were three times more likely to reveal personal information than the clients who did not interact with the ESA.

ii) We suppose that ESA's appearance (i.e., ethnicity, look) and ESA's nonverbal behaviors (i.e., facial expressions, gestures, posture, etc.) do not have a straight effect on perceived credibility. We think that the cause-effect links would be mediated by some variables concerning the receiver, such as receiver's attractiveness toward ESA, receiver's involvement, and receiver's idiosyncratic properties.

This framework, and its consequent methodology, is going to be tested in an on-going experimental study.

References

1. Reeves, B., Nass, C.: The Media Equation. CUP (1996)
2. Ruttkay, Z., Dormann, C., Noot, H.: Embodied conversational agents on a common ground. In: Ruttkay, Z., Pelachaud, C. (eds.) From Brows to trust. Evaluating embodied conversational agents, pp. 27–66. Kluwer Academic Publishers, Dordrecht (2004)
3. Isbister, K., Doyle, P.: The blind men and the elephant revisited. In: Ruttkay, Z., Pelachaud, C. (eds.) From Brows to Trust, pp. 3–26. Springer, Heidelberg (2004)
4. Cowell, A.J., Stanney, K.M.: Manipulation of non-verbal interaction style and demographic embodiment to increase anthropomorphic computer character credibility. International Journal of Human-Computer Studies 62, 281–306 (2005)
5. Nowak, K., Rauh, C.: The Influence of the Avatar on Online Perceptions of Anthropomorphism, Androgyny, Credibility, Homophily, and Attraction. Journal of Computer Mediated Communication 11(1) (2005)
6. Zanbaka, C., Goolkasian, P., Hodges, L.F.: Can a virtual cat persuade you? The role of gender and realism in speaker persuasiveness. In: Computer Human Interaction Conference, April 22-27, pp. 131–139. ACM Press, Canada (2006)
7. ten Ham, R., Theune, M., Heuvelman, A., Verleur, R.: Judging Laura: perceived qualities of a mediated human versus an embodied agent. In: Panayiotopoulos, T., et al. (eds.) Intelligent Virtual Agents 2005, pp. 381–393. Springer, Berlin (2005)
8. Fogg, B.J.: Persuasive Technology. Morgan Kaufmann Publishers, San Francisco (2003)
9. Reeves, X.: (2001), http://www.oddcast.com/home/cases/research/CSLI_Stanford_Study.pdf

Acceptable Dialogue Start Supporting Agent for Avatar-Mediated Multi-tasking Online Communication

Takahiro Tanaka, Kyouhei Matsumura, and Kinya Fujita

Tokyo University of Agriculture and Technology
2-24-16 Nakacho, Koganei, Tokyo, 184-8588, Japan
{takat,kfujita}@cc.tuat.ac.jp, 50004258054 @st.tuat.ac.jp

In recent years, the instant messaging tools have become popular for daily online communication. The feature of the communication style with these tools is multi-tasking online communication. The users of these tools have a problem in recognizing the status of interaction partners. The start of dialogue has a risk of unintended interruption of the partner. The automatic status estimation and ambient display of the status is expected to assist the avoidance of the interruption. Therefore, we prototyped an acceptable dialogue start supporting (ADSS) agent system, for assisting pleasantly-acceptable dialogue start in avatar-mediated multi-tasking communication (AMMCS). The ADSS agent estimates and expresses the user uninterruptibility, and appeals the dialogue request by the communication partner. Fig. 1 shows the overview of ADSS agent and screenshot of the user desktop while using the agent. Each communication partner is displayed as an avatar in a small individual window.

In attempt to overcome the intelligent activity interruption problem, we focused on the application-switching (AS) as a potential intelligent activity discontinuity for uninterruptibility estimation. The preliminary experiments revealed the uninterruptibility reduction effect of AS. The agent estimates the user uninterruptibility using AS, keystroke and mouse click. We also used the avatar's postures and motions including gazing for ambient and intuitive expression of the uninterruptibility and the appeal of the partner's dialogue request. The agent controls the uninterruptible impression using overlapping these non-verbal expressions. Also, we examined the possibility of impression control by motion overlapping using the postures having different body direction distance and face direction.

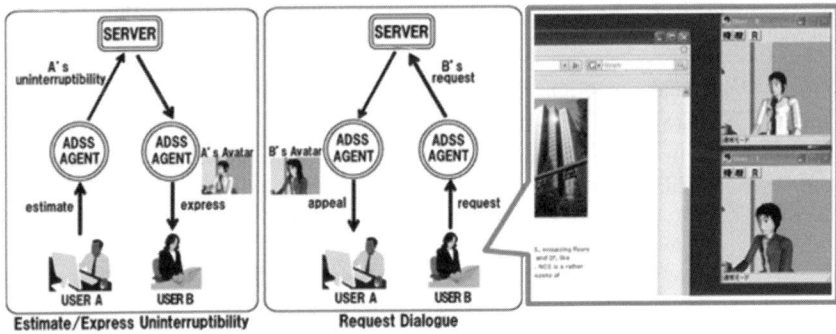

Fig. 1. Overview of the ADSS Agent System

Do You Know How I Feel? Evaluating Emotional Display of Primary and Secondary Emotions

Julia Tolksdorf, Christian Becker-Asano, and Stefan Kopp

Artificial Intelligence Group, University of Bielefeld, 33594 Bielefeld, Germany
{jtolksdo,cbecker,skopp}@techfak.uni-bielefeld.de

1 Motivation and Description of Study

In this paper we report on an empirical study on how well different facial expressions of primary and secondary emotions [2] can be recognized from the face of our emotional virtual human Max [1]. Primary emotions like happiness are more primitive, onto-genetically earlier types of emotions, which are expressed by direct mapping on basic emotion display; secondary emotions like relief or gloating are considered cognitively more elaborated emotions and require a more subtle rendition. In order to validate the design of our virtual agent, which entails devising facial expressions for both kinds of emotion, we tried to find answers to the questions: How well can emotions be read from a virtual agent's face by human observers? Are there differences in the recognizability between more primitve primary and more cognitively elaborated secondary emotions?

In our study, facial expressions of six primary emotions (see Table 1(a)) and seven secondary emotions (see Table 1), i.e. 13 in total, had to be rated on a questionnaire. Stimuli expressions of the secondary emotions were created from pictures and movie actors, as no sufficiently precise specifications was available. Each subject saw a sequence of 15 still pictures of Max's face and had to rate each face for its emotional content by choosing from a total of 15 candidate emotion labels, which were illustrated by an additional German example sentence and could further be weighted by choosing either "maybe", "pretty similar", or "almost perfect". Choosing a weighted second emotion term was optionally

Table 1. The presented primary and secondary emotions (with translations); labels in *italics* indicate a correspondence to one of Ekman's basic emotions [3]

(a) Primary emotions

english	german	facial expr.
happy	erfreut	*happiness*
bored	gelangweilt	bored
concentrated	konzentriert	neutral
annoyed	genervt	*sadness*
sad	traurig	*sadness*
surprised	überrascht	*surprise*
angry	wütend	*anger*

(b) Secondary emotions

english	german
gloating	schadenfroh
ashamed	beschämt
relieved	erleichtert
jealous	neidisch
proud	stolz
frustrated	frustriert
hopeful	hoffnungsvoll

Vol. 4898: M. Kolp, B. Henderson-Sellers, H. Mouratidis, A. Garcia, A.K. Ghose, P. Bresciani (Eds.), Agent-Oriented Information Systems IV. X, 292 pages. 2008.

Vol. 4897: M. Baldoni, T.C. Son, M.B. van Riemsdijk, M. Winikoff (Eds.), Declarative Agent Languages and Technologies V. X, 245 pages. 2008.

Vol. 4894: H. Blockeel, J. Ramon, J. Shavlik, P. Tadepalli (Eds.), Inductive Logic Programming. XI, 307 pages. 2008.

Vol. 4885: M. Chetouani, A. Hussain, B. Gas, M. Milgram, J.-L. Zarader (Eds.), Advances in Nonlinear Speech Processing. XI, 284 pages. 2007.

Vol. 4874: J. Neves, M.F. Santos, J.M. Machado (Eds.), Progress in Artificial Intelligence. XVIII, 704 pages. 2007.

Vol. 4870: J.S. Sichman, J. Padget, S. Ossowski, P. Noriega (Eds.), Coordination, Organizations, Institutions, and Norms in Agent Systems III. XII, 331 pages. 2008.

Vol. 4869: F. Botana, T. Recio (Eds.), Automated Deduction in Geometry. X, 213 pages. 2007.

Vol. 4865: K. Tuyls, A. Nowe, Z. Guessoum, D. Kudenko (Eds.), Adaptive Agents and Multi-Agent Systems III. VIII, 255 pages. 2008.

Vol. 4850: M. Lungarella, F. Iida, J.C. Bongard, R. Pfeifer (Eds.), 50 Years of Artificial Intelligence. X, 399 pages. 2007.

Vol. 4845: N. Zhong, J. Liu, Y. Yao, J. Wu, S. Lu, K. Li (Eds.), Web Intelligence Meets Brain Informatics. XI, 516 pages. 2007.

Vol. 4840: L. Paletta, E. Rome (Eds.), Attention in Cognitive Systems. XI, 497 pages. 2007.

Vol. 4830: M.A. Orgun, J. Thornton (Eds.), AI 2007: Advances in Artificial Intelligence. XIX, 841 pages. 2007.

Vol. 4828: M. Randall, H.A. Abbass, J. Wiles (Eds.), Progress in Artificial Life. XII, 402 pages. 2007.

Vol. 4827: A. Gelbukh, Á.F. Kuri Morales (Eds.), MICAI 2007: Advances in Artificial Intelligence. XXIV, 1234 pages. 2007.

Vol. 4826: P. Perner, O. Salvetti (Eds.), Advances in Mass Data Analysis of Signals and Images in Medicine, Biotechnology and Chemistry. X, 183 pages. 2007.

Vol. 4819: T. Washio, Z.-H. Zhou, J.Z. Huang, X. Hu, J. Li, C. Xie, J. He, D. Zou, K.-C. Li, M.M. Freire (Eds.), Emerging Technologies in Knowledge Discovery and Data Mining. XIV, 675 pages. 2007.

Vol. 4811: O. Nasraoui, M. Spiliopoulou, J. Srivastava, B. Mobasher, B. Masand (Eds.), Advances in Web Mining and Web Usage Analysis. XII, 247 pages. 2007.

Vol. 4798: Z. Zhang, J.H. Siekmann (Eds.), Knowledge Science, Engineering and Management. XVI, 669 pages. 2007.

Vol. 4795: F. Schilder, G. Katz, J. Pustejovsky (Eds.), Annotating, Extracting and Reasoning about Time and Events. VII, 141 pages. 2007.

Vol. 4790: N. Dershowitz, A. Voronkov (Eds.), Logic for Programming, Artificial Intelligence, and Reasoning. XIII, 562 pages. 2007.

Vol. 4788: D. Borrajo, L. Castillo, J.M. Corchado (Eds.), Current Topics in Artificial Intelligence. XI, 280 pages. 2007.

Vol. 4775: A. Esposito, M. Faundez-Zanuy, E. Keller, M. Marinaro (Eds.), Verbal and Nonverbal Communication Behaviours. XII, 325 pages. 2007.

Vol. 4772: H. Prade, V.S. Subrahmanian (Eds.), Scalable Uncertainty Management. X, 277 pages. 2007.

Vol. 4766: N. Maudet, S. Parsons, I. Rahwan (Eds.), Argumentation in Multi-Agent Systems. XII, 211 pages. 2007.

Vol. 4760: E. Rome, J. Hertzberg, G. Dorffner (Eds.), Towards Affordance-Based Robot Control. IX, 211 pages. 2008.

Vol. 4755: V. Corruble, M. Takeda, E. Suzuki (Eds.), Discovery Science. XI, 298 pages. 2007.

Vol. 4754: M. Hutter, R.A. Servedio, E. Takimoto (Eds.), Algorithmic Learning Theory. XI, 403 pages. 2007.

Vol. 4737: B. Berendt, A. Hotho, D. Mladenic, G. Semeraro (Eds.), From Web to Social Web: Discovering and Deploying User and Content Profiles. XI, 161 pages. 2007.

Vol. 4733: R. Basili, M.T. Pazienza (Eds.), AI*IA 2007: Artificial Intelligence and Human-Oriented Computing. XVII, 858 pages. 2007.

Vol. 4724: K. Mellouli (Ed.), Symbolic and Quantitative Approaches to Reasoning with Uncertainty. XV, 914 pages. 2007.

Vol. 4722: C. Pelachaud, J.-C. Martin, E. André, G. Chollet, K. Karpouzis, D. Pelé (Eds.), Intelligent Virtual Agents. XV, 425 pages. 2007.

Vol. 4720: B. Konev, F. Wolter (Eds.), Frontiers of Combining Systems. X, 283 pages. 2007.

Vol. 4702: J.N. Kok, J. Koronacki, R. Lopez de Mantaras, S. Matwin, D. Mladenič, A. Skowron (Eds.), Knowledge Discovery in Databases: PKDD 2007. XXIV, 640 pages. 2007.

Vol. 4701: J.N. Kok, J. Koronacki, R. Lopez de Mantaras, S. Matwin, D. Mladenič, A. Skowron (Eds.), Machine Learning: ECML 2007. XXII, 809 pages. 2007.

Vol. 4696: H.-D. Burkhard, G. Lindemann, R. Verbrugge, L.Z. Varga (Eds.), Multi-Agent Systems and Applications V. XIII, 350 pages. 2007.

Vol. 4694: B. Apolloni, R.J. Howlett, L. Jain (Eds.), Knowledge-Based Intelligent Information and Engineering Systems, Part III. XXIX, 1126 pages. 2007.

Vol. 4693: B. Apolloni, R.J. Howlett, L. Jain (Eds.), Knowledge-Based Intelligent Information and Engineering Systems, Part II. XXXII, 1380 pages. 2007.

Vol. 4692: B. Apolloni, R.J. Howlett, L. Jain (Eds.), Knowledge-Based Intelligent Information and Engineering Systems, Part I. LV, 882 pages. 2007.

Vol. 4687: P. Petta, J.P. Müller, M. Klusch, M. Georgeff (Eds.), Multiagent System Technologies. X, 207 pages. 2007.

Vol. 4682: D.-S. Huang, L. Heutte, M. Loog (Eds.), Advanced Intelligent Computing Theories and Applications. XXVII, 1373 pages. 2007.

Lecture Notes in Artificial Intelligence (LNAI)

Vol. 5221: B. Nordström, A. Ranta (Eds.), Advances in Natural Language Processing. XII, 512 pages. 2008.

Vol. 5208: H. Prendinger, J. Lester, M. Ishizuka (Eds.), Intelligent Virtual Agents. XVII, 557 pages. 2008.

Vol. 5195: A. Armando, P. Baumgartner, G. Dowek (Eds.), Automated Reasoning. XII, 556 pages. 2008.

Vol. 5144: S. Autexier, J. Campbell, J. Rubio, V. Sorge, M. Suzuki, F. Wiedijk (Eds.), Intelligent Computer Mathematics. XIV, 600 pages. 2008.

Vol. 5118: M. Dastani, A. El Fallah Seghrouchni, J. Leite, P. Torroni (Eds.), Languages, Methodologies and Development Tools for Multi-Agent Systems. X, 279 pages. 2008.

Vol. 5113: P. Eklund, O. Haemmerlé (Eds.), Conceptual Structures: Knowledge Visualization and Reasoning. X, 311 pages. 2008.

Vol. 5110: W. Hodges, R. de Queiroz (Eds.), Logic, Language, Information and Computation. VIII, 313 pages. 2008.

Vol. 5108: P. Perner, O. Salvetti (Eds.), Advances in Mass Data Analysis of Images and Signals in Medicine, Biotechnology, Chemistry and Food Industry. X, 173 pages. 2008.

Vol. 5097: L. Rutkowski, R. Tadeusiewicz, L.A. Zadeh, J.M. Zurada (Eds.), Artificial Intelligence and Soft Computing – ICAISC 2008. XVI, 1269 pages. 2008.

Vol. 5078: E. André, L. Dybkjær, W. Minker, H. Neumann, R. Pieraccini, M. Weber (Eds.), Perception in Multimodal Dialogue Systems. X, 311 pages. 2008.

Vol. 5077: P. Perner (Ed.), Advances in Data Mining. XI, 428 pages. 2008.

Vol. 5076: R. van der Meyden, L. van der Torre (Eds.), Deontic Logic in Computer Science. X, 279 pages. 2008.

Vol. 5064: L. Prevost, S. Marinai, F. Schwenker (Eds.), Artificial Neural Networks in Pattern Recognition. IX, 318 pages. 2008.

Vol. 5049: D. Weyns, S.A. Brueckner, Y. Demazeau (Eds.), Engineering Environment-Mediated Multi-Agent Systems. X, 297 pages. 2008.

Vol. 5043: N. Jamali, P. Scerri, T. Sugawara (Eds.), Massively Multi-Agent Technology. XII, 191 pages. 2008.

Vol. 5040: M. Asada, J.C.T. Hallam, J.-A. Meyer, J. Tani (Eds.), From Animals to Animats 10. XIII, 530 pages. 2008.

Vol. 5032: S. Bergler (Ed.), Advances in Artificial Intelligence. XI, 382 pages. 2008.

Vol. 5027: N.T. Nguyen, L. Borzemski, A. Grzech, M. Ali (Eds.), New Frontiers in Applied Artificial Intelligence. XVIII, 879 pages. 2008.

Vol. 5012: T. Washio, E. Suzuki, K.M. Ting, A. Inokuchi (Eds.), Advances in Knowledge Discovery and Data Mining. XXIV, 1102 pages. 2008.

Vol. 5009: G. Wang, T. Li, J.W. Grzymala-Busse, D. Miao, A. Skowron, Y. Yao (Eds.), Rough Sets and Knowledge Technology. XVIII, 765 pages. 2008.

Vol. 5003: L. Antunes, M. Paolucci, E. Norling (Eds.), Multi-Agent-Based Simulation VIII. IX, 141 pages. 2008.

Vol. 5001: U. Visser, F. Ribeiro, T. Ohashi, F. Dellaert (Eds.), RoboCup 2007: Robot Soccer World Cup XI. XIV, 566 pages. 2008.

Vol. 4999: L. Maicher, L.M. Garshol (Eds.), Scaling Topic Maps. XI, 253 pages. 2008.

Vol. 4994: A. An, S. Matwin, Z.W. Raś, D. Ślęzak (Eds.), Foundations of Intelligent Systems. XVII, 653 pages. 2008.

Vol. 4953: N.T. Nguyen, G.S. Jo, R.J. Howlett, L.C. Jain (Eds.), Agent and Multi-Agent Systems: Technologies and Applications. XX, 909 pages. 2008.

Vol. 4946: I. Rahwan, S. Parsons, C. Reed (Eds.), Argumentation in Multi-Agent Systems. X, 235 pages. 2008.

Vol. 4944: Z.W. Raś, S. Tsumoto, D.A. Zighed (Eds.), Mining Complex Data. X, 265 pages. 2008.

Vol. 4938: T. Tokunaga, A. Ortega (Eds.), Large-Scale Knowledge Resources. IX, 367 pages. 2008.

Vol. 4933: R. Medina, S. Obiedkov (Eds.), Formal Concept Analysis. XII, 325 pages. 2008.

Vol. 4930: I. Wachsmuth, G. Knoblich (Eds.), Modeling Communication with Robots and Virtual Humans. X, 337 pages. 2008.

Vol. 4929: M. Helmert, Understanding Planning Tasks. XIV, 270 pages. 2008.

Vol. 4924: D. Riaño (Ed.), Knowledge Management for Health Care Procedures. X, 161 pages. 2008.

Vol. 4923: S.B. Yahia, E.M. Nguifo, R. Belohlavek (Eds.), Concept Lattices and Their Applications. XII, 283 pages. 2008.

Vol. 4914: K. Satoh, A. Inokuchi, K. Nagao, T. Kawamura (Eds.), New Frontiers in Artificial Intelligence. X, 404 pages. 2008.

Vol. 4911: L. De Raedt, P. Frasconi, K. Kersting, S. Muggleton (Eds.), Probabilistic Inductive Logic Programming. VIII, 341 pages. 2008.

Vol. 4908: M. Dastani, A. El Fallah Seghrouchni, A. Ricci, M. Winikoff (Eds.), Programming Multi-Agent Systems. XII, 267 pages. 2008.

possible as well. The order of stimuli was randomized across subjects. Prior to the study it was made clear that there is no *correct* choice, but that we were only interested in each subject's subjective opinion. Participants' age ranged from 18 to 66 (mean value 31.7 years), 67% were male, 33% female, and 28% had prior experiences with our Virtual Human Max.

2 Results and Discussion

The study provided a total of 100 complete data sets. First, we analyzed which emotion label was assigned to which picture and found that the following stimuli were most recognizable: ashamed, happy, concentrated, surprised, sad, and angry. For frustrated, bored, annoyed, relieved, hopeful, jealous, proud, and gloat this does not apply. A correlation analysis revealed a significant relationship between the presented picture and the participant's choice ($\chi^2 = 7087.856$; df = 546; p<0.001). The majority of primary emotions were recognized as expected, with the pictures sad and annoyed correlating with the label angry. Among the secondary emotions, the expressions of gloat, ashamed, relieved, and proud were recognized quite well, in contrast to jealous, frustrated, and hopeful. In total, the labels for the primary emotions happy and sad were chosen very often, while labels for the secondary emotions jealous, proud, and gloat were rarely selected.

We then tested for a correlation between the labels and the pictures they were assigned to. A χ^2 analysis revealed a significant relationship ($\chi^2 = 3715.888$; df = 210; p<0.001) meaning that the most significant amount of choices was given to the "correct" stimulus. The label happy was distributed over a high number different stimuli that showed positive facial expressions (e.g. the secondary emotion hopeful). It turned out that each positive stimulus presented first very often got the label happy. Thereafter, participants tended to choose the label happy less often in the following pictures. The label sad, on the contrary, had a high amount of votes and was mostly assigned to the "correct" picture.

Overall, primary emotions seem to be better recognizable than secondary ones. Moreover, the facial expressions, which are based on four "basic emotions" to express primary emotions, are not only much better recognizable but also chosen more frequently than emotion terms denoting secondary emotions. This supports our assumption that secondary emotions, such as relief or hope, cannot be revealed by facial expressions alone.

References

1. Becker, C., Kopp, S., Wachsmuth, I.: Why emotions should be integrated into conversational agents. In: Nishida, T. (ed.) Conversational Informatics: An Engineering Approach, ch.3, pp. 49–68. Wiley, Chichester (2007)
2. Damasio, A.: Descartes' Error, Emotion Reason and the Human Brain. Grosset/Putnam (1994)
3. Ekman, P.: Facial expressions. In: Handbook of Cognition and Emotion, ch.16, pp. 301–320. John Wiley & Sons, Chichester (1999)

Comparing Emotional vs. Envelope Feedback for ECAs

Astrid von der Pütten[1], Christian Reipen[1], Antje Wiedmann[1],
Stefan Kopp[2], and Nicole C. Krämer[1]

[1] University Duisburg-Essen, Forsthausweg 2, 47057 Duisburg, Germany
[2] University of Bielefeld, Artificial Intelligence Group, 33549 Bielefeld, Germany
vdpuetten@interactivesystems.info,
christian.reipen@stud.uni-due.de,
antje.wiedmann@stud.uni-due.de, skopp@techfak.uni-bielefeld.de,
nicole.kraemer@uni-due.de

Abstract. Opinions in the scientific community differ about what makes an ECA effective. This study investigated whether emotional expressions influence the effectiveness of human-agent-interaction and the participants´ emotional status after the experiment. 70 participants took part in a small talk (10 min.) situation with the ECA MAX. We implemented two different types of feedback: emotional feedback (EMO), which provided a feedback about the emotional state of MAX (including smiles and compliments) and envelope feedback (ENV), which provided a feedback about the comprehension of the participants´ contributions and presents MAX as an attentive listener. We found that smiling was the only nonverbal behaviour which was significantly recognized and that the emotional feedback led to increased feelings of interest in the participants. There were no effects with regard to the evaluation of the effectiveness of the conversation.

Keywords: evaluation study, nonverbal behavior, emotional feedback, envelope feedback, social effects of ECAs.

A great deal of research concentrates on the implementation and evaluation of feedback systems for intelligent virtual agents [1,2,3]. However, it is still an open question how emotional feedback influences the evaluation of the system, users´ feelings during and after the interaction as well as the perceived efficiency of the conversation. Therefore our research concentrated on the following questions: What are the effects of different types of feedback on (a) the self-reported users´ emotional status and (b) the evaluation of the artificial interlocutor? (c) Do different types of feedback increase the perceived effectiveness and efficiency of the conversation?

The study was conducted with the ECA MAX [3]. MAX can express himself multimodally, and is able to respond to natural language input (via a "Wizard of Oz" procedure [4]). To test our research questions, we tested versions of MAX with and without envelope feedback (including vocal backchannels (e.g "Ja" (yes), "mhm") and nonverbal signals (head tilt, nod, frown)), as well as with and without emotional feedback (MAX consistently showed a smile in the beginning and the end of the conversation and gave a compliment ("Your clothes are cool!")). As independent variables we thus varied the existence of emotional and envelope feedback. As dependent variables we used (1) the participants´ emotional status after the interaction, (2) the evaluation of MAX and (3) the evaluation of the conversation. The questionnaire given after the experiment measured the participants' emotional status after the interaction via a 30-item 5 point scale. The questions evaluating the personal perception of MAX included 62 items assessed on a 7 point scale. In addition, the participants were handed questions concerning their evaluation of the conversation (efficiency and enjoyment of the conversation, who controlled the conversation, whether MAX showed nonverbal behavior (e.g. "MAX smiled")).

Seventy persons, ranging from 17 to 48 years age (m=24.09; sd=5.717), participated in the study. In general, MAX was evaluated positive (pacific (m=5.52; sd=.885), friendly (m=5.26; sd=1.112), honest (m=4.84; sd=1.235) and relaxed (m=4.67; sd=1.401)). But he was also described as rather artificial (m=5.63; sd=1.253), stiff (m=4,94; sd=1.433) and wooden (m=4.81; sd=1.406). Factor analysis yielded three factors for the participants´ emotional status: "unhappiness", "stress/strain" and "interest". The factors *"sympathy"*, *"live-likeness"*, *"incompetence"*, *"exertion"*, *"introversion"*, *"dominance"*, *"relaxation"* described the person perception of MAX. Results showed that participants who had experienced emotional feedback (EMO) felt significantly more interested ($F(1;69)$= 4.534; p= .037; partial eta^2= .063). Also there was a non-significant tendency that they felt more stressed or strained ($F(1;69)$= 2.879; p= .094; partial eta^2= .041) talking to MAX in this condition. There were no findings for the third factor *"unhappiness"*. With regard to person perception, MAX was significantly judged as more incompetent ($F(1;69)$= 3.296; p= .074; partial eta^2= .046) within EMO. Participants rated the gestures of MAX as more helpful for the communication in the EMO condition ($F(1;69)$= 2.888; p= .094; partial eta^2= .043). Furthermore, participants correctly perceived more smiles (Chi^2= 9.130; p = .003) in the EMO condition. We observed no effects with regard to the perceived effectiveness of the interaction, and – more surprisingly – not a single main effect for envelope feedback (ENV). In sum, we found that the addition of EMO has an impact, whereas ENV did not affect the feelings and evaluations of the user. In line with the results by Cassell and Thórisson [1] we did not find an influence of emotional feedback on perceived effectivenss and efficiency of the conversation but contrary to their results we could not find an impact of envelope feedback on this dependent variable. We could demonstrate, though, an influence of emotional feedback on numerous dependent variables. In sum, our data stress that the development and systematic evaluation of emotional feedback is worthwhile and should be focused more intensely.

References

1. Cassell, J., Thórisson, K.R.: The Power of a Nod and a Glance: Envelope vs. Emotional Feedback. Animated Conversational Agents. Applied Artificial Intelligence 13, 519–538 (1999)
2. Cassell, J., Bickmore, T., Billinghurst, M., Campbell, L., Chang, K., Vilhjálmsson, H., Yan, H.: Embodiment in Conversational Interfaces: Rea. In: Proceedings of the CHI 1999 Conference, PA, Pittsburgh, pp. 520–527 (1999)
3. Kopp, S., Allwood, J., Ahlsen, E., Grammer, K., Stocksmeier, T.: Modeling Embodied Feedback in a Virtual Human. In: Wachsmuth, I., Knoblich, G. (eds.) Modeling Communication With Humanoids And Robots, Springer, Hamburg (to appear)
4. Dahlbäck, N., Jönsson, A., Ahrenberg, L.: Wizard of Oz studies – why and how. In: Proceedings of the ACM International Workshop on Intelligent User Interfaces (1993)

Intelligent Agents Living in Social Virtual Environments – Bringing Max into Second Life

Erik Weitnauer, Nick M. Thomas, Felix Rabe, and Stefan Kopp

Artifical Intelligence Group, Bielefeld University, Germany
{eweitnau,nthomas,frabe,skopp}@techfak.uni-bielefeld.de

1 Introduction and Background

When developing cognitive agents capable of interacting with humans, it is often challenging to provide a suitable environment in which agent and user are co-situated. This paper presents a straightforward approach to use *Second Life* (SL) [1] as a persistent, "near natural", and socially rich environment for research on autonomous agents in complex surroundings, learning social skills, and how they are perceived by humans.

In SL, with its registered user base of more than 11 millions, human-human interaction is easily found and technically observable. Moreover it is easy to engage into human-agent interaction and thereby offers a promising way for learning and testing social skills in actual interactions. However, our focus is not just on serving and taking advantage of the existing SL user community, but also on how to use the SL technology independently of the official server grid as an advanced and controllable setting testbed for intelligent agent research.

2 Implementation and Architecture

The Artificial Intelligence Group at Bielefeld University has been developing the virtual human Max [2] to study how natural conversational behavior of humans can be modeled and made available for A.I. systems. Aimed at mixed-initiative dialog and collaborative interaction in dynamic virtual environments, Max rests upon a general cognitive architecture that combines abilities for concurrent perception, rational reasoning and deliberation, emotion, and action. The connection of Max's architecture with SL is realized via a dedicated module (libSecondLifeAgent) as shown in figure 1. Building upon libsecondlife, simulated sensors receive data from the server and feed it into Max's perception component. This includes object and avatar positions, status updates, avatar appearances, avatar profiles, chat and instant messaging, as well as changes of friends and inventory status.

Upon these extensions, Max was able to enter the environment and communicate multimodally with human avatars and objects. He employs about 15 of over 100 in-built gestures (including facial expressions), autonomously navigates through the virtual world, follows avatars, and can access all user functions available in the offical SL client.

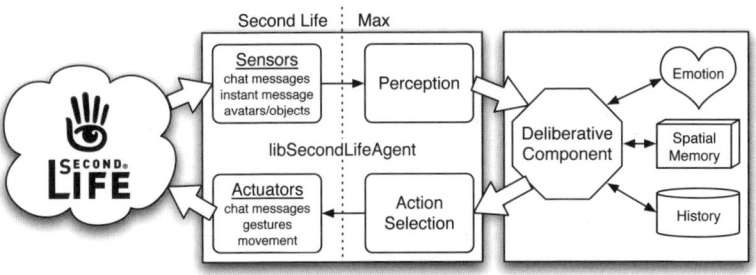

Fig. 1. Software architecture: The libSecondLifeAgent acts as a SL client for Max

3 Max's First Steps into Second Life – Observations

Max's first steps into SL were mainly conducted in two of the biggest German communities in SL, the *Apfelland* and the *Germania* regions. During our trials Max retrieves the position of the nearby resident's avatar, approaches her, initiates a chat, and tries to sustain the conversation as long as possible. Max beliefs the conversation to be finished when either the conversation partner walks away, or expresses her intention to end it. All of Max's first steps were automatically logged to text-files as well as captured on video.

Out of the 15 residents Max approached in total, five responded while the others were already engaged in other activities. The conversations lasted an average of 300 seconds, with an average of 25 turn-switches. In the post-briefing, three of the residents reported they suspected the agent to be controlled by an artificial system at a certain point in the conversation, the other two got frustrated without realizing that they were not interacting with a human. The most common reasons residents gave were repetitive agent behavior, too general or ambiguous answers, or too quick responses. One remarkable concern was that a mixed society, in which human and machine-controlled avatars are so undistinguishable as in SL, would lead to distrust and disharmony in the community.

Summing up this work, it is fair to say that SL offers great opportunities for learning-based approaches to modeling social skills, both by interacting with humans or other agents, as well as by observing other avatars interact. Our first observations are promising in that they show that intelligent agent techniques can be employed with and benefit from these settings, but also do they raise a number of new research issues, e.g., regarding the acceptance of machines that are potentially indistinguishable from humans.

References

1. Second Life, http://secondlife.com/whatis/
2. Kopp, S., Gesellensetter, L., Krämer, N., Wachsmuth, I.: A conversational agent as museum guide — design and evaluation of a real-world application. In: Panayiotopoulos, T., Gratch, J., Aylett, R.S., Ballin, D., Olivier, P., Rist, T. (eds.) IVA 2005. LNCS (LNAI), vol. 3661, pp. 319–343. Springer, Heidelberg (2005)

Author Index

Abe, Yukari 498
Akhter Lipi, Afia 465
Anderson, James 516
Anderson, Nicholas 516
André, Elisabeth 223, 245, 492
Austermann, Anja 308
Aydın, Ali Orhan 468
Aylett, Ruth 73, 348

Bai, Li 511
Baldassarri, Sandra 520
Bao, Haiyan 289
Baylor, Amy L. 208
Beck, Diane 154
Becker-Asano, Christian 15, 548
Bevacqua, Elisabetta 262
Billon, Ronan 470
Bogdanovych, Anton 456
Breitfuss, Werner 472
Breton, Gaspard 545
Bruegmann, Klaus 281
Brusk, Jenny 474
Burden, David 535
Burkitt, Mark 372

Catrambone, Richard 530
Cavazza, Marc 146, 380
Cerekovic, Aleksandra 476
Cerezo, Eva 520
Charfuelan, Marcela 426
Charlton, Daniel 146
Cho, Kenta 364
Chung, Shun-an 479
Corruble, Vincent 524
Courty, Nicolas 215
Cousseau, Aurélie 481

Dautenhahn, Kerstin 59
de Freitas, Sara 535
de Kok, Iwan 176
Deladisma, Adeline 237
de Melo, Celso 484
de Sevin, Etienne 486, 488
Dias, João 348
Duthen, Yves 528

Endrass, Birgit 492
Endres, Christoph 426
Esteva, Marc 456

Fujita, Kinya 547

Gebhard, Patrick 191, 426
Georg, Gersende 380
Gibet, Sylvie 215
Gillies, Marco 89, 494
Gratch, Jonathan 117, 176, 253, 394, 484
Grinberg, Maurice 356

Ha, Eun Young 45
Hahn, Minsoo 526
Hakulinen, Jaakko 146
Hartholt, Arno 117
Hattori, Masanori 364
Haya, Pablo A. 520
Hayashi, Hisashi 364
Heloir, Alexis 215
Hernault, Hugo 139
Heylen, Dirk 270, 543
Hirakawa, Daisuke 505
Ho, Wan Ching 59
Huang, Shwu-Lih 514

Iketani, Naoki 364
Ishii, Ryo 200
Ishizuka, Mitsuru 29, 139, 281, 472, 492
Ito, Jonathan Y. 322

Jack, Mervyn 516
Johnsen, Kyle 237
Jonsdottir, Gudny Ragna 131, 162

Kang, Sin-Hwa 253
Kenny, Patrick 394
Kikuchi, Masaaki 364
Kim, Soyoung 208
Kipp, Michael 191, 215, 426
Koda, Tomoko 245
Komatsu, Takanori 498

Kondo, Toshiyuki 505
Kopp, Stefan 270, 548, 550, 552
Kostadinov, Stefan 356
Krämer, Nicole C. 507, 550
Kriegel, Michael 73
Kruizinga, Edze 543
Kryssanov, Victor V. 509
Kumokawa, Shizuka 509

Lance, Brent 1
Le Pichon, François 481
Lee, Hung-Wei 514
Lee, Jim Jiunde 479
Lee, Jina 117
Lester, James C. 45
Li, Jituo 511
Li, Tsai-Yen 514
Li, Zheng 289
Lim, Mei Yii 348
Lin, Yueh-Hung 514
Lind, David 154
Lind, Scott 237
Liu, Chia-Yang 514
Lok, Benjamin 154, 237
Looije, Rosemarijn 490
Luga, Hervé 528

Maffiolo, Valérie 545
Mancini, Maurizio 262
Mao, Xia 289
Marsella, Stacy C. 1, 117, 270, 322, 334
Matsumura, Kyouhei 547
Matthews, Alexandra 516
Melder, Willem A. 490
Militaru, Dorin 518
Miyawaki, Kenzaburo 97
Montoro, Germán 520
Morency, Louis-Philippe 176

Nagata, Mizue 496
Nakano, Yukiko I. 200, 223, 465
Nakayama, Kenji 533
Nayak, Abhaya 468
Nédélec, Alexis 470
Neerincx, Mark A. 490
Neff, Michael 522
Neviarouskaya, Alena 29
Niewiadomski, Radoslaw 37
Nishida, Toyoaki 223
Nishimura, Keisuke 364

Nivel, Eric 162
Nozawa, Takayuki 505

Ochs, Magalie 37, 524
Ogawa, Hitoshi 509
Ogawa, Kohei 296
Oh, Seungwon 526
Ono, Tetsuo 296
Orgun, Mehmet Ali 468

Paiva, Ana 348
Pammi, Sathish 426
Pan, Xueni 89, 494
Pandzic, Igor S. 476
Panzoli, David 528
Park, Sung 530
Parsons, Thomas D. 394
Pedica, Claudio 104
Pelachaud, Catherine 37, 262, 270, 380
Piattini, Mario 539
Piwek, Paul 139
Pontier, Matthijs 417
Prendinger, Helmut 29, 139, 281, 472, 492
Pynadath, David V. 322, 334

Rabe, Felix 552
Rebolledo Méndez, Jovan David 533
Rebolledo-Mendez, Genaro 535
Rehm, Matthias 223, 245, 465
Reichardt, Dirk M. 537
Reipen, Christian 550
Rizzo, Albert A. 394
Rodriguez, Harold 154
Romano, Daniela M. 372
Romero, Miguel 539
Rossen, Brent 237
Rowe, Jonathan P. 45
Rumpler, Martin 426
Ruttkay, Zsófia 409

Sabouret, Nicolas 524
Sano, Mutsuo 97
Satoh, Ichiro 441
Schröder, Marc 426
Serón, Francisco José 520
Si, Mei 334
Siddiqui, Ghazanfar F. 417
Simoff, Simeon 456
Slater, Mel 89, 494

Printing: Mercedes-Druck, Berlin
Binding: Stein+Lehmann, Berlin

Smid, Karlo 476
Smith, Cameron 146
Sumi, Kaoru 496
Supic, Haris 541
Swartjes, Ivo 543

Tajariol, Federico 545
Tanaka, Takahiro 547
Theune, Mariët 543
Thomas, Nick M. 552
Thórisson, Kristinn R. 131, 162
Tisseau, Jacques 470
Tolksdorf, Julia 548
Traum, David 117
Türk, Oytun 426
Turunen, Markku 146

Ullrich, Sebastian 281

van Doesburg, Willem A. 490
van Velsen, Martin 81
van Welbergen, Herwin 409
Vilhjálmsson, Hannes 104, 270
Vizcaíno, Aurora 539
von der Pütten, Astrid 550

Wachsmuth, Ipke 15
Wang, Ning 253
Wang, Yangsheng 511
Watt, James H. 253
Weitnauer, Erik 552
Wiedmann, Antje 550

Yamada, Seiji 308
Yamaoka, Yuji 465

Zhang, Li 146
Zoric, Goranka 476